AF213569

Textbook of Engineering Chemistry

R. N. Goyal • Harmendra Goel

Textbook of Engineering Chemistry

Third Edition

Springer

Ane Books Pvt. Ltd.

R. N. Goyal
Department of Chemistry
Formerly, IIT Roorkee
Delhi, Delhi, India

Harmendra Goel
Department of Chemistry
G.B. Pant Engineering College
Pauri Garhwal, Uttarakhand, India

ISBN 978-3-032-22067-7 ISBN 978-3-032-22068-4 (eBook)
https://doi.org/10.1007/978-3-032-22068-4

Jointly published with Ane Books Pvt. Ltd.
In addition to this printed edition, there is a local printed edition of this work available via Ane Books in South
Asia (India, Pakistan, Sri Lanka, Bangladesh, Nepal and Bhutan) and Africa (all countries in the African subcontinent).
ISBN of the Co-Publisher's edition: 978-93-84726-03-4

First edition - CRY: 2008
Second edition - CRY: 2011

© The Author(s), under exclusive license to Springer Nature Switzerland AG 2026

This work is subject to copyright. All rights are solely and exclusively licensed by the Publisher, whether the whole or part of the material is concerned, specifically the rights of reprinting, reuse of illustrations, recitation, broadcasting, reproduction on microfilms or in any other physical way, and transmission or information storage and retrieval, electronic adaptation, computer software, or by similar or dissimilar methodology now known or hereafter developed.
The use of general descriptive names, registered names, trademarks, service marks, etc. in this publication does not imply, even in the absence of a specific statement, that such names are exempt from the relevant protective laws and regulations and therefore free for general use.
The publishers, the authors, and the editors are safe to assume that the advice and information in this book are believed to be true and accurate at the date of publication. Neither the publishers nor the authors or the editors give a warranty, express or implied, with respect to the material contained herein or for any errors or omissions that may have been made. The publishers remain neutral with regard to jurisdictional claims in published maps and institutional affiliations.

This Springer imprint is published by the registered company Springer Nature Switzerland AG
The registered company address is: Gewerbestrasse 11, 6330 Cham, Switzerland

If disposing of this product, please recycle the paper.

Preface to the Third Edition

The third revised edition of *Textbook of Engineering Chemistry* is in your hands now. This edition has been completely revised and errors that erupted in the earlier editions have been rectified. Now, the book covers the syllabus prescribed for first year students of technical universities of most of the states. Due to the demand of students, a new chapter on electrical conductivity and ionization has been added in this edition. The chapter on mechanism of organic reactions has been modified and details of Pinacol-Pinacolone rearrangement have been added. In view of the growing demand of polymers, details of biodegradable polymers have also been added in the chapter of polymers, and we regard it as important or interesting information for the students. We believe that the new information added will be useful to the students.

It is hoped that like earlier editions, students and teachers will continuously write us and inform us about their feedback even on this edition. Finally, the enormous support of the families provided to us to complete the work is gratefully acknowledged.

Delhi, India — R. N. Goyal
Pauri Garhwal, Uttarakhand, India — Harmendra Goel

Preface to the First Edition

The title of this book, TEXTBOOK OF ENGINEERING CHEMISTRY was chosen to convey its aim: that is to provide broad overview of the principles of chemistry often needed by the engineers. Learning basics of chemistry is as important as understanding some basic mathematics and biology. It is believed that the book will provide some insight into the chemical aspects of world around us. For example, why is iron important in Indian economy and why does it rust, what is the importance of carbon dioxide, why ozone layer is depleting and why friction is necessary. By finding the answers to these questions, one can come to better general understanding of nature.

The textbook is written with an aim so that students joining engineering education can understand the subject in a simple manner and enjoy reading the contents. The book covers syllabus prescribed by UP Technical University, Uttarakhand Technical University and many more. A variety of problems have been solved to familiarize the students with application part of the topics. In the end of chapters different problems have been given so that students can solve them independently. It is our general observation that students try to manipulate the answers, if given in the book, hence the answers to the problems are purposely not given. It is planned to give the solutions of such problems and many more in a short manual. The details of experiments usually done in chemistry laboratory are described in the end of the book. In addition the details for preparation of general reagents for chemistry laboratory have also been given.

In writing this book we have had support and encouragement of family and of some wonderful friends and students. The help from Ane Books Pvt. Ltd. is also acknowledged. Comments and suggestions from teachers and students are welcome.

R. N. Goyal
Harmendra Goel

Contents

Chapter 1
Chemical Bonding

The valence electrons, located in the outermost shell of an atom, are crucial in the establishment of chemical bonds. These bonds only form when atoms approach sufficiently close to each other, allowing interatomic interactions to take effect, ultimately resulting in molecular formation. This chapter focuses on how such bonding mechanisms influence the three-dimensional configuration of molecules. Notably, theories concerning valency were developed prior to the full understanding of electrons. In 1883, Arrhenius introduced the concept of ions in solution. Shortly after, in 1884, Van't Hoff and Le Bel proposed a tetrahedral structure for carbon atoms. Werner, in 1908, introduced his theory on coordination compounds, while Lewis (1916) sought to explain bonding through the sharing of electrons. Pauling later characterized a chemical bond as a force between atoms that is sufficiently strong to produce a stable grouping, thereby justifying its classification as a distinct molecular species.

1.1 Why Do Atoms Form Bonds?

Atoms tend to achieve a stable electron configuration, typically resembling that of noble gases, through one of three processes: electron loss, electron gain, or electron sharing. In the formation of ionic compounds, atoms lose or gain electrons to attain the nearest noble gas configuration, denoted as ns^2np^6. The number of electrons transferred—whether lost to form a cation or gained to form an anion—defines the *electrovalency* of the atom, which may be either positive or negative depending on the nature of the ion formed.

© The Author(s), under exclusive license to Springer Nature Switzerland AG 2026

R. N. Goyal, H. Goel, *Textbook of Engineering Chemistry*,
https://doi.org/10.1007/978-3-032-22068-4_1

1.2 Types of Chemical Bonds

Based on their ability to donate or accept electrons, elements are generally classified into three categories:

1. **Electropositive elements:** Tend to lose one or more electrons with ease
2. **Electronegative elements:** Prefer to accept electrons
3. **Intermediate elements:** Exhibit minimal tendency to gain or lose electrons

The type of bond that forms depends on the nature of the interacting elements:

- **Ionic bond:** Formed between electropositive and electronegative elements
- **Covalent bond:** Arises when two electronegative elements share electrons
- **Metallic bond:** Results from interaction between two electropositive elements

To understand how atoms utilize their orbitals in forming bonds, two main theoretical frameworks have been proposed:

1. Valence bond theory (VBT)
2. Molecular orbital (MO) theory

1.3 Valence Bond Theory (VBT)

Heitler and London first applied this theory to the hydrogen molecule (H_2). In the H_2 molecule, the possibility of electron exchange between the two identical hydrogen atoms is considered. For the sake of explanation, let us label the two atoms as H_A and H_B. This leads to two possible electron arrangements:

- **Structure I:** H_A with electron 1 and H_B with electron 2
- **Structure II:** H_A with electron 2 and H_B with electron 1:

$$\underset{\text{I}}{H_A^1 H_B^2} \qquad \underset{\text{II}}{H_A^2 H_B^1}$$

The respective wave functions for these configurations are given by

$$\psi_1 = \psi_A^{(1)} \times \psi_B^{(2)} \tag{1.1}$$

$$\psi_2 = \psi_A^{(2)} \times \psi_B^{(1)} \tag{1.2}$$

Here, $\psi_A^{(1)}$, $\psi_B^{(2)}$, $\psi_A^{(2)}$, and $\psi_B^{(1)}$ represent the wave functions of electrons at the respective hydrogen atoms. The actual molecular wave function is expressed as a linear combination:

$$\psi = C_1\psi_1 + C_2\psi_2$$

Fig. 1.1 Plot of the energy E against the intermolecular distance between the atoms of a H_2 molecule

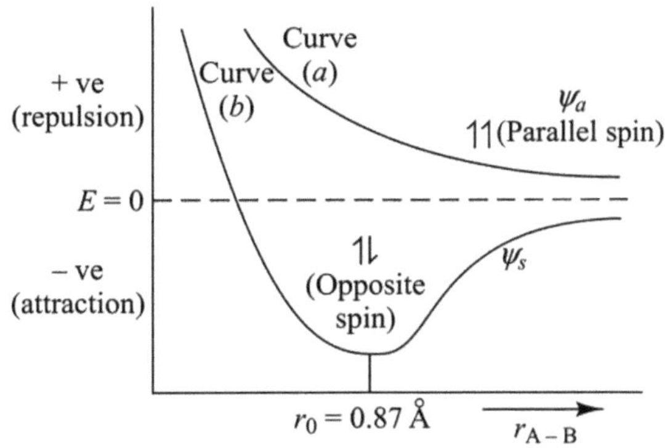

or

$$\psi = C_1\psi_A^{(1)}\psi_B^{(2)} + C_2\psi_A^{(2)}\psi_B^{(1)}$$

When $C_1 = \pm C_2$, two molecular states arise:

$$\psi_s = \psi_A^{(1)}\psi_B^{(2)} + \psi_A^{(2)}\psi_B^{(1)} \quad (1.3)$$

$$\psi_a = \psi_A^{(1)}\psi_B^{(2)} - \psi_A^{(2)}\psi_B^{(1)} \quad (1.4)$$

In Fig. 1.1, which plots energy (E) against intermolecular distance (R_{A-B}) for the H_2 molecule:

- Curve (a), corresponding to ψ_a, represents the case where electron spins are parallel. Here, energy increases as atoms approach, indicating repulsion and no bond formation—this is a nonbonding state.
- Curve (b), associated with ψ_s, corresponds to opposite spin alignment, showing a minimum in energy. This implies an attractive interaction and successful bond formation.

1.3.1 Pauling's Extension of the Model

Pauling refined the Heitler-London approach by introducing two additional ionic configurations for the H_2 molecule, where both electrons reside on the same atom:

- Structure III: Both electrons on atom H_A
- Structure IV: Both electrons on atom H_B

The corresponding wave functions are

$$\psi_{\mathrm{III}} = \psi_A^{(1)} \cdot \psi_A^{(2)}$$

$$\psi_{\mathrm{IV}} = \psi_B^{(1)} \cdot \psi_B^{(2)}$$

Pauling then modified the symmetric wave function as

$$\psi_s = \psi_A^{(1)}\psi_B^{(2)} + \psi_A^{(2)}\psi_B^{(1)} + \lambda\left(\psi_{\mathrm{III}} + \psi_{\mathrm{IV}}\right)$$

or

$$\psi_{\mathrm{s}} = \psi_{\mathrm{cov.}} + \lambda\psi_{\mathrm{ionic}} \tag{1.5}$$

Here, λ quantifies the contribution of ionic character in the primarily covalent bond.

1.3.2 The Resonance Model

Equation (1.5) suggests that the true electronic structure of the hydrogen molecule cannot be depicted by any single Lewis structure. Instead, the actual state is a *resonance hybrid* of covalent and ionic forms (Structures I–IV). The following conditions must be satisfied for valid resonance:

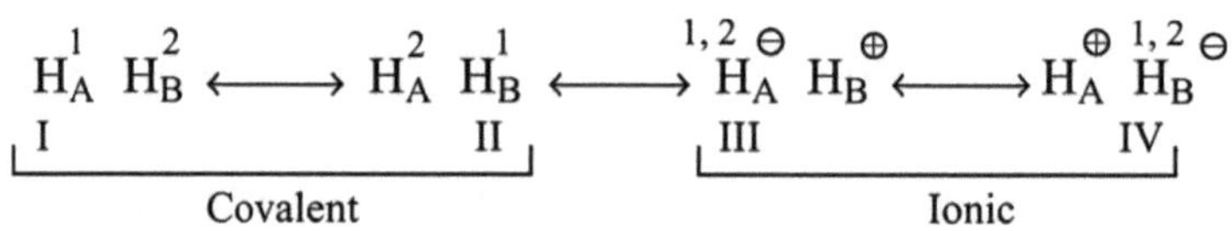

1.3.3 Conditions of Valence Bond Theory

1. All resonance structures must maintain identical atomic positions.
2. Only the arrangement of electrons varies among the contributing forms.
3. The number of unpaired electrons remains constant in all resonance contributors.
4. The resonance hybrid is more stable (lower in energy) than any individual structure—this energy difference is known as *resonance energy*.

1.3.4 Implications/Results of Valence Bond Theory

1. **Atomic identity maintained**: Atoms preserve their individual identities in the molecule.

2. **Covalent bond through overlap**: When two atoms with unpaired electrons come close, their atomic orbitals (AOs) overlap, and the spins pair to form a localized covalent bond. If the unpaired electrons have parallel spins, bond formation is prevented.
3. **Participation of electrons**: Only unpaired electrons are available for bonding. Paired electrons must first become unpaired to engage in bonding. For instance, one unpaired electron forms one bond, two form two, and so on.
4. **Bond strength and overlap**: The stability of the bond depends on the degree of orbital overlap. The overlap integral SSS is defined as

$$S = \int \psi_A \cdot \psi_B dr$$

Here, dr denotes the interatomic separation. When S is positive, a bond can form. A negative S indicates electron density reduction between nuclei, increasing repulsion. If S=0, there is no net bonding interaction. In the case of the H_2 molecule, when the hydrogen atoms move sufficiently close, their $1s$ orbitals partially overlap, resulting in bond formation.

When two atomic orbitals overlap, a maximum of two electrons—each with opposite spin—can be involved in the bond formation. These shared electrons reside within the overlapping region, experiencing the attractive influence of both nuclei, resulting in a stable bond. As the overlap becomes more extensive, the bond strength correspondingly increases.

To understand the energetic variations during bond formation, consider the interaction between two hydrogen atoms. Starting from a large separation, the potential energy is nearly zero. As the atoms approach, the energy initially decreases, reaching a minimum when the nuclei are approximately 70 picometers apart. However, moving closer than this optimal distance results in a steep increase in energy due to repulsion between the positively charged nuclei. This relationship is illustrated in Fig. 1.2, which displays the variation in potential energy relative to intermolecular separation.

For spherical s-orbitals, overlap can occur in any spatial direction, but in the case of directional orbitals like sp, specific orientation is needed to achieve maximum overlap. Such directionally dependent overlaps typically yield stronger *sigma* (σ) *bonds*.

1.3.5 Concept of Hybridization

To address the uniform geometry and bond lengths in certain molecules, Linus Pauling introduced the concept of hybridization. For instance, in methane (CH_4), the carbon atom should theoretically have only two unpaired electrons, yet the molecule forms four equivalent C-H bonds with an angle of $109°28'$ between

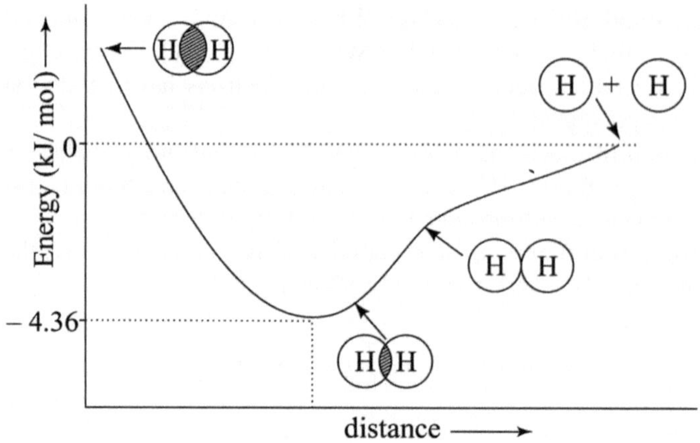

Fig. 1.2 Change in potential energy with the distance between two H atoms

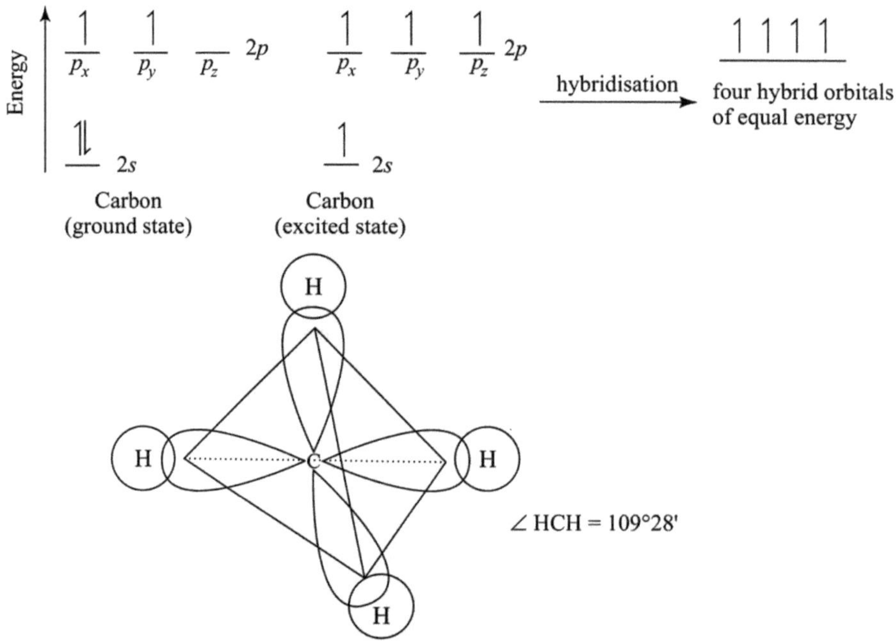

Fig. 1.3 The orbital structure of a CH_4 molecule

them. This observation led to the conclusion that carbon's atomic orbitals undergo hybridization, forming four equivalent hybrid orbitals of identical energy. These orbitals point toward the corners of a tetrahedron, explaining the consistent bond angles. The orbital hybridization involved in methane formation is depicted in Fig. 1.3.

A similar explanation applies to carbon monoxide (CO). The Lewis structure for CO appears as

$$:\overline{\text{C}}\equiv\overset{+}{\text{O}}:$$

indicating the presence of a triple bond. Both carbon and oxygen atoms in CO are *sp hybridized*. Each atom uses one of its *sp* hybrid orbitals to form a σ bond, while the other hybrid orbital holds a lone pair of electrons. The unhybridized *p*-orbitals form two π-bonds through lateral (sidewise) overlap. Hence, CO adopts a linear configuration, resembling acetylene (CH≡CH), except that CO features a lone pair instead of a terminal hydrogen.

1.3.6 Illustrative Example

Example 1.1 Analyze the bonding in acetic acid (CH_3COOH).

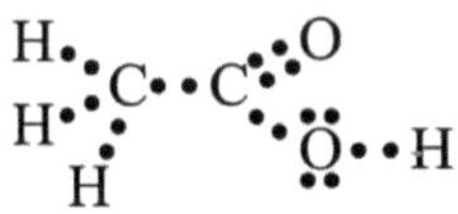

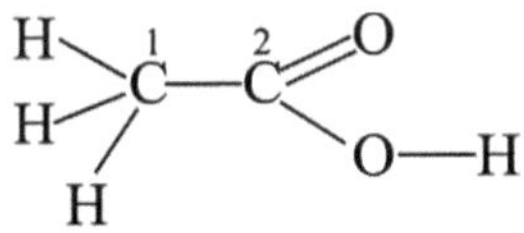

Using the Lewis structure, we analyze bonding beginning with the first carbon atom (C-1), which is connected to three hydrogen atoms and one carbon atom—making a total of four sigma bonds. Therefore, C-1 is sp^3 *hybridized*, forming a tetrahedral arrangement with bond angles around 109°. The second carbon atom (C-2) is doubly bonded to an oxygen atom and also bonded to C-1 and a hydroxyl group (−OH), indicating sp^2 *hybridization* with an approximate bond angle of 120°. The doubly bonded oxygen is also sp^2 hybridized, while the oxygen in the OH group has two lone pairs and forms two bonds, resulting in sp^3 *hybridization*, giving it a roughly tetrahedral geometry.

1.3.7 Limitations of Valence Bond Theory

Although valence bond theory (VBT) successfully explains many features of covalent bonding, it has several significant limitations:

1. It assumes that atoms maintain their individual identities completely within a molecule, which is not always realistic.
2. VBT requires the shared pair of electrons in a covalent bond to originate from two different atoms. This premise fails to explain *coordinate bonds*, where both electrons are supplied by a single atom.

3. The theory incorrectly predicts that the oxygen molecule (O_2) should be *diamagnetic*, as it suggests all electrons are paired. However, experimental evidence reveals that O_2 is *paramagnetic*, implying the presence of unpaired electrons.

1.4 Introduction to Molecular Orbital Theory (MOT)

Molecular orbital theory provides a more accurate explanation of bonding and magnetic properties in molecules such as O_2. For example, the Lewis structure of oxygen—O=O—implies complete electron pairing, which contradicts the observed paramagnetism. According to MOT, atomic orbitals combine to form molecular orbitals in the same manner as described in VBT, but with a critical distinction: the atomic orbitals merge across the entire molecule.

The number of molecular orbitals generated is always equal to the number of atomic orbitals involved. In the simplest case—the hydrogen molecule—two 1s atomic orbitals combine *via linear combination of atomic orbitals (LCAO)* to form two molecular orbitals.

1.4.1 Bonding and Antibonding Molecular Orbitals

MOT outlines two possible modes of combination for atomic orbitals:

1. **Constructive overlap**: When the wave functions of the atomic orbitals combine additively, electron density increases between the nuclei. This produces a *bonding molecular orbital* (σ), which has lower energy than the original atomic orbitals.
2. **Destructive overlap**: When the wave functions subtract from each other, electron density between the nuclei decreases. This forms an *antibonding molecular orbital* (σ), which possesses higher energy than the constituent orbitals.

As the electrons in the molecular orbital of the H_2 molecule are at a lower energy level compared to their positions in the individual atomic orbitals, this energy reduction confirms the molecule's stability.

Let's now examine why the helium molecule (He_2) does not exist under normal conditions. In the case of He_2, the molecular orbital diagram includes four electrons—two occupying the bonding orbital and two placed in the antibonding orbital. While the electrons in the bonding orbital offer stabilizing energy, this benefit is offset by the destabilizing effect of the electrons in the antibonding orbital. Since the rise in energy due to antibonding is marginally greater than the stabilization from bonding, the net effect destabilizes the molecule, thus preventing the formation of a stable He_2 species.

Table 1.1 Molecular orbital occupation of electrons for O_2 and F_2 molecules

Molecular orbital	O_2	F_2
$\sigma^* 2p$	□	□
$\pi^* 2p$	↑ ↑	↑ ↑
$\sigma 2p$	↑↓	↑↓
$\pi 2p$	↑↓ ↑↓	↑↓ ↑↓
$\sigma^* 2s$	↑↓	↑↓
$\sigma 2s$	↑↓	↑↓

In comparison to s-block elements, the energy level diagrams for p-block diatomic molecules are more intricate. Each atom's three p-orbitals contribute differently: one pair undergoes end-to-end overlap forming a sigma (σ) bonding orbital and a corresponding σ^* antibonding orbital, while the remaining two pairs overlap side by side, generating two pi (π) bonding orbitals and their corresponding π^* antibonding orbitals.

Considering a generic second-period diatomic molecule, its energy level diagram (refer to the diagram) illustrates that the bonding and antibonding σ orbitals derived from $2s$ orbitals lie energetically below those formed from $2p$ orbitals. Furthermore, the $\pi 2p$ bonding orbitals are more stable than the $\sigma 2p$ orbital. This is attributed to the sidewise overlap being less efficient than head-on overlap, resulting in greater stabilization of the π bonding orbitals and, correspondingly, greater destabilization of the π^* antibonding orbitals. Table 1.1 presents the molecular orbital occupancy for oxygen (O_2) and fluorine (F_2) molecules.

1.4.2 *Bond Order*

Many molecular characteristics—including internuclear distance and the dissociation energy—can be interpreted through the concept of *bond order*. This metric indicates the number of chemical bonds between a pair of atoms in a molecule. Bond order is defined mathematically as

$$\text{Bond order} = \text{net number of bonding electrons}/2$$

or, expressed in terms of electron pairs

$$= [\text{number of electron pairs in bonding molecular orbitals } (N_b)$$

$-$ number of electron in antibonding molecular orbitals $(N_a)]/2$

where N_b is the total number of electrons in bonding molecular orbitals and N_a is the total in antibonding orbitals.

The molecular orbital filling order follows: *Aufbau, Hund's rule*, and *Pauli's exclusion principle*, analogous to atomic orbitals. Accordingly, the typical sequence of orbital filling is

$$\sigma 1s < \sigma^* 1s < \sigma 2s < \sigma^* 2s < \sigma 2p_x < \pi 2p_y < \pi^* 2p_y < \sigma^* 2p_x$$

$$\pi 2p_z \quad \pi^* 2p_x$$

Based on this, the calculated bond order values for O_2 and F_2 are 2 and 1, respectively. It's also possible for ions to have *fractional bond orders*, as illustrated in the following examples.

1.4.3 Examples Applying Molecular Orbital Theory

1. *H_2 Molecule:*
   Configuration: $1s^1 + 1s^1 \longrightarrow \sigma 1s^2$
   Bond Order (B.O.) = $\frac{1}{2}(2 - 1) = 1$
   Conclusion: Single bond; molecule is stable.
2. *He_2 Molecule:*
   Configuration: $1s^2 + 1s^2 \longrightarrow \sigma 1s^2$ and $\sigma^* 1s^2$
   Bond Order (B.O.) = $\frac{1}{2}(2 - 2) = 0$
   Conclusion: No bond forms; the molecule is unstable.
3. *H_2^+ Ion:*
   Configuration: Only one electron present in $\sigma 1s$
   Bond Order (B.O.) = $\frac{1}{2}(1 - 0) = 0.5$
   Conclusion: Exists as a stable ionic species.
4. *N_2 Molecule (Atomic number 7):*
   Configuration: $\sigma 1s^2 \sigma^* 1s^2 \sigma 2s^2 \sigma 2p_x^2 \sigma^* 2s^2 \pi 2p_y^2 \pi 2p_z^2$
   Bond Order (B.O.) = $\frac{1}{2}(10 - 3) = 4$
   Conclusion: Triple bond, very stable molecule.
5. *O_2 Molecule (Atomic number 8):*
   Configuration: $\sigma 1s^2 \sigma^* 1s^2 \sigma 2s^2 \sigma^* 2s^2 \sigma 2p^2{}_x \pi 2p_y^2 \pi_z 2p^2$
   Bond Order (B.O.) = $\frac{1}{2}(10 - 6) = 2$
   Conclusion: Double bond, stable molecule.
6. *Nitric Oxide (NO):* For nitric oxide (15 electrons):
   Configuration: $\sigma 1s^2 \sigma^* 1s^2 \sigma 2s^2 \sigma^* 2s^2 \sigma 2p^2{}_x \pi 2p_y^2 \pi 2p_z^2 \pi^* 2p_y^1$
   Bond Order (B.O.) = $\frac{1}{2}(10 - 4) = 3$

Conclusion: Thus, a triple bond exists between nitrogen and oxygen, indicating a stable N≡O structure.

1.5 Hybridization Concept

In polyatomic molecules, the bonding patterns are effectively described by the concept of *hybridization*. This refers to the process wherein atomic orbitals within an atom combine to form a new set of orbitals, known as *hybrid orbitals*, which are equal in energy and similar in shape but differ in spatial orientation. These orbitals participate in covalent bond formation via direct overlap. It's important to note that hybrid orbitals do *not* participate in forming π-bonds. The key features of hybridization include the following:

- Hybridization helps explain the bonding and structure of molecules but is *not applicable* to isolated atoms.
- It involves mixing *nonequivalent atomic orbitals* to form hybrid orbitals of uniform energy. For example, mixing an s orbital with a p orbital leads to orbitals with characteristics different from either s or p orbitals alone.
- The total number of hybrid orbitals produced equals the number of atomic orbitals that participate in the process.

1.5.1 Examples of Hybridization

sp Hybridization
An example is *beryllium hydride (BeH_2)*. Beryllium, with electron configuration $1s^2 2s^2$, has both electrons paired in the $2s$ orbital and hence doesn't naturally form covalent bonds. However, upon excitation, one $2s$ electron moves to a $2p$ orbital, resulting in two unpaired electrons. These then hybridize, forming two equivalent sp orbitals oriented 180° apart, leading to a *linear geometry*. Each Be-H bond arises from the overlap of an sp orbital from Be with a $1s$ orbital from H.

sp^2 Hybridization
This is demonstrated in *boron trifluoride (BF_3)*. Boron's valence configuration is $2s^2 2p^1$. When one electron is promoted to an empty $2p$ orbital, three unpaired electrons result. Hybridization of one s and two p orbitals produces three sp^2 orbitals arranged in a plane at 120° angles. The B-F bonds form by overlapping each sp^2 orbital with a $2p$ orbital from fluorine.

sp^3 Hybridization
Seen in *ammonia (NH_3)*, where nitrogen has configuration $1s^2 2s^2 2p^3$. Hybridization of one s and three p orbitals results in four sp^3 hybrid orbitals. Three are used

Table 1.2 Details of shape, angle, and structure of various hybrid orbitals

Hybridisation of the central atom	*Total number of hybrid orbitals*	*Shape*	*Angle*	*Example*	*Structure of hybrid orbitals*
sp^3	4	tetrahedral	109° 28′	NH^+_4	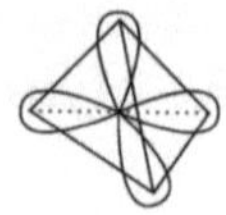
sp^2	3	planar	120°	BF_3	
sp	2	linear	180°	BeH_2	
sp^3d	5	triagonal bipyramidal	120°/90°	$PC1_5$	
sp^3d^2	6	octahedral	90°	SF_6	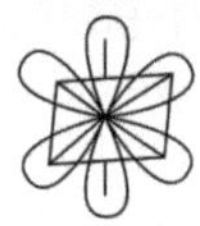

for forming N-H bonds, while the fourth holds a lone pair. Although the ideal bond angle in a tetrahedral structure is $109°28'$, *lone pair-bond pair repulsion* reduces this to approximately $107°20'$. A summary of hybridizations and resulting shapes is given in Table 1.2.

1.5.2 *Hybridization Involving s, p, and d Orbitals*

For elements from the third period onward in the periodic table, hybridization may include *d orbitals*. This enables the formation of geometries like *trigonal bipyramidal* or *octahedral*. For instance, in *sulfur hexafluoride* (SF_6), the valence configuration of sulfur is $3s^2 3p^4$. Due to small energy gaps between $3s$, $3p$, and $3d$ orbitals, electrons can be promoted from the s and p orbitals to the $3d$ orbitals.

Hybridization of one $3s$, three $3p$, and two $3d$ orbitals produces *six equivalent hybrid orbitals* with *octahedral geometry*. Each S-F bond results from the overlap of a hybrid orbital from sulfur with a $2p$ orbital from fluorine.

1.6 Water Molecule (H_2O) Formation

Oxygen, with atomic number 8, has the following electron configuration:

$$1s^2\,2s^2\,2p_x^2\,2p_y^1\,2p_z^1$$

Though oxygen has two unpaired electrons—seemingly enough to form two bonds—this would suggest a *linear* geometry. However, the actual shape of H_2O is *bent* and highly *polar*. This discrepancy is resolved through hybridization.

Here, two lone pairs and two unpaired electrons in oxygen's valence shell undergo hybridization to yield *four sp^3 orbitals*. Two of these orbitals form O-H bonds, while the remaining two hold lone pairs. The resulting shape is *tetrahedral*, with two vertices occupied by hydrogen atoms and two by lone pairs, leading to the observed *angular geometry*.

Experimental studies reveal that the H-O-H bond angle in a water molecule is approximately $104°35'$, which deviates from the expected $109°28'$ angle characteristic of a perfect tetrahedral arrangement. This reduction in angle is primarily due to the strong *repulsive interactions* between the lone pairs of electrons and those involved in bonding.

1.7 Structure of Phosphorus Pentachloride (PCl_5)

Phosphorus (atomic number 15) has the ground-state electronic configuration:

$$1s^2\,2s^2\,2p^6\,3s^2\,3p_x^1\,3p_y^1\,3p_z^1$$

This configuration suggests three unpaired electrons, explaining its trivalency in molecules like PH_3 and PCl_3. However, to account for *pentavalency* in compounds such as PCl_5, phosphorus undergoes *excitation*, promoting one electron to the $3d$ orbital, resulting in

$$3s^1\,3p_x^1\,3p_y^1\,3p_z^1\,3d^1$$

This excited configuration allows *sp^3d hybridization*, involving one s, three p, and one d orbital.

The five resulting hybrid orbitals adopt a *trigonal bipyramidal geometry*, where three orbitals form an equatorial plane at 120° angles, and the remaining two occupy axial positions forming 90° angles with the equatorial plane. Each chlorine atom overlaps its $3p$ orbital with one of the hybrid orbitals, resulting in *five σ-bonds*.

Among these, two *axial bonds* (P-Cl_2) have bond lengths of roughly 2.19 Å, while the three *equatorial (basal)* P-Cl_b bonds are shorter, about 2.04 Å. Due to these bond length variations, the five sp^3d orbitals in PCl_5 are *not strictly equivalent*.

1.8 Structures of SO_2 and SO_3

At room temperature, *sulfur trioxide (SO_3) exists as a gas and exhibits a planar structure. This geometry is attributed to sp^2 hybridization* of sulfur, whose valence electron configuration in the ground state is

$$3s^2\,3p^4$$

Upon excitation, electrons from the $3s$ and $3p$ orbitals are promoted to vacant $3d$ orbitals. Hybridizing one s and two p orbitals forms three sp^2 orbitals that engage in bond formation with oxygen atoms.

The remaining unhybridized p and d orbitals participate in multiple bonding through $p\pi - p\pi$ and $p\pi - d\pi$ overlaps. This leads to a *trigonal planar* molecular shape with an O-S-O bond angle of approximately 120°.

In *sulfur dioxide (SO_2)*, sulfur also undergoes *sp^2 hybridization*. Among the three resulting hybrid orbitals, two form bonds with oxygen atoms, while the third contains a *lone pair of electrons*, leading to a bent geometry with a bond angle of roughly 119°.

Solid SO_3 may exist in polymeric forms—either *cyclic structures* composed of three SO_3 units or *linear chains* where oxygen bridges are shared between adjacent molecules.

1.9 Structure of Xenon Oxides

In *xenon trioxide (XeO_3)*, xenon's outer electron configuration in its ground state is

$$5s^2\,5p^6$$

The structure of *xenon tetroxide (XeO_4)* will also follow from appropriate orbital hybridization and oxidation states (detailed treatment may follow in figures or subsequent pages).

1.10 Metallic Bonding

Metals exhibit compact atomic arrangements where atoms are closely packed. The force that holds these atoms together is referred to as a *metallic bond*, which is distinctive from ionic or covalent interactions.

1.10.1 Characteristics of Metallic Bond

1. This type of bond resembles covalent bonding but lacks saturation, enabling numerous atoms to be linked via shared valence electrons.
2. Metals display *high reflectivity* due to mobile electrons.
3. They conduct *heat efficiently.*
4. They exhibit *excellent electrical conductivity.*
5. Metals have *high absorption coefficients.*
6. Valence electrons are *delocalized*, forming a mobile electron cloud throughout the crystal lattice.
7. Metallic bonds are *weaker* than typical covalent bonds.

1.10.2 Theoretical Models of Metallic Bonding

Free Electron Theory
According to this model, a metallic solid is considered to be composed of *positively charged metal ions* immersed in a "sea" of *free-moving valence electrons.* These electrons are *not confined* to any single atom but move freely within the metallic lattice.

Taking *lithium (Li)* as an example, the metal consists of Li^+ ions surrounded by a cloud of delocalized electrons. Each Li^+ ion exerts an attractive force on multiple nearby electron clouds. Since no single electron is associated with one particular ion, these charge clouds are able to *move freely* among the ions.

This interaction between the metal nuclei and surrounding mobile electrons constitutes the *metallic bond.*

1.10.3 Explaining Metallic Properties Using Free Electron Theory

1. **Optical behavior:** Metals exhibit a characteristic *shine* or *lustre* due to their free electrons. These electrons absorb incoming light across the spectrum and re-emit it almost instantaneously, resulting in *strong reflection.*
2. **Electrical conductivity:** The electrons in a metal are free to move in any direction. When a voltage is applied across a metal, these electrons drift from the negative end toward the positive, resulting in *electric current.*
3. **Thermal conductivity:** Free electrons also serve as carriers of thermal energy, transferring heat throughout the metal and thereby contributing to its *thermal conductivity.*

1.10.4 Limitations of the Free Electron Theory

This theory falls short in explaining why, in some metals, electrical resistance increases *linearly* with temperature, whereas in metals that initially show high resistance, increasing temperature leads to a *nonlinear* decrease in resistance.

1.11 Valence Bond or Resonance Theory

This model is a refined version of the *valence bond theory (VBT)* proposed by Pauling. It views metallic bonding as a form of *covalent bonding*, where delocalized electron pairs resonate between neighboring atoms. Consider a lithium (Li) crystal—each Li atom has eight nearest neighbors but possesses only one valence electron. This limits it to forming a single covalent bond with one adjacent atom.

To maintain a stable metallic structure, it is assumed that the *actual bonding scenario* is a *resonance hybrid* of all possible covalent configurations.

While this theory helps in explaining electrical conductivity by invoking the presence of delocalized charges, it is unable to account for *metallic luster*, *thermal conductivity*, and the *temperature dependence* of electrical resistance.

1.12 Modern View: Band Theory of Metallic Bonding

The *band theory* expands upon earlier ideas by considering all electrons in a metallic crystal to be shared among atoms in the lattice. These electrons occupy *discrete energy bands*, which include the following (Fig. 1.4):

1. **Valence band:** The outermost electrons of atoms occupy this band when the material is in a solid state. It is *completely filled* in metals but *empty* in inert gases.
2. **Conduction band:** When valence electrons gain sufficient energy to leave the valence band, they enter the conduction band and become *free-moving electrons*, contributing to electrical conductivity.
3. **Forbidden energy gap:** The *energy separation* between the valence and conduction bands is termed the forbidden gap. Based on the width of this gap, solids are classified as follows:

   - **Insulators:** Valence band is fully occupied, conduction band is vacant, and the forbidden gap is large.
   - **Conductors (metals):** Either the valence band is partially filled, or it overlaps with the conduction band, with *negligible* or *no gap*.

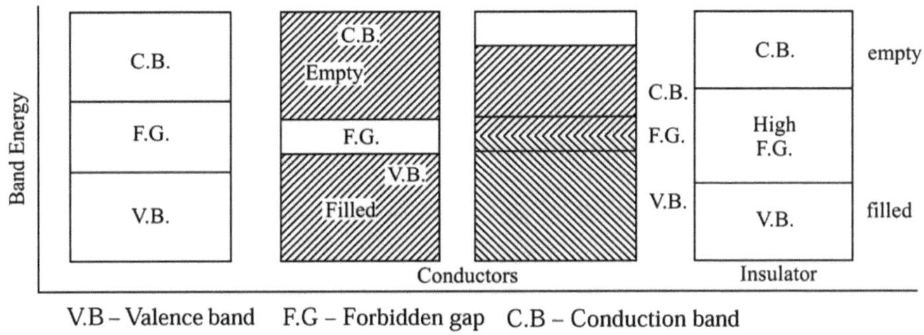

Fig. 1.4 Energy level diagram for insulator, conductors and semiconductors

- **Semiconductors:** Valence and conduction bands are partially filled, separated by a *small energy gap*. Their conductivity is intermediate.

1.13 Bragg's Law and X-Ray Diffraction

In 1913, *M. von Laue* demonstrated that X-rays, having wavelengths on the order of 10^{-3} cm, could be diffracted by crystals due to the similarity with interatomic spacings. W.L. and W.H. Bragg later provided a theoretical basis for this phenomenon using crystal planes.

Crystals consist of parallel atomic layers. When X-rays are directed at a crystal face at an angle θ, they reflect off different planes. If the *reflected waves* are in phase, constructive interference leads to observable diffraction patterns. If *out of phase*, they cancel out via destructive interference.

Let horizontal planes (Fig. 1.5) be spaced at distance d. When incident rays strike at glancing angle θ, those reflecting from successive planes can interfere constructively *only when the path difference* equals an *integer multiple of the wavelength* (λ).

By geometry, the condition for constructive interference becomes

$$2d \sin\theta = n\lambda$$

where:

- n = integer (order of reflection)
- λ = X-ray wavelength
- d = spacing between atomic planes

This relation is known as *Bragg's equation*. If values of n and λ are known, d can be determined.

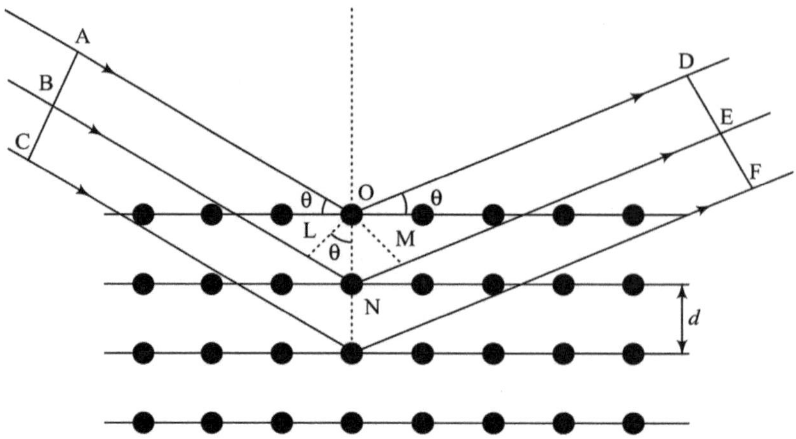

Fig. 1.5 X-ray reflections from a crystal

1.14 Born-Haber Cycle

The *Born-Haber cycle* offers a thermodynamic route to evaluate the *lattice energy* of an ionic solid. Consider the formation of sodium chloride (NaCl):

$$\mathrm{Na(s)} + \frac{1}{2}\mathrm{Cl_2(g)} \rightarrow \mathrm{NaCl(s)}$$

The process involves:

(i) Sublimation of Na:

$$\mathrm{Na(s)} \longrightarrow \mathrm{Na(g)} - \Delta H_\mathrm{S}$$

(ii) Ionization of Na:

$$\mathrm{Na(g)} - \mathrm{e} \longrightarrow \mathrm{Na^+(g)} - \Delta H_\mathrm{I}$$

(iii) Dissociation of Cl_2:

$$\frac{1}{2}\mathrm{Cl_2(g)} \longrightarrow \mathrm{Cl(g)} - \frac{1}{2}\Delta H_\mathrm{D}$$

(iv) Electron affinity of Cl:

$$\mathrm{Cl(g)} + \mathrm{e} \longrightarrow \mathrm{Cl^-(g)} + \Delta H_\mathrm{ea}$$

(v) Formation of NaCl from ions:

$$Na^+(g) + Cl^-(g) \longrightarrow NaCl + u \quad \text{(Lattice energy)}$$

The overall enthalpy change is given by

$$-\Delta H_f = \Delta H_S + \Delta H_I + \frac{1}{2}\Delta H_D - \Delta H_{ea} - u$$

1.15 Imperfections in Solids (Crystal Defects)

An *ideal crystal* is one in which the *unit cell is repeated flawlessly* across the entire structure. At absolute zero, such a perfect lattice exists without any vacancies. However, as temperature rises, atoms can vacate their lattice sites, leading to *defects*. These are *thermodynamic defects*, and their number increases with temperature. The defect concentration n per cmcan be estimated using the following formula:

$$n = Ne^{-W/2RT}$$

where:

- N = total number of potential defect sites per cm^2
- W = energy needed to create one defect
- R = gas constant
- T = absolute temperature

1.15.1 Types of Defects

Crystal imperfections are broadly classified into:

- Stoichiometric defects
- Nonstoichiometric defects

In *stoichiometric defects*, irregularities in crystal structures occur when atoms or ions are either missing from their normal lattice positions or are displaced to other locations. Two well-recognized types of such defects are as follows:

Schottky Defect

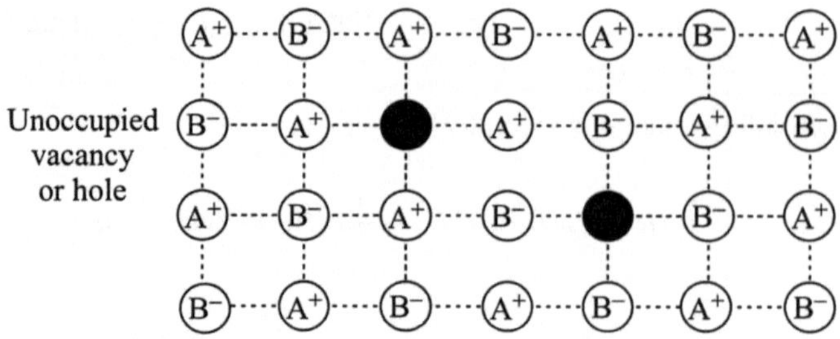

Schottky Defect

This form of defect appears when certain lattice sites remain unoccupied, creating *vacancies* or *holes* within the crystal. Such defects are common in ionic compounds with a *high coordination number* and ions of comparable sizes, such as NaCl and CsCl.

Frenkel Defect

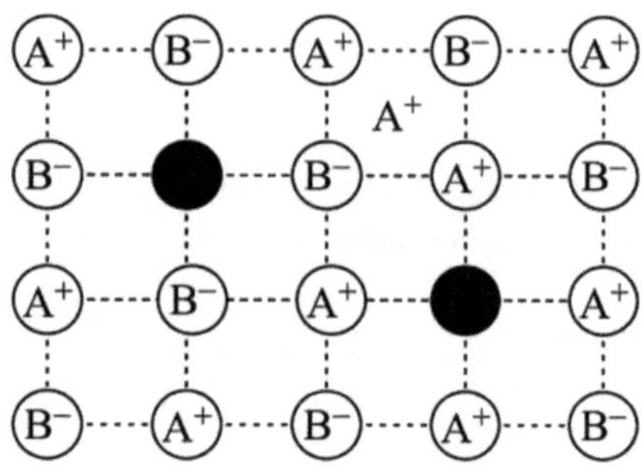

This arises when an ion is displaced from its regular lattice site and occupies an *interstitial position* instead. This generates a vacancy at the original site and an ion in the interstitial space. Typical examples include ZnS and AgBr, where Ag^- ions often leave their normal positions and relocate between other ions in the crystal lattice.

This type of defect is generally found in compounds with:

1. Low coordination numbers
2. A significant difference in the sizes of the constituent ions
3. A *highly polarizing cation* and a *readily polarizable anion*

In contrast, *nonstoichiometric defects* occur when the ratio of constituent atoms does not conform exactly to the ideal whole-number ratios. These defects are often observed in *oxides and sulfides of transition metals*, leading to deviations from the ideal crystal structure.

1.15.2 Metal Excess Defects

There are two notable mechanisms through which excess metal content may be introduced:

1.15.3 F-Centers

In this case, a negatively charged ion is absent from its lattice position, leaving behind a void that is occupied by a free electron. These sites, called F-centers, contribute to the *color* of the compound and confer *paramagnetic properties*, as seen in NaCl and KCl.

1.15.4 Interstitial Ions and Electrons

This defect arises when an *additional positive ion* occupies an interstitial space, and an accompanying *interstitial electron* ensures charge neutrality. Compounds like ZnO and CdO display such characteristics.

Crystals possessing either of these defects contain mobile electrons. When these electrons move, they facilitate *electrical conductivity*, thus classifying such materials as *semiconductors*.

1.15.5 Metal Deficiency Defects

These occur in structures where there is a *shortage of metal ions* and are especially common among *transition metals* due to their multiple oxidation states.

1.15.6 Missing Positive Ions

If a metal ion is missing from a lattice site, its absence can be balanced by an adjacent metal ion that assumes a *higher oxidation state*. Examples include FeO and NiO.

1.15.7 *Excess Interstitial Anions*

While theoretically possible, this defect involves extra negative ions entering interstitial spaces and being charge compensated by neighboring metal ions carrying higher charges. However, because anions are generally larger in size, this type of defect is *rarely observed.*

1.16 Band Theory of Semiconductors

In a crystalline solid, the *electrons in the outer shells* of atoms are influenced by adjacent atoms. This interaction alters their *wave functions*, causing energy levels to *shift and split* into *energy bands.*

When individual atoms aggregate to form a solid, interatomic forces of attraction and repulsion create an *optimal interatomic spacing*. This leads to significant modifications in the *electron energy levels*, ultimately affecting the material's *electrical properties.*

As the atoms come closer, their discrete energy levels expand into *continuous energy bands*. For example, during the formation of a *diamond lattice* from isolated carbon atoms, which have a ground-state electronic configuration of $1s^2 2s^2 2p^2$, the $2s$ and $2p$ orbitals interact and *merge* into a broader $2s$-$2p$ band capable of holding up to *eight electrons.*

As atoms approach their equilibrium spacing, this merged band splits into two distinct regions:

- The *lower energy band* becomes the *valence band.*
- The *upper energy band*, separated by a forbidden energy gap (E_9), becomes the *conduction band.*

Each of these bands can hold up to four spin states. The inner $1s$ *orbitals* remain localized and do not engage in bonding due to their strong attraction to the nucleus.

The remaining electrons from $2s$ and $2p$ orbitals, however, participate in bonding through *hybridization* and the *molecular orbital framework*. These orbitals lose their individual identities and distribute their electrons across both bands (see Fig. 1.6). At *0 Kelvin*, all available states in the *valence band* are filled, whereas the *conduction band* remains vacant.

Each solid has a distinct *band structure*, which determines its *electrical behavior*. Upon the application of an *electric field*, electrons in lower energy levels can be *accelerated* to higher bands, provided there are vacant states. This transition underlies the material's *conductivity*.

The distinction between semiconductors and insulators lies primarily in the size of the *forbidden energy gap (Eg)* between their valence and conduction bands. In *semiconductors*, this energy gap is relatively narrow (approximately 1 eV), enabling electrons to transition from the valence band to the conduction band upon absorbing

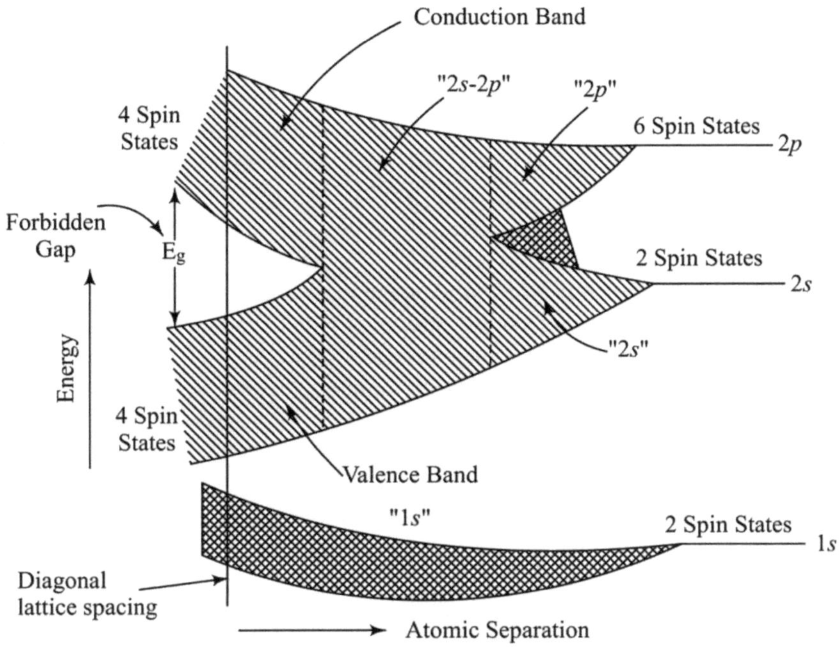

Fig. 1.6 Formation of energy bands as a diamond crystal is formed by bringing together isolated carbon atoms

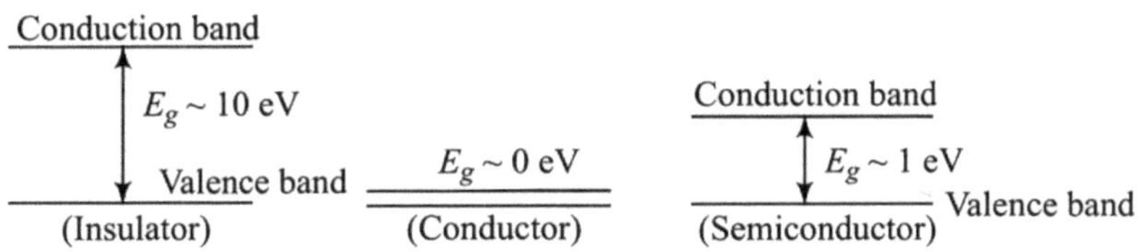

Fig. 1.7 Forbidden gap energy (E_g) levels for different materials

minimal thermal or optical energy. At ambient temperature, this small energy input is sufficient to promote a substantial number of electrons, thus contributing to *electrical conductivity*.

Conversely, in *insulators* like diamond, the energy gap is significantly larger (about 5 eV). The magnitude of this gap prevents thermal or optical energy from enabling electron excitation, resulting in minimal or no conduction. An energy gap as large as 10 eV makes it virtually impossible for electrons to move across bands, rendering such materials excellent electrical insulators.

Figure 1.7 illustrates the comparative energy gap levels observed in insulators, conductors, and semiconductors.

1.17 Intrinsic and Extrinsic Semiconductors

An *intrinsic semiconductor* is a perfect crystal lattice devoid of any impurities or structural flaws. At absolute zero, such a material has no free charge carriers, as the valence band is entirely occupied, while the conduction band remains vacant. However, when temperature increases, thermal energy excites electrons into the conduction band, resulting in the creation of *electron-hole pairs*, which serve as the primary charge carriers.

In contrast, *extrinsic semiconductors* are engineered by intentionally introducing impurities through a process called *doping*, which enhances conductivity by increasing either electron or hole concentration.

There are two primary types of extrinsic semiconductors:

- n-type: electron-dominant
- p-type: hole-dominant

When doping causes the equilibrium concentrations of electrons (n_0) and holes (p_0) to differ from the intrinsic carrier concentration (n_i), the semiconductor is considered extrinsic.

For instance, doping *germanium (Ge)* or *silicon (Si)* with *group V elements* such as phosphorus (P), arsenic (As), or antimony (Sb) introduces impurity levels near the conduction band. These impurity levels—termed *donor levels*—are already occupied at 0 K, and minimal thermal energy is sufficient to elevate these electrons to the conduction band, thereby forming n-type semiconductors. In such materials, $n_0 \gg p_0$.

Similarly, introducing *group III elements* like boron (B), aluminum (Al), gallium (Ga), or indium (In) into Ge or Si results in energy levels near the *valence band.* These levels are devoid of electrons at 0 K, but at slightly elevated temperatures, electrons from the valence band are thermally excited into these impurity levels, leaving behind *holes*. These holes serve as majority charge carriers, and the corresponding impurity levels are known as *acceptor levels*. Materials with dominant hole populations are termed *p-type semiconductors.*

Figures 1.8 and 1.9 respectively depict electron excitation to the conduction band and the formation of holes due to acceptor levels.

1.18 Donor Binding Energy

The energy required to excite the fifth valence electron of a donor atom into the conduction band is known as the *donor binding energy*, which can be approximated by the equation:

$$E = \frac{mq^4}{2K^2h^2}$$

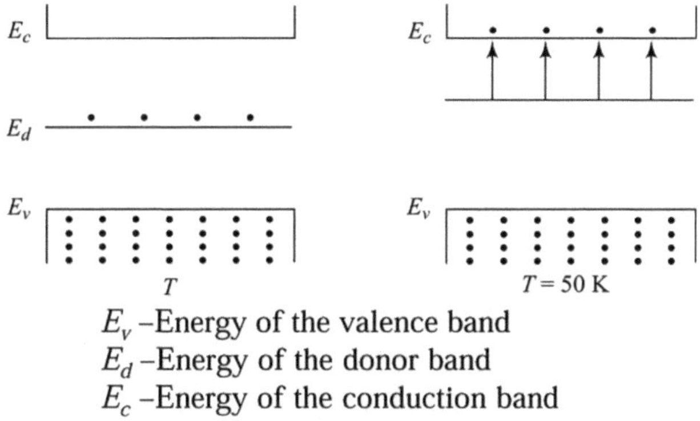

Fig. 1.8 Excitation of electrons to the conduction band

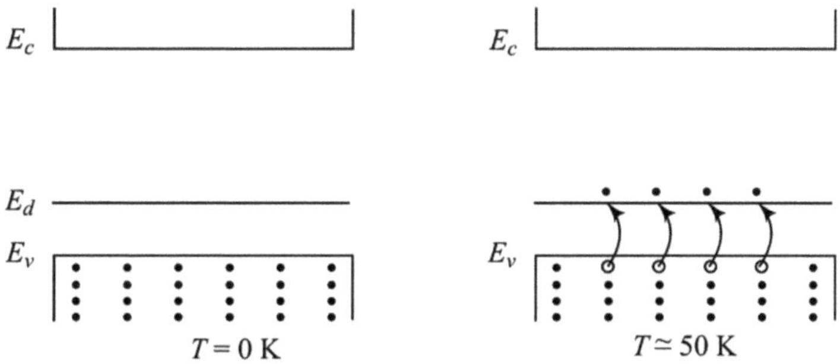

Fig. 1.9 Acceptance of valence band electrons by an acceptor level and the resulting reaction of holes

where:

- m is the effective mass of the electron in the semiconductor
- q is the electronic charge
- h is Planck's constant
- K is modified from the free-space constant to include the relative permittivity:

$$K = 4\pi\varepsilon_0\varepsilon_r$$

For example, in *germanium* where $\varepsilon_r = 16$ and $m_n = 0.12m_0$, substituting into the expression yields

$$E = \frac{m_n q^4}{8\left(\varepsilon_0\varepsilon_r\right)^2 h^2} = \frac{0.12\left(9.11 \times 10^{-31}\right)\left(1.6 \times 10^{-19}\right)^4}{8\left(8.85 \times 10^{-12} \times 10\right)^2\left(6.63 \times 10^{-34}\right)^2}$$

$$= 1.02 \times 10^{-21}\ \text{J} \quad \text{or} \quad 0.0064\ \text{eV}$$

Hence, only a small energy (0.0064 eV) is required to move a donor electron into the conduction band in Ge.

1.19 Hydrogen Bonding

In numerous chemical compounds, hydrogen atoms serve as *bridging elements* between two highly electronegative atoms, forming a *hydrogen bond*. This attractive interaction occurs when hydrogen is covalently bonded to atoms such as *fluorine (F), oxygen (O)*, or *nitrogen (N)*, which significantly *polarize the shared electron pair*, imparting a partial positive charge to the hydrogen atom.

Hydrogen bonding is generally illustrated using dotted lines, for example,

$$\cdots\cdot H - F\cdots H - F\cdots H - F\cdots\cdots$$

There are two main categories of hydrogen bonding:

1. Intermolecular hydrogen bonding:

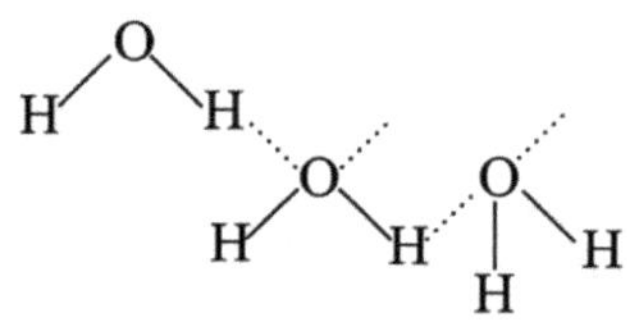

   When the hydrogen atom forms a bridge between two separate molecules. Examples include water (H_2O), ammonia (NH_3), and hydrofluoric acid (HF).
2. Intramolecular hydrogen bonding:

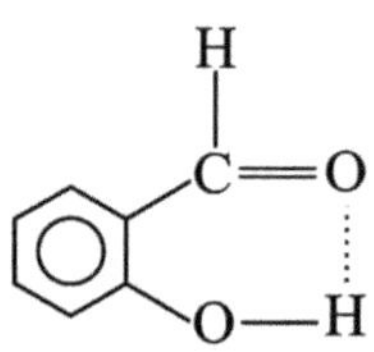

Salicylaldehyde

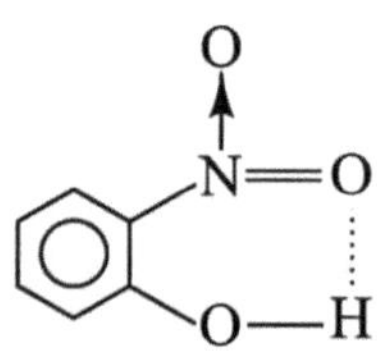

o-nitrophenol

   When the hydrogen bond exists within a single molecule. Classic examples are *o*-nitrophenol and *o*-aminophenol.

1.19.1 Effects of Hydrogen Bonding

Hydrogen bonding significantly influences the physical properties of compounds:

- ***Boiling point elevation:*** Substances exhibiting strong hydrogen bonding tend to have higher boiling points.
- ***Solubility enhancement:*** Organic molecules capable of hydrogen bonding are generally more soluble in water. For example, *dimethyl ether* is water miscible due to hydrogen bonding, whereas *dimethyl sulfide* is not.
- ***Increased viscosity:*** The presence of hydrogen bonds impedes the free movement of molecules, leading to higher viscosity in such liquids.
- ***Anomalous behavior of water:*** Water exhibits unique properties like having *greater density in liquid form* than in solid form (ice), attributed to its extensive hydrogen bonding.

For intramolecular hydrogen bonding:

- The *melting and boiling points* of *para-isomers* are typically higher than those of *ortho-isomers* due to reduced intermolecular attractions in the latter.
- The *solubility* of organic compounds in polar solvents is often improved due to internal hydrogen bonding facilitating interactions with the solvent.

Exercises

1. What do you understand by covalent bond? Give three characteristics of a covalent bond.
2. Discuss the salient features of molecular orbital theory of bonding. How does it differ from valence band theory?
3. What is Heitler-London theory? Write its important features and limitations.
4. Describe Pauling's and Slater's theory of bond formation.
5. Explain clearly the term hybridization.
6. Explain the following:

   (a) O_2 is more reactive than N_2.
   (b) F_2 is more reactive than O_2.
   (c) B_2 exists but Be_2 does not.
   (d) N_2, CN, and CO molecules have triple bond.
   (e) $H_2^+$is more reactive than H_2.

7. Write a short note on hydrogen bonding.
8. On the basis of hydrogen bonding, explain the following:

   (a) Ice is lighter than water.
   (b) *o*-Nitrophenol can be separated from *p*-nitrophenol by steam distillation.

9. Write the energy level diagram from O_2^+ and calculate the bond order.
10. Define the term resonance. What are the necessary conditions for resonance hybrid? Explain applications of resonance with the help of suitable examples.

11. Outline a Born-Haber cycle for the formation of an ionic compound MCl. Define the terms used. Discuss the use of Born-Haber cycle for the calculation of lattice energy.
12. Explain on the basis of MO theory as to why:

    (a) O_2 molecule is paramagnetic while N_2 is diamagnetic
    (b) Hydrogen forms diatomic molecule while helium remains monoatomic
    (c) The bond order of O_2 is less than that in O_2 which, in turn, is less than that in O_2
    (d) The internuclear distances in O – O bonds are as follows:

$$\begin{array}{cccc} O_2^+ & O_2 & O_2^- & O_2^{2-} \\ 1.21\text{Å} & 1.21\text{Å} & 1.30\text{Å} & 1.49\text{Å} \end{array}$$

        (i) Account for the gradation of the internuclear distances in the above series.
        (ii) Which of these species are paramagnetic?

13. Write a short note on:

    (a) *n*-type and *p*-type semiconductors
    (b) Bragg's equation

14. Calculate the bond order in B_2, N_2, O_2 and F_2.
15. State differences between σ (Sigma) and $\pi(pi)$ bonds.
16. Calculate the number of σ (sigma) and $\pi(pi)$ bonds in C_2H_2, C_2H_4, CO_3^-, NH_3, PCl_5.
17. Predict the shape, bond angle, dipole moment, and hybridization of the following compounds on the basis of VSEPR theory:

$$Ag(CN)_2^+, XeO_3, H_2O, PCl_5, XeF_4, XeF_2$$

18. Predict the shape, bond angle, dipole moment, and hybridization of the following compounds on the basis of VSEPR theory: $SO_4^{2-}, NH_4^+, BF_4^-$, $XeO_4, NH_3, XeOF_2$
19. Explain the paramagnetic nature of O_2 and its less stability in comparison to that of N_2.
20. What is meant by "Bond order?" Give a formula to calculate it. Calculate the bond order for NO.
21. Which of the following is paramagnetic. Give reasons in support of your answer:

$$O_2, O_2^- N_2, C_6H_6, CO_3^-$$

22. Explain the order of acidity of $BF_3 < BCl_3 < BBr_3$.

23. Arrange the following in the decreasing order of acidity and C-H bond length: $CH_4, C_2H_6, C_2H_4, C_2H_2$.
24.
   (a) Which types of chemical bonds are involved in the formation of the following compounds? Explain the bond formation on the basis of the electronic structure of the atoms involved in each case:
      (i) NaCl
      (ii) HCl
      (iii) NH_4Cl
   (b) Which of the following compounds has the highest melting point:

$$C_3H_7OH, KCl, C_6H_5COOH$$

25. What is bond order? Calculate the bond order of He_2^+, O_2^- and O_2^{2-} molecules.
Hints:

$$\text{Bond order} = \frac{1}{2}(N_b - N_a)$$

where N_b = Number of electrons in bonding M.O. and N_a = Number of electrons in antibonding M.O.

(i) He_2^+, the configuration is $(\sigma_2 s^2)(\sigma^* 1s^1)$
No of electrons in bonding M.O. = 2
No. of electrons in antibonding M.O. = 1

$$\text{B.O.} = \frac{1}{2}(2-1) = \frac{1}{2}$$

(ii) O_2^- molecule has electronic configuration as $\sigma 2s^2, \sigma^* 2s^2, \sigma 2p_x^2, \pi 2p_y^2, \pi 2p_z^2, \pi 2p_y, \pi^* 2pz$. O_2^- molecule has 17 electrons and the number of electrons in bonding M.O. is = 10, while the number of electrons in antibonding orbital is 7. Hence,

$$\text{B.O.} = \frac{1}{2}(10-7) = \frac{3}{2} = 1.5$$

(iii) O_2^{2-} molecule has electronic configuration as $\sigma 2s^2, \sigma^* 2s^2, \sigma 2p_x^2, \pi 2p_x^2, \pi 2p_y^2, \pi^* 2p_x^2, \pi 2p_y^2$ O_2^{2-} has 18 electrons, the number of electrons in bonding orbital is = 10, while number of electrons in antibonding orbitals is 8. Hence, B.O. = $\frac{1}{2}(10-8) = 1$.

26. Explain in terms of valence bond theory the following observations:
   (a) CCl_4 and $SiCl_4$ are tetrahedral.
   (b) CO_2 molecule is linear and SO_2 is angular.

(c) BF_3 is planar but NH_3 in not.
(d) H-S-H angle in H_2 S is closer to 90° whereas the H–O–H angle in H_2O is 104.5°
(e) [***Hints:*** According to valence shell electron pair repulsion theory, the force of repulsion in the various electron pairs around the center is lone pair-lone pair > lone pair-bond pair > bond pair-bond pair. Therefore, there will be a greater distortion in H_2S as well as in water molecule. The bond angle in case of water is 104.5°, but the size of sulfur atom is greater than oxygen atom. Therefore, in sulfur atom, the two lone pairs of electrons are spread apart and the H-S bond move away from each other. The bond angle between lone pair and bond pair orbital becomes 90°.]

27. The bond energies of H_2 and I_2 are 435.4 kJ mol^{-1} and 150.6 kJ mol^{-1}, respectively. Calculate the bond energy for HI molecule.
Hints: For a diatomic molecule like H_2. The dissociation energy of the substance is identical to the bond energy of the molecule. The bond energy of HI can be calculated from the bond energies of H_2 and I_2 by taking their mean values:

$$\begin{aligned}\text{Bond energy of HI} &= \frac{1}{2}\,(\text{Bond energy of H}_2) + \frac{1}{2}\,(\text{Bond energy of I}_2)\\ &= \frac{1}{2}(435.4) + \frac{1}{2}(150.6)\\ &= 293.0\,\text{kJ mol}^{-1}\end{aligned}$$

Chapter 2
Acids and Bases

The most important reaction in all chemical and biological systems is an acid-base reaction. Before we discuss the definitions of acids and bases, let us consider some of their general properties.

Acids (a) Change blue litmus to red; (b) sour, for example, citrus fruits that contain citric acid; (c) conduct electricity in aqueous solution; and (d) react with carbonates such as $CaCO_3$ or Na_2CO_3 and CO_2 gas is evolved.

Bases (a) Change red litmus to blue, (b) bitter taste, (c) conduct electricity in aqueous solution, and (d) soapy or slippery to touch.

By the end of the eighteenth century, the products of reaction of acids and bases were recognized as salts. For example when HCl reacts with a water-soluble alkali, a salt is formed:

$$2HCl + \underset{\text{slaked lime}}{Ca(OH)_2} \longrightarrow \underset{\substack{\text{Calcium chloride} \\ \text{(salt)}}}{CaCl_2} + 2H_2O$$

2.1 Arrhenius Concept

In the late nineteenth century, Svante Arrhenius, a Swedish chemist (1859–1927) proposed that an acid is a substance that produces H^+ (hydrogen ions) in aqueous solution. Similarly, a base is defined as a substance that gives OH^- (hydroxide ions) in aqueous solution. Thus,

$$HCl(aq) \longrightarrow H^+(aq) + Cl^-(aq)$$

$$HNO_3(aq) \longrightarrow H^+(aq) + NO_3^-(aq)$$

© The Author(s), under exclusive license to Springer Nature Switzerland AG 2026

R. N. Goyal, H. Goel, *Textbook of Engineering Chemistry*,
https://doi.org/10.1007/978-3-032-22068-4_2

and

$$NaOH(aq) \longrightarrow Na^+(aq) + OH^-(aq)$$

$$KOH(aq) \longrightarrow K^+(aq) + OH^-(aq)$$

represent the ionization of acids or bases in aqueous solution.

2.2 Limitations of Arrhenius Concept

The Arrhenius concept satisfactorily explains acids and bases for protonic acids (acids having ionizable H atoms) or metal hydroxides. However, the concept has several limitations:

1. It describes acids and bases only in aqueous solutions, i.e., it refers to ions – H^+ and OH^- derived from water.
2. Many substances that do not have OH^- increase the OH^- concentration when added to water, for example, liquid NH_3 in water.

To overcome the limitations of Arrhenius concept, a more general acid-base concept was proposed by J.N. Bronsted and J.M. Lowry. Before we discuss the concept of Bronsted and Lowry, let us deal with properties of water because we are most concerned with acids and bases in water.

2.3 Water Auto-ionization

The hydrogen ion (H^+) is a base proton having a radius of less than 0.1pm. As it is a tiny particle having positive charge, it is readily attracted by the negative water dipole and thus is hydrated in the aqueous solution:

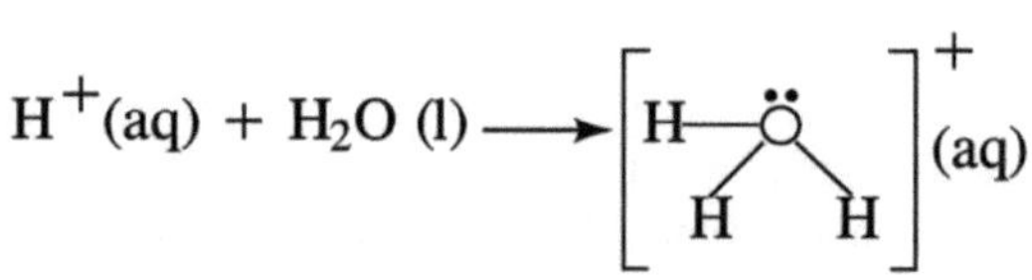

The above reaction indicates that a proton in water can only exist associated with water molecule. The simple proton water complex (H_3O^+) is known as *hydronium ion*. However, it has been established that a proton may be associated with more than one water molecules such as

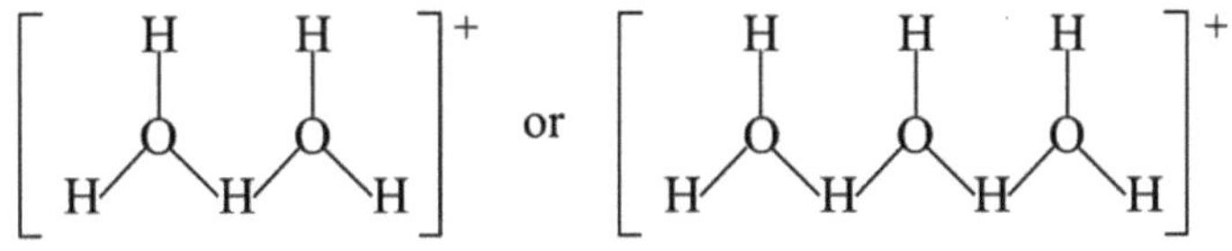

$H_5O_2^+$ $H_7O_3^+$

but the quantitative treatment of acid ionization is not affected by writing either H^+ or H_3O^+ or $H_5O_2^+$ and so on. For the sake of simplicity, we will write proton in aqueous solution as H^+ or H_3O^+.

For hydroxide ion (OH^-) to exist in the water, the presence of an acid such as HCl is not necessary. Two molecules of water can interact to form hydronium ion (H_3O^+) and hydroxide ion and the process is known as *auto-ionization* of water. This is the reason that the purest form of water conducts electricity to a small extent:

$$2H_2O(aq) \rightleftharpoons H_3O^+(aq) + OH^-(aq)$$

2.4 Bronsted-Lowry Concept of Acids and Bases

A new concept of acids and bases was independently suggested by J.N. Bronsted in Copenhagen and J.M. Lowry in Cambridge in 1923. According to them, *an acid is a substance that can donate a proton to other substances. Similarly, a base is a substance capable of accepting a proton.* For example, in the equation:

$$HCl(aq) + H_2O(l) \longrightarrow H_3O^+(aq) + Cl^-(aq)$$

HCl is donating a proton, whereas H_2O is accepting a proton. Thus, the former is an acid and the latter is a base. Acids such as HF, HNO_3, and CH_3COOH, are capable of donating proton. When an acid is capable of donating only one proton, it is called as *monoprotic* acid. However, some acids such as H_2SO_4, H_2CO_3, and H_2S are able to donate more than one proton and hence are called polyprotic acids.

Like acids, there are bases that can accept one or more protons and are respectively called monoprotic or polyprotic bases. Some common polyprotic bases are S^{2-}, $SO_4{}^{2-}$, PO_4^{3-}, HS^- etc.:

$$S^{2-}(aq) + H^+(aq) \longrightarrow HS^-(aq)$$

$$HS^-(aq) + H^+(aq) \longrightarrow H_2S(aq)$$

In contrast to Bronsted and Lowry acids or bases, there are also amphiprotic molecules, ions, or substances that can behave either as acid or base. For example, water can act as acid or base depending upon the conditions employed. With HCl, water accepts a proton and acts as a base:

Table 2.1 Some common amphiprotic ions

Acting as base	$\xleftarrow{H^+}$ Amphiprotic ion $\xrightarrow{-H^+}$	Acting as acid
H_2S	SH^-	S^{2-}
H_2CO_3	HCO_3^-	CO_3^{2-}
$H_2PO_4^-$	$HPO_4{}^{2-}$	PO_4^{3-}
H_3PO_4	$H_2PO_4^-$	HPO_4^{2-}

$$HCl(aq) + H_2O(l) \longrightarrow H_3O^+(aq) + Cl^-(aq)$$

whereas in NH_3, water donates a proton and acts as an acid:

$$NH_3(aq) + H_2O(l) \longrightarrow NH_4^+(aq) + OH^-(aq)$$

A list of some common amphiprotic ions is presented in Table 2.1.

Interestingly, whenever there is a reaction between two substances, one is an acid relative to water and other is a base relative to water:

$$HNO_3(aq) + NH_3(aq) \longrightarrow NH_4^+(aq) + NO_3^-(aq)$$

A substance neither acidic nor basic in character is formed in a reaction, and such a reaction is often called neutralization reaction.

2.5 Conjugate Acid-Base Pairs

The concept of acid-base pair is the extension of Bronsted-Lowry definition, according to which *the conjugate base of an acid is the species that is formed after removal of proton from the acid.* Thus, the conjugate base of HCl

$$HCl(aq) \longrightarrow H^+(aq) + Cl^-(aq)$$

is Cl^-. On the other hand, when proton is added to a base, a conjugate acid is formed. Thus, the conjugate acid of NH_3

$$NH_3(aq) + H_2O(l) \longrightarrow NH_4^+(aq) + OH^-$$

in the equation is NH_4^+. Similarly, we can consider several other compounds and write their conjugate acid or base. Some important acids and their conjugate bases are given in Table 2.2.

The relative strengths of acids and bases can also be explained using Bronsted-Lowry concept. Some acids are better proton donors than others in water; similarly, some bases are better proton acceptors than others. A strong Bronsted acid is the

Table 2.2 Some common acids and their conjugate bases

Acid	Conjugate base
HCl	Cl^- (Chloride ion)
HCN	CN^- (Cyanide ion)
HNO_3	NO_3^- (Nitrate ion)
H_2SO_4	HSO_4^- (Hydrogen sulfate ion)
H_2CO_3	HCO_3^- (Bicarbonate ion)

one that has a strong tendency to lose a proton in water, and a strong base is the one that strongly accepts a proton from water to give OH^-.

Let us consider two acids HCl and acetic acid (CH_3COOH). A solution of HCl in water mainly consists of ions H_3O^+ and Cl^-; thus, HCl completely ionizes in water. On the other hand, dilute solution of acetic acid has very few ions as less than 1% ionization of acetic acid occurs. For example, if we take 0.1 M solution of acetic acid in water, the ions present are nearly 0.001 M as shown below:

$$\underset{0.1\,\text{M}}{CH_3COO(aq)} + H_2O(l) \rightleftharpoons \underset{0.001\,\text{M}}{H_3O^+(aq)} + \underset{0.001\,\text{M}}{CH_3COO^-(aq)}$$

Similarly, oxide ion is a strong base in aqueous solution because it readily accepts proton from water to give OH^-:

$$\underset{\text{oxide ion}}{O^{2-}(aq)} + H_2O(l) \longrightarrow 2OH^-(aq)$$

whereas ammonia is a weak base because it produces very few OH^- in water:

$$NH_3(aq) + H_2O(l) \rightleftharpoons NH_4^+(aq) + OH^-(aq)$$

However, the important aspect of Bronsted-Lowry concept is that if the acid is strong, its conjugate base is weak. Similarly, if the base is strong, its conjugate acid is weak. This can be illustrated by the following examples:

$$HCl(aq) + H_2O(l) \rightleftharpoons H_3O^+(aq) + Cl^-(aq)$$

HCl is a strong acid because its tendency to donate proton to $H_2O(l)$ is higher and conjugate base Cl^- is formed. As H_2O is accepting a proton, it acts as a base to form conjugate acid H_3O^+. The reaction, however, proceeds completely to the right, and therefore at equilibrium, practically no HCl will be left in the solution. This also suggests that acid and base on the left side of the equation are stronger than the acid and base formed on the right side of the equation. Thus, Cl^-, conjugate base of strong acid HCl is weaker than HCl. Similarly, we can write an equation for ionization of acetic acid as:

$$CH_3COOH(aq) + H_2O(l) \rightleftharpoons CH_3COO^-(aq) + H_3O^+(aq)$$

Table 2.3 Relative strengths of some common acids and bases

	Conjugate acid	Conjugate base	
	$HClO_4$	ClO_4^-	
	H_2SO_4	HSO_4^-	
	HCl	Cl^-	
	HNO_3	NO_3^-	
Decreasing acid strength ↓	H_3O^+	H_2O	Decreasing base strength ↑
	H_3PO_4	$H_3PO_4^-$	
	CH_3COOH	CH_3COO^-	
	H_2S	SH^-	
	HCN	CN^-	
	H_2O	OH^-	
	NH_3	NH_2^-	
	CH_3NH_2	CH_3NH^-	

CH_3COOH ionizes to a very small extent and the conjugate base CH_3COO^- has a strong tendency to abstract a proton from H_3O^+; it is expected that at equilibrium, the solution will consist mainly of acetic acid with a very little concentration of CH_3COO^- and H_3O^+. Thus, CH_3COO^- is a strong base than H_2O, and its conjugate acid CH_3COOH is a weak acid.

Thus, it proves that strong acid has a weak conjugate base and strong base has a weak conjugate acid. Table 2.3 provides information about the relative strengths of some common acids and their conjugate bases.

2.6 Limitations of Bronsted-Lowry Concept

1. The theory fails to explain the acidic character of proton-free molecules such as CO_2, SO_2, NO_2.
2. The protonic definition cannot be used to explain reactions occurring in non-protonic solvents such as $COCl_2$, SO_2, and BrF_3.

2.7 Lewis Concept or Electron Pair Donor-Acceptor System

According to the concept, "An acid is any species that can accept an electron pair i.e., make a coordinate bond and a base is any species which donates an electron pair and form coordinate bond."

Some of the examples of Lewis acids and bases have been discussed below:

1. BF_3 is electron deficient, whereas NH_3 contains a lone pair of electrons:

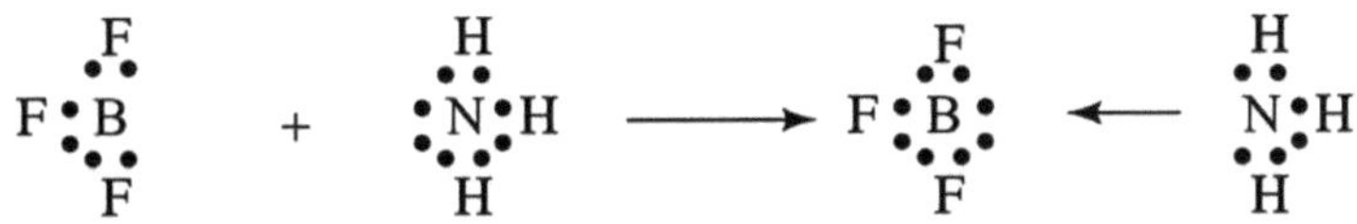

acid (electron deficient) | base (with lone pair) | Lewis Adduct (coordination compound)

2. The heavy metal cations that are electron deficient with empty orbitals act as acids and form Lewis adduct with substances (base) containing a lone pair of electrons:

$$Ag^{+} + 2NH_3 \longrightarrow [NH_3 \longrightarrow Ag \leftarrow NH_3]^{+}$$

$$Cu^{2+} + 2H_2O \longrightarrow [H_2O \longrightarrow Cu \leftarrow H_2O]$$

$$Fe^{2+} + 6CN^{-} \longrightarrow$$

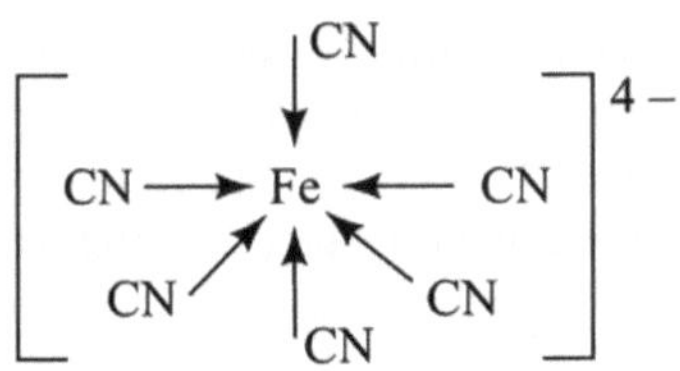

3. Molecules having multiple bonds, between atoms of different electronegativity such as $O = C = O, O = S = O$.

2.8 Applications of Theory

1. The concept is more general in which the behavior is not dependent on an element or a solvent.
2. It explains metallic and nonmetallic compounds and their basic and acidic properties, respectively.
3. The theory explains nonaqueous solvent reactions.

2.9 Limitations

1. The theory explains fast acid-base reactions; however, no explanation of slow reactions is possible.
2. The strength of acids and bases depend on the type of reaction. It is not possible to arrange them in any order. For example, the fluoride complex of Be^{2+} is more stable than that of Cu^{2+}, while the amine complex of Cu^{2+} is more stable than that of Be^{2+}.

2.10 Strength of Acids and Bases

The relative strength of an acid or base can also be expressed quantitatively using equilibrium constant. For example, if an acid HA dissociates into its ions in aqueous solution, we can write

$$HA(aq) \rightleftharpoons H^+(aq) + A^-(aq)$$

and hence

$$K_a = \frac{[H^+][A^-]}{[HA]}$$

where the subscript "a" in K_a represents that it is for acid ionization. Similarly, for a base, we use the term K_b. Higher values of K_a indicate strong acid and lower values indicate weak acid. Similarly, higher values of K_b indicate the base as a strong one. The equilibrium constant or ionization constant or dissociation constant values can vary in the range 10^{-2}–10^{-13} depending upon the acidity or basicity of the acids and bases. The ionization constants of some weak acids and bases is given in Table 2.4 to provide an idea about these values.

The values of K_a and K_b depend on the temperature. The dissociation reaction of acid presented above can also be used for determining the relative strength of acids or bases considering the degree of dissociation (α). For example, the dissociation of acetic acid can be represented by the equation:

$$CH_3COOH \rightleftharpoons CH_3COO^- + H^+ \quad (2.1)$$

Considering the original and equilibrium concentrations, we can write

original conc.	C	0	0
equilibrium conc.	$C(1-\alpha)$	α	α

Table 2.4 Ionization constants of some weak acids and bases

Acid	K_a	Base	K_b
H_2SO_3	1.2×10^{-2}	HSO_3^-	8.3×10^{-13}
HSO_4^-	1.2×10^{-2}	SO_4^{2-}	8.3×10^{-13}
HF	7.2×10^{-4}	F^-	1.4×10^{-11}
HCOOH	1.8×10^{-4}	$HCOO^-$	5.6×10^{-11}
$C_6H_5\overset{+}{N}H_3$	2.6×10^{-5}	$C_6H_5NH_2$	3.8×10^{-10}
C_6H_5COOH	6.3×10^{-5}	$C_6H_5COO^-$	16×10^{-10}
HCN	4.0×10^{-10}	CN^-	2.5×10^{-5}
HS^-	1.3×10^{-13}	S^{2-}	0.077

where the total molar concentration of the acetic acid is C and α is the degree of dissociation. Thus,

$$K_a = \frac{[CH_3COO^-][H^+]}{[CH_3COOH]}$$

or

$$K_a = \frac{[C\alpha][C\alpha]}{[C(1-\alpha)]} = \frac{C\alpha^2}{(1-\alpha)} \tag{2.2}$$

As the value of α is very small for weak acids, we can consider $1-\alpha = 1$ and hence

$$K_a = C\alpha^2$$

$$\alpha = \sqrt{\frac{K_a}{C}} \tag{2.3}$$

If the dissociation constants of two acids at the same concentration are K_{a1} and K_{a2}, respectively, then the ratio of their degree of dissociation will be

$$\frac{\alpha_1}{\alpha_2} = \sqrt{\frac{K_{a_1}}{K_{a_2}}} \tag{2.4}$$

because the term C (concentration) will cancel out in the ratio. Thus, the relative strength of any two weak acids can be represented by the ratio of the square root of their dissociation constant. If the value of α is known, the concentration of $[H^+]$ can also be calculated from Eq. (2.1) as

$$[H^+] = C\alpha$$

$$= C\sqrt{\frac{K_a}{C}} = \sqrt{CK_a}$$

Thus, the concentration of $[H^+]$ in a solution of weak acid is directly proportional to the square root of the dissociation constant of the acid.

Similarly, we can consider dissociation of the weak base:

$$NH_3(aq) + H_2O(l) \rightleftharpoons NH_4^+(aq) + OH^-(aq)$$

and the dissociation constant of NH_3 can be written as

$$K_b = \frac{[NH_4^+][OH^-]}{[NH_3][H_2O]}$$

Since very few water molecules are consumed as compared to the total concentration of H_2O, therefore we can treat the concentration of H_2O as constant; hence,

$$K_b = \frac{[NH_4^+][OH^-]}{[NH_3]}$$

and K_b is known as the base ionization constant.

2.11 Ionic Product of Water

In contrast to weak acids and bases, strong acids and bases lead to larger concentration of $[H^+]$ or $[OH^-]$ ions per mole of substance in water. If the concentrations of $[H^+]$ or $[OH^-]$ could be measured, it would be possible to predict the relative strengths of acids and bases. In this section, we will deal with one way to measure the concentrations of $[H^+]$ or $[OH^-]$.

As discussed earlier, the auto-ionization of water is a common phenomenon, and expressing the hydrated proton as H^+rather than H_3O^+, we can write an equilibrium reaction as

$$H_2O(l) \rightleftharpoons H^+(aq) + OH^-(aq)$$

Since OH^- is a much stronger base than water (see Table 2.3) and H_3O^+ is a stronger acid than water, the equilibrium will be from the right to the left side. It has also been proved that in pure water at 25 °C, only two molecules in 10^9 molecules exist in ionic form at any instant. Thus, the equilibrium constant of the reaction can be expressed as

$$K = \frac{[H^+][OH^-]}{[H_2O]}$$

As very few molecules of water are ionized, we can consider the concentration of water as unchanged and hence

$$K\,[H_2O] = [H^+][OH^-]$$

or

$$K_w = [H^+][OH^-] \tag{2.5}$$

The K_w is expressed as water ionization constant, and in pure water at 25 °C, its value is 1.0×10^{-14}. Also, the concentrations of $[H^+]$ and $[OH^-]$ in pure water are equal and found to be 1.0×10^{-7} M, respectively. Thus,

$$
\begin{aligned}
\left[H^+\right] &= [OH] = \sqrt{K_w} \\
&= \sqrt{1.0 \times 10^{-14}} = 1.0 \times 10^{-7}\,M
\end{aligned}
$$

It must be realized that whenever, in aqueous solution, the concentration of $\left[H^+\right] = \left[OH^-\right]$, the solution is called neutral. Hence, in acidic solution, $\left[H^+\right]$ will be more than 10^{-7} M, whereas in basic solution, $\left[OH^-\right]$ will be more than 10^{-7}; however, the product of $\left[H^+\right]$ and $\left[OH^-\right]$ will always be 1.0×10^{-14}. In other words, we can also say that in acidic solution, the concentration of $\left[H^+\right]$ should be more than 1.0×10^{-7} M, and in basic solution, the concentration of $\left[OH^-\right]$ should be more than 1.0×10^{-7} M.

Example 2.1 A household cleaning solution was analyzed for OH^- ions, and the concentration of OH^- is found to be 0.0025 M. What will be the concentration of H^+ ions? Will the cleaning solution be acidic or basic in nature?

Solution Using Eq. (2.5) for aqueous solutions, we can write

$$K_w = \left[H^+\right]\left[OH^-\right]$$

or

$$\left[H^+\right] = \frac{K_w}{\left[OH^-\right]}$$

The value of K_w is known to be 1.0×10^{-14}; hence,

$$
\begin{aligned}
\left[H^+\right] &= \frac{1.0 \times 10^{-14}}{0.0025\,M} \\
&= 4.0 \times 10^{-12}\,M
\end{aligned}
$$

As the concentration of $\left[H^+\right]$ is lesser than 1.0×10^{-7} M, the concentration of $\left[OH^-\right]$ will be more than 1.0×10^{-7} M. Hence, the cleaning solution will be basic in nature. As the concentration of $\left[H^+\right]$ or $\left[OH^-\right]$ ions is very small, a new convenient method is used to represent these concentrations and is known as pH. The pH of a solution is defined as the negative of base 10 logarithm of the hydrogen ion concentration. Thus,

$$pH = -\log\left[H^+\right]$$

Similarly, the pOH of a solution will be

$$pOH = -\log\left[OH^-\right]$$

For pure water, the concentration of $\left[H^+\right]$ and $\left[OH^-\right]$ is equal and 1.0×10^{-7} M; therefore, for pure water,

$$\begin{aligned} pH &= -\log\left[1.0 \times 10^{-7}\right] \\ &= -\log 1.0 - \log(10^{-7}) = -0 + 7.0 = 7.0 \end{aligned}$$

Also we can similarly deduce that pOH for water is also 7.0. Using Eq. (2.5), we can also write an expression that

$$K_w = \left[H^+\right]\left[OH^-\right] = 1.0 \times 10^{-14}$$

or on taking log of both sides

$$-\log\left[H^+\right]\left[OH^-\right] = -\log\left[1.0 \times 10^{-14}\right]$$

$$(-\log\left[H^+\right]) + (-\log OH^-) = 14$$

$$pH + pOH = 14 \tag{2.6}$$

Thus, at 25 °C, the sum of the pH and the pOH of a solution is equal to 14.

Other useful expressions one must remember from these equations are

$$\left[H^+\right] = 10^{-pH} \tag{2.7}$$

and

$$\left[OH^-\right] = 10^{-pOH} \tag{2.8}$$

The following schematic representation of pH scale and H^+ and OH^- concentrations further makes you understand the above discussion:

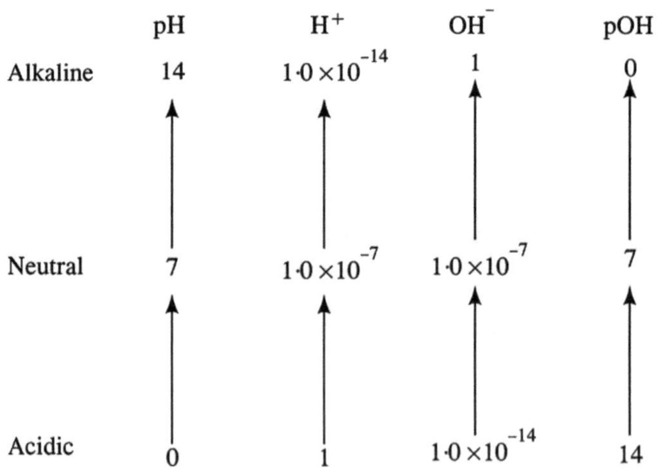

Note One must remember that pH scale is not limited to values ranging between 1 and 14 only. Most of the solutions have pH in this range; however, values lower than 1 and higher than 14 are also possible.

Let us consider few typical examples involving pH calculations.

Example 2.2 The pH of a human blood sample is 7.30; calculate the H^+ and OH^- concentrations.

Solution Using Eq. (2.7), we can write

$$\left[H^+\right] = 10^{-pH}$$

or

$$\begin{aligned}\left[H^+\right] &= 10^{-7.3}\\ &= 5.0 \times 10^{-8}\,M\end{aligned}$$

The value of $\left[OH^-\right]$ can be calculated using Eq. (2.5):

$$\begin{aligned}K_w &= \left[H^+\right]\left[OH^-\right]\\ 1.0 \times 10^{-14} &= \left[5.0 \times 10^{-8}\right]\left[OH^-\right]\end{aligned}$$

Hence,

$$\begin{aligned}\left[OH^-\right] &= \frac{1.0 \times 10^{-14}}{5.0 \times 10^{-8}}\\ &= 0.2 \times 10^{-6}\,M\end{aligned}$$

Alternatively, one can calculate pOH from the expression (2.6)

$$pH + pOH = 14$$

Hence,

$$\begin{aligned}pOH &= 14 - 7.3\\ &= 6.7\end{aligned}$$

Then,

$$\begin{aligned}pOH &= -\log\left[OH^-\right]\\ 6.7 &= -\log\left[OH^-\right]\\ \left[OH^-\right] &= 0.2 \times 10^{-6}\,M\end{aligned}$$

If different volumes of acids and bases of known molarity are mixed, one can calculate the pH of the resulting solution by determining the equivalents of acid or base left after neutralization.

Example 2.3 25 ml of 0.2 N HCl are mixed with 50 ml of 0.25 M NaOH. Calculate the pH of the resulting.

Solution Let us calculate the number of milliequivalents of both acid and alkali:

$$\begin{aligned}\text{milliequivalents of the acid} &= 25 \times 0.2 = 5.0\\ &= 50 \times 0.25\\ &= 12.5\end{aligned}$$

As acid and alkali will neutralize in the mixture, the remaining alkali equivalents (which are higher) will be equal to 12.5 − 5.0 = 7.5 ml in the total volume of 75.0 ml of the mixture.

Thus, the concentration of OH^-per liter in the mixture will be

$$= \frac{7.5 \times 10^3}{75 \times 10^3} = 0.10\,\text{M}$$

Now,

$$\left[H^+\right]\left[OH^-\right] = 1.0 \times 10^{-14}$$

Therefore,

$$\left[H^+\right] = \frac{1.0 \times 10^{-14}}{0.1} = 1.0 \times 10^{-13}$$

And thus pH of the mixture will be

$$\begin{aligned}\text{pH} &= -\log\left[H^+\right]\\ &= -\log 2\left[1.0 \times 10^{-13}\right]\text{pH} = 13\end{aligned}$$

Another important point in pH calculations is to check your answer for logical and reasonable explanation. For example, if you are asked to calculate the pH of 10^{-8} MHCl, your immediate reply will be 8 by using expression

$$\begin{aligned}\text{pH} &= -\log\left[H^+\right]\\ &= -\log\left[10^{-8}\right] = 8.0\end{aligned}$$

Table 2.5 pK_a values of some common acids

Name	Formula	pK_a
Acetic acid	CH_3COOH	4.8
Benzoic acid	C_6H_5COOH	4.2
Formic acid	$HCOOH$	3.8
Trichloroacetic acid	CCl_3COOH	0.6
Trifluoroacetic acid	CF_3COOH	0.2

However, if the answer is 8.0, the solution should be basic in nature. Remember, one cannot prepare a basic solution by adding very dilute solution of acid in water. In such cases, water becomes the species, as it has 10^{-7} M concentration of H^+ ions, which controls the pH, and the solution will be essentially neutral.

The dissociation constants of acids (K_a) and bases (K_b) discussed earlier in this chapter can also be represented by the expressions similar to pH to

$$pK_a = -\log K_a$$

$$pK_b = -\log K_b$$

convert small numerical values of K_a and K_b to convenient numbers. It is important to note that a stronger acid has a larger K_a value and smaller pK_a value, and a stronger base has a smaller K_b value and higher pK_b value. The pK_a values of some common acids are presented in Table 2.5.

Polyprotic Acids Many acids are capable of donating more than one protons such as phosphoric acid (H_3PO_4), sulfuric acid (H_2SO_4), oxalic acid ($H_2C_2O_4$), and hydrosulfuric acid (H_2S). It is noticed that in such acids, the loss of successive proton becomes more difficult. The K_a values for the successive ionization of protons confirm this fact. For example,

$$H_2S(aq) \rightleftharpoons HS^-(aq) + H^+(aq)$$

$$K_{a_1} = 1.0 \times 10^{-7}$$

$$\left.\begin{aligned} HS^-(aq) &\rightleftharpoons S^{2-}(aq) + H^+(aq) \\ K_{a_2} &= 1.3 \times 10^{-13} \end{aligned}\right] \tag{2.9}$$

Similarly, for other acids, it is noticed that each successive loss of proton becomes nearly 10^4–10^6 times difficult. Thus, the first ionization step is basically responsible for producing H^+ ions and hence the pH of these acids. The production of H^+ ions for the successive steps can be neglected due to their weak ionization.

2.12 Buffer Solutions

Before we discuss buffer solutions, let us see what happens if the salt of a weak acid is added to the solution of acid itself, for example, if we add sodium acetate to a solution of acetic acid. As acetic acid is a weak acid, it ionizes very little in water:

$$CH_3COOH \rightleftharpoons CH_3COO^- + H^+ \tag{2.10}$$

When sodium acetate is added, which completely ionizes to sodium and acetate ions,

$$CH_3COONa \longrightarrow CH_3COO^- + Na^+$$

More acetate ions have been added in the solution, and therefore the equilibrium of Eq. (2.10) will shift toward the left and hence the ionization of the acetic acid will further decrease. This phenomenon is in agreement with Le-Chatlier's principle to maintain the equilibrium. Similarly, if NH_4Cl is added to a weak base like NH_4OH, the dissociation is suppressed. This behavior of weak acids or bases is called *common ion effect*, which causes suppression of the dissociation on the addition of its own ion. In such cases, the concentration of $\left[H^+\right]$ or $\left[OH^-\right]$ can be calculated using the expression for ionization:

$$K_a = \frac{\left[H^+\right]\left[A^-\right]}{[HA]}$$

Hence,

$$\left[H^+\right] = \frac{K_a[HA]}{\left[A^-\right]}$$

If we consider CH_3COOH and CH_3COONa mixture, the dissociation will be

$$CH_3COOH \rightleftharpoons CH_3COO^- + H^+ \tag{2.11}$$

Hence,

$$\left[H^+\right] = \frac{K_a\left[CH_3COOH^-\right]}{\left[CH_3COO^-\right]} \tag{2.12}$$

Now $\left[CH_3COO^-\right]$ concentration will be basically from the complete ionization of sodium acetate, and the concentration of unionized acid will be same as that of acid; thus, the expression can be simplified to

$$\left[H^+\right] = \frac{K_a[\text{Acid}]}{[\text{Salt}]} \tag{2.13}$$

As K_a for the acid is constant, the ratio of the concentration of acid and its salt in the mixture controls the $\left[H^+\right]$ concentration of such mixtures. A useful property of such mixtures is that they resist the change in their pH values on the addition of an acid or base and are called *buffer solutions*. Let us see how such mixtures resist the charge in pH on addition of acid or base.

If a strong acid is added to the mixture of CH_3COOH – CH_3COONa, the $\left[H^+\right]$ ions reacts with $\left[CH_3COO^-\right]$ ions produced in Eq. (2.11) to give weak acid (CH_3COOH), which is very slightly dissociated. Hence, the change in pH is practically very little when a strong base is added, as it reacts

$$OH^- + CH_3COOH \rightleftharpoons CH_3COO^- + H_2O$$

with acetic acid to form water and hence does not affect the pH.

The pH of buffer solution can be calculated using Eq. (2.13), which on taking logarithm of both sides gives

$$\log\left[H^+\right] = \log K_a + \log\frac{[\text{Acid}]}{[\text{Salt}]}$$

on multiplying both sides by a negative sign

$$-\log\left[H^+\right] = -\log K_a + \log\frac{[\text{Salt}]}{[\text{Acid}]}$$

as $pK_a = -\log K_a$ from Eq. (2.9)

$$pH = pK_a + \log\frac{[\text{Salt}]}{[\text{Acid}]} \tag{2.14}$$

This equation is called *Henderson's equation* and is useful for calculating pH of the buffer solutions prepared by mixing weak acid and its highly dissociable salt. Similarly, when a mixture possesses weak base and its highly ionizable salt its pOH will be

$$pOH = pK_b + \log\frac{[\text{Salt}]}{[\text{Base}]} \tag{2.15}$$

Example 2.4 A buffer solution possesses 0.2 M acetic acid and 0.2 M sodium acetate per liter. Calculate the pH of the buffer solution. If 1.0 ml of 1 M HCl is added in the buffer, how much change will occur in pH. The dissociation constant of acetic acid is 1.8×10^{-5} at 25 °C

Solution Using Eq. (2.14),

$$pH = pK_a + \log \frac{[Salt]}{[Acid]}$$

as $pK_a = -\log K_a$ from Eq. (2.9)

$$pH = -\log K_a + \log \frac{[Salt]}{[Acid]}$$

$$pH = -\log \left(1.8 \times 10^{-5}\right) + \log \frac{0.20}{0.20}$$

$$= -\log \left(1.8 \times 10^{-5}\right)$$

$$= 4.744$$

$$\text{On adding HCl} = 10\,\text{ml of } 1\,\text{M HCl}$$

$$= 0.001\ \text{mole}$$

which is completely ionized, 0.001 mole of H^+ ions will be increased in the buffer mixture. Thus,

$$H^+ + CH_3COO^- \longrightarrow CH_3COOH$$

The amount of acetic acid will be increased and that of acetate ion will be decreased. The concentrations of CH_3COOH and $\left[CH_3COO^-\right]$ after adding HCl will be

$$[CH_3COOH] = \text{initial conc.} + 0.001$$

$$= 0.2 + 0.001$$

$$= 0.201\,\text{M}$$

$$\left[CH_3COO^-\right] = \text{initial conc.} - 0.001$$

$$= 0.2 - 0.001$$

$$= 0.199\,\text{M}$$

The change in volume of buffer after adding HCl will be negligible. Hence, pH after adding HCl will be

$$pH = pK_a + \log \frac{[Salt]}{[Acid]}$$

$$= 4.744 + \log \frac{[0.199]}{[0.201]} = 4.744 - 0.044$$

$$= 4.740$$

$$\text{Change of pH} = 4.744 - 4.740 = 0.004$$

2.13 pH Indicators

The pH of a solution can be determined by using an instrument called pH meter or chemically, by involving the use of a substance that changes its color with changes in pH in some known range. These substances are known as *indicators*, and they can exist in conjugate acid and base forms and have different colors depending on the form. Litmus paper is well-known for indicating the acidic or basic nature of a solution. It is coated with a natural plant juice having red color in solutions more acidic than pH < 5 and blue in solutions having pH > 8.

A common indicator you must have used in the laboratory is phenolphthalein, an acid-base indicator. The neutral form of phenolphthalein is colorless in acidic solutions, whereas, in alkaline solution, two protons are lost. The conjugate base formed has extended π-conjugation, which imparts a beautiful pink color to the solution:

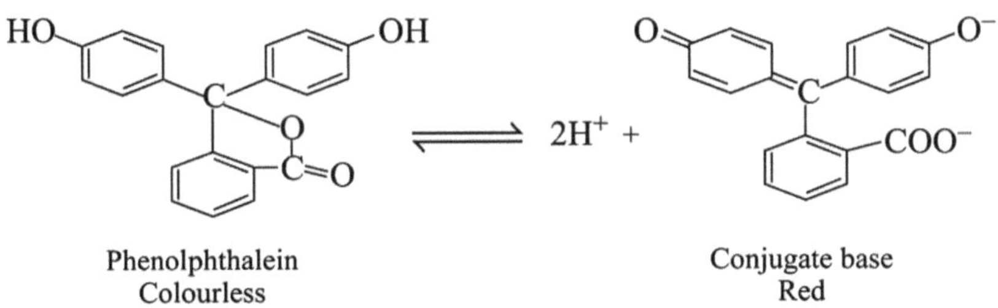

Some other common indicators that show changes in color with pH are shown in Table 2.6.

2.14 The pH in Biological System

The digestive system of a normal human body produces an average of 2–3 L gastric juice per day, which contains mostly hydrochloric acid. The pH of the gastric juice is ~1.5, which corresponds to HCl concentration of 0.03 M. Now, the question arises: where do H^+ ions come from, and what is the purpose of the highly acidic medium in the stomach.

Table 2.6 Some common indicators and their colors in different pH range

Indicator	Color	pH range
Crystal violet	Yellow	0–1
	Blue	1–2
Methyl orange	Red	3.2–4
	Yellow	4–4.4
Phenol	Yellow	6.8–7.2
	Red	7.2–8.2

The H^+ ions in the stomach come from the carbonic acid (H_2CO_3), which is formed by the hydration of CO_2:

$$CO_2(g) + H_2O(l) \rightleftharpoons H_2CO_3(aq)$$
$$H_2CO_3(aq) \rightleftharpoons H^+(aq) + HCO_3^-(aq)$$

These reactions occur in the blood plasma, and H^+ ions move across the membrane into the stomach. Once H^+ ions reach the stomach, an equal number of Cl^- ions also move from the blood plasma to maintain electrical balance. It is also interesting that these ions cannot diffuse back to blood plasma as cell membranes prevent them. The highly acidic medium in the stomach helps in the digestion of food. It also converts the proenzyme pepsinogen to its active form pepsin, which plays an important role in digestion. If the acid content in the stomach becomes excessively high, the H^+ ions begin to move through the membrane into the blood plasma. Such a situation can cause pain, inflammation, and swelling and may even lead to bleeding. Then the H^+ ion concentration in the stomach can be temporarily reduced by taking an antacid, which neutralizes excess HCl in gastric juices. The ingredients of the common antacids are calcium carbonate, magnesium carbonate, sodium bicarbonate etc., which neutralize HCl through simple reactions:

$$CaCO_3 + 2HCl \longrightarrow CaCl_2 + H_2O + CO_2$$
$$MgCO_3 + 2HCl \longrightarrow MgCl_2 + H_2O + CO_2$$
$$NaHCO_3 + HCl \longrightarrow NaCl + H_2O + CO_2$$

In normal conditions, the human body systems have a tendency to keep remarkably constant pH values of various body fluids.

Any alteration in pH values leads to pathological conditions. Also the enzymes that catalyze biochemical reactions function best at optimum pH values. The large number of solved numericals given below will help you practice all kinds of calculations involving concentration and pH of solutions.

1. (a) (i) Calculate the pH of $\frac{N}{1000}$ HCl;
   (ii) $\frac{N}{10000}$ NaOH;
   (iii) $\frac{N}{1000}$ NaOH [assume complete ionization].
   (b) (i) The concentration of

$$H^+ \text{ ion} = \frac{N}{1000} \text{ g equivalent/L}$$
$$= 10^{-3} \text{ g ion/L}$$

So,

$$\text{pH of the solution } = 3$$

(ii) The concentration of OH^- ions $= \frac{1}{10{,}000}$ g ion/L:

$$\left[OH^-\right] = 10^{-3} \text{ g ion/L}$$

As

$$K_w = \left[H^+\right]\left[OH^-\right] = 1.0 \times 10^{-14}$$

$$\left[H^+\right]\left[10^{-3}\right] = 1.0 \times 10^{-14}$$

or

$$\left[H^+\right] = \frac{1.0 \times 10^{-14}}{10^{-3}} = 10^{-11}$$

Hence,

$$\text{pH} = -\log 10^{-11} = 11$$

(iii) The concentration of OH^- ions $= \frac{N}{1000}$ g ion/L $= 10^{-4}$ g ion/L:

$$K_w = \left[H^+\right]\left[OH^-\right] = 1 \times 10^{-14}$$

Therefore,

$$\left[H^+\right]\left[10^{-4}\right] = 1 \times 10^{-14}$$

$$\left[H^+\right] = \frac{1 \times 10^{-14}}{10^{-4}} = 10^{-10}$$

Hence,

$$\text{pH} = \log 10^{-10}$$

or

$$\text{pH} = 10$$

(c) 0.1 N solution of a weak base had pH equal to 9. Calculate the percentage ionization of the base.

The pH of the given base $= 9$; therefore,

$$[H^+] = 10^{-9}$$

Now,

$$[H^+][OH^-] = 10^{-14}$$

$$[OH^-] = \frac{10^{-14}}{10^{-9}} = 10^{-5}\,\text{g ion/L}$$

Hence, $[OH^-]$ in 0.1 N solution of base is given by relationship:

$$[OH^-] = \text{Normality} \times \alpha$$

or

$$10^{-5} = \frac{1}{10} \times \alpha$$

or

$$\alpha = 10^{-5} \times 10 = 10^{-4}\,\text{g ion/L}$$

Percentage dissociation of base in 0.1 N solution

$$= 1.0 \times 10^{-4} = 10^{-2} = 0.01\%$$

2. (a) Find out the pH of a 0.002 N acetic acid solution if it is 2.3% ionized in the given dilution:

$$[H^+] = \text{Normality} \times \alpha$$
$$= 0.002\,\text{N} \times \frac{2.3}{100}$$
$$= 0.002 \times 0.023 = 0.000046$$

Therefore,

$$\text{pH} = -\log[0.000046]$$
$$\text{pH} = 5.6628 = -[-4.3372] = 4.3372$$

Hence, pH value of acid solution $= 4.3372$.

(b) (i) Calculate the pH of 0.01 N solution of an acid, which is ionized to 60% in a solution.

(ii) Calculate the pH value of a decinormal acetic acid, which is 1.8% ionized in a solution.

(iii) Calculate the pH value of 0.04 N HNO_3, 0.001 N, 1.0 N HCl, 0.1 N NaOH and 0.01 N NaOH solution.

(iv) Calculate the pH of a solution containing 4 g of NaOH per liter of water.

Solution:

(i) As

$$\left[H^+\right] = \text{Normality } \times \alpha = 0.01 \times \frac{60}{100} = 0.006$$

Now,

$$\begin{aligned} pH &= -\log[0.006] = -[\bar{3}.7782] \\ &= 2.2219 \end{aligned}$$

Hence, pH value of acid solution = 2.2219.

(ii)

$$\begin{aligned} \left[H^+\right] &= \text{Normality } \times \alpha \\ &= \frac{1}{10} \times \frac{1.8}{100} = 0.0018 \end{aligned}$$

Therefore,

$$\begin{aligned} pH &= -\log[0.0018] \\ &= \bar{3}.2553 = -[-2.7447] \\ &\equiv 2.7447 \end{aligned}$$

Hence, pH value of acetic acid solution = 2.7447.

(iii) Concentration of HNO_3 = 0.04 N; therefore,

$$\left[H^+\right] = 0.04 \text{ g equivalent/L}$$

[because HNO_3 is completely ionized]:

$$\begin{aligned} pH &= -\log\left[H^+\right] = -\log[0.04] \\ &= 1.398 \end{aligned}$$

Concentration of HCl = 0.0001 N.

$$[H^+] = 0.0001 \text{ g equivalent/L}$$

$$pH = -\log[H^-] = -\log\left(1 \times 10^{-4}\right) = 4$$

$1\,NHC1, H^+ = 10^0$; therefore,

$$pH = -\log_{10}\left(10^0\right) = 0$$

In 0.1 N NaOH, $OH^- = 10^{-1}$; therefore

$$H^+ = K_w/[OH^-] = 10^{-14}/10^{-1} = 10^{-13}$$

Therefore,

$$pH = -\log\left(10^{-13}\right) = 13.$$

Similarly in 0.001 NNaOH, $OH^- = 10^{-3}$; therefore

$$H^+ = 10^{-14}/10^{-3} = 10^{-11}$$

Therefore,

$$pH = -\log\left(10^{-11}\right) = 11.$$

(iv) If 40 g of NaOH is dissolved per liter, it will form 1 N solution. Therefore, 4 g of NaOH dissolved per liter forms $\frac{N}{10}$ solution.
In 0.1 NNaOH, $OH^- = 10^{-1}$. Therefore,

$$H^+ = K_w/[OH^-] = 10^{-14}/10^{-1} = 10^{-13}$$

Therefore,

$$pH = -\log_{10}\left(10^{-13}\right) = 13.$$

3. (a) In a 0.1 N solution of acetic acid, the degree of ionization (α) is 0.013. Calculate the hydrogen ion concentration of the solution per liter:

$$\begin{aligned}[H^+] &= \text{Normality} \times \text{degree of ionization} \\ &= 0.1 \times 0.013 \\ &= 0.0013 = 1.3 \times 10^{-3} \text{ g ion/L.}\end{aligned}$$

(b) What will be the pH of 10^{-8} molar HCl at 25 °C. HCl ionizes in H^+ and Cl^- completely; hence, the concentration of H^+ as g ions per liter in the solution is equal to the molarity of the acid:

$$\text{Concentration of } H^+ \text{ ions} = 10^{-8}\ \text{g ion/L}$$

$$pH = -\log\left[H^+\right] = -\log 10^{-8}$$

$$pH = -1 \times -8 = 8$$

Theoretically, the pH should be 8; however, the value will be close to 7 because H^+ ions of water also play a role.

(c) Assuming complete dissociation, calculate the pH values of 0.05 N – HCl solution; 0.05 M H_2SO_4 solution:

$$\text{Conc. of HCl} = 0.05\ \text{N}$$

$$pH = \log\left[H^+\right] = -\log[0.05]$$

$$= -[-1.301] = 1.301$$

$$\text{Conc. of } H_2SO_4 = 0.05\ \text{M}$$

$$pH = -\log[0.05] = 1.301.$$

4. (a) Calculate the pH value of solution whose H^+ ion concentration is 0.005 gm/L:

$$pH = -\log\left[H^+\right]$$

$$= -\log[0.005]$$

$$= -[-2.301]$$

$$= 2.301$$

(b) Calculate the values of pH, pOH, $(OH)^-$for a solution in which H^+ = 3.0×10^{-4} M:

$$pH = -\log\left(3.0 \times 10^{-4}\right) = -[\log 3.0 - 4 \log 10]$$

$$= [-0.48 + 4.00] = 3.52$$

$$pOH = 14.00 - 3.52 = 10.48$$

As

$$OH^- = \frac{K_w}{H^+}$$

$$OH^- = \frac{1.0 \times 10^{-14}}{3.0 \times 10^{-3}} = 3.3 \times 10^{-11}$$

(c) The hydroxyl ion concentration of a solution is $\frac{N}{10}$ g ion/L. Calculate the pH of the solution:

$$[OH^-] = \frac{1}{10}\,\text{g ion/L} = 10^{-1}\,\text{g ion/L}$$

But

$$[H^+][OH^-] = 10^{-14}$$

$$[H^+] = \frac{10^{-14}}{10^{-1}} = 10^{-13}$$

$$\text{Hence pH of solution} = 13$$

(d) The dissociation constant of acetic acid at 180 °C is 1.8×10^{-5}. Calculate the pH of 0.1 N, 0.01 N, and 0.001 N solutions of acetic acid.

According to Ostwald' dilution formula,

$$\frac{\alpha^2}{V} = K \quad \text{or} \quad \alpha^2 = K \times V$$

In decinormal, i.e., 0.1 N solution, V = 10 L; therefore,

$$\alpha^2 = 1.8 \times 10^{-5} \times 10$$

or

$$\alpha = \sqrt{(1.8 \times 10^{-5} \times 10)} = 1.34 \times 10^{-2}$$

Now,

$$\begin{aligned}
[H^+] &= \text{Normality } \times \alpha \\
&= 0.1 \times 1.34 \times 10^{-2} \\
&= 1.34 \times 10^{-3}\,\text{g ion/L} \\
\text{pH} &= -\log[H^+] \\
&= -\log\left(1.34 \times 10^{-3}\right) = -\bar{3}.1271 \\
&= -[2.8720] = 2.873
\end{aligned}$$

In 0.01 N solution, i.e., V = 100 L. Therefore,

$$\alpha^2 = 1.8 \times 10^{-5} \times 100 = 1.8 \times 10^{-3}$$

or

$$\alpha = \sqrt{\left(1.8 \times 10^{-3}\right)} = \sqrt{\left(1.8 \times 10^{-4}\right)} = 4.24 \times 10^{-2}$$

Therefore,

$$\begin{aligned}\left[H^{+}\right] &= 4.24 \times 10^{-2} \times \frac{1}{100} \\ &= 4.24 \times 10^{-4}\,\text{g ion/L}\end{aligned}$$

Hence,

$$\begin{aligned}\text{pH} &= -\log\left[H^{+}\right] = -\log\left[4.24 \times 10^{-4}\right] \\ &= -(\bar{4} \cdot 6275) = +3.3726 \\ &= 3.3726\end{aligned}$$

In 0.001 N solution, i.e., V = 1000 L:

$$\alpha^{2} = 1.8 \times 10^{-5} \times 1000 = 1.8 \times 10^{-2}$$

or

$$\alpha = \sqrt{\left(1.8 \times 10^{-2}\right)} = 1.34 \times 10^{-1}$$

$$\left(H^{+}\right) = 1.34 \times 10^{-1} \times \frac{1}{1000}$$

$$\left(H^{+}\right) = 1.34 \times 10^{-4}\,\text{g ion/L}$$

Hence,

$$\begin{aligned}\text{pH} &= -\log\left(H^{+}\right) = -\log\left(1.34 \times 10^{-4}\right) \\ &= -[\bar{4}.1271] = 3.8729.\end{aligned}$$

5. (a) Calculate the pH value of a 0.01 N solution of a weak monobasic acid $\left[K = 1.0 \times 10^{-8}\right]$:

$$K = \frac{\alpha^{2}}{V}$$

The solution of monobasic acid is 0.01 N; hence, $V = 100$ L:

$$1 \times 10^{-8} = \frac{\alpha^2}{100} \quad \Rightarrow \quad \alpha^2 = 1 \times 10^{-8} \times 100 \quad \Rightarrow \alpha = 10^{-3}$$

$$\begin{aligned} H^+ &= \text{Normality } \times \alpha \\ &= 0.01 \times 10^{-8} = 10^{-5}\,\text{g ion/L} \\ pH &= -\log\left[H^+\right] = -\log\left[10^{-5}\right] \\ &= -1 \times -5 = 5. \end{aligned}$$

(b) At 20 °C, 0.1 N solution of an acid is 4% dissociated. Calculate the dissociation constant of the acid.

In 0.1 N solution, V = 10 L α = 4% = 0.04. According to Ostwald's dilution formula,

$$K = \frac{\alpha^2}{V}$$

Therefore,

$$K = \frac{0.04 \times 0.04}{10} = 1.6 \times 10^{-4}$$

(c) Calculate the pH value of (i) 0.1 N HCl, (ii) N/50HCl, and (iii) N/100 H_2SO_4 solution.

(i) Assume complete ionization of HCl in 0.1 solution:

$$\begin{aligned} \text{Conc. of } H^+ \text{ ions} &= 0.1 \text{ gm ions/L} \\ &= 10^{-1} \end{aligned}$$

Therefore,

$$pH = 1$$

(ii) Assume complete ionization of HCl in N/50 solution:

$$\begin{aligned} \text{Conc. of } H^+\text{ions} &= 1/50\,\text{g ion/L} \\ &= 0.02\,\text{g ion/L} \end{aligned}$$

Therefore,

$$\begin{aligned} pH &= -\log(0.02) = -[\bar{2}.30] \\ &= [-1.7] = 1.7 \end{aligned}$$

(iii) N/100 solution of H_2SO_4 will contain $1/100$ H^+g ions:

$$[H^+] = 1/100 = 10^{-2}$$

Therefore,

$$pH = \log [H^+] = \log_{10} \frac{1}{[10^{-2}]}$$

$$= \log_{10} 10^2 = 2$$

6. Calculate the pH value of a mixture containing 50 ml of NHCl and 30 ml NNaOH solutions. Assume both to be completely dissociated in solution.

   When 50 ml of N HCl is added to 30 ml NNaOH solution, the latter is neutralized giving in 50–30 = 20 ml NHCl in a total of (50 + 30 = 80 ml) solution:

$$20\,\text{ml N HCl} = \frac{20 \times 1}{1000}\ \text{g equivalent}/80\,\text{ml}$$

$$= \frac{1}{50}\ \text{g equivalent}/80\,\text{ml}$$

$$= \frac{1}{50} \times \frac{1000}{80} = \frac{1}{4}\ \text{g equivalent/L}$$

$$[H^+] = \frac{1}{4}\ \text{g ion/L}$$

Now,

$$pH = -\log H^+ = -\log \frac{1}{4}$$

$$= \log 4 = 0.602$$

7. Assuming complete dissociation of all solutes, determine the pH of a solution obtained by mixing equal volume of $N/10$ NaOH and $N/20$ HCl.

   Let us assume that 1 L of NaOH and 1 L of HCl are mixed so that the total solution on mixing becomes 2 L. Now since NaOH is stronger than HCl, the resultant solution will contain excess of NaOH:

$$1\ \text{L of N/10 NaOH} \equiv 2\ \text{L of N/20 NaOH}$$

Therefore,

$$V_1 \times N_1 \equiv V_2 \times N_2$$

Furthermore, 1 L of N/10

$$HCl \equiv 1 \text{ liter of } N/10NaOH$$

Hence, the resultant 2 L solution after neutralization will contain 1 L of N/20NaOH. Thus, 1 L of N/20NaOH in 2 L solution:

$$\equiv \frac{1}{20} \text{ g equivalent of NaOH}/2\,L$$
$$\equiv \frac{1}{20} g \times \frac{1}{2} \text{ g equivalent/L}$$
$$\equiv \frac{1}{40} \text{ g OH ion/L}$$

But

$$\left[H^-\right]\left[OH^-\right] = 10^{-14}$$

Hence,

$$\left[H^+\right] = \frac{10^{-14}}{1/40} = 4 \times 10^{-13} \text{ g ion/L}$$
$$pH = -\log 4 \times 10^{-13} = -[0.6028 - 13]$$
$$= -[12.3972] = 12.4.$$

8. Calculate the pH of a solution having 2.0 g of sodium hydroxide dissolved in 1 L of water:

$$\text{Normality of NaOH solution} = \frac{2}{40} = \frac{1}{20}$$

Therefore,

$$\text{concentration of } OH^- \text{ions} = \frac{1}{20} \text{ g ion/L}$$

Now,

$$\left[H^-\right]\left[OH^-\right] = 10^{-14}$$
$$\left[H^+\right]\left[\frac{1}{20}\right] = 10^{-14}$$
$$\left[H^+\right] = 20 \times 10^{-14} = 2 \times 10^{-13}$$
$$pH = -\log\left[H^+\right] = -\log\left[2 \times 10^{-13}\right]$$
$$= -[0.3010 - 13] = 12.7$$

9. Calculate the pH of a 0.001 molar solution of benzoic acid. The dissociation constant of benzoic acid being 7.3×10^{-5} at 95 °C.

   The solution of benzoic acid is $\frac{N}{1000}$; hence, $V = 1000$ L (Eq. wt. is the same as mol. wt.).

   Benzoic acid is a weak acid; hence, $K = \frac{\alpha^2}{V}$; on substituting the value, we get

$$7.3 \times 10^{-5} = \frac{\alpha^2}{1000}$$

$$\alpha^2 = 7.3 \times 10^{-5} \times 1000$$

or

$$\alpha^2 = 73 \times 10^{-2}$$

$$\alpha = 2.696 \times 10^{-1} = 0.2696$$

$$\text{H}^+\text{concentration} = \frac{\alpha}{V} = \frac{0.2696}{1000}$$

$$= 2.696 \times 10^{-4}\ \text{g ion/L}$$

10. Calculate the hydrogen ion concentration of a solution of formic acid containing 0.092 gm of the acid per liter. The dissociation constant of the formic acid at 25 °C is 2.4×10^{-4}.

$$\text{Normality of formic acid} = \frac{0.092}{46} = 0.002\,\text{N or } \frac{2\,\text{N}}{1000}$$

Therefore,

$$V = 500\,\text{L}$$

$$\alpha^2 = 2.4 \times 10^{-4} \times 500$$

$$= 12 \times 10^{-2}$$

or

$$\alpha = 3.464 \times 10^{-1}$$

As

$$[\text{H}^+] = \text{Normality} \times \text{degree of dissociation}$$

$$= 0.002 \times 3.464 \times 10^{-1}$$

$$[\text{H}^+] = 6.928 \times 10^{-4}.$$

11. The dissociation constant of NH_4OH is 2.5×10^{-5} at room temperature. Calculate the hydroxyl ion concentration in N/100 solution of the base. Also determine the concentration of hydroxyl ion if a hundredth part $(1/100)$ of a g mol. of NH_4Cl are dissolved in a liter of this solution.

Ammonium hydroxide is a weak base. It dissociates as

$$\underset{1-\alpha}{NH_4OH} \rightleftharpoons \underset{\alpha}{NH_4^+} + \underset{\alpha}{OH^-}$$

Applying Ostwald's dilution law,

$$K = \frac{\alpha^2}{(1-\alpha)V} = \frac{\alpha^2}{V} \quad [\text{as } \alpha \text{ is very small}]$$

or

$$\alpha = \sqrt{KV} = \sqrt{(2.5 \times 10^{-5} \times 100)}$$
$$= \sqrt{(2.5 \times 10^{-5})} = \sqrt{(25 \times 10^{-4})} = 5 \times 10^{-2} = 0.05$$

or

$$\text{degree of dissociation} = 0.05$$

$$[OH^-] = \text{normality} \times \text{degree of dissociation}$$
$$= \frac{1}{100} \times 0.05 = 0.0005$$

or

$$\text{conc. of } [OH]^- \text{ ions} = 0.0005\,\text{g ion/L}$$

When $\frac{1}{100}$ g mol. of NH_4Cl is dissolved in a liter of $\frac{N}{100}NH_4OH$ solution, there will be common ion effect:

$$NH_4Cl \rightleftharpoons NH_4^+ + Cl^-$$

$$\underset{(1-\alpha_1)}{NH_4OH} \rightleftharpoons \underset{(\alpha_1)}{NH_4^+} + \underset{(\alpha_1)}{OH^-}$$

Therefore,

$$[NH_4^+] = \frac{\alpha_1}{100} + 0.01 = \frac{\alpha_1 + 1}{100}$$

$$[OH^-] = \frac{\alpha_1}{100}; \quad [NH_4OH] = \frac{1-\alpha_1}{100} \text{ and } K = 2.5 \times 10^{-5}$$

$$K = \frac{(NH_4^+)\,(OH^-)}{(NH_4OH)}$$

$$\frac{\left(\frac{\alpha_1+1}{100}\right)\frac{\alpha_1}{100}}{\frac{1-\alpha_1}{100}} = 2.5 \times 10^{-5}$$

or

$$\frac{(\alpha_1 + 1)\,(\alpha_1)}{(1 - \alpha_1)\,100} = 2.5 \times 10^{-5}$$

or

$$\frac{(\alpha_1 + 1)\,\alpha_1}{100} = 2.5 \times 10^{-5}$$

(as α is very small, $1 - \alpha_1$ may be taken as 1). Or

$$\alpha_1^2 + \alpha_1 - 2.5 \times 10^{-3} = 0$$

or

$$\alpha_1 = \frac{-1 \pm \sqrt{(1^2 + 4 \times 2.5 \times 10^{-3})}}{2}$$
$$= \frac{-1 \pm 1.00498}{2} = 0.0025$$

Hence, the concentration of OH^- ions in the presence of NH_4Cl

$$= 0.0025 \times \frac{1}{100} = 0.000025 = 2.5 \times 10^{-5}\ \text{g ion./L}$$

Alternative Method In a solution, NH_4OH ionizes as

$$NH_4OH \rightleftharpoons NH_4^+ + OH^-$$

In the presence of highly ionized NH_4Cl, ammonium hydroxide is practically unionized. Thus, all NH_4^+ ions are obtained from the dissociation of NH_4Cl, i.e.,

$$\left[NH_4^+\right] = 0.01\ \text{g ion/L}$$
$$[NH_4OH] = 0.01\ \text{g mol./L}$$

Substituting these values in

$$\frac{[\mathrm{NH_4}]\left[\mathrm{OH^-}\right]}{[\mathrm{NH_4OH}]} = 2.5 \times 10^{-5}$$

or

$$\frac{0.01\left[\mathrm{OH^-}\right]}{0.01} = 2.5 \times 10^{-5}$$

or

$$\left[\mathrm{OH^-}\right] = 2.5 \times 10^{-5}$$

Now,

$$\left[\mathrm{OH^-}\right] = \text{Normality} \times \text{ degree of dissociation.}$$

Therefore,

$$2.5 \times 10^{-5} = \frac{1}{100} \times \alpha$$

or

$$\alpha = 2.5 \times 10^{-5} \times 10^2 = 0.0025$$

12. Calculate the concentration of OH^- in 0.01 N NH_4OH solution. If 0.05 mol of solid NH_4Cl is added, what change would occur in OH^- concentration. The dissociation constant of NH_4OH is 1.7×10^{-5}.

    The concentration of OH ions before addition of NH_4Cl is given by

$$K = \frac{\alpha^2}{(1-\alpha)V} = \frac{\alpha^2}{V} \text{ (as } \alpha \text{ is very small)}$$

$$1.7 \times 10^{-5} = \frac{\alpha^2}{100}$$

or

$$\alpha^2 = 1.7 \times 10^{-5} \times 10^2 = 1.7 \times 10^{-3}$$

or

$$\alpha = \sqrt{\left(1.7 \times 10^{-3}\right)} = 4.12 \times 10^{-3}$$

Therefore,

$$[OH^-] = 4.12 \times 10^{-2} \times \frac{1}{100}$$
$$= 4.12 \times 10^{-4} \text{ g ion/L.}$$

The concentration of hydroxyl ions after adding NH_4Cl.

In the presence of highly ionized NH_4Cl, ammonium hydroxide will remain practically unionized. Thus, all $NH_4{}^+$ ions will be obtained from dissociation of NH_4Cl, i.e.,

$$[NH_4{}^+] = 0.05 \text{ g ion/L}, \ [NH_4OH] = 0.01 \text{ g mol/L}.$$

Therefore,

$$\frac{[NH_4^+][OH^-]}{[NH_4OH]} = 1.7 \times 10^{-5}$$

or

$$\frac{0.05 \times [OH^-]}{0.01} = 1.7 \times 10^{-5}$$

or

$$[OH^-] = \frac{1.7 \times 10^{-5} \times 0.01}{0.01} = \frac{1.7}{5} \times 10^{-5}$$
$$= 3.4 \times 10^{-6} \text{ g ion/L.}$$

A comparison of OH^- concentration under two conditions indicates that hydroxyl ion concentration is decreased by addition of ammonium chloride.

13. The dissociation constant of acetic acid at 18 °C is 1.8×10^{-5}. Calculate the degree of dissociation and H^+ ion concentration in:

(a) 0.25 N acetic acid solution.

For weak acids, the degree of dissociation $\alpha = \sqrt{KV}$. Normality

$$0.25\,N = \frac{25}{100} \text{ or } V = \frac{100}{25} = 4\,L.$$

or

$$\alpha^2 = 1.8 \times 10^{-5} \times 4$$
$$= 7.2 \times 10^{-5} = 72 \times 10^{-6}$$
$$= \sqrt{(72 \times 10^{-6})} = 8.484 \times 10^{-3}$$

As

$$\left[H^+\right] = \text{Normality} \times \text{degree of dissociation}$$
$$= 0.25 \times 8.484 \times 10^{-3}$$
$$= 2.121 \times 10^{-3}\ \text{g ion/L.}$$

(b) 0.25 N acetic acid solution containing 0.25 N sodium acetate. Assume complete dissociation of sodium acetate.

Acetic acid ionizes in a solution as

$$CH_3COOH \rightleftharpoons CH_3COO^- + H^+$$

In the presence of completely dissociated sodium acetate, acetic acid will be practically unionized. Thus, all the acetate ions are obtained from dissociation of sodium acetate. Hence,

$$\left[CH_3COO^-\right] = 0.25\ \text{g ion/L}$$
$$[CH_3COOH] = 0.25\ \text{g mol/L}$$
$$\left[H^+\right] = ?$$

We know

$$\frac{\left[CH_3COO^-\right]\left[H^+\right]}{[CH_3COOH]} = K$$

or

$$\frac{\left[H^+\right] \times 0.25}{0.25} = 1.8 \times 10^{-5}$$

or

$$\left[H^+\right] = 1.8 \times 10^{-5}$$

As

$$\left[H^+\right] = \text{Normality} \times \text{degree of dissociation}$$
$$1.8 \times 10^{-5} = 0.25 \times \alpha$$
$$\alpha = \frac{1.8 \times 10^{-5}}{0.25} = 7.2 \times 10^{-5}$$

14. A buffer solution is made by mixing 0.06 moles of acetic acid and 0.04 moles of sodium acetate per liter. Calculate its pH. K for acetic acid $= 1.8 \times 10^{-5}$.

$$\left[CH_3COO^-\right] = 0.04 \text{ g ion/L}$$

$$[CH_3COOH] = 0.06 \text{ g mol./L}$$

This is due to the fact that sodium acetate is completely dissociated, while acetic acid remains practically undissociated in its presence; therefore, acetate ions are mainly obtained from sodium acetate:

$$\frac{\left[H^+\right]\left[CH_3COO^-\right]}{[CH_3COOH]} = K$$

$$\frac{\left[H^+\right] \times 0.4}{0.06} = 1.8 \times 10^{-5}$$

or

$$\left[H^+\right] = 2.7 \times 10^{-5}$$

Now,

$$\begin{aligned} pH &= -\log\left[H^-\right] \\ &= -\log 2.7 \times 10^{-5} \\ &= 5\log 10 - \log 2.7 \\ &= 5 - 0.4314 = 4.5686 \end{aligned}$$

15. Calculate the pH value of a buffer solution containing 0.2 M acetic acid and 0.2 M sodium acetate. (Assume dissociation constant for acetic acid as 1.8×10^{-5}.)

    It is a sodium acetate-acetic acid buffer; hence, Henderson's equation can be used:

$$pH = pK_a + \log\frac{\text{salt}}{\text{acid}}$$

$$\begin{aligned} pK_a &= -\log K_a \\ &= -\log 1.8 \times 10^{-5} = 4.7447 \end{aligned}$$

Given salt $= 0.2$ M and acid $= 0.2$, therefore

$$pH = 4.7447 + \log\left[\frac{\text{salt}}{\text{acid}}\right]$$

$$pH = 4.7447 + \log\frac{0.2}{0.2}$$

$$pH = 4.7447.$$

16. (a) Calculate H^+ ion concentration of $\frac{N}{100}$ acetic acid solution. In $\frac{N}{100}$ solution, i.e.,

$$V = 100$$

therefore

$$\alpha^2 = 1.8 \times 10^{-5} \times 100$$

or

$$\alpha = \sqrt{(1.8 \times 10^{-3})} = 4.24 \times 10^{-2}$$

Therefore,

$$[H^+] = 4.24 \times 10^{-2} \times \frac{1}{100} = 4.24 \times 10^{-4}\ \text{g ion/L}$$

(b) What will be the change in H^+ ion concentration on adding 0.01 mole of solid sodium acetate to 1 L of the above solution (K_a for acetic acid $= 1.8 \times 10^{-5}$):

$$[CH_3COO^-] = 0.01\ \text{g ion/L}$$

$$[CH_3COOH] = 0.01\ \text{g mol/L}$$

Now,

$$\frac{[CH_3COO^-][H^+]}{[CH_3COOH]} = K$$

$$\frac{0.01 \times [H^+]}{0.01} = 1.8 \times 10^{-5}$$

or

$$[H^+] = 1.8 \times 10^{-5} = 0.000018\ \text{g ions/L.}$$

17. Calculate the H^+ in a solution prepared by diluting 0.10 mole of HCN and 0.10 mole of NaCN to 10 L with water. (K_a for HCN $= 7.2 \times 10^{-10}$):

$$HCN \rightleftharpoons H^+ + CN^-$$

$$\frac{[H^+][CN^-]}{[HCN]} = K_a = 7.2 \times 10^{-10}$$

$$[Na^+] + [H^+] = [OH^-] + [CN^-]$$

$$Na^+ = 0.10, [HCN] + [CN] = 0.20$$

$$CN^- = 0.10 + H^+ - OH^-$$

$$= 0.10$$

$$\frac{[H^+](0.10)}{(0.10)} = 7.2 \times 10^{-10}, 7.2 \times 10^{-10} H^+$$

$$OH^- = \frac{1 \times 10^{-14}}{7.2 \times 10^{-10}} = 1.4 \times 10^{-5}$$

18. A buffer solution is prepared from an acid of dissociation constant 1.7×10^{-5} mole/L and its sodium salt. Calculate the ratio of salt to the acid in buffer pH of 3.77.
[Given $\log 1.7 = 0.23$]

$$pH = -\log K_a + \log\left[\frac{salt}{acid}\right]$$

$$3.77 = -\log 1.7 \times 10^{-5} + \log\left[\frac{salt}{acid}\right]$$

$$3.77 = +4.77 + \log\left[\frac{salt}{acid}\right]$$

$$\log\left[\frac{salt}{acid}\right] = -4.77 + 3.77 = -1$$

$$-\log\left[\frac{salt}{acid}\right] = 1$$

$$\log\left[\frac{acid}{salt}\right] = 1$$

$$\left[\frac{acid}{salt}\right] = \text{antilog } 1 = 10$$

19. Determine the degree of dissociation of 0.01 molar acetic acid if its dissociation constant is 1.8×10^{-5}. What is the pH of the solution?

For weak acid, the degree of dissociation $\alpha^2 = KV$

$$\text{Molarity} = 0.01 = \frac{1}{100} \text{ or } V = 100\,\text{L}$$

or

$$\alpha^2 = 1.8 \times 10^{-5} \times 100 = 1.8 \times 10^{-3}$$

$$\alpha = \sqrt{1.8 \times 10^{-3}} = \sqrt{18 \times 10^{-4}}$$

$$\alpha = 4.24 \times 10^{-2}$$

$$\text{H}^+ = \frac{\alpha}{V}$$

$$\text{H}^+ = \frac{1}{100} \times 4.24 \times 10^{-2} = 4.24 \times 10^{-4}\ \text{g ion/L}$$

Hence,

$$\text{pH} = -\log \text{H}^+ = -\log\left(4.24 \times 10^{-4}\right)$$

$$\text{pH} = 3.3726$$

20. A buffer solution contains 0.2 g molecules of acetic acid and 0.25 g molecules of sodium acetate per liter. A total of 0.5 mL of N hydrochloric acid is added in the buffer. Calculate the pH of the resulting liquid. The ionization constant of acetic acid at room temperature is 1.8×10^{-5}.

$$\text{pH} = \text{p}K_\text{a} + \log\left[\frac{\text{salt}}{\text{acid}}\right]$$

$$\text{p}K_\text{a} = -\log K_\text{a} = -\log\left(1.8 \times 10^{-5}\right) = 4.7447$$

pH value of the solution before adding HCl will be given by the equation:

$$\text{pH} = \text{pK}_a + \log\frac{0.25}{0.20}$$

$$\text{pH} = 4.7447 + 0.0969 = 4.8416$$

$$\text{Amount of HCl added} = 0.5\ \text{ml of N.HCl}$$

$$= 0.0005\ \text{moles}$$

Assuming HCl to be completely ionized, the amount of H^+ ions added will be $= 0.0005$ mole. The addition of H^+ ions will result in the following reaction:

$$CH_3COO^- + H^+ \longrightarrow CH_3COOH,$$

i.e., the amount of acetic acid will increase and that of the salt will correspondingly decrease by 0.0005 moles. Therefore, after adding HCl,

$$\begin{aligned}[CH_3COOH] &= 0.2 + 0.0005 \\ &= 0.2005 \text{ moles/L} \\ [CH_3COONa] &= 0.25 - 0.0005 \\ &= 0.2495 \text{ moles/L} \\ pH &= pK_a + \log\left[\frac{\text{salt}}{\text{acid}}\right] \\ &= 4.7447 + \log\frac{0.2495}{0.2005} \\ &= 4.7447 + 0.0950 = 4.8397 \\ \text{Change of pH} &= 4.8416 - 4.8397 = 0.0019\end{aligned}$$

21. Calculate the hydroxyl ion concentration in a liter of $\frac{N}{10}$ solution of NH_4OH before and after adding 2 g of NH_4Cl. Assume complete dissociation of the salt. The dissociation constant of NH_4OH is 18×10^{-6}.

Ammonium hydroxide is a weak base and ionizes as

$$NH_4OH \rightleftharpoons NH_4^+ + OH^-$$

$$\frac{[NH_4^+][OH^-]}{[NH_4OH]} = K$$

Now,

$$\frac{\alpha^2}{(1-\alpha)V} = \text{K, assuming } (1-\alpha) = 1 \text{ (as } \alpha \text{ is small)}$$

$$\frac{\alpha^2}{V} = K$$

$$\alpha^2 = 18 \times 10^{-6} \times 10$$

$$\alpha = \sqrt{(1.8 \times 10^{-4})} = 1.342 \times 10^{-2}$$

As

$$[OH^+] = \frac{\text{degree of dissociation}}{\text{dilution}} = \frac{\alpha}{V},$$

therefore

$$\left[OH^-\right] = \frac{1.342 \times 10^{-2}}{10}\left(\because \text{ solution is } \frac{N}{10}\text{i.e.,} V = 10 \text{ liters}\right)$$

$$\left[OH^-\right] = 1.342 \times 10^{-3}\,\text{g ion/L}$$

$$\text{On addition of } 2\,\text{g}NH_4C1 = \frac{24}{53}\,\text{g mol.}$$

$$= 0.037\,\text{g mol./L}$$

$$\left[NH_4^+\right] = 0.037; [NH_4OH] = 0.1\,\text{g mol./L}\left(\text{since solution is } \frac{N}{10}\right)$$

Therefore,

$$\frac{\left[OH^-\right][.037]}{0.1} = 18 \times 10^{-6}$$

or

$$[OH^-] = \frac{18 \times 10^{-6} \times 0.1}{0.037}$$

or

$$[OH^-] = 48.6 \times 10^{-5}$$

$$\left[CH_3COO^-\right] = 0.02\,\text{g ion/L}$$

$$[CH_3COOH] = 0.2\,\text{g mol./L}$$

Therefore,

$$\frac{\left[H^+\right] \times 0.02}{0.20} = 1.8 \times 10^{-5}$$

or

$$\left[H^+\right] = 1.8 \times 10^{-4}$$

Hence,

$$\text{pH} = +\log\left(1.8 \times 10^{-4}\right)$$

$$= +[4.2553]$$

$$= 3.7447$$

22. The equivalent conductivity of 0.02 acetic acid solution at 20 °C is 11.92 mho. The equivalent ionic conductances of an infinite dilution of hydrogen and acetate ion are 360 mho and 40 mho, respectively. Calculate the dissociation constant of acetic acid. What will be the pH of solution when 8.2 g of sodium acetate are added per liter of solution?

The dissociation constant of acetic acid:

$$\lambda_y = 11.92\,\text{mho}$$

$$\lambda_\infty = \lambda CH_3COO^- + \lambda H^+ = 40 + 360 = 400\,\text{mho}$$

$$\alpha = \frac{\lambda v}{\lambda_\infty} = \frac{11.92}{400}$$

According to Ostwald's dilution formula,

$$K = \frac{\alpha^2}{V} = \frac{11.92 \times 11.92}{400 \times 400} \times \frac{1}{50} = \mathbf{1.78 \times 10^{-6}}$$

(Since the solution is 0.02 N, $V = 50\,L$)

pH of solution in the presence of 8.2 g sodium acetate:

$$\text{mol. wt. of } CH_3\ COONa = 82$$

Therefore, g mol. of sodium acetate $= \frac{8.2}{82} = \frac{1}{10} = 0.1$ g mol./L $\frac{N}{50}$ acetic acid contains $\frac{1}{50}$ g mol. of CH_3COOH

$$= 0.02\,\text{g mol. of } CH_3COOH$$

In the presence of highly ionized sodium acetate, all the acetate ions are obtained from sodium acetate only. Hence,

$$\left[CH_3COO^-\right] = 0.1\,\text{g ion/L}$$

$$[CH_3COOH] = 0.02\,\text{g mol./L}$$

Now,

$$\frac{\left[CH_3COO^-\right]\left[H^+\right]}{[CH_3COOH]} = K$$

or

$$\frac{0.1 \times \left[H^+\right]}{0.02} = 1.78 \times 10^{-5}$$

or

$$[H^+] = \frac{1.78 \times 10^{-5} \times 0.02}{0.1} = \left[3.56 \times 10^{-6}\right]$$

Therefore,

$$\begin{aligned} pH &= -\log H^+ = -\log\left[3.56 \times 10^{-6}\right] \\ &= -\log 0.00000356 \\ &= -[6.5514] = 3.45 \end{aligned}$$

Exercises

1. Explain the reasons for the following:
   (a) HCl is stronger acid than H_2S.
   (b) NH_3 is more basic than H_2O.
   (c) The order of acid strength of the acids is $H_2SO_4 > H_3PO_4 > H_3BO_3$
   (d) HI acid is stronger than HBr.
   (e) $HClO_4$ acid is stronger than HClO.
2. Write short notes on:
   (a) Bronsted concept of acid and base
   (b) Buffer value
3. What do you mean by pK value? The pK_a and pK_b values for formic acid and phenol are 3.95 and 9.92, respectively. Find their dissociation constant.
4. Calculate the pH values of the following solutions assuming complete ionization:
   (a) 4.9×10^{-4} N acid
   (b) 0.0016 N base
5. Find out the pH of a 0.0016 acetic acid solution if it is 2.3 percent ionized at this dilution.
6. Calculate the pH of a buffer solution containing 0.1 mole of acetic acid and 0.01 mole of potassium acetate per liter. The dissociation constant of acetic acid is $1.84'10^{-5}$.
7. 50 ml of 0.2 M acetic acid is mixed with 50 ml of 0.2 m sodium acetate solution. What will be the pH of the mixture $\left[K_a = 1.85 \times 10^{-5}\right]$?
8. A buffer solution contains 0.015 moles of ammonium hydroxide and 0.025 moles of ammonium chloride. Calculate the pH value of the solution. The dissociation constant of NH_4OH at room temperature is 1.80×10^{-5} and $K_w = 1.0 \times 10^{-14}$.

9. The pH value of a buffer solution containing 0.1M acetic acid and 0.01M sodium acetate per liter is 3.735. Calculate the change in pH value when 1 ml of N HCl is added to 1 L of this buffer solution. $(K_a = 1.84 \times 10^{-5}$; volume change on adding HCl may be neglected).
10. Calculate the pH value of a 0.15M solution of ammonium chloride if the dissociation constant for ammonia is 1.75×10^{-5}.
11. Calculate the pH value of a solution of ammonium acetate. $[K_a = 1.75 \times 10^{-5}, K_b = 1.8 \times 10^{-5}$ and $K_w = 1.0 \times 10^{-4}]$.
12. Calculate the pH of N/1000 HCl and N/1000 NaOH, assuming both to be completely ionized.
13. Calculate the pH of 0.001 $mBa(OH)_2$ solution, assuming complete ionization.
14. Calculate the pH of 0.001 N acetic acid, if it is 10% ionized.
15. Calculate the pH of a solution containing 4 g/L of NaOH.
16. Calculate the pH of a solution containing 2.9 L of NaOH(Na = 23, $\log 10^2 = 0.3010$).
17. Calculate the number of sulfate ions, $SO_4{}^{2-}$ in 100 ml of $0.001MH_2SO_4$ solution and also the pH of the solution.
18. A buffer solution contains 0.29 moles CH_3COOH and 0.25 g moles potassium acetate per liter. Calculate the change in pH value on mixing 0.5 ml N HCl to the above solution. The dissociation constant of acetic acid is 1.8×10^{-5}.
19. Calculate pH of $0.005MCa(OH)_2$ assuming complete dissociation.
20. Calculate the pH of a 1.0×10^{-8}MHCl solution.
21. What is the pH of 1.0×10^{-8}MNaOH solution?
22. The pH of a 0.10M hydrocyanic acid solution is 5.2. What is the value of K_a for hydrocyanic acid?
23. (a) Calculate the pH of a solution obtained by mixing 150 ml of 0.1 N NaOH and 150 ml of 0.2 NHCl.
    (b) Calculate the pH of the solution produced by mixing equal volumes of solutions of pH = 5 and pH = 3.
24. How may grams of NaOH must be dissolved in 1 L of the solution to give it a pH value of 10?
25. Calculate the pH of a solution obtained by mixing 10 ml of 0.1MHCl and 40 ml of 0.2 ml of H_2SO_4.
26. (a) Calculate the pH of a buffer solution containing 0.2M acetic acid and 0.02M sodium acetate. (The pK_a of acetic acid at 18 °C is 4.47.)
    (b) What will be the change in pH when 1 ml of INHCl is added to 1 L of this buffer?
    (c) Compare the pH change when the same amount of acid is added to 1 L of water.
27. Calculate the $[H^+]$ of a solution obtained by mixing 100 ml of 0.3 N NH_4OH and 50 ml of 0.6 N NH_4Cl. Dissociation constant of water is 1×10^{-14}

Chapter 3
Chemical Kinetics and Catalysis

Chemical kinetics focuses on the rate at which chemical reactions occur. It helps determine not only whether a reaction is feasible but also how rapidly it initiates, progresses, and ultimately reaches completion. The speed of a reaction is influenced by several factors, which include the following:

- **Physical state and particle size of the reactants:** These influence how quickly a reaction proceeds. Typically, reactants are dissolved in water or another solvent to ensure full molecular interaction, thereby accelerating the reaction.
- **Reactant concentration:** An increase in the concentration of reactants usually leads to a faster reaction rate.
- **Temperature:** In most cases, raising the temperature increases molecular motion and enhances the rate of reaction.
- **Catalysts:** The presence of a catalyst accelerates the reaction without undergoing permanent chemical change.

Let us now take an example to understand how the rate of a chemical reaction can be determined. In an aqueous solution, bromine reacts with formic acid according to the following equation:

$$Br_2 + HCOOH \longrightarrow 2Br^- + 2H^+ + CO_2$$

Since bromine has a distinct red coloration, the progress of the reaction can be monitored by observing the fading of this red color, which corresponds to the disappearance of Br_2.

A straightforward technique to determine the concentration of bromine involves UV-visible spectroscopy. Bromine exhibits strong absorption at 390 nm, which allows its concentration to be measured at various time intervals by recording the absorbance at this wavelength. As the reaction proceeds, the bromine concentration decreases over time, producing a graph of Br_2 concentration versus time (as depicted

© The Author(s), under exclusive license to Springer Nature Switzerland AG 2026

R. N. Goyal, H. Goel, *Textbook of Engineering Chemistry*,
https://doi.org/10.1007/978-3-032-22068-4_3

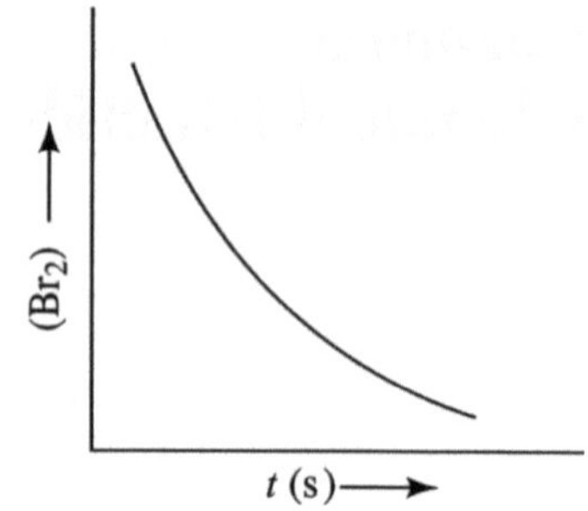

Fig. 3.1 Variation of reaction rate with time based on change in concentration of reactants and products

Table 3.1 Variation of Br_2 concentration with time at 25 °C

Time/s	$[Br_2]$ mol/l	Δt	ΔBr_2	Rate (M/s)	k (l/s)
0	0.0120	–	–	–	–
50	0.0100	50	0.0020	4.0×10^{-5}	4.00×10^{-3}
100	0.0085	50	0.0015	3.0×10^{-5}	3.56×10^{-3}
200	0.0060	100	0.0025	2.5×10^{-5}	4.16×10^{-3}
300	0.0042	100	0.0017	1.7×10^{-5}	4.07×10^{-3}
400	0.0030	100	0.0012	1.2×10^{-5}	4.00×10^{-3}

in Fig. 3.1). The reaction rate can then be expressed as the change in bromine concentration over time:

$$\text{rate} = -\frac{\Delta\,[Br_2]}{\Delta t}$$

The negative sign indicates that the concentration of Br_2 is decreasing over time, and the Δ symbol signifies the change in its concentration. In a similar manner, the formation of CO_2 can be tracked, where the concentration increases with time:

$$\text{rate} = \frac{\Delta\,[CO_2]}{\Delta t}$$

The units of the reaction rate depend on the units of concentration (typically mol/L) and time (usually seconds), which means the unit becomes M/s.

It's important to note that when the rate is measured at a specific point in time using this method, it is referred to as the instantaneous rate rather than an average rate. The instantaneous rate is influenced by the reactant's concentration at that particular moment.

Experimental observations for the reaction described above are presented in Table 3.1.

From the table, it's evident that the rate during the first 50 s was 4.0×10^{-5} M/s, in the next 50 s it dropped to 3.0×10^{-5} M/s, and for the following 100 s, it further reduced to 2.5×10^{-5} M/s. In the last 100-s interval, the rate reached 1.2×10^{-5} M/s. To determine the average reaction rate, a plot of rate versus bromine concentration

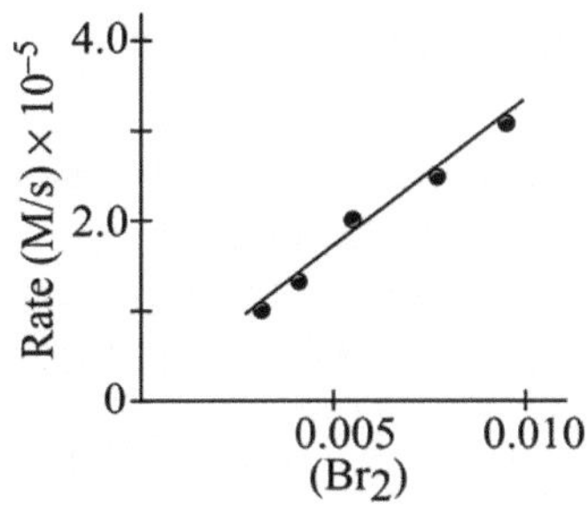

Fig. 3.2 Rate vs. bromine concentration [Br]

(shown in Fig. 3.2) is constructed, and it clearly shows that the rate is directly proportional to the Br_2 concentration:

$$\text{Rate} \propto [Br_2] \quad \text{or} \quad \text{Rate} = k\,[Br_2]$$

Here, k is the proportionality constant and is referred to as the rate constant. Its value can be calculated using the expression:

$$k = \frac{\text{Rate}}{[Br_2]}$$

If we compute this ratio across different bromine concentrations, it remains nearly constant over the entire range, as shown in Table 3.1. This confirms that the rate constant k depends only on the temperature at which the reaction is conducted.

In the example above, each reactant participates in a 1:1 ratio. However, if more than one mole of a reactant is involved, the scenario changes. For instance, if $2A \rightarrow B$, then the rate must be expressed accordingly.

Since two moles of A are consumed for every mole of B produced, the rate can be written as

$$\text{rate} = -\frac{1}{2}\frac{\Delta[A]}{\Delta t} \quad \text{or} \quad \frac{\Delta[B]}{\Delta t}$$

This accounts for either the disappearance of A or the appearance of B. More generally, for a reaction of the form

$$aA + bB \longrightarrow cC + dD$$

The rate is expressed as

$$\text{rate} = -\frac{1}{a}\frac{\Delta[A]}{\Delta t} = -\frac{1}{b}\frac{\Delta[B]}{\Delta t} = \frac{1}{c}\frac{\Delta[C]}{\Delta t} = \frac{1}{d}\frac{\Delta[D]}{\Delta t}$$

3.1 Rate Law

As discussed earlier, the rate of a chemical reaction is influenced by the concentration of the reactants involved. For instance, in Table 3.1, the observed reaction rate is directly related to the concentration of [Br_2]. More generally, the rate of a reaction is often found to be proportional to the concentrations of the reactants raised to specific powers.

Consider a general reaction:

$$aA + bB \longrightarrow cC + dD$$

The rate of the forward reaction can be represented as

$$\text{rate} \propto [A]^p[B]^q \quad \text{or} \quad \text{rate} = k[A]^P[B]^q \tag{3.1}$$

Here, p and q are experimentally determined exponents that describe how the rate depends on the concentrations of A and B. These values cannot be inferred directly from the stoichiometric coefficients of the balanced chemical equation; instead, they must be determined through experimentation.

Equation (3.1) is known as the rate law or rate equation. It quantitatively relates the reaction rate to the rate constant k and the concentrations of the reactants involved. The sum of the exponents p and q is referred to as the overall order of the reaction. Therefore, for Eq. (3.1), the reaction order is equal to $(p + q)$.

To apply the rate law effectively or determine the overall order of a reaction, it is crucial to first determine the individual values of p and q. This is typically done by conducting a series of controlled experiments.

Experiment 1 In the first experimental setup, the concentration of one reactant is kept constant while the other is varied, and the reaction rate is recorded. Let's suppose the concentration of $[B]$ is maintained at 0.20 M, and the results are as shown.

Initial concentration (M)		
A	B	Rate mol/L . h
0.50	0.20	0.50×10^{-2}
0.75	0.20	0.50×10^{-2}
1.00	0.20	0.50×10^{-2}
2.00	0.20	0.50×10^{-2}

Since the reaction rate remains unchanged despite variations in $[A]$, it indicates that the rate is independent of $[A]$. Hence, the exponent $p = 0$, and the reaction is said to be zero order with respect to A.

Experiment 2 In this experiment, $[A]$ is fixed, and the concentration of $[B]$ is altered. The corresponding reaction rates are then measured.

Initial concentration (M)		
A	B	Rate mol/L . h
0.50	0.20	0.50×10^{-2}
0.50	0.40	1.00×10^{-2}
0.50	0.60	1.50×10^{-2}
0.50	1.60	4.00×10^{-2}

From the data, it is clear that doubling $[B]$ doubles the rate, tripling $[B]$ triples the rate, and so on. This directly proportional relationship implies that the rate is first order with respect to B; hence, $q = 1$.

Therefore, the rate law for this reaction is expressed as

$$\text{rate} = k[A]^0[B]^1 = k[B] \tag{3.2}$$

Given that $p = 0$ and $q = 1$, the overall order of the reaction becomes $p + q = 0 + 1 = 1$. This leads to a revised definition of the rate law: it is the number of molecules whose concentrations affect the rate during the course of a reaction.

3.2 First-Order Reaction

Many reactions involving the transformation of *cis* isomers to *trans* forms follow first-order kinetics. Typical examples include the isomerization of *cis*-2-butene to *trans*-2-butene, and *cis*-azobenzene to *trans*-azobenzene:

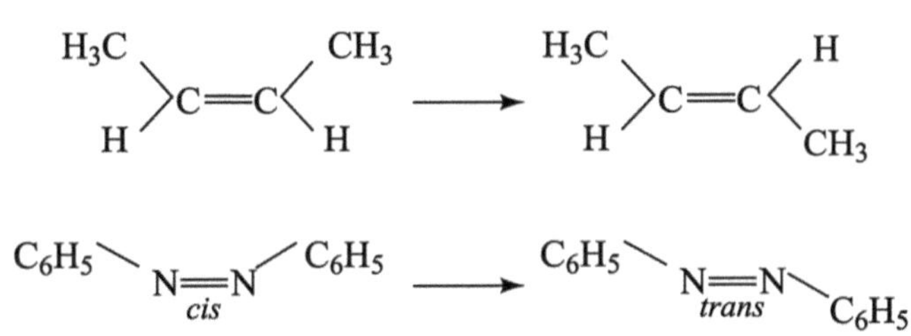

A first-order reaction is one in which the rate depends on the concentration of a single reactant raised to the power of one. That means $p + q = 1$, or if only one substance is reacting, then $p = 1$.

Let us examine a reaction:

$$A \longrightarrow \text{Product}$$

As previously established, the rate of this reaction can be expressed as

$$\text{rate} = -\frac{\Delta A}{\Delta t} \tag{3.3}$$

From the rate law, we also know that

$$\text{rate} = k[A]^1$$

Equating both expressions, we get

$$\frac{-\Delta[A]}{\Delta t} = k[A] \tag{3.4}$$

This can also be rearranged as

$$k = \frac{-\Delta[A]}{[A]} \cdot \frac{1}{\Delta t} \tag{3.5}$$

The units of k are then

$$\frac{\text{M}}{\text{M} \cdot \text{s}} = \frac{1}{\text{s}} = \text{s}^{-1}$$

We can rewrite Eq. (3.4) in its differential form:

$$\frac{-dA}{dt} = k[A]$$

Integrating both sides from time 0 to t, we get

$$\int_0^t \frac{dA}{[A]} = K \int_0^t dt$$

Solving the integral, we obtain

$$\ln\left(\frac{[A]_t}{[A]_0}\right) = -kt$$

Or in logarithmic form:

$$\log \frac{[A]_0}{[A]_t} = \frac{kt}{2.303} \tag{3.6}$$

Here, $[A]_0$ is the initial concentration of A, and $[A]_t$ is its concentration at time t.

Equation (3.6) can also be rearranged in the format of a straight-line equation $y = mx + c$, yielding

$$\log[A]_0 - \log[A]_t = \frac{kt}{2.303} \quad (3.7)$$

Or

$$\log[A]_t = \frac{kt}{2.303} + \log[A]_0 \quad (3.8)$$

Therefore, a plot of $\log[A]_t$ versus time will give a straight line with slope $-k/2.303$, which can be used to determine the value of the rate constant k. Both Eqs. (3.6) and (3.7) have several practical uses in calculating k, t, or $[A]_t$.

Example 3.1 In the isomerization of cyclopropane to propene in the gas phase, the rate constant k is reported to be $6.70 \times 10^{-4}\ \mathrm{s}^{-1}$ at 25 °C. If the initial concentration of cyclopropane is 0.25 M, how long will it take for the concentration to fall to 0.15 M? Using Eq. (3.6),

$$\log\frac{[A]_0}{[A]_t} = \frac{kt}{2.303}$$

$$\log\frac{0.25}{0.15} = \frac{6.70 \times 10^{-4}t}{2.303}$$

Solving for t, we get

$$t = 7.62 \times 10^2\ \mathrm{s}$$

From Eq. (3.8), we can also derive a general form of the rate constant:

$$k = \frac{2.303}{t}\log\frac{[A]_0}{[A]_t}$$

Or more commonly expressed as

$$k = \frac{2.303}{t}\log\frac{[A]_0}{[A-x]}$$

or

$$k = \frac{2.303}{t}\log\frac{a}{a-x}$$

Here, a is the initial concentration of A, and x is the amount decomposed after time t, so the concentration at that time becomes $(a - x)$.

3.2.1 Half-Life ($t_{1/2}$) of a First-Order Reaction

The half-life of a reaction, denoted $t_{1/2}$, is the time required for the concentration of a reactant to reduce to half its initial value. Applying the rate equation,

$$t = \frac{2.303}{t} \log \frac{a}{a - x}$$

By definition, when $t = t_{1/2}$, then $(a - x) = a/2$. Substituting into the equation,

$$t_{1/2} = \frac{2.303}{k} \log \frac{a}{(a/2)} = \frac{2.303}{k} \log 2 = \frac{0.693}{k}$$

3.3 Second-Order Reactions

A reaction is classified as second order when its rate depends on the square of the concentration of a single reactant or on the product of the concentrations of two separate reactants, each raised to the power one. For instance, consider the reaction:

$$2A \longrightarrow \text{Products}$$

Its rate can be expressed as

$$\text{rate} = -\Delta[A]/\Delta t = k[A]^2$$

So,

$$-\Delta[A]/\Delta t = k[A]^2$$

From this, the rate constant k is given by

$$k = -(\Delta[A]/\Delta t) \cdot (1/[A]^2) \tag{3.9}$$

The unit of k is determined as follows:

$$\frac{\text{mol/L}}{(\text{mol/L})^2 \cdot \text{s}} = \frac{\text{L/mol}}{\text{s}} = L \cdot \text{mol}^{-1} \cdot \text{s}^{-1}$$

Now consider another type of second-order reaction:

$$A + B \longrightarrow \text{Product}$$

In this case,

$$\text{rate} = -\frac{\Delta[A]}{\Delta t} = -\frac{\Delta[B]}{\Delta t}$$

And from the rate law,

$$\text{rate} = k[A][B]$$

So,

$$-\frac{\Delta[A]}{\Delta t} = \frac{-\Delta[B]}{\Delta t} = k[A][B]$$

This indicates that the rate of disappearance of A is the same as that of B. Since the rate depends linearly on both $[A]$ and $[B]$, the reaction is first order in A and first order in B, giving a total reaction order of 2. The unit of k in this case is again $\text{L·mol}^{-1}\text{·s}^{-1}$.

Now, consider a second-order reaction of the type:

$$A \longrightarrow \text{Product}$$

Let $[A]_0$ be the initial concentration of the reactant and $[A]_t$ its concentration at time t. By integrating Eq. (3.9) from 0 to t, we obtain

$$\frac{1}{[A]_t} = \frac{1}{[A]_0} + kt \tag{3.10a}$$

Rearranging gives

$$k = \frac{1}{t}\left(\frac{1}{[A]_t} - \frac{1}{[A]_0}\right)$$

Alternatively, if the initial concentration of A is denoted as a, and after time t, the concentration drops to $(a - x)$, the expression becomes

$$k = \frac{1}{t}\left(\frac{1}{(a-x)} - \frac{1}{a}\right) = \frac{1}{t} \cdot \frac{x}{(a-x)} \tag{3.10b}$$

This is known as the second-order rate equation.

3.3.1 Half-Life of a Second-Order Reaction

To determine the half-life ($t_{1/2}$) of a second-order reaction, substitute $[A]_t = [A]_0/2$ into Eq. (3.10a):

$$1/([A]_0/2) = 1/[A]_0 + kt_{1/2}$$

Solving gives

$$t_{1/2} = 1/(k[A]_0)$$

A key distinction from first-order reactions is that in second-order reactions, the half-life depends on the initial concentration of the reactant. This property is often used to differentiate between first- and second-order kinetics.

Example 3.2 For the reaction,

$$A + 2B \longrightarrow C$$

The following experimental data were obtained at 25 °C. The objective is to determine the rate law and calculate the rate constant.

Experiment	$[A]$	$[B]$	Initial rate (M/s)
1	0.100	0.100	5.50×10^{-6}
2	0.200	0.100	2.20×10^{-5}
3	0.400	0.100	8.80×10^{-5}
4	0.100	0.300	1.65×10^{-5}
5	0.100	0.600	3.30×10^{-5}

Solution Let us assume the general rate law:

$$\text{rate} = k[A]^x[B]^y$$

From Experiments 1 and 2, where $[B]$ remains constant, doubling $[A]$ leads to a fourfold increase in the reaction rate. Similarly, in Experiments 2 and 3, doubling $[A]$ again results in the same fourfold increase. This clearly indicates that the reaction is second order in $A(x = 2)$.

From Experiments 4 and 5, where $[A]$ is held constant and $[B]$ is doubled, the rate also doubles. This shows that the reaction is first order in $B(y = 1)$.

Thus, the rate law becomes

$$\text{rate} = k[A]^2[B]$$

The overall reaction order is $2 + 1 = 3$.

To calculate the rate constant k, we use the formula:

$$k = \frac{\text{rate}}{[A]^2[B]}$$

Using the data from Experiment 1,

$$k = \frac{5.50 \times 10^{-6}}{(0.100)^2(0.100)} = 5.50 \times 10^{-3}\ \text{M}^2/\text{s}$$

Reactions Involving More Than One Step

While reactions that proceed through a single step are relatively straightforward to analyze, many reactions—especially those occurring in biological systems—take place through multiple steps. Understanding such complex mechanisms requires careful study to determine how each step contributes to the overall process and to deduce the reaction order.

A representative example is the decomposition of hydrogen peroxide (H_2O_2) in the presence of iodide ions (I^-) under basic conditions. The reaction occurs in the following two steps:

$$H_2O_2 + I^- \xrightarrow{k_1} H_2O + OI^- \quad (\text{rate constant} = k_1)\ (\text{step 1})$$
$$H_2O_2 + OI \xrightarrow{k_2} H_2O + O_2 + I^- \quad (\text{rate constant} = k_2)\ (\text{step 2})$$

The overall chemical equation for this reaction becomes

$$2H_2O_2 \longrightarrow 2H_2O + O_2 \tag{3.11}$$

When the rate law is determined experimentally for this reaction, it is found to be

$$\text{rate} = k\,[H_2O_2]\left[I^-\right]$$

This shows that the reaction is first order with respect to both hydrogen peroxide and iodide ion, even though iodide does not appear in the overall balanced equation. This occurs because the iodide ion functions as a catalyst, being consumed and regenerated in the course of the reaction.

In this mechanism, the second step proceeds significantly faster than the first ($k_2 \gg k_1$), meaning the first step controls the overall reaction rate. This slower step is referred to as the rate-determining step—the step that limits the speed of the entire reaction sequence.

In this case, the species OI^-, which is generated in Step 1 and consumed immediately in Step 2, is known as a reaction intermediate. Intermediates like OI^- may be identified using spectroscopic or other analytical techniques, but they do not appear in the final reaction products since they are used up before the reaction completes.

Because the intermediate OI^- does not accumulate and is quickly consumed, the rate of the overall reaction is governed entirely by the slow first step:

$$\text{rate} = k\,[H_2O_2]\,[I^-]$$

3.4 Complex Reactions

The kinetics of chemical reactions becomes more challenging to analyze when the reaction pathways involve multiple interconnected processes. Some common types of complex reactions include:

- Reversible reactions
- Parallel (or competing) reactions
- Consecutive (or sequential) reactions

3.4.1 *Reversible Reactions*

Many chemical reactions do not proceed exclusively in one direction but instead occur in both forward and reverse directions at rates that may be comparable. A familiar example of such a process is the hydrolysis of esters, where an ester reacts with water to produce an acid and an alcohol, while the reverse reaction reforms the ester:

$$CH_3COOC_2H_5 + H_2O \leftrightharpoons CH_3COOH + C_2H_5OH$$

In this reaction, k_1 and k_2 represent the rate constants for the forward and reverse reactions, respectively.

At equilibrium, a significant amount of the ester remains, indicating that both the forward and backward reactions are occurring simultaneously. At the beginning of the reaction, the concentrations of the acid and alcohol are nearly zero, so the reverse reaction is negligible. As the reaction proceeds, the concentrations of these products increase, and the reverse reaction becomes more prominent.

To determine the net rate of the reaction, the rate of the reverse process is subtracted from that of the forward process. Assuming the initial concentration of the ester is a, and no acid or alcohol is present initially, let x be the amount of ester converted at time t. The rate of the reaction is then given by

$$\frac{dx}{dt} = k_1(a - x)^2 - k_2x^2 \tag{3.12}$$

At equilibrium, the forward and reverse reaction rates become equal. If x_1 is the concentration of the acid and alcohol formed at equilibrium, then

$$k_1(a - x)^2 = k_2x_1^2$$

Substituting this into Eq. (3.12), we get

$$\frac{dx}{dt} = k_1(a - x)^2 - \frac{k_1\,(a - x_1)^2}{x_1^2}x^2 \tag{3.13}$$

This expression can be used to compute the value of the forward rate constant k_1 using experimental data.

Furthermore, the equilibrium constant (K) for the reaction can be related to the forward and reverse rate constants by the expression:

$$K = \frac{k_f}{k_b}$$

where k_f and k_b are the rate constants for the forward and backward reactions, respectively.

3.4.2 Parallel Reactions

In several chemical processes, a single reactant may simultaneously participate in multiple reactions, producing different products. The proportions in which these products form depend on the reaction conditions. However, under a fixed set of conditions, both products are typically observed—one usually in a major amount and the other in a minor amount. A common example of this is the reaction between chlorine (Cl_2) and toluene:

- At low temperatures and in the absence of light, the major products are ortho- and para-chlorotoluene.
- At elevated temperatures, benzyl chloride is the main product.

At a fixed temperature, these two pathways compete, and the dominant product is determined by the relative rates of the competing reactions.

To analyze such systems, consider a case where a reactant A undergoes two parallel reactions forming products B and C, with rate constants k_1 and k_2, respectively.

Assume the initial concentration of $[A]$ is a, which reduces to $(a - x)$ after time t. The rate at which A is consumed can be written as

$$\frac{dx}{dt} = k_1(a - x) + k_2(a - x) = (k_1 + k_2)\,(a - x)$$

Integrating this expression from $t = 0$ to t, we obtain a relation similar to a first-order equation:

$$k_1 + k_2 = \frac{1}{t} \cdot 2.303 \log \frac{a}{(a - x)} \tag{3.14}$$

The individual values of k_1 and k_2 can also be determined independently. Since

$$\text{rate of formation of } B = k_1(a - x)$$

$$\text{rate of formation of } C = k_2(a - x)$$

The ratio of the two rate constants can be written as

$$\frac{k_1}{k_2} = \frac{[B]}{[C]} \tag{3.15}$$

Here, $[B]$ and $[C]$ refer to the concentrations of products B and C at any given time. Therefore, Eqs. (3.14) and (3.15) allow us to calculate both k_1 and k_2 individually from experimental data.

3.4.3 Consecutive Reactions

In various chemical processes, an intermediate product formed in one reaction step may further react to yield the final product. For example, consider a case where reactant A first converts into B, and then B undergoes further transformation to form C, with respective rate constants k_1 and k_2:

$$A \xrightarrow{k_1} B \xrightarrow{k_2} C$$

In such a scenario, the slower step among the two controls the overall rate and is referred to as the rate-determining step.

Suppose the initial concentration of A is a, while the concentrations of B and C are zero at the start. As time progresses, the concentration of A decreases, and the concentrations of B and C increase. Let x, y, and z be the concentrations of A, B, and C at time t, respectively. Then,

$$a = x + y + z$$

The rate at which A is consumed at any time t can be written as

$$\frac{-dx}{dt} = k_1 x \tag{3.16}$$

Since $x = a$ when $t = 0$, integration yields

$$x = ae^{-k_1 t} \tag{3.17}$$

For intermediate B, its concentration at time t depends on both the rate of its formation from A and its consumption in forming C:

$$A \xrightarrow{k_1} B$$

$$B \xrightarrow{k_2} C$$

Hence, the rate of change of B is the difference between its rate of formation and its rate of decomposition:

$$\frac{dy}{dt} = k_1 x - k_2 y \tag{3.18}$$

Substituting x from Eq. (3.17), we get

$$\frac{dy}{dt} = k_1 a e^{-k_1 t} - k_2 y$$

Solving this differential equation by integration gives

$$y = \frac{a k_1}{(k_2 - k_1)} \left[e^{-k_1 t} - e^{-k_2 t} \right] \tag{3.19}$$

Using Eqs. (3.17) and (3.19), the concentrations of A and B at any time t can be calculated. Once these concentrations are experimentally measured, the values of k_1 and k_2 can also be determined.

If the concentrations of A, B, and C are plotted as functions of time, the resulting graph (refer to Fig. 3.3) shows the following:

- A decreases exponentially over time.
- B increases initially, reaches a peak, and then declines.
- C increases continuously as the final product accumulates.

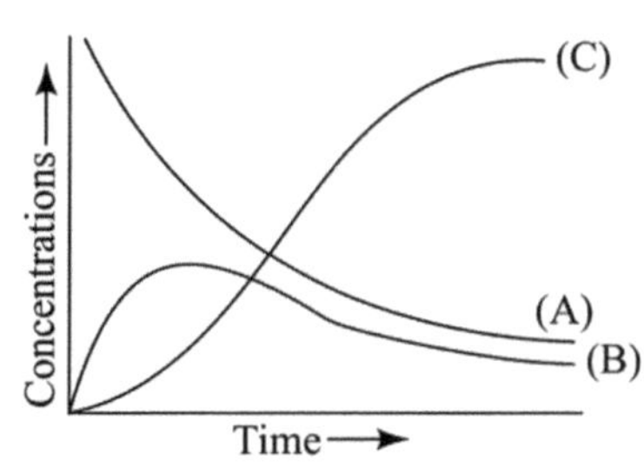

Fig. 3.3 Dependence of concentrations of A, B, and C with time

3.5 Arrhenius Equation

The rate at which a chemical reaction occurs is not only influenced by energy considerations but also by the number and nature of collisions between reacting molecules. To explain this behavior quantitatively, Arrhenius introduced the following equation:

$$k = Ae^{(-E/RT)^*} \tag{3.20}$$

where:

- k = rate constant of the reaction
- A = frequency factor, with units of 1/time; this depends on the total number of collisions and the fraction of them with proper orientation
- R = universal gas constant $= 8.30 \times 10^{-3}$ kJ/mol·K
- E^* = activation energy in kJ/mol
- T = absolute temperature in Kelvin (K)

The exponential factor $e^{(-E/RT)^*}$ represents the fraction of molecules that have sufficient energy (equal to or greater than E^*, the activation energy) to react. This factor is always less than one and is highly sensitive to temperature changes—it increases notably at higher temperatures, which explains the acceleration of reaction rates with heat.

Taking the natural logarithm of both sides of Eq. (3.20) gives

$$\ln k = \ln A - E/RT^*$$

Or

$$\ln k = -\frac{E}{R}\left(\frac{1}{T}\right) + \ln A^* \tag{3.21}$$

This is in the form of a straight-line equation: $y = mx+c$, where the slope is $-E/R^*$ and the intercept is $\ln A$. Therefore, if $\ln k$ is plotted against $1/T$, a straight line is obtained. The slope of this line allows us to determine the value of the activation energy E^*.

To express the equation using common logarithms instead of natural logs, it can be rewritten as

$$\log \mathrm{k} = \frac{-E}{2.303R}\left(\frac{1}{T}\right) + \log A^* \tag{3.22}$$

Furthermore, comparing the rate constants at two different temperatures T_1 and T_2 gives

$$\log \frac{k_2}{k_1} = \frac{\mathrm{E}}{2.303R} \left\{ \frac{1}{T_1} - \frac{1}{T_2} \right\}^* \tag{3.23}$$

Example 3.3 The rate constants for the thermal decomposition of an organic compound were measured at various temperatures, and the following data were recorded. The objective is to determine the activation energy (E^*) for the reaction.

To analyze the data, we first convert the measurements into the required form for plotting $\log k$ versus $1/T$:

$T\,(k)$	$1/T$	k	$\log k$
700	1.43×10^{-3}	0.011	−1.96
730	1.37×10^{-3}	0.035	−1.46
760	1.32×10^{-3}	0.105	−0.979
790	1.27×10^{-3}	0.343	−0.465

Solution Upon plotting $1/T$ against $\log k$, a straight line is obtained with a slope approximately equal to -9.9×10^3 K.

Using the slope to calculate the activation energy,

$$\text{Slope} = -\frac{E}{2.303R^*}$$

Rearranging:

$$\begin{aligned} E^* &= 2.303 \times R \times \text{Slope} \\ &= 2.303 \times 8.30 \times 10^{-3}\ (\text{kJ/K.mol}) \times 9.9 \times 10^3\ (\text{K}) \\ &= 189.2\ \text{kJ/mol} \end{aligned}$$

Thus, the activation energy for the reaction is 189.2 kJ/mol.

3.6 Catalysis

As discussed earlier, one straightforward approach to accelerate a chemical reaction is by raising the reaction temperature. However, this method can sometimes lead to the formation of undesirable by-products due to enhanced side reactions. To avoid such complications, an alternative strategy is often employed: the use of a catalyst.

A catalyst is a substance that increases the rate of a chemical reaction without requiring a rise in temperature. It functions by offering an alternative reaction pathway with a lower activation energy, thereby enhancing the reaction speed.

A catalyst is formally defined as a substance that facilitates a chemical reaction by increasing its rate while remaining chemically unchanged at the end of the reaction.

For example, consider the reaction:

$$A + B \longrightarrow C + D$$

If the rate constant in the absence of a catalyst is denoted as k_a, and the rate constant with a catalyst is k_c, then

$$k_c > k_a$$

This indicates that the reaction proceeds more rapidly in the presence of a catalyst.

A notable feature of a catalyst is that it may temporarily combine with a reactant to form an intermediate complex, but it is ultimately regenerated at the conclusion of the reaction. Because it is not consumed, only a small amount of catalyst is typically required to significantly accelerate the reaction rate.

3.7 Activation Energy and Catalysis

Catalysts, regardless of their specific chemical nature, function based on the same underlying principle. They promote the proper alignment and proximity of reactant molecules, enabling more effective molecular collisions. By lowering the activation energy barrier, catalysts make reactions proceed more rapidly.

This effect can be visualized using energy versus reaction progress curves, as illustrated in Fig. 3.4. Even though the overall energy of the reactants and products remains unchanged, the activation energy (E_a) required to initiate the reaction is reduced when a catalyst is present. In this context, the activation energy E_c in the presence of a catalyst is always less than the activation energy required without it.

For example,

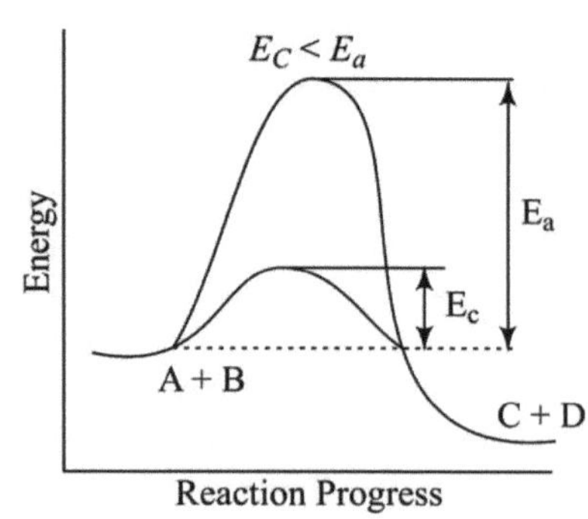

Fig. 3.4 Energy profile diagram for the reaction

$$A + B \longrightarrow C + D \text{ (in both catalyzed and uncatalyzed pathways)}$$

Types of Catalysis Catalysis can generally be classified into three main types.

3.7.1 Homogeneous Catalysis

In homogeneous catalysis, the reactants, products, and catalyst all exist in the same phase, typically either gas or liquid. Several important reactions fall into this category.

One common example is the acid-catalyzed hydrolysis of esters. In the absence of a catalyst, ester hydrolysis follows first-order kinetics and is described by

$$\text{rate} = k\,[\mathrm{CH_3COOC_2H_5}]$$

However, when a small quantity of acid such as HCl is added, the rate law becomes

$$\text{rate} = k_c\,[\mathrm{CH_3COOC_2H_5}]\left[\mathrm{H^+}\right]$$

Here, the H^+ ion temporarily attaches to the carbonyl group (>C=O) of the ester via the oxygen's lone pair, forming a more reactive intermediate. This facilitates a faster attack by water molecules, leading to quicker breakdown of the ester into alcohol and acid. A similar mechanism is observed in many base-catalyzed reactions as well.

Another significant example of homogeneous catalysis in the gaseous phase is found in the lead chamber process used in the production of sulfuric acid. The key reactions involved are

$$\begin{aligned} \mathrm{S + O_2} &\longrightarrow \mathrm{SO_2} \\ \mathrm{2SO_2 + O_2} &\longrightarrow \mathrm{2SO_3} \\ \mathrm{SO_3 + H_2O} &\longrightarrow \mathrm{H_2SO_4} \end{aligned} \tag{3.24}$$

Among these, reaction (3.24)—the conversion of SO_2 to SO_3—proceeds very slowly without a catalyst. When nitrogen dioxide (NO_2) is used as a catalyst, the process becomes significantly more efficient.

NO_2 facilitates the following sequence of reactions:

$$\begin{aligned} \mathrm{2SO_2 + 2NO_2} &\longrightarrow \mathrm{2SO_3 + 2NO} \\ \mathrm{2NO + O_2} &\longrightarrow \mathrm{2NO_2} \\ \hline \mathrm{2SO_2 + O_2} &\longrightarrow \mathrm{2SO_3} \end{aligned}$$

As shown, NO_2 is regenerated at the end of the process, satisfying the requirement that a catalyst must not be consumed during the reaction. Moreover, because all species involved—including the catalyst—are gases, this is a clear example of homogeneous catalysis in the gas phase.

3.7.2 *Heterogeneous Catalysis*

In heterogeneous catalysis, the catalyst exists in a different phase than the reactants and products. Typically, the catalyst is a solid, while the reactants and products are gases or liquids. This type of catalysis plays a vital role in industrial chemistry, where it facilitates the large-scale production of many essential compounds.

A classic example of heterogeneous catalysis is the synthesis of ammonia (NH_3) from nitrogen (N_2) and hydrogen (H_2) gases:

$$N_2 + H_2 \longrightarrow NH_3$$

Under standard conditions, this reaction proceeds at an extremely slow rate, making it unsuitable for industrial-scale ammonia production. Although increasing the temperature speeds up the reaction, it also promotes the reverse decomposition of NH_3 into N_2 and H_2, limiting overall yield.

In 1905, scientist Fritz Haber discovered that the use of a solid iron (Fe) catalyst—enhanced with potassium and aluminum oxides—significantly increases the reaction rate at around 450 °C. In the presence of this catalyst, nitrogen and hydrogen molecules dissociate on the catalyst's surface into individual atoms. These atoms, now adsorbed onto the metal surface, become highly reactive and readily combine to form ammonia:

$$N + 3H \longrightarrow NH_3$$

It has been experimentally shown that hydrogen dissociates into atoms on the catalyst at approximately 196 °C, while nitrogen requires around 450 °C to undergo similar dissociation.

Another notable application of heterogeneous catalysis is in automobile catalytic converters, which are widely used today to minimize harmful exhaust emissions. These devices enable the complete combustion of hydrocarbons and convert toxic gases into less harmful forms. For example:

1. Carbon monoxide (CO) is oxidized to carbon dioxide (CO_2)
2. Unburned hydrocarbons are converted into CO_2 and H_2O
3. Nitric oxide (NO) is reduced to nitrogen (N_2) and oxygen (O_2)

All of these reactions are catalyzed by solid materials (such as platinum, palladium, and rhodium), while the reactants and products are in the gaseous phase, fulfilling the criteria for heterogeneous catalysis:

$$2CO + O_2 \xrightarrow{\text{Catalyst}} 2CO_2$$

$$2C_8H_{18} + 25O_2 \xrightarrow{\text{Catalyst}} 16CO_2 + 18H_2O$$

$$2NO \xrightarrow{\text{Catalyst}} N_2 + O_2$$

3.7.3 Enzymic Catalysis

An enzyme is a biologically produced complex organic catalyst, typically synthesized by living cells. These are high molecular weight proteins characterized by intricate structures that contain specific regions with unique conformations, known as active sites.

Enzymes are mainly composed of amino acids, and their general structure is shown above. The side chain (R group) differs for each amino acid. Some common R group values for typical amino acids are listed in Table 3.2.

A notable feature of enzymic catalysis is the ability of enzymes to facilitate reactions at around 37 °C—the average human body temperature—and under physiological pH conditions. Nearly all biochemical transformations in the human body are enzyme driven. What makes enzymes remarkable is that they accomplish these reactions efficiently under such mild conditions, whereas performing the same reactions in a lab using chemical catalysts would require much more extreme conditions, like high temperatures or pressures, and even then, the rate may be significantly slower.

Enzymes that function within the cell that produces them are referred to as intracellular enzymes, while those that act outside the producing cell are called extracellular enzymes. Enzymes enhance reaction rates by lowering the activation energy required for the reaction to proceed. The mechanism of enzymic catalysis was long ago explained by the Lock and Key theory, which suggests that the substrate fits into the enzyme's active site much like a key fits into a lock. This results in the formation of a complex, which is typically illustrated in the diagram shown below.

Table 3.2 Name and structure of some common amino acids

R	*Name*
$-CH_3$	Alanine
$-CH_2COOH$	Aspartic acid
$-H$	Glycine
$-CH_2-CH(CH_3)_2$	Leucine
$-CH(CH_3)-C_2H_5$	Isoleucine
$-CH_2-CH_2-CONH_2$	Glutamine

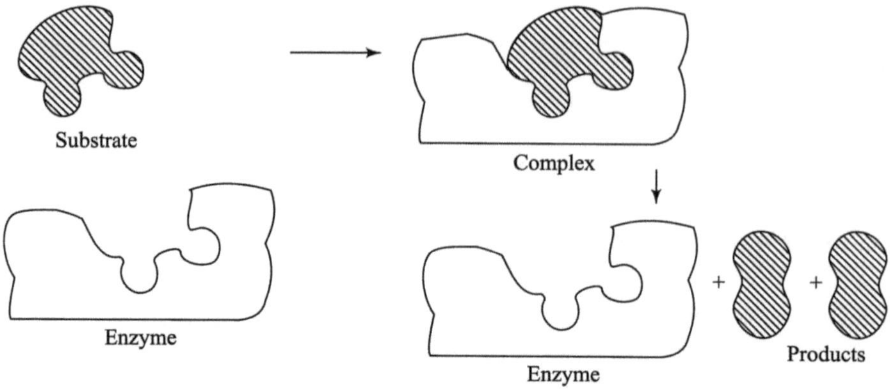

The substrate aligns precisely with the enzyme's active site, undergoes transformation to yield products, and the enzyme itself is regenerated unchanged. These enzyme-substrate complexes are usually stabilized by weak interactions, such as electrostatic forces, hydrogen bonds, and occasionally covalent bonds. The equilibrium constants for these complexes typically fall within the range of 10^{-2}–10^{-8}, corresponding to a free energy of activation between −3 and −12 kcal/mol.

A number of factors influence enzyme-catalyzed reactions.

Temperature

As temperature increases, the rate of enzyme-catalyzed reactions also increases, up to a point. Beyond a certain limit, the enzyme starts to denature and the reaction rate declines. The temperature at which the reaction rate is highest is known as the optimum temperature, which for most enzymes is close to 35 °C. At 0 °C or 100 °C, enzyme activity is nearly zero. This temperature dependence explains why boiling is used in medical sterilization. A typical graph illustrating this temperature-rate relationship is shown.

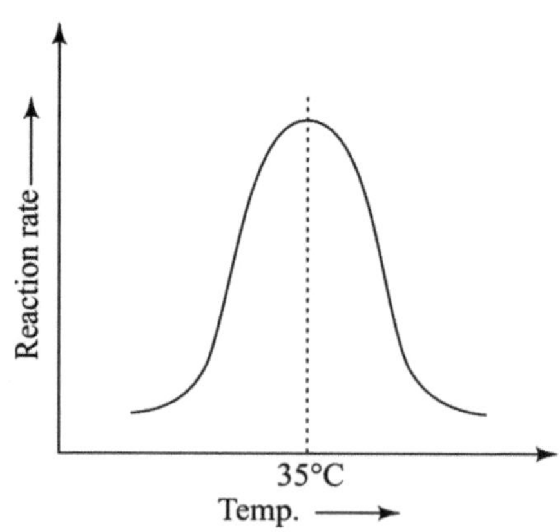

pH

Enzyme activity is also sensitive to the pH of the environment. Highly acidic or strongly alkaline conditions can denature the protein structure of the enzyme. In most cases, maximum enzyme activity is observed between pH 6.0 and 8.0, which aligns with the pH of most bodily fluids—except for gastric juices. When the pH falls below 6.0 or rises above 8.0, enzyme activity declines, producing a curve similar to that of the temperature dependence.

Concentration of Enzyme

When all other conditions in an enzyme-catalyzed reaction—such as temperature, pH, and substrate concentration—are held constant, increasing the enzyme concentration results in a proportional increase in reaction rate. A linear relationship between enzyme concentration and rate is typically observed, as shown in the corresponding figure.

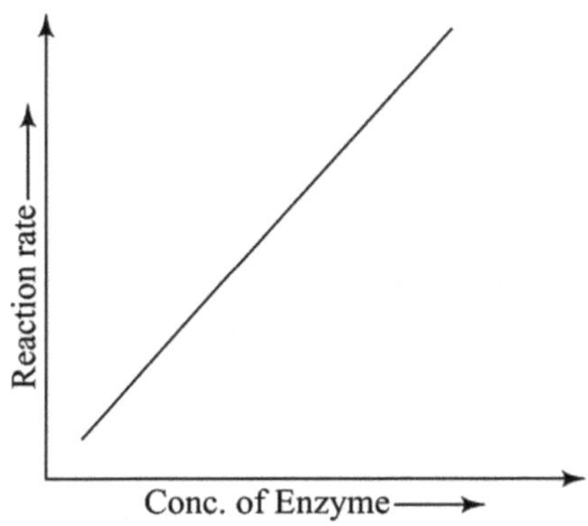

To better understand the significance of enzymes in the human body, consider the metabolic breakdown of adenosine monophosphate (AMP)—a nucleic acid component. This degradation pathway, which involves a series of enzymatic steps, ultimately leads to the formation of uric acid from AMP via intermediate conversion to adenosine.

Since uric acid is only sparingly soluble in water, its excessive buildup in joints can lead to a painful condition known as gout.

Enzymes exhibit remarkable specificity due to the highly selective nature of their active sites. This specificity can be categorized into four main types:

(a) **Absolute specificity:** These enzymes act on only one specific substrate, showing no activity toward similar molecules. An example is urease, which catalyzes the hydrolysis of urea to produce ammonia and carbon dioxide, but has no effect on related compounds like methyl urea or phenyl urea:

$$NH_2CONH_2 \xrightarrow[H_2O]{\text{urease}} 2NH_3 + CO_2$$

(b) **Stereospecificity:** Enzymes in this category catalyze reactions involving only one stereoisomer of a compound. For instance, fumarase adds water exclusively to fumaric acid (the *trans*-isomer), but not to its *cis*-isomer, maleic acid. These enzymes are known for their extremely high specificity.
(c) **Group specificity:** These enzymes are moderately selective and act on molecules with similar functional groups. For example, pepsin hydrolyzes any peptide bond (—CO-NH—) that is adjacent to an aromatic amino acid, regardless of the rest of the molecular structure.
(d) **Linkage specificity:** This type of enzyme targets a specific type of chemical bond, regardless of the surrounding molecular structure. As such, they exhibit the broadest activity range among the four types. A classic example is lipase, which catalyzes the hydrolysis of ester linkages found in lipids.

3.8 Characteristics of Catalytic Reactions

Catalytic reactions possess several distinctive features, which can be outlined as follows:

- Catalysts are not consumed during the reaction. A small amount is sufficient to significantly accelerate the process. While the catalyst participates in the reaction mechanism, it is regenerated at the end. In many homogeneous catalyzed reactions, the rate is influenced by the concentration of the catalyst—increasing the amount of catalyst often results in a higher reaction rate. In contrast, in heterogeneous catalysis, the reaction rate tends to increase with greater catalyst surface area, making finely powdered catalysts much more effective than coarser forms.
- Catalysts remain chemically unchanged after the reaction. Although there may be changes in physical properties, such as color or particle size, the chemical identity of the catalyst remains intact from start to finish.
- Catalysts cannot initiate a chemical reaction on their own. Their role is to speed up an already feasible reaction by lowering its activation energy. This principle is evident in enzyme-catalyzed reactions. However, it's important to note that the final products remain the same whether or not a catalyst is used. For instance, potassium chlorate decomposes to form KCl and O_2 both in the presence and absence of MnO_2—but the reaction occurs significantly faster when MnO_2 is used.
- Catalysts exhibit specificity. The same reactants may produce different products depending on the catalyst employed. For example, the reaction of carbon monoxide (CO) with hydrogen (H_2) gives varying products under different catalytic conditions:

$$CO + 2H_2 \rightarrow CH_3OH \qquad (ZnO + Cr_2O_3 \text{ catalyst})$$

$$CO + H_3 \rightarrow HCHO \qquad (\text{ Cu catalyst})$$

$$Cl + 3H_2 \rightarrow CH_4 + H_2O \qquad (\text{Ni catalyst})$$

- Catalytic activity can be enhanced by promoters. These are substances that, although not catalysts themselves, boost the effectiveness of a catalyst when present in small amounts. For instance, in the Haber process for synthesizing ammonia, a mixture of iron with potassium and aluminum oxides serves as the catalyst. Adding molybdenum further enhances the catalytic activity of iron:

$$N_2 + 3H_2 \xrightarrow[\text{(molybdenum)}]{\text{Fe}} 2NH_3$$

  Another example is the hydrogenation of unsaturated oils using nickel as a catalyst to produce saturated fats with improved shelf stability and flavor. Copper or tellurium, when added in trace amounts, acts as a promoter and increases the reaction rate.
- Catalysts can be deactivated by poisons. Certain substances, even in very small quantities, can completely inhibit catalyst activity. These are known as poisons. A well-known case is the inhibition of platinum catalysts by arsenic oxide during the conversion of SO_2 and O_2 to SO_3 in the contact process for manufacturing sulfuric acid. Arsenic oxide adsorbs more strongly than the reactants on the catalyst's surface, effectively blocking active sites and thereby reducing the reaction rate significantly.

3.9 Autocatalysis

In some chemical reactions, one of the products formed acts as a catalyst for the same reaction, causing the reaction rate to increase as it progresses. Such reactions are referred to as autocatalytic reactions. The observed acceleration in reaction rate is attributed to the catalytic role played by the product once it accumulates.

A typical example is the hydrolysis of esters, which produces an acid and an alcohol:

$$CH_3COOC_2H_5 + H_2O \longrightarrow CH_3COOH + C_2H_5OH$$

When this reaction is carried out in pure water, it proceeds at a very slow rate. However, as acetic acid (CH_3COOH) forms, it releases H^+ ions, which begin to catalyze the reaction, significantly increasing the rate of ester hydrolysis.

Another illustrative example is the reaction of oxalic acid with potassium permanganate ($KmnO_4$) in the presence of sulfuric acid (H_2SO_4). In this redox process, oxalic acid is oxidized to form carbon dioxide (CO_2) and water (H_2O). One of the reaction products, Mn^{2+} ions, functions as a catalyst and accelerates the oxidation, which in turn leads to the rapid decolorization of $KMnO_4$.

At the beginning of the reaction, the consumption of $KMnO_4$ is slow, but as Mn^{2+} accumulates, the reaction speed increases and $KMnO_4$ is used up much more rapidly.

The reactions proceed as follows:

$$2KMnO_4 + 3H_2SO_4 \longrightarrow K_2SO_4 + 2MnSO_4 + 3H_2O + 5[O]$$

$$MnSO_4 \longrightarrow Mn^{2+} + SO_4^{2-}$$

$$5[O] + 5(COOH)_2 \longrightarrow 10CO_2 + 5H_2O$$

3.10 Negative Catalysis

In certain reactions, the addition of specific substances results in a decrease in the reaction rate. This phenomenon is known as negative catalysis, and the substances responsible are referred to as negative catalysts or inhibitors. Slowing down a reaction in this way can be advantageous, especially when there is a need to regulate the reaction speed to avoid undesired outcomes.

An example of this is the oxidation of sodium sulfite (Na_2SO_3) to sodium sulfate (Na_2SO_4) upon exposure to air. This reaction normally proceeds at a noticeable rate. However, introducing a small amount of alcohol (C_2H_5OH) significantly reduces the rate of oxidation, effectively acting as a negative catalyst.

The reactions can be represented as

$$Na_2SO_3 \xrightarrow{\text{atmosphere}} Na_2SO_4$$

$$Na_2SO_3 \xrightarrow[C_2H_5OH]{\text{atmosphere}} Na_2SO_4 \text{ (very slow)}$$

3.11 Theory of Catalysis

As previously discussed, a catalyst functions by reducing the activation energy required for a reaction, thereby increasing the reaction rate. For example, in the Haber process for synthesizing ammonia (NH_3), the activation energy without a catalyst is approximately 326 kJ/mol. However, when finely divided iron is used as a catalyst, the required activation energy drops to about 150 kJ/mol. This demonstrates how a catalyst offers an alternative reaction pathway that leads to the same products with less energy input.

To illustrate this concept, the catalyst's effect is often compared to tunneling through a mountain instead of climbing over it—both routes lead to the same destination, but the tunnel (like the catalyst) provides a quicker and more efficient path. To explain how catalysts function at the molecular level, two main theories have been proposed. These are briefly discussed in the following section.

3.11.1 Theory of Intermediate Compound

This theory proposes that the catalyst temporarily reacts with the reactants to form an intermediate compound. This intermediate is typically highly reactive and unstable, and it subsequently breaks down to yield the final products while regenerating the catalyst. The formation of such intermediate species has been confirmed in many catalytic processes. Below are a few representative examples:

(a) Production of oxygen from potassium chlorate ($KclO_3$):

When manganese dioxide (MnO_2) is used as a catalyst, the reaction appears overall as

$$2KClO_3 \xrightarrow{MnO_2} 2KCl + 3O_2$$

However, the actual reaction occurs in multiple steps involving intermediate species:

$$2KClO_3 + 2MnO_2 \longrightarrow 2KMnO_4 + Cl_2 + O_2$$
$$2KMnO_4 \longrightarrow K_2MnO_4 + MnO_2 + O_2$$
$$K_2MnO_4 + Cl_2 \longrightarrow 2KCl + MnO_2 + O_2$$

Experimental observations confirm the formation of intermediates like Cl_2 and $KMnO_4$. For instance, the oxygen gas generated by this process is often found to contain traces of chlorine. Additionally, sudden cooling of the reaction vessel during the process leads to the appearance of a purple coloration, indicating the presence of $KMnO_4$.

(b) Decomposition of formic acid in an acidic environment:

This reaction is thought to proceed via the formation of an intermediate. Initially, H^+ ions react with formic acid to form an unstable species:

$$HCOOH + H^+ \longrightarrow (HCOOH_2)^+$$
$$(HCOOH_2)^+ \longrightarrow H^+ + CO + H_2O$$

(c) Formation of ether from ethanol using sulfuric acid (H_2SO_4):

This is another example of catalytic intermediate formation. In this case, H_2SO_4 acts as a catalyst. The first step involves the generation of ethyl hydrogen sulfate ($C_2H_5 \cdot HSO_4$):

$$C_2H_5OH + H_2SO_4 \longrightarrow C_2H_5 \cdot HSO_4 + H_2O$$

Then, this intermediate reacts with another ethanol molecule to form diethyl ether, and H_2SO_4 is regenerated:

$$C_2H_5 \cdot HSO_4 + C_2H_5OH \longrightarrow C_2H_5OC_2H_5 + H_2SO_4$$

This theory also successfully explains enzymatic catalysis, where intermediate complexes form between the enzyme and substrate. These enzyme-substrate complexes have been characterized using various spectroscopic techniques. A common example is the hydrolysis of sucrose by sucrase, where the reaction proceeds as follows:

$$\text{Sucrose} + \text{Sucrase} \longrightarrow [\text{Sucrose} - \text{sucrase}] \quad (\text{Intermediate})$$

$$[\text{Sucrose} - \text{sucrase}] \longrightarrow \text{Sucrase} + \text{Glucose} + \text{Fructose}$$

While the theory of intermediate compounds effectively accounts for many catalytic mechanisms, it does have its limitations. Notably, *it fails* to explain the effects of promoters and poisons, which can enhance or inhibit catalytic activity without forming direct intermediates.

3.11.2 Theory of Adsorption

The adsorption theory of catalysis shares similarities with the mechanism of enzymatic reactions. In this theory, the catalyst surface is considered to possess active sites, much like the active sites in enzymes. These specific regions on the surface can attract and hold gas or liquid molecules—a phenomenon known as adsorption. Adsorption can take place through two main mechanisms.

Physical Adsorption

This type of adsorption occurs due to van der Waals forces, which are relatively weak intermolecular attractions. In physical adsorption, the catalyst surface does not form chemical bonds with the adsorbed species. The extent of gas adsorption depends on the type of gas and is generally greater at lower temperatures. As the temperature decreases, the amount of adsorbed gas increases. Physical adsorption is typically reversible, meaning the gas can be released if external conditions (such as pressure or temperature) are altered. Because physical adsorption occurs on the catalyst's surface, a larger surface area enhances adsorption efficiency. This is why finely powdered catalysts are preferred—they provide more surface area than bulk materials.

Chemisorption

Unlike physical adsorption, chemisorption involves actual chemical interactions between the catalyst and the reactants. This kind of adsorption is highly specific and occurs when the catalyst can chemically bond with the reactant molecules. For example, gases such as oxygen may form surface oxides, or hydrogen gas can dissociate into individual atoms and form metal hydrides with the catalyst.

Chemisorption is governed by valence forces and is accompanied by the release of significant heat. It can occur at higher temperatures and is generally irreversible. Increasing the pressure can also enhance chemisorption, which is why industrial catalytic reactions are often conducted under high-pressure conditions to improve catalyst performance.

A common example involves nickel used as a catalyst in hydrogenation reactions. Hydrogen gas interacts with the nickel surface, where the H-H bonds weaken, and individual hydrogen atoms attach to nickel atoms. Since chemical bonds are formed during this process, it is classified as chemisorption. The resulting structure is referred to as an activated complex, and it forms specifically at the catalyst's surface.

Because molecules are more concentrated at the catalyst surface than in the surrounding phase (gas or liquid), they are more likely to react, leading to efficient product formation. The products then desorb from the catalyst surface, making the active sites available again for new reactant molecules.

A well-known example of this process is the oxidation of sulfur dioxide (SO_2) to sulfur trioxide (SO_3), using platinum (Pt) as a catalyst:

$$SO_2 + O_2 \xrightarrow{Pt} SO_3$$

Initially, SO_2 and O_2 are adsorbed onto the platinum surface. The molecules then acquire enough energy to chemically bind to the surface, leading to chemisorption. An activated complex forms and subsequently breaks down to produce SO_3, which is then released (desorbed) from the catalyst surface, leaving it ready for further adsorption cycles.

The adsorption theory is particularly effective in explaining the impact of promoters and poisons on catalyst activity:

- Promoters improve catalytic performance by modifying the surface to create more active sites, enhancing the catalyst's ability to form bonds with reactant molecules and thus increasing the reaction rate.
- Poisons, on the other hand, selectively adsorb onto the catalyst surface, thereby blocking active sites and hindering reactant adsorption. As a result, the reaction rate is greatly reduced or completely halted. If this poisoning occurs via physical adsorption, the substance can often be removed by heating or mechanically

cleaning the catalyst. However, if the adsorption is chemical in nature, the catalyst is generally irreversibly deactivated.

3.12 Solved Problems

1. The amount of CaO after equilibrium for the reaction

$$CaCO_3 \rightleftharpoons CaO + CO_2(g)$$

can be increased by:

(i) Adding soda lime water
(ii) Adding more $CaCO_3$
(iii) Removing some $CaCO_3$
(iv) Decreasing the temperature

Solution Out of the options given, (i) is correct because lime water will absorb CO_2 and the equilibrium will shift toward theright.

2. The decomposition of H_2O_2 by $KMnO_4$ is a first-order reaction. Different volumes of $KMnO_4$ were added to a fixed volume of H_2O_2, and the time required for complete decomposition was found as follows:

Time (min)	0	10	20	30	40
mL of $KMnO_4$	25	20	15.7	12.5	9.6

Calculate the value of the rate constant.

Solution For the first-order reaction,

$$k = \frac{2.303}{t} \log \frac{a}{a - x}$$

At

$$t = 10 \text{ min k} = \frac{2.303}{t} \log \frac{25}{20}$$
$$= 0.02228/\text{min}$$

At

$$t = 20 \text{ min k} = \frac{2.303}{20} \log \frac{25}{15.7}$$
$$= 0.02323/\text{min}$$

similarly calculate at other t values and you will see that k values are almost constant.

3. Twenty-five percent of a reactant decomposed in 40.5 min. Assuming the decomposition as a first-order reaction, calculate the rate constant of the reaction.

Solution Reactant decomposed in 40.5 min—25%. Therefore, remaining reactant —75%. Thus, at $t = 40.5$ min$(a - x) = 0.75a$. Using first-order reaction,

$$\begin{aligned} k &= \frac{2.303}{t} \log \frac{a}{(a-x)} \\ &= \frac{2.303}{40.5} \log \frac{a}{0.75a} \\ &= 7.10 \times 10^{-3}/\text{min} \end{aligned}$$

4. The following data was obtained for the hydrolysis of benzene diazonium chloride:

$$C_6H_5\,\mathrm{N=N-Cl + H_2O} \longrightarrow C_6H_5\mathrm{OH + HCl + N_2\ (g)}$$

Time (min)	0	2	8	16	24	50	∞
Pressure of (p) of N_2 at constant volume (arbitary units)	0	1.6	6.2	11.2	15.5	24.4	34.0

Assuming first-order kinetics, calculate the rate constant.

Solution In this case, evidently, $a \propto (p_\infty - p_0)$ and $(a - x) \propto (p_\infty - p_t)$

Time (min), t	2	8	16	24	50
$(a - x)$	32.4	27.8	22.8	18.5	9.6

$$k_1 = \frac{1}{t} \ln \frac{a}{(a-x)}$$

Incorporating the various values of t, a, and $a - x$, we get following values of k_1 at different intervals of time:

The average value of *rate constant* thus comes out to be 2.498×10^{-2} **min**$^{-1}$.

5. The following results were obtained for the saponification of ethyl acetate:

$$\mathrm{(CH_3COOC_2H_5) + NaOH \longrightarrow CH_3COONa} + C_2H_5\mathrm{OH}$$

Time (min)	k_1 per min
2	2.41×10^{-2}
8	2.52×10^{-2}
16	2.50×10^{-2}
24	2.53×10^{-2}
50	2.53×10^{-2}

Assume concentrations of ester and alkali are equal.

t (min)	0	4.89	10.07	23.66	∞
mL of alkali used	47.65	38.92	32.62	22.58	11.84

Show that the reaction is second order.

Solution The equation for the second order:

$$k_2 = \frac{1}{t}\frac{x}{a(a-x)}$$

In the present case,

$$a = 47.65$$

The value of k_2 is calculated as follows:

t	$(a-x)$	x	$\frac{1}{t}\frac{x}{a(a-x)}$	$k_2\ (\text{dm}^3\ \text{mol}^{-1}\ \text{min}^{-1})$
4.89	38.92	8.73	$\frac{1}{4.89} \times \frac{8.73}{47.65 \times 38.92}$	$= 0.000962$
10.07	32.62	15.03	$\frac{1}{10.07} \times \frac{15.03}{47.65 \times 32.62}$	$= 0.000960$
23.66	22.58	25.07	$\frac{1}{23.66} \times \frac{25.07}{47.65 \times 22.58}$	$= 0.000983$

Almost constant values of k_2 indicate that the reaction is second order.

6. The radioactive decay of $_{88}Ra^{226}$ is a first-order reaction. If the half-life of $_{82}Ra^{226}$ is 1590 years, calculate the value of decay constant.

Solution For first-order reactions,

$$k = \frac{0.693}{t_{1/2}}$$
$$= \frac{0.693}{1590}$$
$$= 0.000436/\text{year}$$

7. Prove that the time required for 99.9% completion of a first-order reaction is ten times than required for 50% completion.

Solution Using first-order reaction,

$$t = \frac{2.303}{k} \log \frac{a}{a-x}$$

We can write

$$t(99.9\%) = \frac{2.303}{k} \log \frac{100}{100 - 99.9}$$

and

$$t(50\%) = \frac{2.303}{k} \log \frac{100}{100 - 50}$$

On dividing the two equations,

$$\frac{t(99.9\%)}{t(50\%)} = \frac{\log 100/0.1}{\log 2}$$
$$= \frac{\log 1000}{\log 2}$$
$$= \frac{3}{0.310}$$
$$= 10$$

8. The catalyzed decomposition of hydrogen peroxide in aqueous solution, which is of the first order, is followed by titrating equal volumes of samples of the solution with potassium permanganate at different times. The following data was obtained:

Time (min)	0	5	15	25	45
cm^3 of $KMnO_4$	37.0	29.8	19.6	12.3	5.0

Calculate the rate constant of the reaction.

Solution For first-order reaction,

$$k = \frac{2.303}{t} \log \frac{a}{(a - x)}$$

Here, $a = 37.0\,\text{cm}^3$; $(a - x) = 29.8, 19.6, 12.3$ and $5.0\,\text{cm}^3$ at $t = 5, 15, 25$ and 45 min, respectively.

The values of k at different times are

$$k = \frac{2.303}{5} \log \frac{37.0}{29.8} = 4.329 \times 10^{-2}\,\text{min}^{-1}. \quad (\text{At } t = 5\text{ min})$$

$$= \frac{2.303}{5} \log \frac{37.0}{19.6} = 4.236 \times 10^{-2}\,\text{min}^{-1}. \quad (\text{At } t = 15\text{ min})$$

$$= \frac{2.303}{25} \log \frac{37.0}{12.3} = 4.407 \times 10^{-2}\,\text{min}^{-1}. \quad (\text{At } t = 25\text{ min})$$

$$= \frac{2.303}{45} \log \frac{37.0}{5.0} = 4.447 \times 10^{-2}\,\text{min}^{-1}. \quad (\text{At } t = 45\text{ min})$$

As the different values of k are nearly same, the reaction is of *first order*. The average value of k is $4.355 \times 10^{-2}\,\mathbf{min}^{-1}$.

9. At 45 °C, the first-order decomposition of N_2O_5 in CCl_4 has a rate constant of 6.0×10^{-4}/min. Calculate (a) the rate constant in terms of seconds and (b) the half-life of reaction in minutes.

Solution (a)

$$k = \frac{6.0 \times 10^{-4}}{\text{min}} \times \frac{1\text{ min}}{60\,\text{s}}$$
$$= 1.0 \times 10^{-5}/\text{s}$$

(b)

$$t_{1/2} = \frac{0.693}{k}$$
$$= \frac{0.693}{6.0 \times 10^{-4}/\text{min}}$$
$$= 1.1 \times 10^{3}\text{ min}$$

10. During decomposition of N_2O_5 dissolved in carbon tetrachloride, at 35 °C, the following data was obtained. Demonstrate that the reaction is first order.

Time (min)	0	40	80	100	160	240	∞
Volume of oxygen collected at constant pressure (mL)	0	15.6	27.6	7.7	45.8	58.3	84.6

Solution The rate equation for the first-order reaction is

$$k_1 = \frac{1}{t}\frac{a}{(a-x)}$$

The initial concentration of N_2O_5 in the solution $= 84.6 - 0 = 84.6$, i.e., $a = 84.6$. The values of $(a - x)$ at different intervals of time are as follows:

Time (min), t	40	80	120	160	240
$(a - x)$	69.0	57.0	46.9	36.8	26.3

On substituting the values of t, a and $(a - x)$, the values of k_1 at different time intervals come out to be as follows:

Time (min)	k_1 per min
40	5.03×10^{-3}
80	4.94×10^{-3}
120	4.92×10^{-3}
160	5.20×10^{-3}
240	4.87×10^{-3}

As k_1 is fairly constant, the reaction is *first order*.

11. Alkaline hydrolysis of ethyl nitro-benzoate proceeds according to the reaction:

$$NO_2 \cdot C_6H_4COOC_2H_5 + OH \longrightarrow NO_2C_6H_4COO^- + C_2H_5OH$$

The date obtained at 15 °C, for the equal (0.05 M) concentration of the reactants, are as follows:

Time, s	120	240	530	600
Extent of reaction (%)	32.95	48.8	69.0	70.35

Calculate the second-order rate constant for the reaction.

Solution Using Eq. (3.10b),

$$k = \frac{1}{t}\frac{x}{a(a-x)}$$

The rate expression can be written as

$$k = \frac{1}{at}\frac{x/a}{(1-x/a)}$$

where x/a is the fractional extent of the reaction for $a = 0.05\,\text{M}$.

At $t = 120\,\text{s}$,

$$k = \frac{1}{120} \times \frac{1}{0.05}\left(\frac{0.3295}{0.6705}\right) = 8.19 \times 10^{-2}$$

$t = 240\,\text{s}$

$$k = \frac{1}{240} \times \frac{1}{0.05}\left(\frac{0.488}{0.512}\right) = 7.94 \times 10^{-2}$$

$t = 530\,\text{s}$

$$k = \frac{1}{530} \times \frac{1}{0.05}\left(\frac{0.690}{0.310}\right) = 8.40 \times 10^{-2}$$

$t = 600\,\text{s}$

$$k = \frac{1}{600} \times \frac{1}{0.05}\left(\frac{0.7035}{0.2965}\right) = 7.91 \times 10^{-2}$$

Average the value of $k = 8.1 \times 10^{-2}\,\text{dm}^3\,\text{mol}^{-1}\,\text{s}^{-1}$.

12. For the reaction,

$$CH_3 - N = N - CH_3(g) \longrightarrow N_2(g) + C_2H_6(g)$$

The decomposition of azomethane is studied by monitoring the partial pressure of the reactant at different times.

Time (s)	0	100	150	200	250	300
Partial pressure of azomethane (mm Hg)	284	220	193	170	150	132

Justify that the above data fit first-order kinetics.

Solution The partial pressure of azomethane at anytime will be directly proportional to its concentration (in mol/L), because it is the product of its mole fraction and the total pressure of the system. Thus, the higher the concentration, the greater its mole fraction. Therefore, Eq. (3.8) can be rewritten as

$$\log P_t = -\frac{-kt}{2.303} + \log P_o$$

where P_o and P_t are partial pressures of azomethane at $t = 0$ and $t = t$. This equation is in the form $y = m_x + C$, and a plot of $\log P_t$ versus t should exhibit a straight line with slope $\frac{-kt}{2.303}$.

Plot the above data and prove that is a first order kinetics.

13. A flask containing 100 ml of 0.1 M HCl was placed in a thermostatic water both at 30 °C. 5.0 ml of ethyl acetate was added to it, and the kinetics of hydrolysis was monitored by taking 5.0 ml of the solution at different times and its titration against standard alkali solution. The data shown below was obtained.

Time (min)	0	75	119	183	∞
Alkali used (mL)	9.62	12.10	13.10	14.75	21.05

From the above data, show that the hydrolysis of ethyl acetate is a first-order reaction.

Solution The hydrolysis of ethyl acetate will be a first-order reaction if the above data conform to the equation:

$$k = \frac{1}{t} \ln \frac{V_\infty - V_o}{V_\infty - V_t}$$

where V_o, V_t and V_∞ represent the volumes of alkali used at the commencement of the reaction, after time t and at the end of the reaction, respectively. Hence,

$$V_\infty - V_0 = 21.05 - 9.62 = 11.43 \text{ ml}$$

The values of k_1 at different times can be calculated as follows:

Time	$V_\infty - V_t$(ml)	$\frac{1}{t}\ln\frac{V_\infty - V_o}{V_\infty - V_t} = k_1$
75 min	$21.05 - 12.10 = 8.95$	$\frac{1}{75}\ln\frac{11.43}{8.95} = 0.003259\ \text{min}^{-1}$
119 min	$21.05 - 13.10 = 7.95$	$\frac{1}{119}\ln\frac{11.43}{7.95} = 0.003264\ \text{min}^{-1}$
183 min	$21.05 - 14.75 = 6.30$	$\frac{1}{183}\ln\frac{11.43}{6.30} = 0.003254\ \text{min}^{-1}$

An almost constant value of k_1 shows that the hydrolysis of ethyl acetate is a *first-order* reaction.

14. The half-life period $(t_{1/2})$ of the first-order reaction is 15 min. Calculate the rate constant and the time taken to complete 80% of the reaction.

Solution

$$k = \frac{0.693}{t_{1/2}} = \frac{0.693}{15\ \text{min}} = 0.0462\ \text{min}^{-1}$$

Now $k = 0.0462\ \text{min}^{-1}$; $a = 100$, then $x = 80$ or $(a - x) = 100 - 80 = 20$; and $t = ?$
Therefore,

$$\begin{aligned} t &= \frac{2.303}{k}\log\frac{a}{(a-x)} \\ &= \frac{2.303}{0.0462\ \text{min}^{-1}}\log\frac{100}{20} \\ &= \frac{2.303\ \text{min}}{0.0462}\log 5 \\ &= 49.85\ \text{min} \times 0.6990 \\ &= 34.85\ \text{min} \\ &= 2091\ \text{s} \end{aligned}$$

15. The optical rotation of a sucrose solution in 0.9 M HCl at various times is observed as follows. Show that the reaction is of the first order.

Time (min)	0	6.18	18.00	27.05	∞
Rotation (degrees)	+ 24.09	+ 21.4	+ 17.7	+ 15.0	– 10.74

Solution Here,

$$a = r_0 - r_\infty = 24.09^\circ - (-10.74) = 34.83^\circ$$

$$(a - x) = r_t - r_\infty = r_t + 10.74^\circ$$

For the first order,

$$k = \frac{2.303}{t}\log\frac{a}{(a-x)} = \frac{2.303}{t}\log\frac{34.83^\circ}{(r_t + 10.74^\circ)}$$

At 6.18 min,

$$k = \frac{2.303}{6.18}\log\frac{34.83^\circ}{21.4^\circ + 10.74^\circ} = \frac{2.303}{6.18}\log\frac{34.83^\circ}{32.14^\circ} = 11.505\ \text{min}^{-1}$$

At 18.00 min,

$$k = \frac{2.303}{18}\log\frac{34.83^\circ}{17.7^\circ + 10.74^\circ} = \frac{2.303}{18}\log\frac{34.83^\circ}{28.44^\circ} = 11.262\ \text{min}^{-1}$$

At 27.05 min,

$$k = \frac{2.303}{27.05}\log\frac{34.83^\circ}{15.0^\circ + 10.74^\circ} = \frac{2.303}{27.05}\log\frac{34.83^\circ}{25.74^\circ} = 11.183\ \text{min}^{-1}$$

Since k values are nearly same, the reaction is of first order.

16. The optical rotations of a sucrose solution in 0.5 M HCl at 35 °C at various time intervals are as follows:

Time (min)	0	10	20	30	40	∞
Rotation (degrees)	+32.4	+28.8	+25.5	+22.4	+19.6	−11.1

Show that the reaction is first order.

Solution The inversion of sucrose will be a first-order reaction if the above data conform to the equation:

$$k = \frac{1}{t}\ln\frac{r_0 - r_\infty}{r_t - r_\infty}$$

where r_0, r_t, and r_∞ represent the optical rotations at the commencement of the reaction, after time t° at the completion of the reaction, respectively.

In this case,

$$a_0 = r_0 - r_\infty = +32.4 - (-11.1) = +43.5$$

The values of k_1 at different times are as follows:

Time	r_t	$r_t - r_\infty$	$\frac{1}{t}\ln\frac{r_0 - r_\infty}{r_t - r_\infty}$
10 min	−28.8	39.9	$\frac{1}{10}\ln\frac{43.5}{39.9} = 0.008625\ \text{min}^{-1}$
20 min	+25.5	36.6	$\frac{1}{20}\ln\frac{43.5}{36.6} = 0.008625\ \text{min}^{-1}$
30 min	+ 22.4	33.5	$\frac{1}{30}\ln\frac{43.5}{33.5} = 0.008694\ \text{min}^{-1}$
40 min	+19.6	30.7	$\frac{1}{40}\ln\frac{43.5}{30.7} = 0.008717\ \text{min}^{-1}$

The more as less constant values of k_1 indicates that the inversion of sucrose is a *first-order* reaction.

17. The rate of a chemical reaction doubles with 10 °C rise in temperature. If the reaction is carried out at 27 °C, what will be the activation energy of the reaction?

Solution Using Eq. (3.20), we can write

$$\frac{k_{305}}{k_{295}} = \frac{e^{-E_a/RT_{305}}}{e^{-E_a/RT_{295}}} = 2$$

$$E_a = \frac{(\ln 2)(8.314\ \text{JK}^{-1}\ \text{mol}^{-1})(305\ \text{K})(295\ \text{K})}{10\ \text{K}}$$

$$= 51851\ \text{J mol}^{-1}$$

$$= 51.85\ \mathbf{kJ\ mol^{-1}}$$

18. A first-order reaction has a k value of $4.5 \times 10^3\ \text{s}^{-1}$ at 1 °C and an activation energy of $58\ \text{kJ mol}^{-1}$. At what temperature would k be $1.00 \times 10^4\ \text{s}^{-1}$?

Solution Using Eq. (3.22), we can write

$$\log k = \log A - \frac{E_a}{2.303RT}$$

$$\log A = \log k + \frac{E_a}{2.303RT}$$

$$= \log\left(4.5 \times 10^3\right) + \frac{58 \times 10^3\ \text{J/mol}}{2.303 \times 8.31\ \text{J mol}^{-1}\ \text{K}^{-1} \times 274\ \text{K}}$$

$$\log A = 14.71$$

On substituting the value of $\log A$ in the above equation for unknown temperature,

$$\log\left(1.00 \times 10^4\right) = 14.71 - \frac{58 \times 10^3\ \text{J/mol}}{2.303 \times 8.31\ \text{J mol}^{-1}\ \text{K}^{-1} \times T}$$

or

$$4.00 = 14.71 - 3,031/T$$

and

$$T = 3,031/10.71 = \mathbf{283\,K}$$

19. At 300 K, a first-order reaction completes 50% in 20 min. At 350 K, the same reaction completes 50% in 5 min. Calculate the energy of activation of the reaction.

Solution k_1 at

$$300\ \text{K} = 0.6932/t_{1/2} = 0.6932/20\ \text{min}^{-1}$$

k_2 at

$$350\ \text{K} = 0.6932/t_{1/2} = 0.6932/5\ \text{min}^{-1} \text{so}\ k_2/k_1 = 4$$

Now,

$$\log\frac{k_2}{k_1} = \log 4 = \frac{E_a}{2.303 \times 8.314\ \text{J}}\left[\frac{350 - 300}{300 \times 350}\right] = 0.6020$$

Hence,

$$E_a = \frac{0.6020 \times 2.303 \times 8.314\ \text{J} \times 300 \times 350}{50}$$
$$= \mathbf{24,206\,J\,mol^{-1}}$$
$$= 24.2\,\text{kJ mol}^{-1}$$

20. The rate constant for a first-order reaction is gives by the expression:

$$\log k\left(\text{in s}^{-1}\right) = 14.34 - \frac{1.25 \times 10^4\ \text{K}}{T}$$

Calculate the value of (i) activation energy and (ii) k at 670 K.

Solution (i) Using Eq. (3.22),

$$\log k = \log A - \frac{E_a}{2.303RT}$$
$$= 14.34 - \frac{1.25 \times 10^4 \text{ K}}{T}$$

therefore

$$\frac{E_a}{2.303RT} = \frac{1.25 \times 10^4 \text{ K}}{T}$$

$$E_a = \left(1.25 \times 10^4 \text{ K}\right)(2.303)8.31 \text{ J mol}^{-1} \text{ K}^{-1}$$
$$= 239 \times 10^3 \text{ J mol}^{-1}$$
$$= \mathbf{239\ kJ\ mol^{-1}}$$

(ii) Now,

$$\log k = 14.34 - \frac{1.25 \times 10^4 \text{ K}}{670 \text{ K}} = -4.32$$
$$k = \text{Antilog } (-4.32)$$
$$= 4.8 \times 10^{-5} \text{ s}^{-1}$$

21. Calculate the activation energy of a reaction whose rate constant is tripled by a 10 °C rise at a temperature in the vicinity of 27 °C.

    Solution The given temperature is $27°C = 300$ K. The second temperature due to 10 °C rise is 305 K. The two temperatures, evidently, are 295 and 305 K:

$$k = Ae^{-Ea/\text{RT}}$$

$$\frac{k_{305}}{k_{295}} = \frac{e^{-E_a/RT_{305}}}{e^{-E_a/RT_{295}}}$$
$$= 3$$

Taking logs on both sides,

$$\ln 3 = -\frac{E_a}{R}\left[\frac{1}{305} - \frac{1}{295}\right]$$
$$= -\frac{E_a}{R}\left[\frac{-10}{305 \times 295}\right]$$
$$E_\text{a} = \frac{(\ln 3)(8.314\text{JK}^{-1}\ \text{mol}^{-1})(305\ \text{K})(295\ \text{K})}{10\ \text{K}}$$
$$= 82{,}182\ \text{J mol}^{-1}$$
$$= \mathbf{82.18\ kJ\ mol^{-1}}$$

22. Indicate whether each of the following statements is true or false. Correct the false statement and rewrite it to make it true:

    (a) In stepwise reaction, the rate-determining step is the slow one.
    (b) It is possible to change the rate constant for a reaction by changing the temperature.
    (c) At constant temperature, when the first-order reaction proceeds, the rate remains constant.
    (d) The rate constant for a reaction is independent of reactant concentrations.

    Solution (a) True
    (b) True
    (c) False
    (d) True
    For the first-order reaction as the reactant decreases, the rate will decrease.

Exercises

1. How can the rate constant k_1 and k_2 be determined for the reaction?

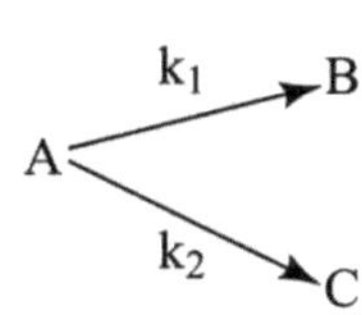

2. Derive the equations for determining k_1 and k_2 for the following reactions:

$$A \underset{k_2}{\overset{k_1}{\rightleftharpoons}} B$$

3. Differentiate between:

(a) Order and molecularly
(b) First-order and pseudo-first-order reaction
(c) Rate constant and equilibrium constant
(d) Heat of reaction and activation energy

4. For reaction:

$$A + 2B \longrightarrow C + D$$

The following data are given:

[A] mols l^{-1}	[B] mols l^{-1}	Rate mole l^{-1} s^{-1}
0.01	0.10	0.0032
0.03	0.10	0.0096
0.01	0.20	0.0064

Calculate the order with respect to A, B, and the total order of reaction.

5. For reaction:

$$2NO + Cl_2 \longrightarrow 2NOCl$$

It was found that on doubling the conc. of both reactants, the rate increases eightfold. But on doubling the conc. of Cl_2 alone, the rate only doubles itself. What is the order of reaction with respect to NO? Justify your answer.

6. In a reversible reaction:

$$A \underset{k_2}{\overset{k_1}{\rightleftharpoons}} B,$$

both the forward and reverse reactions are of first order with half lives of 10.0 hr at 400 K. If one starts with 1.0 M of A, how much will be left after 10 h at 400 K?

7. Derive the equation for a second-order reaction. Write its characteristic and examples.
8. Derive an equation for a first-order reaction. Write its characteristics and examples.
9. The rate for the reaction:

$$2NO + O_2 \longrightarrow 2NO_2 \text{ is : } k[NO]\,[O_2]$$

What is the possible mechanism for this reaction?

10. Why does a reaction speed up with an increase in temperature?
11. What is Arrhenius equation? Explain it.
12. Explain the following with examples:

(a) Autocatalysts
(b) Negative catalysts
(c) Catalyst poisons
(d) Catalyst promoters

13. Why does the energy of activation change by the use of catalyst?
14. What is the role of catalyst in the adsorption theory of catalysis?
15. What do you know about enzyme catalysis? Describe its important characteristics and uses in industry.
16. State the intermediate compound formation theory and adsorption theory of catalysis, and show how they explain characteristics of a catalytic reaction.
17. What are the order of reaction for:

(a) Rate $= k[A]^{1/2}[B]$
(b) Rate $= k\,[Cl_2]\,[NO]^2$

18. What is the order of reaction, whose mechanism is as follows:

$$NO_2(g) + NO_2(g) \longrightarrow NO_3(g) + NO(g) \quad \text{(slow)}$$

$$NO_3(g) + CO(g) \longrightarrow NO_2(g) + CO_2(g) \quad \text{(fast)}$$

19. For the reaction: $A + B \longrightarrow C + D$, rate $= k[A]^2[B]^{1/2}$. Find the order when A is present in large excess.
20. Discuss the difference between the order of a reaction and the molecularity of a reaction.
21. In the hydrolysis of an ester of a monobasic organic acid at 298 K using 0.5 N HCl as the catalyst, it was found that 10 cm^3 of the reaction mixture requires 24.36, 29.32 and 47.15 cm^3 of a standard alkali at the start, after 75 min, and after completion, respectively. Calculate the rate constant of the reaction.
22. The rate constant for a second-order reaction is $8.00 \times 10^{-5}\,M^{-1}\,min^{-1}$. How long will it take for a 1.0 M solution to be reduced to 0.5 M in the reactant? How long will it take from that point until the solution is 0.25 M in the reactant?
23. In the saponification of ethyl acetate by sodium hydroxide 25 cm^3, portions of the reaction mixtures were removed from time to time and titrated with 0.1 N hydrochloric acid. The following results were obtained:

t, min	0	4.1	10.4	28.3	∞
Titrate value (cm^3)	6.55	5.55	4.55	3.20	1.60

Show that the reaction is of second order and calculate the rate constant in $dm^3\,mole^{-1}\,s^{-1}$ units.

24. Optical rotation of sucrose in 1 N – HCl at various times was found as shown below:

Time (min)	0	7.18	18.0	27.5	∞
Rotation (degree)	+24.09	+21.7	+17.7	+15.0	−10.74

Show that inversion of sucrose is a first-order reaction.

25. Find the two-third life $(t_{2/3})$ of a first-order reaction in which $k = 5.48 \times 10^{-4}\ \mathrm{s}^{-1}$. ($\log 3 = 0.4771$; $\log 2 = 0.3010$)
26. The rate constant of a reaction with respect to A is 6 min^{-1}. If we start with $[A] = 0.8$, when would [A] reach the value of 0.08 mol L^{-1}?
27. In a first-order reaction, the time taken for 50% change is 16 min. How much time will it take to complete the reaction for 87.5% change?
28. Find the three-fourth time $(t_{3/4})$ of a first-order reaction, which has a rate constant (k) equal to $6.2 \times 10^{-4}\ \mathrm{s}^{-1}$. ($\log 2 = 0.3010$)
29. A first-order reaction takes 30 min for 60% of the reaction to be completed. What is the velocity constant for the reaction?
30. Show that in the case of a first-order reaction, the time required for 99.9% of the reaction to take place is ten times that required for half of the reaction.
31. The reaction

$$N_2O_5 \longrightarrow 2NO_2 + \frac{1}{2}O_2$$

is of first order in N_2O_5. Its rate constant is $6.2 \times 10^{-6}\ \mathrm{s}^{-1}$. If in the beginning $[N_2O_5]$ is 15 mol L^{-1}, calculate the rate of reaction in the beginning.

32. A reaction that is of the first order with respect to reactant "x" has a rate constant 6 min^{-1}. If we start with $[x] = 5\ \mathrm{mol\,L}^{-1}$, when will $[x]$ reach the value 0.05 mol L^{-1}? [$\log 5 = 0.6990$].
33. Show that the time required for the completion of three-fourths of a first-order reaction is twice the time required for completion of the half reaction.
34. Thermal decomposition of a compound is of first order. If 50% of the compound is decomposed in 120 min, how long will it take for 90% of the compound to decompose?
35. The rate of a particular reaction quadruples when the temperature changes from 293 K to 313 K. Calculate the activation energy for such a reaction.
36. The decomposition of $N_2O_5(g)$ according to the reaction

$$2\,N_2O_5(g) \longrightarrow 4NO_2(\,g) + O_2(g)$$

is a "first-order" reaction. After 30 min from the start of the decomposition, in a closed vessel, the total pressure developed is found to be 284.5 mm of Hg, and on complete decomposition, the total pressure is 284.5 mm of Hg. Calculate the "rate constant" of the reaction.

37. The experimental data for the decomposition of nitrogen pentoxide in the gaseous phase at 318 K is as follows:

$$2\,N_2O_5(g) \longrightarrow 4NO_2(g) + O_2(g)$$

Time (seconds)	$[N_2O_5]$ (mol L^{-1})
0	1.63×10^{-2}
400	1.36×10^{-2}
800	1.14×10^{-2}
1200	0.93×10^{-2}
1600	0.98×10^{-2}
2000	0.64×10^{-2}
2400	0.53×10^{-2}
2400	0.43×10^{-2}
3200	0.35×10^{-2}
3600	0.29×10^{-2}

(a) Plot $[N_2O_5]$ *vs.* time.
(b) Find the time taken for the concentration of N_2O_5 to change from 1.63×10^{-2} mol L^{-1} to half the value.
(c) Plot $\log [N_2O_5]$ *vs.* time.
(d) What is the rate law?
(e) What is the rate constant (k) (calculated from the slope of the line in (C)?
(f) Calculate the half-life from k and verify it with answer (b).

38. The reaction:

$$C_2H_5I \longrightarrow C_2H_4 + HI$$

is of first order and its rate constants are $3.20 \times 10^{-4}\ s^{-1}$ at 600 K and $1.60 \times 10^{-2}\ s^{-1}$ at 1200 K. Calculate the energy of activation for the reaction (given $R = 8.3\ J\ K^{-1}\ mol^{-1}$).

39. In the Arrhenius equation for a certain reaction, the value A and E_a (activation energy) are $4 \times 10^{13}\ s^{-1}$ and 98.6 kJ mol^{-1}, respectively. If the reaction is of first order, at what temperature will its half-life period be 10 min?

Chapter 4
Solid State

Gases exhibit significant compressibility due to the large intermolecular distances and minimal intermolecular forces among their particles. In contrast, solids possess a fixed shape and volume, are virtually incompressible, and exhibit considerable mechanical strength and rigidity. This is because the constituent atoms or molecules in a solid are tightly packed in a well-organized, orderly arrangement.

4.1 Classification of Solids

Solids are broadly categorized into two types: crystalline and amorphous. Crystalline solids, such as sugar and sulfur, not only exhibit rigidity and resistance to compression but also feature a well-defined, repeating arrangement of their constituent particles. In contrast, amorphous solids—including materials like rubber, plastic, and glass—while also rigid and incompressible, lack a regular internal structure and do not possess a defined geometric form.

Crystalline materials melt sharply at a specific temperature, whereas amorphous substances do not exhibit a distinct melting point. For instance, when glass is heated, it softens gradually and begins to flow without undergoing a clear transition to the liquid phase. X-ray diffraction studies reveal that amorphous solids lack a long-range ordered atomic or molecular arrangement, which is why they are often considered to behave like supercooled liquids across all temperature ranges.

In a more precise context, the term "solid state" typically refers to the crystalline form, which is recognized as the true representation of a solid.

© The Author(s), under exclusive license to Springer Nature Switzerland AG 2026

R. N. Goyal, H. Goel, *Textbook of Engineering Chemistry*,
https://doi.org/10.1007/978-3-032-22068-4_4

4.1.1 Difference Between Crystalline and Amorphous Solids

Property	Crystalline solids	Amorphous solids
Symmetry	Exhibit a well-defined symmetry along specific axes	Lack any definite symmetry in their internal structure
Geometric structure	Possess a highly ordered and periodic arrangement of particles	Display an irregular or partially disordered arrangement of atoms or molecules
Cleavage planes	When cut with a sharp tool, they split cleanly along distinct planes, producing flat surfaces	Breaks occur irregularly and not along specific planes, resulting in uneven surfaces
Melting characteristics	Melt sharply at a specific temperature	Gradually soften upon heating and begin to flow without a precise melting point
Anisotropy/isotropy	Show anisotropic behavior, meaning physical properties such as electrical conductivity and refractive index vary with direction	Exhibit isotropic properties; physical characteristics remain uniform in all directions
Cooling curve	The temperature versus time graph during solidification shows distinct breaks, indicating crystallization events	The cooling curve is continuous and smooth, lacking sharp transitions

4.2 Crystal Structure

A crystal is a uniform segment of a crystalline material characterized by a consistent and repeating arrangement of its structural components, resulting in a definite geometric shape.

It was observed by Haüy that a well-known crystal like calcite ($CaCO_3$) can be broken down into many smaller fragments, each maintaining the same external shape as the original crystal. The smallest repetitive portion that retains the structural identity of the crystal is called the *unit cell.*

The *crystal lattice*, also known as the *space lattice*, refers to a systematic three-dimensional network of points that represents the spatial distribution of atoms, ions, or molecules within the crystal. The concept of a lattice was introduced by Bravais, who proposed that each point in the lattice could represent a particle located at the center of a repeating structural unit within the crystal. Therefore, a space lattice visually conveys the orderly and periodic arrangement of the constituents in a crystalline solid.

4.2.1 *Laws of Crystallography*

The unit cell of a crystal is defined by the lengths of its edges and the angles between its faces. Crystals made of the same substance and maintained at identical temperatures show consistent interfacial angles between their faces, even if their external forms differ. This observation is known as the *first law of crystallography*, introduced by D. Guglielmini in 1688.

Therefore, it is not necessary to measure the exact dimensions of a unit cell when analyzing crystal structures, as the ratio of the unit cell's dimensions remains constant throughout the entire crystal. For example, in the case of copper sulfate ($CuSO_4$), the unit cell parameters are

$$a : b : c = 0.571 : 1 : 0.558$$

$$\text{alpha} = 82^\circ 16', \quad \text{beta} = 107^\circ 26', \quad \text{gamma} = 102^\circ 40$$

Here, a, b, and c are the unit cell edges, and alpha, beta, and gamma are the angles between them.

The *second law of crystallography*, called the *law of rational indices*, states that the intercepts of crystal faces on the crystallographic axes occur in simple whole-number ratios.

Consider a three-dimensional system with x, y, and z axes. Let a crystal plane intersect these axes at

$$OA = a, \quad OB = b, \quad \text{and} \quad OC = c$$

Now, suppose another face intersects at

$$OL = l \times a, \quad OM = m \times b, \text{ and } ON = n \times c$$

(where l, m, and n are integers)

The coefficients l, m, and n are known as the *Weiss indices* of the plane. To determine the *Miller indices*, take the reciprocals of the Weiss indices ($1/l$, $1/m$, $1/n$) and multiply all reciprocals by the smallest number that converts them into whole numbers.

For example, if the Weiss indices of a plane are

$$a : b : (3/2)c,$$

then multiply all terms by 2:

$$2a : 2b : 3c$$

So, the Miller indices become (2 2 3).

Example 4.1 A plane within a crystal intersects the crystallographic axes at $3a$, $2b$, and $3c$. Determine the *Weiss indices* and the *Miller indices* for this plane.

Solution: Intercepts along the axes:

$$3a : 2b : 3c$$

Weiss indices: 3 : 2 : 3

Reciprocals of Weiss indices: 1/3 : 1/2 : 1/3

To eliminate fractions, multiply each term by the least common multiple of the denominators (LCM = 6):

$$\left(\frac{1}{3} \times 6\right) : \left(\frac{1}{2} \times 6\right) : \left(\frac{1}{3} \times 6\right) = 2 : 3 : 2$$

Miller Indices = (2 3 2)

The *primary application* of Miller indices is to represent the spacing between parallel crystal planes, denoted as $d(hkl)$.

The general formula for interplanar spacing in a crystal is

$$\frac{1}{(d(hkl))^2} = \left(\frac{h^2}{a^2}\right) + \left(\frac{k^2}{b^2}\right) + \left(\frac{l^2}{c^2}\right)$$

where:

- h, k, and l are the Miller indices of the plane
- a, b, and c are the unit cell dimensions along the x, y, and z axes, respectively

In a *cubic system*, where $a = b = c$, the formula simplifies to

$$d(hkl) = \frac{a}{\sqrt{(h^2 + k^2 + l^2)}}$$

Example 4.2 Calculate the interplanar spacing for the (100), (110), and (111) planes in a *cubic lattice*.

Solution: Using the formula for cubic crystals:

$$d(hkl) = \frac{a}{\sqrt{(h^2 + k^2 + l^2)}}$$

For (100):

$$d(100) = \frac{a}{\sqrt{(1^2 + 0^2 + 0^2)}} = \frac{a}{\sqrt{1}} = a$$

For (110):

$$d(110) = \frac{a}{\sqrt{(1^2 + 1^2 + 0^2)}} = \frac{a}{\sqrt{2}}$$

For (111):

$$d(111) = \frac{a}{\sqrt{(1^2 + 1^2 + 1^2)}} = \frac{a}{\sqrt{3}}$$

Ratio of interplanar spacings:

$$\begin{aligned} d(100) : d(110) : d(111) &= a : \left(a/\sqrt{2}\right) : \left(a/\sqrt{3}\right) \\ &= 1 : 1/\sqrt{2} : 1/\sqrt{3} \\ &\approx 1 : 0.707 : 0.577 \end{aligned}$$

4.3 Crystal Symmetry

Crystals can be categorized based on their *symmetry elements*. Although more than 200 different crystal shapes have been observed, they are grouped into *32 symmetry classes*, which are further organized into *seven crystal systems*. The basic types of symmetry found in crystals are referred to as *elements of symmetry*, which include:

1. Center of symmetry
2. Axis of symmetry
3. Plane of symmetry

4.3.1 Center of Symmetry

A *center of symmetry* is a specific point within the crystal such that any straight line drawn through it will intersect the crystal surface at equal distances in opposite directions. A crystal that has a center of symmetry may or may not possess planes or axes of symmetry.

4.3.2 Axis of Symmetry

An *axis of symmetry* is an imaginary line passing through a crystal such that the crystal can be rotated around this line and still appear identical more than once during a full 360° rotation. If a crystal looks the same *n times* during one complete turn, the axis is referred to as an *n-fold axis of symmetry*.

Common types of rotational symmetry in crystals include:

- *2-fold (diad) symmetry*: repeated every 180°
- *3-fold (triad) symmetry*: repeated every 120°
- *4-fold (tetrad) symmetry*: repeated every 90°
- *6-fold (hexad) symmetry*: repeated every 60°

The axis of symmetry is usually perpendicular to a crystal face and aligned parallel to an edge between two faces.

Examples

- The **H_2O** *molecule* exhibits *twofold rotational symmetry*, as it appears the same every 180°.
- The **BCl_3** *molecule* has a *threefold rotational symmetry*, repeating its appearance every 120°. Since $360°/120° = 3$, this confirms a threefold symmetry axis.

4.3.3 Plane of Symmetry

A *plane of symmetry* is an imaginary plane that divides a crystal into two halves that are exact mirror images of one another. Depending on its structure, a crystal can possess multiple planes and axes of symmetry. However, a crystal can have *only one center of symmetry*, if present.

4.4 Bravais Lattices

In 1848, *Auguste Bravais* demonstrated that all possible crystal structures can be formed from just *14 distinct three-dimensional arrangements*, known as *Bravais lattices*. These fundamental lattice types are the building blocks for over *230 different crystal configurations* observed in nature.

Each Bravais lattice represents a unique way of arranging points in space to describe the *unit cell*—the smallest repeating unit in a crystal. These 14 lattices are grouped into *seven crystal systems*, based on their symmetry elements and geometric characteristics.

The seven crystal systems are:

1. Cubic (or Isometric) system
2. Hexagonal system
3. Tetragonal (or quadratic) system
4. Orthorhombic (or rhombic) system
5. Trigonal (or rhombohedral) system

6. Triclinic (or asymmetric/anorthic) system
7. Monoclinic (or monosymmetric) system

The *geometric parameters* and *symmetry characteristics* of these systems determine how the atoms, ions, or molecules are arranged within each unit cell. Further details regarding the *classification and structural properties* of each Bravais lattice type within these seven systems are typically summarized in a reference table (see Table 4.1).

4.4.1 Regular or Cubic or Isometric System

This crystal system is defined by *three mutually perpendicular axes*, all of which are *equal in length*. The angles between the axes are all *90 degrees*, forming a perfectly symmetrical three-dimensional shape.

There are *three variations* within this system based on how particles are arranged:

- Simple cubic
- Body-centered cubic (BCC)
- Face-centered cubic (FCC)

Substances that crystallize in this system typically exhibit high symmetry and dense packing.

Examples include *KCl, NaCl, diamond, AgCl, gold, copper, silver*, and other similarly structured compounds.

4.5 Close Packing of Particles in Crystals

In crystalline solids, each atom, ion, or molecule strives to *maximize contact with surrounding particles*. This arrangement lowers the *potential energy* of the system and increases the *packing efficiency*, thereby leading to *greater density*.

There are two main types of *close packing arrangements*:

1. **Square close packing (SCC)/face-centered close packing (FCC):**
   In this arrangement, particles in one row align *directly above* those in the row beneath, resulting in a *grid-like* pattern. The particles are arranged in both horizontal and vertical lines, forming *square-shaped voids* between them.
2. **Hexagonal close packing (HCC):**
   In this configuration, particles in the second row are positioned so that each sphere *sits in the gap between two particles* in the first row. As a result, each sphere touches two particles from the previous row, forming a *triangular arrangement* and creating *hexagonal symmetry*.

Table 4.1 Characteristics of different crystal systems

System	Axial property	Angles	Space lattices	Common examples
1. Regular or cubic	$a = b = c$	$\alpha = \beta = \gamma = 90°$	3	NaCl, KCl, CsCl
2. Hexagonal	$a = b \neq c$	$\alpha = \beta = 90°, \gamma = 120°$	1	CdS, HgS, SiO_2
3. Tetragonal or quadratic	$a = b \neq c$	$\alpha = \beta = \gamma = 90°$	2	TiO_2, SnO_2, $ZrSO_4$
4. Rhombic or orthorhombic	$a \neq b \neq c$	$\alpha = \beta = \gamma = 90°$	4	$CdSO_4$, $BaSO_4$, KNO_3, AgBr
5. Trigonal or rhombohedral	$a = b = c$	$\alpha = \gamma = 90°, \beta \neq 90°$	1	$CaCO_3$, As, Sb, Bi
6. Triclinic	$a \neq b \neq c$	$\alpha \neq \beta \neq \gamma \neq 90°$	1	$NaHSO_4$, $CuSO_4 \cdot 5H_2O$, $K_2Cr_2O_7$
7. Monoclinic	$a \neq b \neq c$	$\alpha = \gamma = 90°, \beta \neq 90°$	2	$NaHCO_3$, $CaSO_4 \cdot 2H_2O$

From these two packing types, it is evident that *hexagonal close packing* offers *greater efficiency* in space utilization compared to square close packing, due to the reduced amount of empty space between particles.

4.6 Calculation of Atoms/Unit Cells in Crystal

The number of atoms effectively present within a unit cell depends on their positions—whether they are located at the *corners*, *faces*, or the *center* of the unit cell. Since atoms located on the edges or corners are shared among multiple adjacent unit cells, only a *fractional contribution* is considered for each.

4.6.1 Simple Cubic Cell

In a *simple cubic (SC)* arrangement, atoms are positioned *only at the eight corners* of the cube. Each corner atom is shared among *eight neighboring unit cells*, contributing only *1/8* of an atom to the given unit cell. Therefore, the total number of atoms per unit cell is $= 8 \times (1/8) = 1$ atom.

4.6.2 Body-Centered Cubic Cell

A *body-centered cubic (BCC)* structure contains *one atom at the center* of the cube in addition to the *eight corner atoms*. The center atom is *entirely contained* within the unit cell, while each corner atom contributes *1/8* as in the simple cubic cell. So, the total atoms per unit cell $= 1$ (center) $+8 \times (1/8) = 1 + 1 = 2$ atoms.

4.6.3 Face-Centered Cubic Cell

In a *face-centered cubic (FCC)* unit cell, atoms are located *at all eight corners* and *at the centers of the six faces*. Each *face-centered atom* is shared between *two adjacent unit cells*, contributing *1/2* to a single unit cell.

Total atoms in the unit cell: $= 6 \times (1/2) + 8 \times (1/8) = 3 + 1 = 4$ atoms.

4.7 Calculation of Mass of the Unit Cell

The mass of a unit cell is determined by multiplying the *mass of a single atom* by the *number of atoms present within the unit cell*:

$$\text{Mass of unit cell} = \text{Number of atoms per unit cell} \times \text{Mass of one atom}$$

The *mass of a single atom* can be calculated by dividing the *molar mass* (usually in g/mol) of the substance by *Avogadro's number (NA)*:

$$\text{Mass of one atom} = \frac{\text{Molar mass (M)}}{N_A}$$

Therefore,

$$\text{Mass of unit cell} = \frac{M \times Z}{N_A}$$

where:

M = molar mass of the element or compound (in g/mol)
Z = number of atoms in the unit cell
N_A = Avogadro's number ($6.022 \times 10^{23}\ \text{mol}^{-1}$)

4.8 Atomic Radius of Cubic System

The *atomic radius* (r) is defined as half the distance between the centers of two adjacent atoms in a unit cell. The value of r depends on how the atoms are arranged within the crystal lattice. The *edge length* (a) refers to the distance between centers of atoms at adjacent corners of the cube.

4.8.1 Atomic Radius of Simple Cubic Cell (SCC)

In a simple cubic structure, atoms are in direct contact along the edge of the cube.

Relation between the edge length and atomic radius:

$$a = 2r$$

Therefore,

$$r = \frac{a}{2}$$

4.8.2 Atomic Radius of Face-Centered Cubic Cell (FCC)

In a face-centered cubic unit cell, atoms touch along the face diagonal of the cube. A face diagonal spans across two corner atoms and the face-centered atom, equivalent to four atomic radii.

Length of face diagonal:

$$\sqrt{2} \times a = 4r$$

Solving for r:

$$r = \frac{(a)}{(2\sqrt{2})}$$

4.8.3 Atomic Radius of Body-Centered Cubic Cell (BCC)

In a body-centered cubic structure, atoms are in contact along the body diagonal, which spans across two corner atoms and the central atom, equal to four atomic radii.

Length of body diagonal:

$$\sqrt{3} \times a = 4r$$

Solving for r:

$$r = \frac{(\sqrt{3} \times a)}{4}$$

4.9 Radius Ratio

The *radius ratio* is a key geometric parameter in the study of ionic crystals. It is defined as the ratio of the radius of the *cation* to the radius of the *anion* in an ionic compound. This ratio helps predict the *coordination number* and *geometric arrangement* of ions in the crystal lattice.
Formula:

$$\text{Radius ratio} = \frac{r^+}{r^-}$$

where:

r^+ = radius of the cation
r^- = radius of the anion

4.10 Coordination Number

The *coordination number* refers to the number of nearest neighboring atoms or ions that are directly surrounding a given atom or ion in a crystal structure. In simple terms, it is the count of closest contacts an atom has in the lattice.

This value depends on both the *crystal structure* and the *radius ratio* of the constituent ions. Larger radius ratios usually allow more ions to fit around a central ion, leading to a higher coordination number.

Some typical coordination numbers for different unit cell types are:

- Simple cubic: 6
- Body-centered cubic (BCC): 8
- Face-centered cubic (FCC): 12

The coordination number also correlates with the *radius ratio* (r^+/r^-) of the cation to the anion. The following table shows how different radius ratios correspond to different coordination numbers and resulting geometries:

Coordination Number	Radius ratio Range	Geometrical Arrangement
2	0.00–0.155	Linear
3	0.155–0.225	Triangular
4	0.225–0.414	Tetrahedral
6	0.414–0.732	Octahedral
8	0.732–1.000	Cubic
12	1.000	FCC or HCP

4.10.1 Calculation of Radius Ratio of Different Coordination Numbers

The *radius ratio* is the ratio of the radius of a cation (r^+) to the radius of an anion (r^-). It is a crucial factor in determining the geometry and coordination number of ions in a crystal structure. The following are calculations for various coordination numbers:

(i) **Coordination number 3 (triangular geometry)**

Let the radius of the cation be r^+ and the radius of the anion be r^-.
From triangle EBD: Let

$$ED/BD = \tan(30^\circ) \Longrightarrow ED = \frac{r^-}{\sqrt{3}}$$

In triangle EBD (right-angled triangle),

$$(r^+ + r^-)^2 = r^{-2} + \left(\frac{r^-}{\sqrt{3}}\right)^2$$

Expanding the equation,

$$(r^+)^2 + 2r^+r^- + (r^-)^2 = \frac{r^{-2} + r^{-2}}{3}$$

$$r^{+2} + 2r^+r^- + r^{-2} = \left(\frac{4}{3}\right) r^{-2}$$

Multiply both sides by 3:

$$3r^{+2} + 6r^+r^- + 3r^{-2} = 4r^{-2}$$

$$3r^{+2} + 6r^+r^- - r^{-2} = 0$$

Divide the whole equation by r^{-2}:

$$\left(\frac{r^+}{r^-}\right)^2 + 2\left(\frac{r^+}{r^-}\right) - 1 = 0$$

Solve the quadratic:

$$x^2 + 2x - 1 = 0$$

where $x = r^+/r^-$.
Using the quadratic formula,

$$x = \frac{-2 \pm \sqrt{(4+4)}}{2} = \frac{[-2 \pm \sqrt{8}]}{2} = \frac{[-2 \pm 2\sqrt{2}]}{2} = -1 \pm \sqrt{2}$$

Take the positive root:

$$r^+/r^- = \sqrt{2} - 1 \approx 0.155$$

(i) **Coordination number 6 (octahedral geometry)**
In an octahedral structure, the geometry resembles a square cross section. From triangle ABC: Let

$$\cos(45°) = AB/BC \Longrightarrow BC = \frac{AB}{\sqrt{2}}$$

Now, since $AB = r^+ + r^-$, we substitute

$$r^- = \frac{(r^+ + r^-)}{\sqrt{2}}$$
$$\Longrightarrow \sqrt{2} \times r^- = r^+ + r^-$$
$$\Longrightarrow r^+ = \left(\sqrt{2} - 1\right) \times r^-$$

Thus,

$$\frac{r^+}{r^-} = \sqrt{2} - 1 \approx 0.414$$

(iii) **Coordination number 8 (cubic geometry)**
In a cubic geometry, ions touch along the body diagonal of the cube. Let the body diagonal be $(2r^+ + 2r^-)$ and edge length be $2r^-$. Using the Pythagorean theorem in triangle ADC,

$$AC^2 = AD^2 + DC^2$$
$$(2r^+ + 2r^-)^2 = (\sqrt{2} \times 2r^-)^2 + (2r^-)^2$$
$$\Longrightarrow (2r^+ + 2r^-)^2 = 8r^{-2} + 4r^{-2} = 12r^{-2}$$
$$\Longrightarrow 4r^{+2} + 8r^+r^- + 4r^{-2} = 12r^{-2}$$
$$\Longrightarrow 4r^{+2} + 8r^+r^- - 8r^{-2} = 0$$

Divide by 4:

$$r^{+2} + 2r^+r^- - 2r^{-2} = 0$$

Now, divide by r^{-2}:

$$\left(\frac{r^+}{r^-}\right)^2 + 2\left(\frac{r^+}{r^-}\right) - 2 = 0$$

Solve the quadratic:

$$x^2 + 2x - 2 = 0$$

where $x = r^+/r^-$. Using the quadratic formula,

$$x = \frac{[-2 \pm \sqrt{(4+8)}]}{2} = \frac{[-2 \pm \sqrt{12}]}{2} = \frac{[-2 \pm 2\sqrt{3}]}{2} = -1 \pm \sqrt{3}$$

Take the positive root:

$$r^+/r^- = \sqrt{3} - 1 \approx 0.732$$

4.11 Volume, Mass, and Density of Unit Cell

4.11.1 Volume of the Unit Cell

The volume of a unit cell can be determined using data from X-ray diffraction analysis of the crystal. Let the edge length of the unit cell be

$$a\ pm = a \times 10^{-10}\,\text{cm}$$

Then the volume of the unit cell is

$$\text{Volume} = (a \times 10^{-10})^3 = a^3 \times 10^{-40}\,\text{cm}^3$$

4.11.2 Mass of the Unit Cell

The mass of a unit cell is calculated as

$$\text{Mass of unit cell} = \text{Number of atoms in unit cell} \times \text{Mass of one atom}$$

Let:

- Z = number of atoms per unit cell
- M = molar mass (in g/mol)
- N_0 = Avogadro's number

Mass of one atom:

$$m = \frac{M}{N_0}\ \text{grams}$$

So,

$$\text{Mass of unit cell} = Z \times \left(\frac{M}{N_0}\right) = \frac{(Z \times M)}{N_0} \text{ grams}$$

4.11.3 *Density of the Unit Cell*

Density (ρ) is defined as the mass of the unit cell divided by its volume:

$$\begin{aligned}\text{Density } (\rho) &= \frac{\text{Mass of unit cell}}{\text{Volume of unit cell}} \\ &= \frac{(Z \times M)}{(N_0 \times a^3 \times 10^{-30})}\end{aligned}$$

Multiply the numerator and denominator by 10^{30} to simplify

$$\rho = \frac{(Z \times M \times 10^{30})}{(a^3 \times N_0)} \text{ g/cm}^3$$

4.12 Atomic Packing Factor

The atomic packing factor (APF) is defined as the ratio of the total volume actually occupied by atoms in a unit cell to the total volume of the unit cell.

4.12.1 *Atomic Packing Factor for Simple Cubic Cell (SCC)*

A simple cubic unit cell contains eight corner atoms, each shared by eight neighboring unit cells. Therefore, the effective number of atoms per unit cell is

$$= 8 \times (1/8) = 1 \text{ atom}$$

Volume occupied by one atom:

$$= (4/3) \times \pi r^3$$

Volume of the unit cell: $= a^3$, and for SCC, $a = 2r$. So,

$$a^3 = (2r)^3 = 8r^3$$

Packing factor:

$$APF = \frac{[(4/3) \times \pi r^3]}{(8r^3)}$$
$$= \frac{\pi}{6} \approx 0.5236 \text{ or } 52.36\%$$

4.12.2 Atomic Packing Factor for Face-Centered Cubic Cell (FCC)

An FCC unit cell contains:

- eight corner atoms $\times(1/8) = 1$ atom
- six face atoms $\times(1/2) = 3$ atoms. So, total atoms $= 1 + 3 = 4$ atoms

Volume occupied by four atoms:

$$= 4 \times (4/3) \times \pi \times r^3 = (16/3) \times \pi \times r^3$$

In FCC, the relation between a and r is

$$a = 2\sqrt{2} \times r$$

so

$$a^3 = (2\sqrt{2} \times r)^3 = 16\sqrt{2} \times r^3$$

Packing factor:

$$APF = \frac{[(16/3) \times \pi \times r^3]}{(16\sqrt{2} \times r^3)}$$
$$= (\pi/3\sqrt{2}) \approx 0.7405 \text{ or } 74.05\%$$

4.12.3 Atomic Packing Factor for Body-Centered Cubic Cell (BCC)

A BCC unit cell has:

- eight corner atoms $\times(1/8) = 1$ atom
- one atom at the body center $= 1$ atom. So, total atoms $= 2$ atoms

Volume occupied by two atoms:

$$= 2 \times (4/3) \times \pi \times r^3 = (8/3) \times \pi \times r^3$$

In BCC, the relation between a and r is

$$a = \frac{(4r)}{\sqrt{3}}$$

so

$$a^3 = \left[\frac{(4r)}{\sqrt{3}}\right]^3 = \left(\frac{64}{3\sqrt{3}}\right) \times r^3$$

Packing factor:

$$\begin{aligned} APF &= \frac{[(8/3) \times \pi \times r^3]}{[(64/3\sqrt{3}) \times r^3]} \\ &= \frac{(\pi\sqrt{3})}{8} \approx 0.6802 \text{ or } 68.02\% \end{aligned}$$

4.13 X-Ray Diffractions of Crystals

In 1913, Laue proposed that crystals could diffract X-rays, due to the fact that the wavelength of X-rays is on the same order of magnitude as the spacing between atoms in a crystal, typically around 1 angstrom (1 Å).

When an X-ray beam is directed at a crystal, a diffraction pattern is observed. This pattern contains valuable information about the arrangement of atoms within the crystal and forms the foundation of a technique called *X-ray crystallography.*

The first detailed analysis using this technique was conducted by *Bragg* on a sodium chloride crystal. His findings led to the development of what is now known as *Bragg's law.*

4.14 Bragg's Equation

Bragg observed that X-ray scattering from a crystal can be treated similarly to the reflection of light. Specifically, the X-rays reflected from successive atomic planes interfere constructively under certain conditions. The relationship among the X-ray wavelength (λ), the interplanar spacing (d), and the angle of reflection (θ) is defined by Bragg's law:

$$n\lambda = 2d \sin \theta$$

where:

- n is an integer (the order of diffraction)
- λ is the wavelength of the incident X-rays
- d is the distance between adjacent crystal planes
- θ is the angle between the incident X-rays and the crystal planes

This principle is foundational for determining crystal structures and is further elaborated in *Chap. 1*.

4.14.1 Bragg's Spectrophotometer for Elucidation of Sodium Chloride Structure

In this method, a sharp and narrow beam of X-rays is reflected from one face of a crystal. These X-rays are generated in a chamber (A), as shown in the diagram, using a suitable metal target such as chromium (Cr), molybdenum (Mo), or copper (Cu), which emit X-rays with wavelengths of 2.29 Å, 0.79 Å, and 1.54 Å, respectively.

The beam is made monochromatic by passing it through an appropriate filter. It is then collimated using slits (S and L) to produce a sharp beam, which is directed onto one face of the crystal. The crystal is mounted on a turntable (C) that can be rotated to any desired angle of incidence (θ) using a calibrated circular scale.

The X-ray beam reflected from the crystal enters an ionization chamber (I) filled with a gas—typically sulfur dioxide (SO_2). This gas gets ionized by the reflected X-rays, and the extent of ionization, which is proportional to the intensity of the X-rays, is measured using an electrometer as electric current.

By rotating the turntable, the glancing angle θ of the incident X-ray is gradually increased. The reflected intensity reaches a maximum at certain angles that satisfy Bragg's condition. This procedure is repeated for various crystallographic planes like (100), (110), and (111). For NaCl, the first-order reflection angles are:

- (100) $\rightarrow$ 5.9°
- (110) $\rightarrow$ 8.4°
- (111) $\rightarrow$ 5.2°

The condition for maximum reflection is given by *Bragg's law*:

$$n\lambda = 2d \sin \theta$$

where:

- n = order of reflection
- λ = wavelength of X-rays
- d = spacing between atomic planes
- θ = angle of incidence

When n, λ, and θ are known, the interplanar distance d can be calculated. Rearranging Bragg's law,

$$1/d = 2\sin\theta/n\lambda$$

This implies that

$$1/d \propto \sin\theta$$

or

$$d \propto 1/\sin\theta$$

Using this relationship for the three planes,

$$\begin{aligned} d_{100} : d_{110} : d_{111} &= \frac{1}{\sin(5.9^\circ)} : \frac{1}{\sin(8.4^\circ)} : \frac{1}{\sin(5.2^\circ)} \\ &\approx \frac{1}{0.103} : \frac{1}{0.146} : \frac{1}{0.091} \\ &\approx 9.73 : 6.84 : 11.04 \\ &\approx 1 : 0.7 : 1.1 \end{aligned}$$

This ratio matches the pattern of a *face-centered cubic (FCC)* lattice structure.

4.15 Two- and Three-Dimensional Solids

Carbon solids can exist in both two-dimensional and three-dimensional crystalline forms. These different structural arrangements lead to unique properties—some carbon solids are extremely hard, while others are soft; some conduct electricity well, while others do not. Carbon exhibits *three primary allotropic forms*: *diamond*, *graphite*, and *fullerene*.

4.15.1 Diamond

Diamond consists of carbon atoms arranged in a *three-dimensional tetrahedral network*. Each carbon atom is *sp*^{3} *hybridized* and forms four sigma bonds with other carbon atoms in a compact, giant covalent structure.

- Bond angle (C-C): 109°28′.
- Bond length (C-C): *1.54 Å*.

- Structure type: *face-centered cubic (FCC)*, with carbon atoms occupying alternate tetrahedral holes in the lattice.
- Density: *3.52 g/cm*
- Properties: Due to strong covalent bonding throughout the lattice, diamond is *extremely hard* and possesses a *very high melting point*. It is *electrically insulating*.

4.15.2 Graphite

Graphite is made up of *sp^2 hybridized carbon atoms* arranged in a *layered two-dimensional hexagonal lattice*. Each carbon atom forms three sigma bonds with neighboring atoms, creating flat sheets of hexagonal rings:

- Bond length (C-C in hexagon): *1.42 Å*
- Interlayer distance (van der Waals gap): *3.4 Å*
- Density: *2.25 g/cm^3*

The fourth valence electron in each carbon atom remains unhybridized and contributes to a *delocalized π-electron cloud* above and below each hexagonal layer. These mobile electrons make graphite a *good conductor of electricity*.

The layers are held together by *weak van der Waals forces*, allowing them to slide over one another easily. This makes graphite *soft and slippery*, suitable for use as a *solid lubricant*.

Types of Graphite

(a) **Natural graphite**: Extracted from natural mineral deposits. It is a *good conductor of electricity*, *highly crystalline*, and used in applications like solid lubricants and refractory materials due to its *high thermal resistance*.
(b) **Artificial graphite**: Produced from *fancoke* or *petroleum pitch tar*. It is *less crystalline* compared to natural graphite and is used in industrial applications where pure crystalline structure is not essential.

Applications of Graphite

1. **Refractory use**: Graphite is widely utilized in manufacturing *refractory materials* such as *crucibles* due to its high thermal stability.
2. **Blast furnace lining**: It serves as a *protective lining material* inside *blast furnaces*, withstanding extreme temperatures and chemical corrosion.
3. **Electrodes and electrolysis**: Graphite functions as an *arc electrode* in *electric arc furnaces* and is also employed as the *anode* in various *electrolytic processes*.
4. **Electrical machinery**: It is used in *carbon brushes* for *electric motors*, *generators*, and *turbines*, providing good conductivity and wear resistance.

5. **Nuclear applications**: Owing to its *high melting point* and ability to *absorb neutrons*, graphite is used as a *moderator* and *shielding material* in *nuclear reactors*.
6. **Miscellaneous uses**: *Amorphous graphite* is used in the production of *pencils*, *solid lubricants*, *paints*, and *coatings*, offering both lubricating and conductive properties.

4.16 Fullerenes

Fullerene was discovered in 1985 by Robert Curl, Harold Kroto, and Richard Smalley at the University of Sussex and Rice University. The name was inspired by Richard Buckminster Fuller, as the structure resembled his geodesic domes. The molecule was first identified as a product of graphite vaporization using a laser, a discovery that earned the Nobel Prize in Chemistry in 1996. Subsequently, fullerenes were also found in residues from carbon arc lamps, lightning strikes, and other high-energy carbon events. The most common fullerene, C_{60}, has a structure known as a truncated icosahedron, composed of five- and six-membered rings of sp^2-hybridized carbon atoms. This differs from graphite, which consists solely of fused hexagonal rings of sp^2 carbons arranged in a flat, planar sheet. Larger fullerenes also consist of similar five- and six-member carbon rings. The truncated icosahedron has 60 vertices and 32 faces, including 12 pentagons and 20 hexagons.

The van der Waals diameter of a C_{60} molecule is approximately 1 nanometer (nm), while the distance from nucleus to nucleus is around 0.7 nm. Other fullerenes formed in trace amounts contain 70, 76, 84, 90, and 96 carbon atoms. The general chemical characteristics of fullerenes include the following:

- Fullerenes act as electrophiles and readily undergo nucleophilic addition reactions. The resulting carbanion intermediate is stabilized by capturing an electrophile. Typical nucleophiles include Grignard and organolithium reagents. For instance, C_{60} reacts with methylmagnesium chloride to form a stable penta-adduct, where methyl groups cluster around a cyclopentadienyl anion that is later protonated. The Bingel reaction is another example of a nucleophilic addition to fullerenes.
- Fullerene reacts with chlorobenzene and aluminum chloride in a Friedel-Crafts-type alkylation. This hydroarylation yields a 1, 2-addition product (Ar-CC-H).
- The [6,6] bonds in fullerenes can behave as dienes or dienophiles in cycloaddition reactions, such as the Diels-Alder reaction. Four-membered rings may be formed *via* [2+2] cycloadditions with compounds like benzyne. The Prato reaction is an example of a 1,3-dipolar cycloaddition to a five-membered ring.
- Fullerenes are easily hydrogenated by various methods. Commonly studied hydrofullerenes include $C_{60}H_{18}$ and $C_{60}H_{36}$. However, a fully saturated $C_{60}H_{60}$ molecule remains hypothetical due to excessive molecular strain. Highly

hydrogenated fullerenes are generally unstable; extended hydrogenation at high temperatures causes the cage to collapse into polycyclic aromatic hydrocarbons.

- Though more challenging than reduction, oxidation of fullerenes is feasible with agents like oxygen and osmium tetroxide.
- Fullerenes can also participate in electrophilic addition. For instance, bromination can introduce up to 24 bromine atoms onto the C_{60} sphere. The fluorinated compound $C_{60}F_{48}$ holds the record for fluorine addition. Computational predictions suggest that the elusive $C_{60}F_{60}$ might exist with some fluorine atoms pointing inward (endo positions), giving it a tubular rather than a spherical shape.
- Reactions with carbenes form methanofullerenes. A well-known example is the reaction between C_{60} and bromo(dimethoxyphosphoryl) acetate in the presence of sodium hydride in toluene, producing alkoxycarbonyl (dimethoxyphosphoryl)methanofullerene.

Fullerenes do occur naturally, albeit in small quantities, during combustion processes such as fires and lightning discharges. However, they were first synthetically created in laboratory conditions by vaporizing graphite with a laser, resulting in soot that contained fullerenes. The earliest large-scale production method was developed in 1990 and is known as the arc discharge method, which involves graphite electrodes. This technique primarily produces C_{60} and C_{70} but can also yield higher fullerenes when more porous electrodes are used. Solvent extraction, particularly with toluene, can purify C_{60} to nearly 100%. Fullerenes have also been synthesized *via* pyrolysis of different aromatic compounds. Advanced methods such as sputtering and electron beam evaporation have been shown to preferentially produce larger fullerenes like C_{70}, C_{76}, C_{78}, and C_{84}.

Fullerenes exhibit remarkable electrical properties, prompting their exploration for use in electronic applications ranging from data storage to photovoltaic devices. For example, scientists at Virginia Tech in the USA have employed ultrathin fullerene layers as electron acceptors in flexible organic solar cells. They are also under investigation for use in X-ray photodetectors. The unique electronic behavior of fullerenes is further being explored for fuel cell technologies. Fullerene-based materials hold potential in photonic devices as well, owing to their strong nonlinear optical response, which makes them promising candidates for telecommunications technologies. Their ability to neutralize free radicals, which are harmful to biological tissues, has sparked interest in their use in skin care cosmetics and in preventing neuronal damage caused by oxidative stress—a potential shown in early rat studies. Since the size of the C_{60} molecule closely matches that of several biologically active compounds, including steroid hormones and pharmaceuticals, it could serve as a template for creating a broad range of bioactive derivatives. Notably, buckyballs exhibit high chemical and physical affinity for the active site of HIV protease—an enzyme critical to HIV replication—and can inhibit its function. Although existing AIDS treatments target the same enzyme, drug resistance has become a concern due to their similar mechanisms. In contrast, buckyballs act *via* a different mechanism, suggesting they may bypass current resistance pathways. Certain fullerenes and their chemical derivatives have demonstrated superconducting properties, although only at extremely low temperatures. They may also serve as precursors for advanced materials like diamond-like coatings or carbon nanotubes. In a more fundamental context, fullerenes have been utilized in quantum mechanics research: C_{60} molecules are the largest particles to date that have displayed wave-particle duality, as seen in double-slit experiments. Due to their affinity for organic compounds, C_{60} and its derivatives can be coated onto piezoelectric crystals and used as sensors for detecting organic molecules.

The electrochemical reduction of fullerenes in an aqueous potassium hydroxide (KOH) medium involves multiple electron transfer steps, with three electrons reducing each C_{60} molecule. Transient intermediates such as KC_{60} and K_2C_{60} exhibit ion-exchange properties, while the final product, K_3C_{60}, is electrically conductive. Films of partially reduced fullerene C_{60} have been found to possess sufficient conductivity to serve as electrodes in various electrochemical processes. These films also show notable electrocatalytic behavior and good biocompatibility. However, further reduction to form highly hydrogenated or potassium-saturated structures like $C_{60}H_{60}$ or $K_{60}C_{60}$ is unfeasible due to the extreme strain imposed on the molecular framework:

$$C_{60} \xrightarrow{e^-} KC_{60} \xrightarrow{e^-} K_2C_{60} \xrightarrow{e^-} K_3C_{60}$$

4.17 Carbon Nanotubes

Carbon nanotubes (CNTs) are highly ordered carbon allotropes with an exceptionally high aspect ratio, which have attracted widespread research interest since their identification in 1991. They were first observed by Iijima in 1991 using a transmission electron microscope. CNTs are considered among the most intriguing materials due to their distinctive electronic, mechanical, and chemical characteristics. Composed of sp^2-hybridized carbon atoms, these tubes have diameters of a few nanometers and can extend several micrometers in length. Their synthesis is achieved through methods such as arc discharge, laser ablation, and chemical vapor deposition. Depending on their diameter and helicity, CNTs can act as either metals or semiconductors. The carbon atoms on the sidewalls and ends of the tubes differ in reactivity, akin to the basal and edge planes of highly oriented pyrolytic graphite (HOPG).

These cylindrical carbon structures possess novel properties that make them promising for a variety of applications across nanotechnology, electronics, optics, and advanced material sciences, with potential architectural uses as well. They combine exceptional mechanical strength, thermal conductivity, and unique electrical characteristics. However, concerns regarding their potential toxicity could influence their eventual range of applications. Their integrated electrical, mechanical, optical, thermal, and structural attributes make them valuable candidates for use in optoelectronics, high-performance materials, and biomedical engineering. Due to their structure, CNTs can serve as both conductors and semiconductors, opening up wide-ranging uses including gas storage systems, battery electrodes, and field emission displays. Their subtle electronic behavior allows them to facilitate electron-transfer processes when employed as electrode materials in electrochemical systems. These remarkable qualities also suggest that CNTs could lead to the development of highly efficient electrochemical sensors, ranging from amperometric enzyme electrodes to label-free DNA biosensors.

Carbon nanotubes are broadly classified into two types: single-walled nanotubes (SWNTs) and multi-walled nanotubes (MWNTs). SWNTs consist of a single graphene sheet rolled into a seamless cylinder, effectively forming a molecular wire with all atoms exposed on the surface. MWNTs, in contrast, are composed of multiple such tubes concentrically nested like the rings of a tree trunk. Single-walled carbon nanotubes typically have diameters ranging from 0.4 nm to 2 nm, while MWNTs consist of graphite cylinders with diameters ranging from 2 nm to 100 nm. SWNTs are more susceptible to chemical reactions, whereas MWNTs display greater resistance. CNTs often group into bundles comprising dozens or even hundreds of parallel tubes in physical contact. They retain their electrical conductivity and stability even in harsh chemical environments. Both SWNTs and MWNTs possess extraordinary tensile strength, are extremely lightweight, and demonstrate exceptional chemical and thermal resilience. These properties, along with their conductive or semiconductive nature, have led to numerous proposed applications. SWNTs are particularly important because they exhibit electronic behaviors not found in MWNTs, making them strong candidates for advancing electronics beyond current microelectromechanical systems. In such miniaturized devices, where the electric wire serves as the fundamental unit, SWNTs may function as excellent conductors.

All forms of carbon nanotubes are expected to be efficient thermal conductors along their longitudinal axis through a mechanism known as "ballistic conduction," but they act as good insulators in the direction perpendicular to the axis. The combination of high strength and flexibility makes CNTs suitable for manipulating other nanoscale structures, positioning them as key materials in nanotechnology engineering. The maximum tensile strength recorded for an individual MWNT is around 63 PGa. CNTs display a wide array of remarkable properties, from their nanoscale dimensions to their unusual current conduction behavior, making them highly suitable for use in electronic circuits. Additionally, they have been proposed for biomedical applications such as gene delivery systems and, when paired with radiofrequency radiation, as tools for selectively destroying cancer cells.

4.18 Solved Problems

1. A unit cell of NaCl has *FCC* structure, and the edge length is 0.651 nm. What is the density of NaCl?

Solution: $Z = 4$ (for *FCC*)

$$V = a^3 = (0.651 \times 10^{-7})^3\,\text{cm}^3 = 0.276 \times 10^{-21}$$
$$= 2.76 \times 10^{-22}\,\text{cm}^3$$
$$M = 58.5\,\text{g mol}^{-1}$$

$$\text{Density } \rho = \frac{Z \times M}{V \times N_o}$$
$$= \frac{4 \times 58.5}{2.76 \times 10^{-22} \times 6.023 \times 10^{23}}$$
$$= 1.408\ \text{g/cm}^3$$

2. Ionic radii of some ions are given below:

Ions	K^+	Cs^+	Mg^{2+}	Sr^{2+}	Br^-	O^{2-}	S^{2-}
Radius (pm)	133	169	66	112	196	140	184

Give the most probable structures of KBr, CsBr, MgO, and SrS.
Solution:

(i) KBr

$$\frac{r_+}{r_-} = \frac{133}{196} = 0.678 < 0.732$$

Octahedral coordination no. 6.

(ii) CsBr

$$\frac{r_+}{r_-} = \frac{169}{196} = 0.863 > 0.732$$

Cubic coordination no. 8.

(iii) MgO

$$\frac{r_+}{r_-} = \frac{66}{140} = 0.472 > 0.414$$

but <0.732 Octahedral coordination no. 6.

(iv) SrS

$$\frac{r_+}{r_-} = \frac{112}{184} = 0.609 > 0.414$$

but <0.732 Octahedral coordination no. 6.

3. If two ions A^+ and B^- have radii 88 and 200 pm, respectively, in a compound AB, find out the coordination number of A^+.
   Solution:

$$\frac{r_+}{r_-} = \frac{88}{200} = 0.44 > 0.414$$

   but <0.732 Octahedral coordination no. 6.
4. Calculate the value of Avogadro's number from the data. The density of NaCl = 2.165 g cm^{-3}, and the distance between ions = 281 pm.
   Solution: For NaCl FCC structure $Z = 4$,

$$\rho = 2.165 \text{ g cm}^3$$

$$M = 58.5 \text{ g mol}^{-1}$$

$$a = 2 \times 281 \times 10^{-10} \text{ cm} = 5.62 \times 10^{-8} \text{ cm}$$

   So

$$\rho = \frac{ZM}{a^3 \times N_o} \Longrightarrow N_o = \frac{ZM}{\rho a^3}$$

$$N_o = \frac{4 \times 58.5}{\left(5.62 \times 10^{-8}\right)^3 \times 2.165} = 6.09 \times 10^{23} \text{ mol}^{-1}$$

5. A compound has a cubic structure with elements A and B. The elements A occupy the corners of the cube, while the elements B occupy alternate opposite faces. Find the formula of the compound.
   Solution:

$$\text{Number of } A \text{ in unit cell} = 8 \times \frac{1}{8} = 1 \text{ atom}$$

$$\text{Number of } B \text{ atoms} = 3 \times \frac{1}{2} = \frac{3}{2}$$

   Formula of the compound:

$$= \frac{A}{B} = \frac{1}{3/2} = \frac{2}{3}$$

i.e.,

$$A = 2 \text{ and } B = 3$$

So, the formula is A_3B_2.

6. Calculate the density of an *FCC* lattice with a unit cell length of = 0.409 nm and atomic weight of 107.88 g/mol^{-1}.
   Solution: Given $a = 0.409\,\text{nm} = 4.09 \times 10^{-10}\,\text{m} = 4.09 \times 10^{-8}\,\text{cm}$ for FCC $Z = 4$ atoms and $M = 107.88\,\text{g mol}^{-1}$

$$\rho = \frac{ZM}{a^3N_o} = \frac{4 \times 107.88}{\left(4.09 \times 10^{-8}\right) \times 6.023 \times 10^{23}}$$
$$= \frac{431.52}{412.08 \times 10^{-1}}$$
$$= 10.47\,\text{g cm}^{-3}$$

7. Calculate the density of CsCl for a *BCC* lattice. The cell edge of the unit cell is 412 pm (atomic wt: Cs = 133, Cl = 35.5).
   Solution: Given Z = 2

$$a = 412\text{ pm} = 4.12 \times 10^{-8}\,\text{cm}$$
$$V = \left(4.12 \times 10^{-8}\right)^3 = 6.993 \times 10^{-23}\,\text{cm}^3$$
$$M = 133 + 35.5 = 168.5\,\text{g mol}^{-1}$$
$$N_o = 6.023 \times 10^{23}$$
$$\rho = \frac{Z \times M}{V \times N_o}$$
$$= \frac{2 \times 168.5}{6.993 \times 10^{-23} \times 6.023 \times 10^{23}}$$
$$= \frac{337.0}{42.12} = 8.0\,\text{g cm}^{-3}$$

8. A *BCC* structure has a cell edge 314 pm and density 10.3 g cm^{-3}. Calculate the atomic mass of element ($N_o = 6.023 \times 10^{23}\,\text{mol}^{-1}$).
   Solution: Given $a = 314\text{ pm} = 3.14 \times 10^{-8}\,\text{cm}$

$$\rho = 10.3\,\text{g cm}^{-3}$$

$$Z = 2$$

$$\rho = \frac{ZM}{a^3 \times N_o}$$

$$10.3 = \frac{2 \times M}{\left(3.14 \times 10^{-8}\right) \times 6.023 \times 10^{23}}$$

$$= \frac{2 \times M}{186.5 \times 10^{-1}}$$

$$M = \frac{10.3 \times 186.5 \times 10^{-1}}{2} = 96.04\text{ g mol}^{-1}$$

9. A *BCC* structure has density 1.51 g cm^{-3}, and the radius of the atom is 248 pm. If the atomic mass of the atom is 85.5, calculate Avogadro's number.
Solution: Given $\rho = 1.51\,\text{g cm}^{-3}$

$$\rho = 1.51\text{ g/cm}^3$$

$$Z = 2$$

$$M = 85.5\text{ g mol}^{-1}$$

$$r = 248\text{ pm} = 2.48 \times 10^{-8}\text{ cm}$$

for *BCC*

$$a = \frac{4r}{\sqrt{3}} = \frac{4 \times 2.48 \times 10^{-8}}{\sqrt{3}}$$

$$= 5.73 \times 10^{-8}\text{ cm}$$

$$V = a^3 = \left(5.73 \times 10^{-8}\right)^3$$

$$= 1.88 \times 10^{-22}\text{ cm}^3$$

so

$$\rho = \frac{ZM}{V \times N_o}$$

or

$$N_o = \frac{ZM}{V\rho} = \frac{2 \times 85.5}{1.88 \times 10^{-22} \times 1.51}$$

$$= 6.023 \times 1023$$

10. The density of KBr is 2.73 g/cm^3. The cell edge is 654 pm; find the value of Z if the atomic weights of K = 39 and Br = 80.
Solution: Given

$$\rho = 2.37\,\text{g/cm}^3$$

$$a = 654\text{ pm} = 6.54 \times 10^{-8}\text{ cm}$$

$$M = 39 + 80 = 119\,\text{g mol}^{-1}$$

$$N_o = 6.023 \times 10^{23}\,\text{mol}^{-1}$$

$$Z = \frac{\rho a^3 N_o}{M}$$

$$= \frac{2.73 \times \left(6.54 \times 10^{-8}\right)^3 \times 6.023 \times 23}{119}$$

$$= \frac{2.73 \times 279.73 \times 10^{-24} \times 6.023 \times 10^{23}}{119}$$

$$= 3.86 \cong 4$$

11. An element has *BCC* structure. Its unit cell edge is 400 pm. Calculate the density of the unit cell and the number of unit cells in 100 g of an element (atomic mass = 100, $N_o = 6.023 \times 10^{23}\,\text{mol}^{-1}$).
Solution: $Z = 2$

$$a = 400\,\text{pm} = 4 \times 10^{-8}\text{ cm}$$

$$M = 100\,\text{g mol}^{-1}$$

$$V = a^3 = (4 \times 10^{-8})^3$$

$$= 6.4 \times 10^{-23}\text{ cm}^3$$

$$N_o = 6.023 \times 10^{23}\,\text{mol}^{-1}$$

So, density

$$\rho = \frac{Z \times M}{V \times N_o}$$

$$= \frac{2 \times 100}{6.4 \times 10^{-23} \times 6.023 \times 10^{23}}$$

$$= 5.29\,\text{g/cm}^3$$

The amount of unit cell in 100 g of atom:

$$\text{weight of unit cell} = Z \times \text{weight of one atom}$$

and

$$\text{weight of atom} = \frac{\text{Atomic weight}}{N_o} = \frac{100}{6.023 \times 10^{23}}$$

so

$$\text{weight of unit cell} = \frac{2 \times 100}{6.023 \times 10^{23}}$$

and number of unit cell in 100 g

$$= \frac{100}{\frac{2\times 100}{6.023\times 10^{23}}} = \frac{6.023 \times 10^{23}}{2}$$

Number of unit cell in 100 g = 3.0115×10^{23} atoms.

12. 208 g of a *FCC* structure has 4.28×10^{24} atoms. Calculate the cell edge of the unit cell if the density of the unit cell is 7.2 g cm^{-3}.
Solution: Given $\rho = 7.2\,\text{g cm}^{-3}$

$$Z = 4$$

If the cell edge is a, then the volume = a^3

$$\frac{M}{N_o} = \frac{208}{4.28 \times 10^{24}}$$

Now,

$$\text{Density } \rho = \frac{Z \times M}{V \times N_o}$$

$$7.2 = \frac{4 \times \frac{208}{4.28\times 10^{24}}}{a^3}$$

or

$$a^3 = \frac{4 \times 208}{7.2 \times 4.28 \times 10^{24}}$$

$$a^3 = 2.7 \times 10^{-25}\ \text{cm}^3$$

$$a = 3 \times 10^{-8}\ \text{cm} = 300\,\text{pm}$$

13. In a cubic structure, one atom is at each corner and two atoms on one diagonal. The cell edge is 289 pm and the density is $7.2\,\text{g/cm}^3$. Calculate number of atoms in 200 g of an element.
Solution: Given no. of atom $Z = \frac{1}{8} \times 8 + 2 = 3$

$$a = 289 \text{ pm} = 2.89 \times 10^{-8} \text{ cm}$$

$$V = a^3 = \left(2.89 \times 10^{-8}\right)^3 = 24.1 \times 10^{-24} \text{ cm}^3$$

$$\rho = 7.2\,\text{g/cm}^3$$

$$\text{mass of 1 atom} = \frac{V \times \rho}{Z}$$

$$= \frac{24.1 \times 10^{-24} \times 7.2}{3}$$

$$= 5.7 \times 10^{-24} \text{ g}$$

Number of atoms in 200 g $= \frac{200}{5.7 \times 10^{-24}}$:

$$= 3.4 \times 10^{24} \text{ atoms}$$

14. For a *FCC* structure, the cell edge is 100 pm and the density is $10\,\text{g m}^{-3}$. How many atoms are present in 100 g?
Solution: Given $a = 100 \text{ pm} = 10^{-8}$ cm

$$V = a^3 = (10^{-8})^3 = 10^{-24} \text{ cm}^3$$

$$Z = 4$$

$$M = 100\,\text{g}$$

$$\rho = 10\,\text{g m}^{-3}$$

$$= 10 \times 10^{-6}\,\text{g/cm}^3$$

$$\text{Number of atoms} = \frac{Z \times M}{a^3 \times \rho}$$

$$= \frac{4 \times 100}{10^{-24} \times 10 \times 10^{-6}}$$

$$= 4 \times 10^{31} \text{ atoms.}$$

15. An X-ray beam is diffracted at an angle 20° for second-order spectrum. If the wavelength of X-ray is 2.6 Å, then find the distance of an atom in a crystal.
Solution: Given $\theta = 20°$

$$\lambda = 2.6\ \text{Å} = 2.6 \times 10^{-10}\ \text{m}$$

$$n = 2$$

Then $n\lambda = 2d \sin\theta$

$$d = \frac{n\lambda}{2\sin\theta} = \frac{2 \times 2.6 \times 10^{-10}}{2 \times \sin 20}$$

$$d = \frac{2.6 \times 10^{-10}}{0.3420}$$

$$= 7.602 \times 10^{-10}\ \text{m} = 7.6\ \text{Å}$$

16. An X-ray beam ($\lambda = 150$ pm) gives first-order diffraction (110) on KCl plane. The cell edge of the unit cell is 305 pm. Find out the glancing angle of diffraction.
Solution: Given $\lambda = 150$ pm

$$a = 305\ \text{pm}$$

$$n = 1$$

$$d_{110} = \frac{a}{\sqrt{1^2 + 1^2 + 0}} = \frac{a}{\sqrt{2}}$$

$$= \frac{305}{\sqrt{2}} = 215.7\ \text{pm}$$

Now,

$$n\lambda = 2d \sin\theta$$

$$1 \times 150 = 2 \times 215.7 \times \sin\theta$$

or

$$\sin\theta = \frac{150}{2 \times 215.7} = 0.3477$$

$$\theta = \sin^{-1} 0.3477 = 20.35^\circ$$

17. KCl has a simple cubic lattice, and for 100 plane, the spacing is 315.2 pm. At what angle will the first-order reflection from the 100 plane be observed (λ of X-ray = 157.3 pm)?
Solution: Given $a = 315.2$ pm for 100 plane

$$\lambda = 153.7 \text{ pm}, n = 1$$

$$d_{100} = \frac{a}{\sqrt{1^2 + 0^2 + 0^2}} = a = 315.2 \text{ pm}$$

$$n\lambda = 2d \sin\theta$$

$$1 \times 153.7 = 2 \times 315.2 \times \sin\theta$$

$$\sin\theta = \frac{153.7}{630.4} = 0.2438$$

$$\theta = \sin^{-1} 0.2438 = 10°07'$$

18. A beam of X-ray ($\lambda = 0.842$ Å) is falling at an angle of $8°35'$ for I order reflection. Calculate the glancing angle for the third-order reflection.
Solution: Given $\lambda = 0.842$ Å

$$\theta_1 = 8°35'$$

$$n_1 = 1$$

$$n_2 = 3$$

Case I:

$$n_1\lambda = 2d \sin\theta_1$$

$$1 \times 0.842 = 2d \sin 8°35' \tag{4.1}$$

Case II:

$$n_3\lambda = 2d \sin\theta_3$$

$$3 \times 0.842 = 2d \sin\theta_3 \tag{4.2}$$

From Eqs. (4.1) and (4.2) after dividing

$$\frac{1}{3} = \frac{\sin 8°35'}{\sin\theta_3}$$

or

$$\sin\theta_3 = 3 \times \sin 8°35' = 3 \times 0.15 = 0.45$$

$$\theta_3 = \sin^{-1}(0.45) = 26.7°$$

19. For KCl crystal using X-ray ($\lambda = 154$ pm) for planes 100, 110, and 111, the glancing angles are $11°27'$, $16°30'$, and $20°07'$, respectively. Determine the structure of KC.
Solution:

$$d_{110} : d_{110} : d_{111} = \frac{1}{\sin\theta_{100}} : \frac{1}{\sin\theta_{110}} : \frac{1}{\sin\theta_{111}}$$

$$= \frac{1}{\sin 11°27'} : \frac{1}{\sin 16°30'} : \frac{1}{\sin 20°07'}$$

$$= 5.03 : 3.52 : 2.90$$

$$d_{100} : d_{110} : d_{111} = 1 : 0.7 : 0.6$$

20. For a second-order diffraction, the diffraction angle is 8.4°, and the interplaner distance is 0.4 nm. Calculate the wavelength of X-ray.
Solution: Given $n = 2$

$$d = 0.4\,\text{nm} = 4 \times 10^{-10}\,\text{m}$$

$$\theta = 8.4°$$

$$n\lambda = 2d\sin\theta$$

$$2 \times \lambda = 2 \times 4 \times 10^{-10} \times \sin 8.4$$

$$\lambda = 0.584 \times 10^{-10}\,\text{m}$$

$$= 0.584\,\text{Å}$$

Exercises

1. What is a unit cell? Draw a unit cell of NaCl.
2. Define and explain the law of interfacial angle.
3. What is the difference between Weiss and Miller indices?
4. Describe the different types of cubic lattices.
5. Derive the relation $n\lambda = 2d\sin\theta$ in crystallography.
6. A solid AB has NaCl structure. If the radius of a cation is 200 pm, find out the radius of an anion.
7. What is the coordination number of $CaCl_2$ if $Cl^- = 181$ pm and $I^- = 221$ pm.
8. A compound A has cubic structure and density 3.4 g cm^{-3}. Calculate the length of the unit cell?
9. A solid has NaCl structure. If an anion has a radius of 250 pm, calculate the radius of a cation. Can a cation of radius 180 pm be occupied in the tetrahedral site of the crystal?

10. A primitive cubic structure may be thought of as eight spheres of radius and just touching each other. Calculate the volume of a unit cell and the packing factor of a unit cell.
11. A unit cell has FCC structure with an edge of 200 pm. Calculate the density of a unit cell if there are 24×10^{24} atoms in 200 g of the substance.
12. An *FCC* element has cell edge 400 pm. What is its density (atomic mass = 60 g mol^{-1})?
13. Calculate Avogadro's number if the distance between Na^+ and Cl^- is 281 pm and the density of NaCl is 2.165 g cm^{-3}.
14. In an experiment, elemental silver was crystalized to form a face-centered cubic lattice with a unit cell length of 4.986 Å. Calculate the atomic radius and the density of the elemental silver. The atomic mass of silver is 107.8, and Avogadro's number is 6.02×10^{23}.
[*Hint:* Calculate the density by using $d = \frac{4M}{V_M}$ and the length of the face diagonal $= \sqrt{2a^2}$]
15. Explain the rotating crystal method for determining crystal structures. Also, explain why X-rays are diffracted by crystals.
16. Fill in the blanks:

(i) Fullerene polygon possesses ________ vertices and ________ faces. Out of ________ faces, ________ are pentagonal and ________ are hexagonal.
(ii) The common hydrogenated species of fullerenes are ________ and ________.
(iii) On reacting with carbenes, fullerenes give ________.
(iv) Fullerene on reacting with chlorobenzene in the presence of $AlCl_3$ gives ________.
(v) The reaction of fullerene with bromine can add up to ________ atoms of bromine, whereas with fluorine, the addition can give ________.
(vi) Sputtering can be used to prepare fullerenes like ________.
(vii) In KOH medium, the electrochemical reduction of fullerenes gives ________, ________, and ________.
(viii) KC_{60} and K_2C_{60} are well-known ________.
(ix) Carbon nanotubes can be synthesized by ________.
(x) CNT behaves as metals and semiconductors depending on ________.
(xi) The diameter of MWNT is usually in the range ________.
(xii) Out of SWNT and MWNT, the one which is highly resistant to chemical attack is ________.
(xiii) Carbon nanotubes promote ________ when used as electrode materials.
(xiv) The potential use of carbon nanotubes is limited by their ________.

17. In a silver crystal, Ag atoms are 2.8×10^{-8} cm apart. Calculate the density of silver assuming that silver crystallizes in *FCC* form.
[*Hint:* The unit cell of silver for *FCC* will have $\frac{1}{8} \times 8$ corner $+ \frac{1}{2} \times 6$ face-centered atoms, that is, 4Ag atoms. Thus, $\frac{\sqrt{2}a}{2} = 2.8 \times 10^{-8}$ and $a = 4.0 \times 10^{-8}$ cm]

18. Fill in the blanks:

(a) Fullerene C_{60} has ________ pentagons and ________ hexagons.
(b) The state of hybridization of carbon in fullerenes is ________.
(c) Fullerenes on reacting with C_6H_5Cl in the presence of $AlCl_3$ chiefly give ________.
(d) Complete hydrogenation of C_{60} to $C_{60}H_{60}$ is not possible due to ________.
(e) Fullerenes react with carbenes to give ________.
(f) Reduction of fullerene C_{60} in KOH chiefly gives ________.
(g) Carbon nanotubes are categorized as ________.
(h) Carbon nanotubes possess ________ tensile strength and ______ weight.
(i) Carbon nanotubes are ________ thermal conductors.

Chapter 5
Electrochemistry

Electrolytes such as NaCl and KCl play an important role in human system. Human cells possess KCl, which are bathed by a solution having sodium chloride. The K^+ ions migrate across cell membranes and the electrical balance of membrane changes, for example, the membranes of the muscle cells of the heart. Due to movement of K^+ ions, the electrical balance of the membrane changes and it starts acting as an electric generator. A detectable amount of current is actually noticed. In the case of fracture, electric currents are generated, which are believed to control the healing process. These are some simple phenomena, which demonstrate the use of electrochemistry in biological systems. Basically electrochemistry is the study of the relationship between chemical and electrochemical phenomena. Electrochemistry deals with electrolysis and chemical reactions caused by electrolysis and the reverse that is production of electricity from chemical reactions. This aspect is the basis of batteries, dry cells, and more recently fuel cells. It has its origin from the first experiment reported by Luigi Galvani in 1786 on the effect of two different metals on the muscles of dead frog. He found that if the frog was placed on copper sheet attached to iron sheet, the legs of the frog exhibited contraction of muscles. He explained this observation due to the passage of fluids from the nerves to muscles. Later on in 1796, another scientist explained that whenever two different metals come in contact with each other, electricity is produced by chemical reaction.

A common experiment showing electricity from chemical reactions is galvanic cell or more commonly a cell. If we place zinc rod in the solution of $ZnSO_4$ and copper rod in the solution of $CuSO_4$, and connect the two solutions through a porous plug, the reactions shown in Fig. 5.1 occur, and oxidation of zinc to zinc ions and reduction of copper ions to copper take place.

The following reaction takes place:

$$Zn + CuSO_4(aq) \longrightarrow Cu + ZnSO_4$$

© The Author(s), under exclusive license to Springer Nature Switzerland AG 2026

R. N. Goyal, H. Goel, *Textbook of Engineering Chemistry*,
https://doi.org/10.1007/978-3-032-22068-4_5

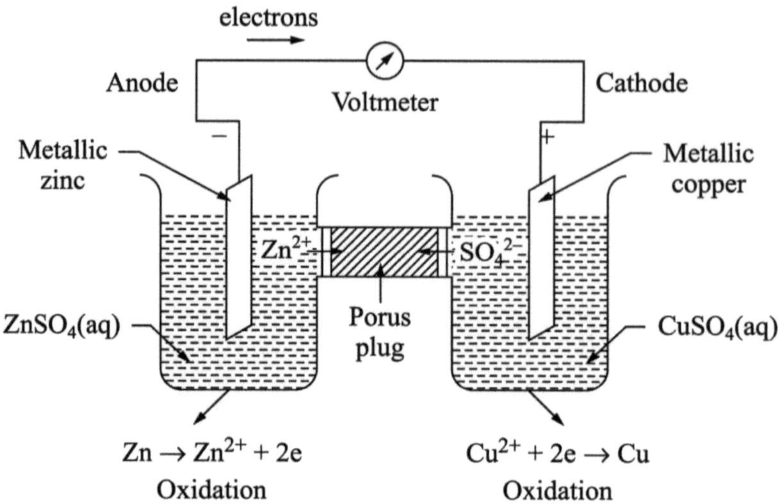

Fig. 5.1 A typical galvanic cell

or

$$Zn + Cu^{2+} \longrightarrow Cu + Zn^{2+}$$

The reaction is exothermic and indicates that if Zn is added to Cu^{2+} solution, heat is evolved but electricity is not produced. To obtain electricity zinc should be placed in $ZnSO_4$ solution and copper in $CuSO_4$ solution, and both should be connected by a wire. The common reactions that take place in such an arrangement are shown in Fig. 5.1, and electrons flow from the zinc electrode to the copper electrode and a voltage of 1.1 V is obtained. The flow of electrons constitutes electricity which can run a motor or light a bulb. The source of electrons in this cell is oxidation of zinc to zinc ions that is

$$Zn - 2e \longrightarrow Zn^{2+} \quad \text{(Oxidation)}$$

and the electrons generated then flow through wire and are consumed by Cu^{2+}:

$$Cu^{2+} + 2e \longrightarrow Cu \quad \text{(Reduction)}$$

The addition of the two electrode reactions gives overall cell reaction as

$$Zn + Cu^{2+} \longrightarrow Cu + Zn^{2+}$$

This is an oxidation-reduction reaction and is the basis of galvanic cell. The electrode at which oxidation reaction takes place is called as *anode*, and the one at which reduction takes place is called as *cathode*. The negative sign for anode is

used as source of electrons. Negative ions and positive ions which move through porous plug are called as anions and cations, respectively.
[Students must remember that anions always move toward anode and cations always move toward cathode.]
The migration of ions is important as no migration means no electricity is produced.

Example 5.1 In a laboratory the following reaction is used to produce electricity:

$$Mg + CdCl_2 \longrightarrow MgCl_2 + Cd$$

Write the electrode reactions and identify anode and cathode.

Solution

$$Mg \longrightarrow Mg^{2+} + 2e \quad \text{oxidation}$$

$$Cd^{2+} + 2e \longrightarrow Cd \quad \text{reduction}$$

Thus, source of electrons is Mg; hence, it should be anode and Cd should be cathode.

5.1 Redox Reactions

The chemical reaction which involves oxidation and reduction processes simultaneously is called as redox reaction. For example, if we place zinc metal in $CuSO_4$ solution, then the following redox reaction will take place:

$$Zn(s) + Cu^{2+}(aq) \longrightarrow Zn^{2+}(aq) + Cu(s)$$

Here zinc, oxidized to Zn^{2+} by losing electrons and the Cu^{2+}, reduced to Cu atom by gaining electrons. The overall reaction is called as redox or oxidation-reduction reaction.

5.2 Electrode Potential

When a metal is dipped in its own salt, the positive ions in the metal come into equilibrium with +ve ions of solution, leaving behind an equivalent number of electrons on the metal (Fig. 5.2). Thus, the metal gets −ve charge and salt solution gets +ve charge. Due to this opposite charge development in metal and solution, a constant potential difference is established between metal and solution. This potential difference is called as *electrode potential* of a metal, which is the measure

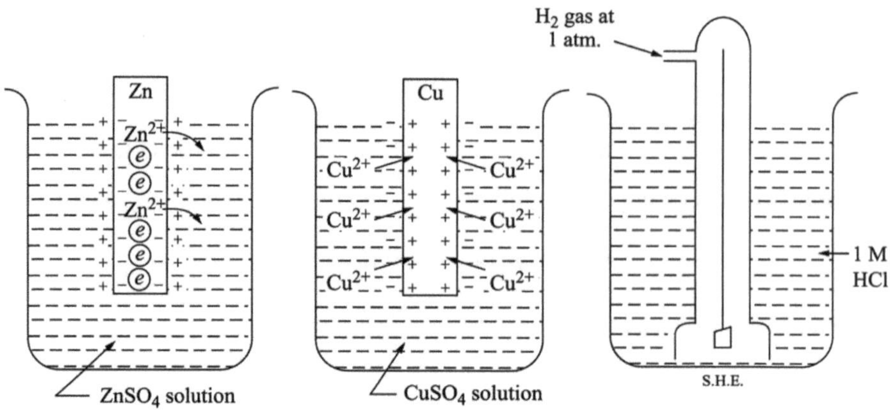

Fig. 5.2 Standard hydrogen electrode and electrode potential generated for zinc and copper dipped in the solutions of the respective salts

of tendency of an electrode to lose or gain electrons, in its own salt of unit molar concentration at 25 °C.

The tendency of metal to lose electrons is a measure to get oxidized and is called as *oxidation potential*. Similarly, the tendency of a metal to gain electrons is a measure to get reduced and is called as reduction potential. The oxidation potential of electrode is +ve and reduction potential is −ve.

5.2.1 Measurement of Electrode Potential

The measurement of potential difference between an electrode and its surrounding solution is not possible. Although we can measure the relative value of electrode potential, we fix the potential of one of the electrodes. For this purpose we use a hydrogen gas electrode by fixing H_2 gas (by continuous bubbled, H_2 gas) over a platinum electrode at atmospheric pressure and dipped it in 1 M HCl solution (Fig. 5.2). This new electrode is called as standard hydrogen electrode (SHE), whose potential has been arbitrarily fixed at *zero*. This electrode is represented as

$$\text{Pt, } H_2(1\,\text{atm}) : \; H^+(1\,\text{M})$$

In order to measure the electrode potential of an electrode into its salt, we couple this electrode with SHE through salt bridge (Fig. 5.3) and measure the electrode potential taken SHE as zero.

Here Zn electrode acts as an anode with respect to SHE, so electrode potential of Zn is assigned a −ve sign. Similarly if copper electrode is connected, it acts as cathode with respect to SHE; hence, electrode potential of Cu is assigned as +ve as (+0.34 V).

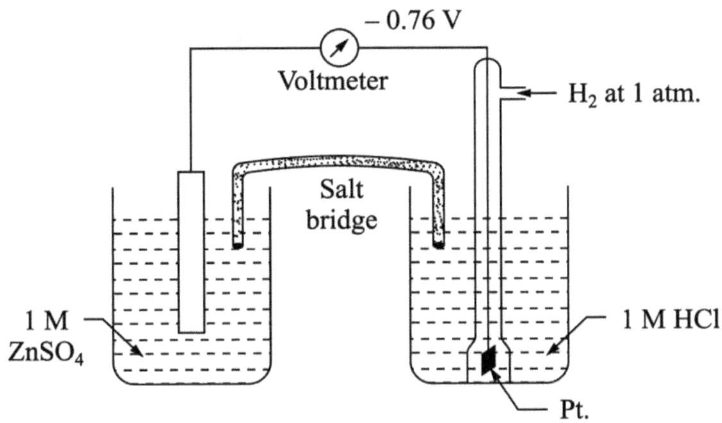

Fig. 5.3 Electrochemical cell for measuring electrode potential relative to SHE

Electrochemical Series This is the arrangement of standard electrode potential of elements in increasing order (Table 5.1).

5.3 Electrochemical Cell or Galvanic Cells

Electrochemical cell is an arrangement of two electrodes dipped in an electrolyte, which generates current due to the chemical reaction or produced chemical reaction due to the passing of current.

Galvanic Cell or Daniell Cell It is a voltaic cell which generates electric current due to chemical redox reaction.

Consider a redox reaction:

$$Mn^{+}(aq) + ne^{-} \rightleftharpoons M(s)$$

For a reversible reaction, the free energy change will be, as per Van't Hoff reaction isotherm

$$\Delta G = RT \ln K + RT \ln \left(\frac{\text{Right}}{\text{Left}} \right)$$

$$\Delta G = \Delta G^{\circ} + RT \ln \left(\frac{\text{Right}}{\text{Left}} \right) \tag{5.1}$$

where ΔG° is the standard free energy change at unit concentration of reactant and products.

The value of free energy will be equal to

Table 5.1 Standard reduction potentials in aqueous solution at 25 °C*

	Reduction half-reaction		$E°(V)$ vs. S.H.E.	
	$F_2(g) + 2e^-$	$\longrightarrow 2F^-$	+2.87	
	$H_2O_2(aq) + 2H^-(aq) + 2e^-$	$\longrightarrow 2H_2O(l)$	+1.77	
	$PbO_2(s) + SO_4^{2-}(aq) + 4H^+(aq) + 2e^-$	$\longrightarrow PbSO_4(s) + 2H_2O(l)$	+1.685	
	$Au^{3+}(aq) + 3e^-$	$\longrightarrow Au(s)$	+1.50	
	$Cl_2(g) + 2e^-$	$\longrightarrow 2Cl^-(aq)$	+1.360	
	$O_2(g) + 4H^+(aq) + 4e^-$	$\longrightarrow 2H_2O(l)$	+1.229	
	$Br_2(l) + 2e^-$	$\longrightarrow 2Br^-(aq)$	+1.08	
	$Hg^{2+}(aq) + 2e^-$	$\longrightarrow Hg(l)$	+0.855	
Increasing strength of oxidizing agents ↑	$Ag^+(aq) + e^-$	$\longrightarrow Ag(s)$	+0.80	Increasing strength of reducing agents ↓
	$Hg_2^{2+}(aq) + 2e^-$	$\longrightarrow 2Hg(l)$	+0.789	
	$Fe^{3+}(aq) + e^-$	$\longrightarrow Fe^{2+}(aq)$	+0.771	
	$O_2(g) + 2H_2O(l) + 4e^-$	$\longrightarrow 4OH^-(aq)$	+0.40	
	$Cu^{2+}(aq) + 2e^-$	$\longrightarrow Cu(s)$	+0.337	
	$Sn^{4+}(aq) + 2e^-$	$\longrightarrow Sn^{2+}(aq)$	+0.15	
	$2H^+(aq) + 2e^-$	$\longrightarrow H_2(g)$ Reference	0.00	
	$Sn^{2+}(aq) + 2e^-$	$\longrightarrow Sn(s)$	−0.14	
	$Ni^{2+}(aq) + 2e^-$	$\longrightarrow Ni(s)$	−0.25	
	$PbSO_4(s) + 2e^-$	$\longrightarrow Pb(s) + SO_4^{2-}(aq)$	−0.356	
	$Cd^{2+}(aq) + 2e^-$	$\longrightarrow Cd(s)$	−0.40	
	$Fe^{2+}(aq) + 2e^-$	$\longrightarrow Fe(s)$	−0.44	
	$Zn^{2+}(aq) + 2e^-$	$\longrightarrow Zn(s)$	−0.763	
	$2H_2O(l) + 2e^-$	$\longrightarrow H(g) + 2OH^-(aq)$	−0.8277	
	$Al^{3+}(aq) + 3e^-$	$\longrightarrow Al(s)$	−1.66	
	$Mg^{2+}(aq) + 2e^-$	$\longrightarrow Mg(s)$	−2.37	
	$Na^+(aq) + e^-$	$\longrightarrow Na(s)$	−2.714	
	$Ca^{2+} + 2e$	$\longrightarrow Ca$	−2.90	
	$K^+ + e$	$\longrightarrow K$	−2.93	
	$Li^+(aq) + e^-$	$\longrightarrow Li(s)$	−3.045	

$$\Delta G = -nEF \text{ and } \Delta G^\circ = -nE^\circ F$$

where E is electrode potential, E° is standard electrode potential and F is Faraday constant.

Putting the value of ΔG and ΔG° in Eq. (5.1)

$$-nEF = -nE^\circ F + RT \ln\left(\frac{M}{M^{n+}}\right) \text{ or } \left(\frac{\text{Right}}{\text{Left}}\right)$$

$$E = E^\circ - \frac{RT}{nF} \ln\left(\frac{\text{Right}}{\text{Left}}\right)$$

$$\boxed{E = E^\circ + \frac{RT}{nF}\left(\ln M^{n+}\right) = E^\circ + \frac{2.3RT}{nF} \log\left(M^{n+}\right)}$$

This is called as Nernst's equation for electrode potential at 298 K.

From Nernst's equation it is near that the electrode potential E is proportional to conc. of solution $\left(M^{n+}\right)$.

The electrochemical series provides several information about element as given below:

1. The elements at the top of electrochemical series are strongly electropositive and they lose electors quickly.
2. The lower position of element, the more tendency to reduce, for example, F_2 reduced to F^-has highest standard electrode potential (+2.87 V) so F^- ions are very easily reduced to F_2 or F_2 oxidized to F^- not so easily.
   In the reverse case, Li^+ to Li has the lowest standard electrode potential (−3.05 V), so Li^+ is reduced with difficulty to Li or Li will oxidize to Li^+easily.
3. The elements with higher value of reduction potential can replace the lower value reduction potential elements. For example,

   $$Cu^{2+}/Cu = +0.34\,V$$

   and

   $$Zn^{2+}/Zn = -0.76\,V$$

   So Cu^{2+} has higher tendency to convert Cu than Zn^{2+} to Zn. It means if we dip zinc metal in $CuSO_4$ solution, then copper will be deposited over Zn metal, and Zn goes to solution as $ZnSO_4$.
4. The possibility/spontaneity of a redox reaction can be predicted from the emf of the cell. If emf is +ve, then redox reactions will be spontaneous; otherwise, reaction is not possible.
5. The metals above hydrogen can replace H_2 from acids if they will displace H_2 by reactions with acids.

6. The equilibrium constant of cell reaction can be calculated by using Nernst's equation:

$$\boxed{E^\circ = \frac{RT}{nF} \ln K_{eq}}$$

where K_{eq} is the equilibrium constant.

5.3.1 *Representation of Galvanic Cell*

The galvanic cell is represented as

$$Zn + Cu^{2+} \longrightarrow Zn^{2+} + Cu$$

To represent the above cell, anode reaction will be on the left and cathode reaction will be on the right. The anode half-cell will be represented as Zn/Zn^{2+} (1 M), and cathode half-cell will be represented as $Cu^{2+}/Cu(1\,M)$.

Now to represent the whole cell, a salt bridge is indicated by two vertical lines, separating the two half cells which may be represented as

$$Zn; \quad Zn^{2+}(1\ M)\|Cu^{2+}(1\ M)Cu$$

5.4 Determination of Electromotive Force (EMF) of a Cell

The electrochemical cell is the combination of two half-cells, i.e., the sum of oxidation half and reduction half cells. So the emf of the cell will be the algebraic sum of two half-cells, emfs. The emf of whole cell (E°) will be

$$E^\circ_{\text{cell}} = E_{\text{right}} - E_{\text{left}}$$

or

$$E^\circ_{\text{cell}} = E_{\text{cathode}} - E_{\text{anode}}$$

or

$$E_{\text{reduction}} - E_{\text{oxidation}}$$

where

$$E_{\text{right}} \text{ or } E_{\text{cathode}} = \text{reduction potential of right-side electrode}$$

and

$$E_{\text{left}} \text{ or } E_{\text{anode}} = \text{reduction potential of left-side electrode}$$

The cell reaction is only feasible if the value of E° cell is +ve.

Example 5.2 Calculate the E° for the following cells, in which the cell reaction occurs as follows:

(i) $CaF_2(aq) + Cl_2(g) \longrightarrow CaCl_2(aq) + F_2(g)$
(ii) $Ca + 2H^+ \longrightarrow Ca^{2+}(aq) + H_2(g)$
(iii) $2KBr(aq) + Cl_2 \longrightarrow 2KCl(aq)Br_2(l)$
(iv) $Zn(s) + Sn^{2+}(aq) \longrightarrow Sn(s) + Zn^{2+}(aq)$
(v) $Cd^{2+}(aq) + Cu(s) \longrightarrow Cd(s) + Cu^{2+}(aq)$

Solution

(i) As written, equation (i) indicates that F^- is oxidized to F_2, whereas Cl_2 is reduced to Cl^-. The involved reactions are

$$2\,F^- \rightarrow F_2 + 2e \quad \text{Oxidation}$$

$$Cl_2 + 2e \longrightarrow 2Cl^- \quad \text{Reduction}$$

$$E^\circ_{\text{cell}} = E_{\text{reduction}} - E_{\text{oxidation}}$$

Now considering E° values from Table 5.1

$$= 1.36 - 2.87 = -1.51\,\text{V}$$

(ii) Equation (ii) indicates that H^+ ions are reduced to H_2 and Ca is oxidized to Ca^{2+}. Hence,

$$\begin{aligned} E^\circ_{\text{cell}} &= E_{\text{reduction}} - E_{\text{oxidation}} \\ &= 0.0 - (2.76) = -2.76\,\text{V} \end{aligned}$$

(iii) Equation (iii) indicates that the actual reaction is

$$2Br^- + Cl_2 \longrightarrow 2Cl^- + Br_2$$

Hence, Br^- is oxidized to Br_2 and Cl_2 is reduced to Cl^-:

$$\begin{aligned} E^\circ_{\text{cell}} &= E_{\text{reduction}} - E_{\text{oxidation}} \\ &= 1.360 - (1.08) = 0.280\,\text{V} \end{aligned}$$

(iv)

$$\text{Zn} \longrightarrow \text{Zn}^{2+} + 2\text{e} \quad \text{Oxidation}$$

$$\text{Sn}^{2+} + 2\text{e} \longrightarrow \text{Sn} \quad \text{Reduction}$$

$$\begin{aligned} E^{\circ}_{\text{cell}} &= E_{\text{reduction}} - E_{\text{oxidation}} \\ &= -0.14 - (-0.76) = 0.62\,\text{V} \end{aligned}$$

(v)

$$\text{Cd}^{2+} + 2\text{e} \longrightarrow \text{Cd} \quad \text{Reduction}$$

$$\text{Cu} \longrightarrow \text{Cu}^{2+} + 2\text{e} \quad \text{Oxidation}$$

$$\begin{aligned} E^{\circ}_{\text{cell}} &= E_{\text{reduction}} - E_{\text{oxidation}} \\ &= -0.40 - (0.34) = -0.74\,\text{V} \end{aligned}$$

5.5 Reference Electrode

Standard hydrogen electrode is considered as primary reference electrode but suffers from several disadvantages such as difficulty in preparation and operation as it requires hydrogen gas at 1 atm. pressure. In addition many arsenic, mercury, and many more compounds poison the platinum electrode. Hence, few other reference electrodes were developed and commonly used in laboratories. Two such electrodes are as follows: (i) calomel electrode and (ii) silver/silver chloride electrode.

(i) ***Calomel Electrode:*** It is most widely used due to its ease of preparation and constant potential and is prepared by using calomel [mercury grinded with mercurous chloride] paste. Little mercury is then placed in the glass vessel and then the paste is added. It is covered with known concentration of KCl, and the side arm is filled with agar-agar to provide a contact with solution as shown in Fig. 5.4. The electrical contact is made using platinum wire dipped in mercury. The usual concentrations of KCl are 0.1 and 1.0 or saturated, and accordingly the calomel electrodes are known as decimolar, molar, or saturated electrode. Once prepared the electrode is left overnight to attain equilibrium. The potentials of these electrodes have been determined with respect to hydrogen electrode at 25 °C and have been found as 0.339, 0.282, and 0.244 V for 0.1, 1.0, and saturated KCl electrode, respectively. The electrode reaction that takes place in this half-cell can be written as

Fig. 5.4 A typical calomel electrode

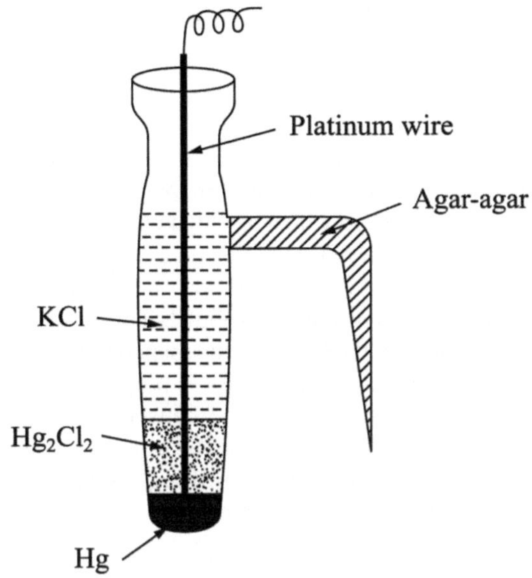

$$Hg_2Cl_2 + 2e \rightleftharpoons 2Hg(l) + 2Cl^-$$

and the potential of the electrode is due to reduction of mercurous ions to mercury. The EMF of the cell is also equal to this potential as the potential of hydrogen electrode is zero. Thus, we can represent the cell for saturated KCl as

$$Pt, H_2(g, 1\text{ atm})\left|H^+(1\text{ M})\right||KCl\text{ (saturated)}\,|\,Hg_2Cl_2, Hg^+$$

Usually the electrode is properly closed from top so that the concentration of KCl does not change. Nowadays, compact calomel electrodes are marketed by several companies. As mercury salts are poisonous, precautions should be taken while preparing such electrodes.

(ii) ***Silver-Silver Chloride Electrode:*** Silver-silver chloride is prepared by using a silver-coated platinum wire. The wire is electrochemically coated with a layer of silver chloride and is then dipped in a solution of KCl of known concentration. The usual concentrations are 0.1, 1.0, and saturated KCl. The potential of the electrode is controlled by the activity of chloride ions present in the KCl solution. At 25 °C the potential of this electrode for saturated KCl with reference to hydrogen electrode is 0.199 V, whereas, for 1.0 M KCl the value is 0.227 V. It is represented as

$$Ag,\ AgCl/Cl^-$$

and is stable for several months.

5.6 Indicator Electrode

A common electrode you must have seen is glass electrode, which is used for determining the pH of the solutions. Thus, a glass electrode gives response to H^+ ions. It has an advantage that the equilibrium is rapidly achieved and is easy to use. The electrode is prepared by using a bulb of glass membrane made up of high-quality glass which has high electrical conductivity and relatively low melting point. The bulb is filled with 0.1 NHCl solution as shown in Fig. 5.5, and I silver wire coated with AgCl is dipped in it. The bulb is now immersed in the solution of which pH is to be determined. The two solutions (one inside the bulb and the second outside the bulb) are thus separated by a thin glass membrane, and a potential difference is developed, which is proportional to the difference in pH value. As the internal HCl is maintained at constant concentration, the potential developed for AgCl coated silver wire will be constant. The only potential that will change is between the outer surface of the glass membrane and the solution in which it is dipped and can be represented as

$$E_g = E_g^\circ + 0.0592\,\text{V pH}$$

where E_g is potential and E_g° is standard electrode potential of glass. The value of E_g° can be determined by taking a solution of known pH. Nowadays glass electrodes are available as combination electrodes with a reference electrode usually Ag/AgCl or calomel electrode. Such a combination is easy to handle and has both indicator and reference electrodes. Figure 5.5 presents a combination electrode in which bulb of the glass electrode is fused to a wide tube and has saturated KCl solution in which a silver wire coated with AgCl is dipped. The electrode is sealed from the top so that

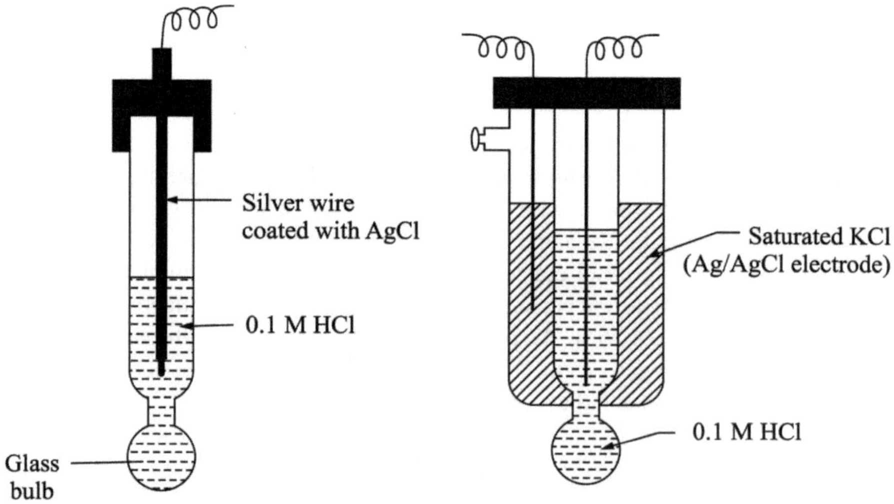

Fig. 5.5 A glass electrode and a combination electrode

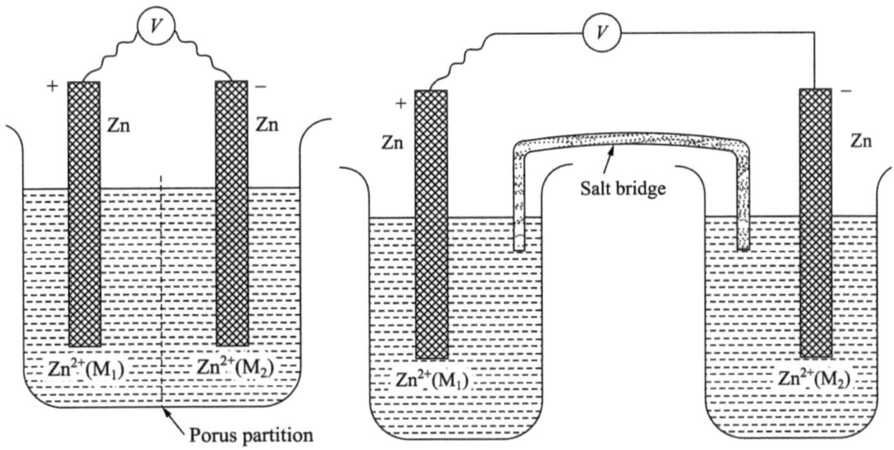

Fig. 5.6 Concentration cell in which two half-cells are joined with porous partition or salt bridge

the concentration of KCl does not change. The system forms a complete cell and its emf is given by

$$E_{\text{cell}} = E_{\text{right}} - E_{\text{left}} \text{ or } E_{\text{reduction}} - E_{\text{oxidation}}$$
$$= 0.2412 - \left[E_g^\circ + 0.0592\,\text{V pH}\right]$$

Concentration Cell It is the cell which produces electrical energy by transfer of solute from higher concentration to lower concentration.

The concentration cell is different from galvanic cell in which electrical energy (nEF) is produced from decrease in free energy, when a chemical reaction takes place in the cell. However, there is no chemical reaction in concentration cell figure.

The concentration cell is made up of two half-cells having electrodes dipped in same electrolytes with different concentration as shown in Fig. 5.6. The half-cells are joined by a salt bridge. The concentration cells are of two types:

(i) In which the two electrodes are the same but dipped in the same solution of different concentration.
(ii) In which the two electrodes are different but dipped in the same solution of different concentration. For example,

$$\text{Zn}\left|\text{Zn}^{2+}\,(\text{M}_1)\right|\left|\text{Zn}^{2+}\,(\text{M}_2)\right|\text{Zn}$$

and

$$\text{H}_2(\text{g})(1\text{Atm})\left|\text{H}^+\,(\text{M}_1)\right|\left|\text{H}^+\,(\text{M}_2)\right|\text{H}_2\text{g}(1\text{Atm})$$

5.7 Calculation of EMF of Concentration Cell

For example, a cell is formed as

$$\mathrm{Zn}\left|\mathrm{Zn}^{2+}\,(M_1)\right|\left|\mathrm{Zn}^{2+}\,(M_2)\right|\mathrm{Zn}$$

where $M_2 > M_1$ and electrode potentials of each half-cell are E_1 and E_2 then as per Nernst's equation:

$$E_1 = E^\circ + \frac{0.0591}{n}\log M_1$$

and

$$E_2 = E^\circ + \frac{0.0591}{n}\log M_2$$

For the above cell, $n = 2$ for Zn ion.
The EMF of cell $E = E_2 - E_1$

$$E = \frac{0.0591}{2}\log M_2 - \frac{0.0591}{2}\log M_1$$

or

$$\boxed{E = \frac{0.0591}{2}\log\frac{M_2}{M_1}}$$

At left-hand side reaction is

$$\mathrm{Zn} \rightleftharpoons \mathrm{Zn}^{2+}\,(\mathrm{M}_1) + 2\mathrm{e}^-$$

At right-hand side reaction is

$$2\mathrm{e}^- + \mathrm{Zn}^{2+}\,(\mathrm{C}_2) \rightleftharpoons \mathrm{Zn}$$

overall reaction is

$$\mathrm{Zn}^{2+}\,(C_2) \rightleftharpoons \mathrm{Zn}^{2+}\,(C_1)$$

which shows that there is no chemical change; only movement of Zn^{2+} ions from higher concentration to lower concentration takes place.

5.8 Applications

1. In determination of solubility of sparingly soluble salts, e.g., the concentration cell

$$\text{Ag} \mid \text{N}/100\ \text{AgNO}_3 \parallel \text{saturated AgBr} \mid \text{Ag}.$$

We have to measure the solubility of AgBr. Then the emf

$$E = \frac{0.0591}{1} \log \frac{0.01}{C_{\text{Ag}}}$$

By calculating E from voltmeter, we can calculate C_{Ag}, and by multiplying with mol weight of AgBr, the solubility of AgBr can be determined.
2. In determination of valency of cation.
3. In determination of hydrolysis of salt.
4. In determination of ionization constant of water.
5. In determination of pH of a solution.

5.9 Solved Problems

1. If

$$E^\circ \left(\text{Ag}/\text{Ag}^{2+}\right) = 0.80\,\text{V}$$

and

$$E^\circ \left(\text{Zn}^{2+}/\text{Zn}\right) = -0.76\,\text{V}$$

predict whether the reaction

$$2\text{Ag} + \text{Zn}^{2+} \longrightarrow 2\text{Ag}^{+} + \text{Zn}$$

is possible or not.

Solution Half-cells are:
Anode

$$2\text{Ag} \longrightarrow 2\text{Ag}^{+} + 2\text{e}^{-}$$

Cathode

$$Zn^{2+} + 2e^- \longrightarrow Zn(s)$$

$$E^\circ_{cell} = -0.76 - 0.80 = -1.56\,V$$

as E°_{cell} is −ve so the above reaction is not feasible.

2. For a cell

$$Zn\left|Zn^{2+}(0.01\,M)\right|\left|Ni^{2+}(0.5)\right|Ni$$

the SEP of Zn | Zn^{2+} = −0.76 V and for Ni^{+2} | Ni = −0.25 V. Find out the EMF of cell.

Solution

$$E = E^\circ + \frac{0.0591}{2}\log\left(\frac{Ni^{2+}}{Zn^{2+}}\right)$$

$$E^\circ = E_{right} - E_{left} = -0.25 - (-0.76) = 0.51\,V$$

$$E = 0.51 + \frac{0.0591}{2}\log\frac{0.5}{0.01}$$

$$= 0.51 + 0.0503 = 0.5603\,V.$$

3. Calculate the emf of the following cell at 25 °C:

$$Zn \mid Zn^{2+}(0.002\,M) \mid\mid Cu^{2+}(0.1\,M) \mid Cu$$

Assume E° of the cell as 1.2 V.

Solution

$$E_{cell} = E^\circ + \frac{0.0591}{n}\log\frac{Cu^{2+}}{Zn^{2+}}$$

$$= 1.2 + \frac{0.0591}{2}\log\left(\frac{0.1}{0.002}\right) = 1.2 + \frac{0.0591}{2}\log 50$$

$$= 1.2 + 0.05 = 1.25\,V.$$

4. From the following half-cells, calculate the standard electrode potential (E°) of cell and emf of cell:

$$PbSO_4 + 2e^- \longrightarrow Pb + SO_4^{2-} \quad E^\circ = -0.31\,V$$

$$PbO_2 + 4H^+ + SO_4^{2-} \longrightarrow PbSO_4 + 2H_2O \quad E^\circ = 1.7\,V$$

The concentrations of $H^+$and SO_4^{2-} are 0.2 M and 2.1 M, respectively.

Solution

$$\text{SEP} = E^\circ_{\text{right}} - E^\circ_{\text{left}} = 1.7 - (-0.31) = 2.01\,\text{V}$$

$$\begin{aligned} E &= E^\circ + \frac{0.0591}{n} \log\left[\left(\text{H}^+\right)^4 \left(\text{SO}_4^{-2}\right)^2\right] \\ &= 2.01 + \frac{0.0591}{2} \log\left[(0.2)^4 \cdot (2.1)^2\right] \\ &= 2.01 + \frac{0.0591}{2} \log 0.0071 \\ E &= 1.947\,\text{V}. \end{aligned}$$

5. Calculate the concentration of Ni^{2+} in the following cell:

$$\text{Ni}\left|\text{Ni}^{2+}\,(\text{M}_1)\right|\left|\text{Cu}^{2+}(0.8)\right|\text{Cu}$$

Given: $E^\circ \text{Ni}^{2+}/\text{Ni} = -0.251\,\text{V}$, $E^\circ \text{Cu}^{2+}/\text{Cu} = 0.3\,\text{V}$ and $E_{\text{cell}} = 0.6\,\text{V}$

Solution

$$\begin{aligned} E^\circ &= 0.3 - (0.251) \\ &= 0.0551\,\text{V} \\ E &= E^\circ + \frac{0.0591}{2} \log\left(\frac{\text{Cu}^{2+}}{\text{Ni}^{2+}}\right) \\ 0.6 &= 0.551 + \frac{0.0591}{2} \log\left(\frac{0.8}{M_1}\right) \\ \frac{0.049 \times 2}{0.0591} &= \log \frac{0.8}{M_1} \\ \frac{0.8}{M_1} &= \text{anti}\log(1.6582) = 45.5204 \end{aligned}$$

or

$$M_1 = \frac{0.8}{45.5204} = 0.0176$$

6. E° of some elements are given below:

(i) $I_2 + 2e^- \longrightarrow 2I^-$ $\quad E^\circ = +0.54\,\text{V}$
(ii) $MnO_4^- + 8H^+ + 5e^- \longrightarrow Mn^{2+} + 4H_2O$ $\quad E^\circ = +0.52\,\text{V}$

(iii) $Fe^{3+} + e^- \longrightarrow Fe^{2+} \quad E^\circ = +0.77\,V$
(iv) $Sn^{4+} + 2e^- \longrightarrow Sn^{2+} \quad E^\circ = +0.1\,V$

(a) Find out the strongest and weakest reductant
(b) Find out the strongest and weakest oxidant
(c) Also by using the above data, select the spontaneous/feasible reactions from the following reactions:

$$Sn^{4+} + 2Fe^{2+} \longrightarrow Sn^{2+} + 2Fe^{3+}$$

$$2Fe^{2+} + I_2 \longrightarrow 2Fe^{3+} + 2I^-$$

$$Sn^{4+} + 2I^- \longrightarrow Sn^{2+} + I_2$$

$$Sn^{2+} + I_2 \longrightarrow Sn^{4+} + 2I^-$$

Solution (a) the strongest reductant and weakest oxidant is (ii) due to the high value of E° in the given data.

(b) the strongest oxidant and weakest reductant is (iv) due to the lowest E° in the given data.

(c) The spontaneous reactions/feasible reactions are those whose emf value of complete cell reaction is positive, i.e., the substance (element) with lower reduction potential can displace another element with higher reduction potential from it's solution spontaneously:

(i) $Sn^{4+}2Fe^{2+} \longrightarrow Sn^{2+}+2Fe^{3+} \longrightarrow \text{emf} = E^\circ_{red}-E^\circ_{ox} = 0.1-0.77 = -0.67$
(ii) $2Fe^{2+} + I_2 \longrightarrow 2Fe^{3+} + 2I^- \longrightarrow \text{emf} = 0.77 - 0.54 = +0.23\,V$
(iii) $Sn^{4+} + 2I^- \longrightarrow Sn^{2+} + I^2 \longrightarrow \text{emf} = 0.1 - 0.54 = -0.44\,V$
(iv) $Sn^{2+} + I_2 \longrightarrow Sn^{4+} + 2I^- \longrightarrow \text{emf} = 0.54 - 0.1 = +0.44\,V$

So Equations (ii) and (iv) are feasible.

7. For a redox reaction

$$AgCl(s) + \frac{1}{2}H_2(g) = HCl + Ag \downarrow$$

Write the cell reactions. Calculate the E° of cell. Find out the activity (conc.) of HCl, if the EMF of the cell is

$$0.2\,V. \left(H^+ \longrightarrow H_2 \quad E^\circ = 0.0\,V, Ag \mid Ag^+ = -0.2224\,V\right)$$

Solution Cell reaction will be

$$\text{Reduction}\quad H^+(aq) + e^- \longrightarrow \frac{1}{2}H_2(g) \quad E^\circ = 0\,V$$

$$\text{Oxidation}\quad Ag(s) \longrightarrow Ag^+ + e^- \quad E^\circ = -0.2224\,V$$

$$\text{Redox Reaction} = Ag(s) + H^+(aq) = Ag^+(aq) + \frac{1}{2}H_2(\,g)$$

$$E^\circ_{cell} = 0.0 - (0.2224) = +0.2224\,V.$$

$$E_{cell} = 0.20\,V, E_{H_2} = 0 - \frac{0.0592}{n}\log\frac{1}{(H^+)}$$

Again

$$E_{cell} = E_{right} - E_{left} = E_{\mu_2} - E_{Ag+/Ag}$$

$$0.20 = \frac{+0.0592}{1}\log H^+ - (-0.2224)$$

$$0.20 = +0.0592\left(\log H^+\right) + 0.2224$$

$$\log H^+ = \frac{-0.2224 + 0.20}{0.0592} = -\frac{0.0224}{0.0592} = -\bar{0}.3784$$

$$H^+ = \text{antil}\log(-0.3784)$$

$$H^+ = \text{anti}\log(-\bar{1}.6216) = 0.4184\,\text{mol}l^{-1}$$

8. For a cell

$$Cd \mid CdCl_2\text{ (aq sat.)} \mid\mid AgCl \mid Ag(s)$$

The emf of the cell at 0 °C is 0.6915 V and at 25 °C 0.7653 V. Calculate the heat of reaction ΔH, free energy, ΔG, and entropy ΔS at 25 °C.

Solution According to the Gibbs-Helmholtz equation

$$\text{emf} = -\frac{\Delta H}{nF} + T\left(\frac{\Delta E}{\Delta T}\right)_\rho$$

here $n = 2$, $F = 96{,}500C$, $\Delta E = (0.7653 - 0.6915)V = 0.0738\,V$

$$\Delta T = 25\,^\circ C \text{ or } 298\,K, \text{ emf at } 298 = 0.7653\,V$$

$$0.7653 = \text{emf} = -\frac{\Delta H}{2 \times 96{,}500} + 298\left(\frac{0.0738}{298}\right)$$

$$0.7653 = -\frac{\Delta H}{193{,}000} + 0.0738$$

$$\Delta H = (0.0738 - 0.7653) \times 193{,}000$$

$$= -0.6915 \times 193{,}000 = -133{,}459.5\,\text{J} = -133.5\,\text{kJ}$$

9. A H_2-electrode is dipped in HCl solution of unknown strength. The emf is 0.29 V at 25 °C. Calculate the activity and pH of acid.

Solution

$$E_{\text{cell}} = E^\circ_{\text{H}_2} - E^\circ_{\text{H}_2} = 0 - \left[-\frac{0.0592}{1}\log\left(\frac{1}{\text{H}^+}\right)\right]$$

$$0.29 = +0.0592\left(-\log \text{H}^+\right) = 0.0592\text{pH}$$

$$\text{pH} = \frac{0.29}{0.0592} = 4.9$$

$$\text{Activity} = -\log\left(\text{H}^+\right) = 4.9$$

$$\text{H}^+ = \text{anti}\log(-4.9) = \text{anti}\log(\bar{5}.1000)$$

$$\text{H}^+ = 1.2589\,\text{mol}l^{-1}.$$

10. Calculate the equilibrium constant of a cell at 25 °C:

$$\text{Cu}^{2+} \mid \text{Cu} \| \text{Zn}^{2+} \mid \text{Zn}$$

given

$$\left(\text{Zn}^{+2} \mid \text{Zn}E^\circ = 0.765\,\text{V}, \text{Cu}^{2+}/\text{Cu} = -0.337\,\text{V}\right)$$

$$E^\circ_{\text{cell}} = 0.765 - (-0.337) = 1.102\,\text{V}$$

$$E = E^\circ - \frac{RT}{nF}\log\left(\frac{\text{oxi}}{\text{red}}\right)$$

$$0 = E^\circ - \frac{RT}{nF}\log \text{K}$$

at equilibrium

$$\log K = \frac{E^\circ nF}{RT} = \frac{1.102 \times 2}{0.0592} = 37.23$$

$$K = \text{anti }\log(37.23) = 1.697 \times 10^{37}$$

11. For cell Mg $|Mg^{2+}|$ $|Ag|Ag$, calculate maximum work of cell given

$$E^\circ_{Mg^{2+}/Mg} = -2.37\,V,\ E^\circ Ag^+/Ag = 0.80\,V$$

Solution

$$\begin{aligned} E^\circ_{cell} &= E_{ox} - E_{red} \\ &= 0.80 - (-2.37) = 3.17\,V \\ \text{Work} &= -\Delta G^\circ = nFE^\circ_{cell} \\ &= 2 \times 96500 \times 3.17 = 611810\,J = 611.8\,kJ \end{aligned}$$

12. Calculate the valency of Hg^+ ion using the following cell:

$$Hg \left| \begin{matrix} 0.001N\ HgNO_3 \\ 0.1\,N\ HNO_3 \end{matrix} \right\| \left| \begin{matrix} 0.01N\ HgNO_3 \\ 0.1\,N\ HNO_3 \end{matrix} \right| Hg$$

emf observed at 18 °C is 0.029 V(F = 96, 500)

Solution

$$E = 2.303\frac{RT}{nF}\log\frac{C_2}{C_1}$$

At 18°$T = 291$ K, $R = 8.318 \times 10^7$, $F = 96,500$, $n = ?$

$$\begin{aligned} \frac{C_2}{C_1} &= \frac{0.01}{0.001} = 10 \\ E &= \frac{0.058}{n}\log 10 \\ 0.029 &= \frac{0.058}{n} \times 1 \\ \Rightarrow \quad n &= \frac{0.058}{0.029} = 2 \end{aligned}$$

13. Calculate the emf of cell at 25 °C:

Solution

$$\text{Cu}\left|\text{Cu}^{2+}(0.01\text{ M})\right|\left|\text{Cu}^{+2}(0.1\text{ M})\right|\text{Cu}$$

$$E = \frac{0.0592}{2}\log\frac{0.1}{0.01}$$

$$E = 0.0295\text{ V}$$

14. What will be the potential of half-cell represented by

$$2\text{Cr}^{3+}(0.2\text{ M}) + 7\text{H}_2\text{O} \longrightarrow \text{Cr}_2\text{O}_7^{2-}(0.3\text{ M}) + 14\text{H}^+(2\text{ M}) + 6\text{e}^-$$

given $E^\circ = -1.33$ V.

Solution

$$E = E^\circ - \frac{0.0591}{n}\log\frac{\left(\text{Cr}_2\text{O}_7^{2-}\right)\left(\text{H}^+\right)^{14}}{\left(\text{Cr}^{3+}\right)\text{h}^2}$$

$$= -1.33 - \frac{0.059}{6}\log\frac{0.3\times(2.0)^{14}}{(0.2)^2}$$

$$= -1.33 - 0.05 = -1.28\text{ V}$$

Exercises

1. What is the difference between voltage and emf?
2. Write the half-cell reaction on iron, when it is connected with copper dipped in dil. H_2SO_4 solution.
3. Explain standard electrode potential and what are its uses.
4. Derive Nernst's equation.
5. What are redox electrodes? Explain it with suitable example.
6. What is the use of salt bridge in formation of electrochemical cells? Why NH_4NO_3 and KCl are used as salt bridge?
7. Explain single electrode potential. How can it be measured?
8. Write a concentration cell with transference. Give the mathematical expression for its emf
9. Calculate the equilibrium constant for the following reaction at 25 °C:

$$\text{Cu(s)} + 2\text{Ag}^+(\text{aq}) \longrightarrow \text{Cu}^{2+}(\text{aq}) + 2\text{Ag(s)} \quad \left(E^\circ_{\text{cell}} = 0.46\text{ V}\right)$$

10. What is the equilibrium constant of the following reaction at 25 °C?

$$Ni_{(s)} \left| Ni^{2+}_{(aq)} \right| /Ag^{+}_{(aq)} Ag_{(s)} \left(E^{\circ}_{cell} = 1.05\,V\right)$$

11. Calculate the maximum work of the following cell:

$$2Al(s) + 3Fe^{2+}(aq) \longrightarrow 2Al^{3+}(aq) + 3Fe(s)$$

conc. of

$$Al^{3+} = (0.1\,M)$$

conc. of

$$Fe^{2+}(aq) = 0.2\,M$$

$$E^{\circ}_{Al^{3+}/Al} = -0.44\,V,\ E^{\circ}_{Fe^{2+}/Fe} = -1.66\,V$$

12. (i) Can we use a copper vessel to store $AgNO_3$ solution?

$$\left\{E^{\circ}_{Ag^{+}/Ag} = 0.80\,V,\ E^{\circ}_{cell^{2+}/Cu} = +0.34\,V\right\}$$

Ans. Yes.

$$\left(E^{\circ}_{Ag^{2+}/Ag} > E^{\circ}_{Cu^{2+}/Cu}\right)$$

(ii) Can we use a nickel rod in $CuSO_4$ solution?

$$\left(E^{\circ}_{Ni^{2+}/Ni} = 0.025\,V,\ E^{\circ}_{Cu^{2+}/Cu} = 0.34\,V\right)$$

13. The standard reduction potential of Cu^{2+}/Cu and Ag^{+}/Ag are 0.337 V and 0.799 V, respectively. Construct a galvanic cell using these electrodes so that its E°_{cell} is positive. For what conc. of Ag^{+} the emf of the cell will be zero if the conc. of Cu^{2+} is 0.01 M?
14. Determine the numerical value of the equilibrium constant of the reaction.

$$2Fe^{2+} + I_2 \longrightarrow 2Fe^{3+} + 2I^{-}$$

given

$$E^{\circ} = -0.771\left(Fe^{3+} \longrightarrow Fe^{2+}\right)$$

$$E^{\circ} - 0.536\left(I_2 \longrightarrow I^{-}\right)$$

Chapter 6
Environmental Chemistry

Broadly speaking, environment means surroundings—air, water, food, necessities of all living things, etc. The temperature, direction of wind, and many other parameters keep the environment changing, and hence the environment for any system is never constant. During the last few decades, the air and water have been polluted due to industrialization and discharge of toxic wastes in water bodies. The environmental chemistry is a branch of science involving chemistry, physics, life sciences, medical science, public health, and many more areas of human concern. In this chapter we will deal with atmospheric chemistry in some detail:

> *Environment refers to the sum of the conditions which surround man at a given place in space and time.*
>
> *or*
>
> *Environment is the sum of all social, economical, physical, chemical or biological factors which constitute the surrounding of man, who is both the creator and moulder of his environment.*

Environment has never been constant or static. It has been changing always, sometimes slowly and at other times rapidly. This change affects living beings, and some of them cannot cope up with the change due to which they have since destroyed or vanished, for example, dinosaurs which become extinct due to the tendency of tiny primitive mammals to eat their eggs. The environment, that time, favored those mammals and not the dinosaurs.

Man has been exploiting the environment to fulfill his needs first and then to improve the quality of his life and eventually to fulfill his desires. By using latest technologies, man is able to modify his immediate environment to suit the needs. Deforestation has been done to a very larger extent. Many animal species have become extinct and many others are endangered. This has happened either due to the massive killing of animals for their skin, etc. or their habitats being destroyed.

© The Author(s), under exclusive license to Springer Nature Switzerland AG 2026
R. N. Goyal, H. Goel, *Textbook of Engineering Chemistry*,
https://doi.org/10.1007/978-3-032-22068-4_6

The main objective of environmental chemistry is to determine the nature as well as the quantity of specific pollutants in the environment. Therefore, its study is of utmost importance today.

6.1 Components of Environment

There are three components of environment:

1. **Abiotic or nonliving components** It involves solid, liquid, and gaseous parts of the environment. It has three parts:
   (a) Lithosphere refers to solid part
   (b) Hydrosphere refers to liquid part
   (c) Atmosphere refers to gaseous part
2. **Biotic or living component** It includes living beings and has two parts:
   (a) Flora (vegetation) which involves all trees, plants, etc.
   (b) Fauna (animals) which involves all animals including man.
3. **Energy components** It involves different types of energies such as solar, geochemical, thermal, hydro, and nuclear radiation.

6.2 Factors Affecting Environment

There are four types of environmental factors which affect the ecology of the environment:

1. **Topographic factors:** The environment changes with altitudes. Plains have different types of vegetation, animals, sea, valleys, mountain, etc.
2. **Climatic factors:** Which include atmosphere, light, humidity, etc.
3. **Edaphic factors:** Which include soil and lithosphere.
4. **Biotic factors:** It includes all types of vegetation, animals, plants, i.e., flora and fauna.

Life of living being is affected by the combination of all these ecological factors, which are interrelated with each other. Thus, the ecological factors under natural conditions operate in combination.

6.3 Types of Environment

It has two categories:

1. **Natural environment:** It is a self-regulating system. Any change in ecosystem is counterbalanced by changes in other components of the environment. Thus, there exists a reciprocal relationship between the components of environment such as air, water, soil, land, trees, animal, flora, and fauna.
2. **Man-made environment:** This is a very diversified environment full of modern technologies to fulfill the needs of humans. It includes latest technologies, housing, industries, transportation space satellites, the Internet, biotechnologies, new medical methods, etc.

6.4 Segments of Environment

There are four segments of environment—atmosphere, hydrosphere, lithosphere, and biosphere.

6.4.1 Atmosphere

The word atmosphere describes the cover of thick gases that surround the earth. This cover not only makes life possible on earth but also saves it from harmful effects of outer atmosphere. The thick gases that surround the earth can be broadly classified in three categories, viz., major compounds, minor compounds, and trace-level components. The major components are nitrogen (78%) and oxygen (21%) with water vapors varying in the range of 0.1–5%. The minor components are carbon dioxide and argon. Their percentage differs from place to place depending upon the extent of pollution. Under normal conditions their percentages are 0.03 and 0.9 respectively. The trace components include CO, NO_2, SO_2, O_3, CH_4, He, Ne, Xe, etc. (Fig. 6.1).

Their percentage is usually very low and of the order of 10^{-4}%. The atmosphere extends up to about 1500 km from the surface of the earth, and density, temperature, and pressure vary with altitude.

For example, atmospheric pressure drops drastically when we move upward from sea level. Thus, the pressure of 1 atmosphere goes down to 3×10^{-7} atmosphere at 100 km above sea level. Similarly the density of the atmosphere decreases with increase in altitude. In broad terms, atmosphere can be divided into five layers:

- The troposphere
- The stratosphere
- The mesosphere
- The heterosphere or ionosphere or thermosphere, and
- The exosphere.

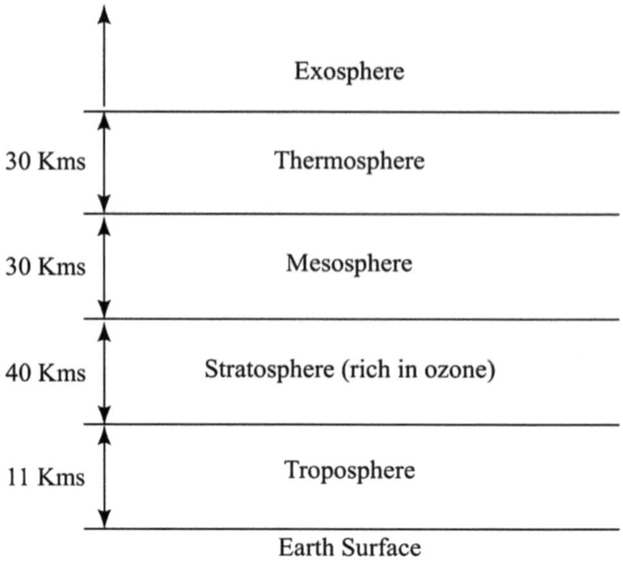

Fig. 6.1 Five layers of atmosphere

The Troposphere

Troposphere is the first layer of gases when one moves from the earth's surface. It extends up to a height of about 11 km and may also vary in some cases. The air is not uniform in this region, and about 75–80% of the total mass of atmosphere is because of the troposphere. With increase in altitude, the temperature decreases systematically by an average rate of 1 °C for 166 m. The upper layer of troposphere has a temperature of nearly −56 °C, and this cold layer of troposphere is called as *tropopause*.

The Stratosphere

The next layer after troposphere is known as stratosphere which extends up to about 40 km from the troposphere layer. In contrast to troposphere, the temperature in this region rises with increase in altitude and reaches to about 5 °C at the top of stratosphere layer. As the temperature in this region is very low, it has no clouds, dust, or water vapors, and hence this sphere is called as inactive sphere. The stratosphere plays an important role in human life as it is rich in ozone, which is formed by photochemical reaction of oxygen. Thus, oxygen molecule takes energy from the sunlight and converts to atomic oxygen, which rapidly combines with another molecule of oxygen to form ozone. The ozone content of stratosphere remains sufficiently constant because of equilibrium between oxygen and ozone:

$$O_2 \overset{hv}{\rightleftharpoons} 2O$$
$$O_2 + O \overset{hv}{\rightleftharpoons} O_3$$

The ozone present in this region absorbs harmful ultraviolet rays present in the sunlight and thus protects us from harmful effects. The UV radiations absorbed by ozone are transformed into heat and thus cause increase in temperature. This is the reason that the temperature of stratosphere increases from −56 °C to about 5 °C at the top of stratosphere.

The Mesosphere

The next zone extended to about 30 km is called as mesosphere. In this region the amount of ozone present is very small and hence absorption of UV rays does not occur much. The result is that the temperature falls with increase in altitude and drops to about −90 °C at the top layer of mesosphere. The chemical species present in mesosphere are N_2, O_2^+, O^+, and NO^+.

The Thermosphere

The layer above the mesosphere is called as thermosphere which again extends up to about 30 km, and the temperature again rises in this region with increase in altitude. The maximum temperature in thermosphere can go up to 1200 °C. Due to absorption of solar energy, most of the gases particularly oxygen and nitric oxide ionize in thermosphere, and charged particles are present in this region ((NO^+, O^+, ${O_2}^+$, etc.). These charged particles are useful because they reflect radio waves back to earth and make communication possible. The radio messages transmitted by radio stations are heard in the world because of reflection of radio waves by the particles of this region.

The Exosphere

The uppermost layer of the atmosphere is called as exosphere or outer space which extends up to about 500 km. The temperature in this region is very high and it possesses only hydrogen and helium gases.

Atmospheric Oxygen Oxygen present in the atmosphere plays an important role. Apart from being used in breathing, it helps in energy-yielding reactions. For example, fuels can burn only when oxygen is available. The LPG used for cooking food in kitchen requires oxygen for burning. The fossil fuels on burning gives CO_2 and H_2O as products, whereas aerobic organisms also use oxygen for the degradation of organic matter present in wastes:

$$CH_4 + 2O_2 \longrightarrow CO_2 + 2H_2O$$

The CO_2 and H_2O so produced are utilized by the green plants for the synthesis of carbohydrates. The process is called as *photosynthesis* in which oxygen is generated:

$$6CO_2 + 6H_2O \xrightarrow{hv} C_6H_{12}O_6 + 6O_2$$

As we move up from the earth's surface, the effects of UV radiations become more, and oxygen exists in different forms. Atomic oxygen is rapidly formed and is very reactive and stable in thermosphere. Thus,

$$O_2 + hv \longrightarrow O + O$$

at very high attitude, molecular oxygen does not exist. The atomic oxygen has been assigned partially responsible for the air glow during which three atomic oxygen combine to give excited oxygen atom and oxygen molecule:

$$O + O + O \xrightarrow{hv} O_2 + O^{\bullet}$$

In the presence of UV-radiations, the atomic oxygen also gives positively charged oxygen atom, and an electron is emitted:

$$O \xrightarrow{hv} O^+ + e$$

The O^+formed is responsible for producing a variety of positively charged species in the atmosphere. Thus,

$$O_2 + O^+ \longrightarrow O_2^+ + O$$

$$N_2 + O^+ \longrightarrow NO^+ + O$$

$O_2{}^+$can also be formed from oxygen in the mesosphere and thermosphere regions by absorbing energy:

$$O_2 \xrightarrow{hv} O_2^+ + e$$

The formation of ozone from oxygen primarily occurs in the stratosphere region by the reactions shown above.

6.4.2 Hydrosphere

More than 75% of the earth's surface is covered with water. Water is present in seas, rivers, ponds, oceans, etc. The world gets supply of water by evaporation of water from water reservoirs to form clouds; on saturation it rains and water returns to the ground. It is called as hydrological cycle.

Water is the best solvent which can dissolve most of the salts into it. It has a high viscosity and surface tension which helps aquatic organisms swim and survive.

O_2 is soluble in water and is used by most of the aquatic living creatures supporting the underwater life.

CO_2 is obtained by decomposition of organic matter and the respiratory activity of aquatic plants and animals. It combines with water to form carbonic acid and changes the pH of water. CO_2 combines chemically with water in the presence of chlorophyll of green plants and light to form carbohydrates.

6.4.3 Lithosphere

It consists of rocks, soil, etc. It has three components, crust, mantle, and outer-inner core. It relates with edaphic factor and includes mainly of soil, earth, rocks, mountains, etc. The core is about 2500 −km diameter from the earth's center. It is composed of Ni − Fe. The mantle extends about 300 km above the core having liquid molten state magma. The crust is about 8–40 km above mantle having soil and rocks. This part of lithosphere is the most significant part of the earth. It is used for agriculture, construction, industrial purposes, and many more. Soil microbes such as bacteria, algae, fungi, and protozoa bind with soil and decompose the nutrients to supply it to the plants through their roots (Fig. 6.2).

Crust too has mountains and rocks. They are made of different kinds of minerals and give rise to different kinds of rocks.

There are three kinds of rocks:

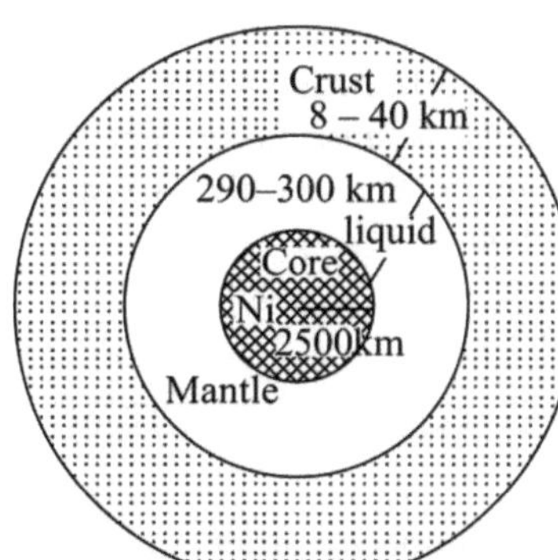

Fig. 6.2 Lithosphere

1. **Igneous Rocks or Fire-Like Rocks:** These are formed by cooling of molten magma obtained from volcanos. These are of three types—granites, diorite, and basalt—which is due to different chemical compositions.
2. **Sedimentary Rocks:** These are formed by deposition of weathered minerals which are derived from igneous rocks. These are also of three types—shales, sandstone, and limestone.
3. **Metamorphic or Changed Rocks:** These are formed by change of igneous and sedimentary rocks due to change of heat and pressure. As the earth's crust moves, rocks are changed by heat and pressure or by chemical action and convert to new kinds of rocks called as metamorphic rocks.

Soil formation takes place due to physical, chemical, and biological weathering process due to which rocks disintegrated into tiny particles.

6.4.4 Biosphere

It is the combination of living beings like flora and fauna. The natural plants and living beings influence each other's life. Green plants by photosynthesis liberate oxygen in atmosphere which animals use in respiration and liberate out CO_2 which is used again in photosynthesis to form a food cycle.

The interaction between living beings is of two types:

1. **Symbiotic interaction:** Symbiotic means "living together"; both the species mutually benefit each other and neither is harmed.
2. **Antagonistic interaction:** Where at least one of the species is harmed.

Radiation Balance of Earth: One may think that earth receives large energy from the sun daily, then how it keeps a balance, for which the earth should also emit energy. The question arises: does earth emit energy also? The answer is yes; the earth absorbs radiation in the visible region and radiates almost at the same rate in the infrared region. The infrared radiations are absorbed by the water vapors and CO_2, and the combined effect leads to the greenhouse effect.

In general the earth absorbs about 66% of the total solar light falls on it, whereas it reflects back about 30%. The loss of heat from the earth's surface primarily occurs through air circulation and in interaction of atoms or molecules to give desired products. Photosynthesis requires a large amount of energy for the synthesis of carbohydrates.

6.5 Greenhouse Effect

As mentioned earlier most solar radiations are absorbed by the earth's surface, and some are reflected by the atmosphere and earth. The absorbed radiations by earth's surface warm it. The infrared radiations emitted from the earth's surface have frequency in the range of 4000–600 cm^{-1}. Some of these radiations are absorbed by CO_2, and some other gases of the atmosphere are partly re-emitted to the earth's surface. This re-emission causes warming of the earth surface and the phenomenon is called as *greenhouse effect*. Carbon dioxide molecules are transparent to sunlight and hence sunlight passes through them, but these molecules readily absorb heat radiations and re-emit part of them to cause warming.

The major gases which cause absorption of infrared radiations in the atmosphere are CO_2, CH_4, N_2O, and chlorofluorocarbons. All these gases are called as greenhouse gases. Among these carbon dioxide is an important greenhouse gas and is of serious environmental concern. If greenhouse gases increase in the environment because of pollution or other reasons, more energy is radiated back by them and consequently causes increase in temperature of the earth. This phenomenon is called as greenhouse effect or global warming. The main source of CO_2 in the atmosphere is respiratory process of living beings, combustion, fermentation, etc., whereas N_2O is mainly contributed by biomass burning and combustion of fossil fuels. Chlorofluorocarbons are released in the atmosphere because they are used as coolants and propellents. The small quantity of chlorofluorocarbons causes drastic changes in the climatic conditions.

6.5.1 Greenhouse Gases and Their Sources

Carbon Dioxide

A major greenhouse gas CO_2 is evolved from the factories due to burning of fuels like coal, oil, natural gas, etc. The exhaust from the vehicle contributes CO_2 in the atmosphere to a significant extent. Another major contributor of CO_2 to the atmosphere is cement industry which has been assessed to contribute about 7% of the global CO_2 output. This higher percentage by cement industry is due to two reasons:

- Burning of fossil fuel to provide heat for roasting.
- Due to the decomposition of $CaCO_3$ into CaO and CO_2 inside the chamber.

The deforestation has also contributed CO_2 to significant extent because plants reduce CO_2 percentage by converting them to carbohydrates. It has been reported that there is a net increase in CO_2 by nearly 10^9 tons per year which is an alarming situation. Scientists believe that many diseases like skin cancer, heart disease, etc. are due to increase in CO_2 in atmosphere (Fig. 6.3).

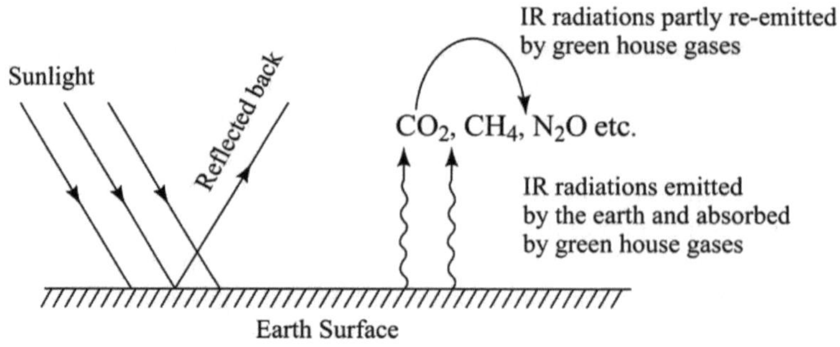

Fig. 6.3 Greenhouse phenomenon

Methane

The major source of methane in atmosphere is biomass combustion which causes 1–5% methane emission. Methane traps about 25 times more heat than a CO_2 molecule, and also it stays in the atmosphere for a short duration in comparison with other gases. A group of anaerobic bacteria also produce methane in swamps, marshes, wetlands, etc., and the process is called as *methanogenesis*, which is largely responsible for producing methane gas in these areas.

Nitrogen Oxides

Three nitrogen oxides are usually found in the atmosphere and they are:

- nitrous oxide (N_2O)
- nitric oxide (NO), and
- nitrogen dioxide (NO_2)

collectively represented as NO_x. The formation of these oxides in atmosphere is possible by several routes. Nitrogen and oxygen gases in the atmosphere break up into fragments at high temperature and pressure due to lightning flash, and the combination of these fragments produces nitrogen oxides. Out of the three oxides of nitrogen, the first one, N_2O, lasts in the atmosphere for about 150 years and has little effect at lower altitudes. However, at higher altitudes it causes severe damage by depleting ozone. Thus, several reactions occur in stratospheric region:

$$N_2O + h\nu \longrightarrow N_2 + O$$

$$N_2O + O \longrightarrow 2NO$$

$$NO + O_3 \longrightarrow NO_2 + O_2$$

In normal air, the level of N_2O is about 0.225 ppm, but it increases in polluted air. N_2O is also formed when nitrogen compounds are formed by chemosynthetic bacteria. The conversion of nitrogen and oxygen to nitrites and nitrates is called as *nitrification*, and their decomposition to free nitrogen is called as *denitrification*. A large quantity of N_2O is produced during denitrification and depends on soil moisture, temperature, and presence of nitrogen compounds in soil. As N_2O also has tendency to deplete ozone, it warms the atmosphere about 200 times more than CO_2.

Bacterial decay of organic waste produces other oxides of nitrogen in the atmosphere. Nitric oxide (NO) reacts with $OH^{\bullet}$ to form HNO_2, whereas NO_2 ultimately converts to HNO_3, and both these acids are removed from the atmosphere as acid rains. The basic reactions involved in these conversions can be represented as follows:

$$2NO + O \longrightarrow 2NO_2$$

$$NO_2 \xrightarrow{h\nu} NO + O$$

$$NO + O_3 \longrightarrow NO_2 + O_2$$

$$NO + OH^{\bullet} \longrightarrow HNO_2$$

$$NO_2 + OH^{\bullet} \longrightarrow HNO_3$$

One of the major contributors of nitrogen oxides in the atmosphere is nuclear explosion carried out by different countries time to time. N_2O released from these explosions depletes ozone layer and thus causes damage to atmosphere and makes it more exposed for UV radiations.

Chlorofluorocarbons (CFCs)

Compounds like trichlorofluoromethane (CCl_3F), dichlorodifluoromethane, (CCl_2F_2) etc., are called as chlorofluorocarbons, and they enter in the atmosphere because of their use as refrigerants and in aerosol sprays. These compounds are inactive in troposphere region; however, they slowly enter into stratosphere region, where free radicals are generated in the presence of UV radiations. Thus,

$$CFCl_3 \xrightarrow{h\nu} CFC1_2 + Cl^{\bullet}$$

$Cl^{\bullet}$ (chlorine free radical) is generated by the decomposition of $CFCl_3$, which reacts with ozone causing its depletion in the following reactions:

$$Cl^{\bullet} + O_3 \longrightarrow O_2 + ClO^{\bullet}$$

$$ClO^{\bullet} + O \longrightarrow Cl^{\bullet} + O_2$$

Thus, a chain reaction is initiated in which $Cl^{\bullet}$ is produced which again reacts with ozone causing its depletion. A single $Cl^{\bullet}$ is sufficient to destroy 10^5 molecules of ozone in such chain reactions before being removed in reactions like

$$Cl^{\bullet} + CH_4 \longrightarrow CH_3^{\bullet} + HCl$$

or

$$ClO^{\bullet} + NO_2 \longrightarrow ClONO_2$$

The greenhouse warming caused by greenhouse gases is expressed by Global Warming Potential (GWP), and it has been reported that GWP of these gases is in the order:

$$CFC > N_2O > CH_4 > CO_2$$

CFC is approximately 10^4 times greater than N_2O, which is 150 times greater than CH_4, which has 25 times greater GWP than CO_2.

6.6 Ozone Hole or Depletion of Ozone Layer

The formation of ozone and its importance in protecting us from harmful UV radiations in stratosphere region have already been discussed earlier in this chapter. In early 1979 the photographs taken by US satellite indicated that ozone layer has a hole over Antarctica which was found to increase by about 3% in the year 1986. The attempts were made to find the reasons for ozone layer depletion, and it was suspected that the probable cause is the use of chlorofluorocarbons (CFCs) which were introduced in the market in the 1930s. Their stable, nontoxic nature soon increased their use as coolants in air conditioners and refrigerators and in cleaning electronic circuits.

On realizing the role of CFCs in atmosphere in ozone depletion, efforts were made to reduce the use of CFCs because it was suggested that each 1% decrease in ozone layer will increase about 6% cases in skin cancer due to exposure to UV radiations. In 1987, 34 countries signed an agreement in Montreal for banning the use of chlorofluorocarbons, and the companies involved in their production were asked to replace CFCs by compounds having at least one hydrogen, i.e., hydroflourochlorocarbons. Nevertheless, the continuous monitoring of ozone layer depletion continued. In 1992, it was reported by Japanese scientists that ozone hole has increased about 13 times during the last 10 years, and in view of this, a total ban on the use of CFCs was imposed in many countries. Other countries agreed to discontinue the production of CFCs by the end of year 2000.

The depletion of ozone layer was found more severe in polar regions (North and South poles) in comparison with other regions. One of the probable causes of such an observation is the cold climate in polar regions, due to which nitrogen oxides present freeze into droplets and no longer available for deactivating the $Cl^{\bullet}$ or ClO°. The thickness of ozone layer is measured using Dobson spectrometer and is presented in Dobson units. In India, the ozone layer has been found between 240 and 300 DU in different parts of the country.

6.7 Air Pollution

Air around us always possesses one or more undesirable constituents that are injurious to humans, plants, or animal life. Emission of gases such as CO, SO_2, and H_2 S by industries or vehicle emission continuously pollutes the air around us. Smoke, dust particles, or tiny particles of liquids or solids are distributed in air by winds and thus cause pollution. In a broader sense, air pollution is defined as the presence of one or more substances in it that are injurious to living beings. More than 90% of air pollution is caused by the following:

- Carbon monoxide (CO)
- Sulfur dioxide (SO_2)
- Nitrogen oxides (NO_x)
- Hydrocarbons, and
- Small solid or liquid particles.

Most of these pollutants exist generally up to the region of about 1 km above the ground level. We have already discussed nitrogen oxides earlier; hence, we will deal with other pollutants in this section.

6.7.1 Carbon Monoxide

The major source of CO production is the incomplete combustion of fuels like coal, gasoline, etc. In such cases carbon monoxide is produced in place of CO_2. Carbon monoxide

$$2C + O_2 \longrightarrow 2CO$$

is a colorless, odorless, and tasteless gas and is insoluble in water. It reduces oxygen-carrying capacity of blood in human system and if inhaled for a long time may lead to death.

Dissociation of CO_2 or its reaction with carbon or carbon-containing compounds at high temperature also produces carbon monoxide:

$$CO_2 \longrightarrow CO + O$$

$$CO_2 + C \longrightarrow 2CO$$

Two wheelers and four wheelers contribute significant amount of CO in the atmosphere. The annual contribution of CO in the atmosphere by the human beings is about 275 million tons per year and is expected to increase per year. In addition to this, natural phenomena like volcanic eruption, electric discharge during thunderstorms, etc. also contribute CO in the atmosphere but to a lesser extent.

Many soil microorganisms consume CO of the air and thus help in controlling pollution due to carbon monoxide; however, such CO sinks are not available uniformly. In general, it is observed that such microorganisms are less available in high CO-polluted areas. To deal with CO pollution in air, the government has made it compulsory to get the automobiles examined for CO emission level. In the case of higher CO emission, the carburetors are to be adjusted to allow proper ratio of fuel and oxygen to provide proper combustion of fuel. Still more efforts are made to reduce CO emission by automobiles. One of the serious efforts is the use of catalytic converter in automobiles. The catalytic convertor is a chemical method for converting incomplete combustion products of fuels to less toxic completely oxidized products. The process involves two steps:

- reduction in the presence of finely divided platinum, and
- oxidation in the presence of air.

Thus, in the first stage, nitrogen oxides of the exhaust are reduced by platinum to N_2 and NH_3, and the platinum oxide formed oxidizes CO and hydrocarbons to CO_2 and H_2O in the presence of oxygen. As lead or any other heavy metal poisons the platinum catalyst, hence the lead-free fuel should be used in automobiles equipped with catalytic convertor. The reactions that occur in catalytic convertor are shown in Fig. 6.4 where CH-represents hydrocarbons.

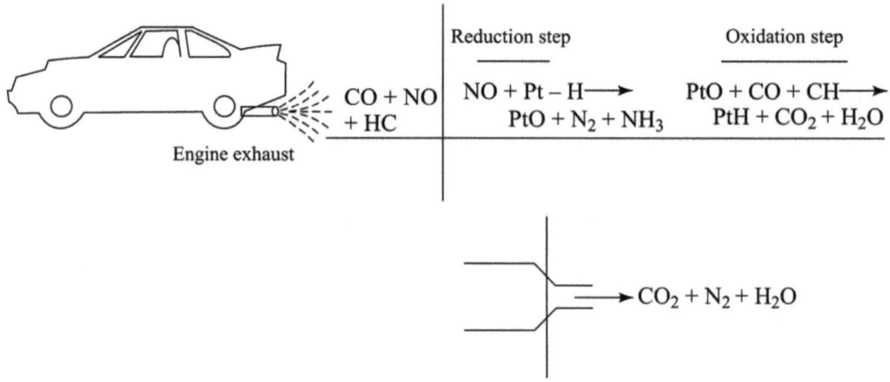

Fig. 6.4 Reactions normally occur in automobiles equipped with catalytic convertor

Hence, in the absence of catalytic converter, the main gases in exhaust are CO, NO, and hydrocarbons, whereas in automobiles equipped with converter, the exhaust has CO_2, N_2, and H_2O.

6.7.2 Toxicity of Carbon Monoxide

CO combines with hemoglobin and reduces it oxygen-carrying capacity which affects respiratory activity and metabolism. It causes headache and blurred vision and in acuteness it may cause unconsciousness and even death.

In blood hemoglobin has $Fe^{(d6)}$ atom in heme of octahedral configuration having four unpaired electrons. The four N-atoms of organic chelate occupy four positions in place. One of the perpendicular positions is occupied by globin protein molecule and the other is available for the coordination of oxygen gas as shown in Figure 4.3. CO has 200 times more affinity than oxygen for occupying the coordination position. Thus, CO replaces oxyhemoglobin (HbO_2) to carboxy hemoglobin (HbCO) due to which replacement of O_2 from the lungs causes respiratory problems:

$$HbO_2 + CO \rightleftharpoons HbCO + O_2$$

This a reversible reaction; thus, the remedy of CO poisoning is supply of pure O_2 (Fig. 6.5).

Fig. 6.5 Structure of oxyhemoglobin (HbO_2)

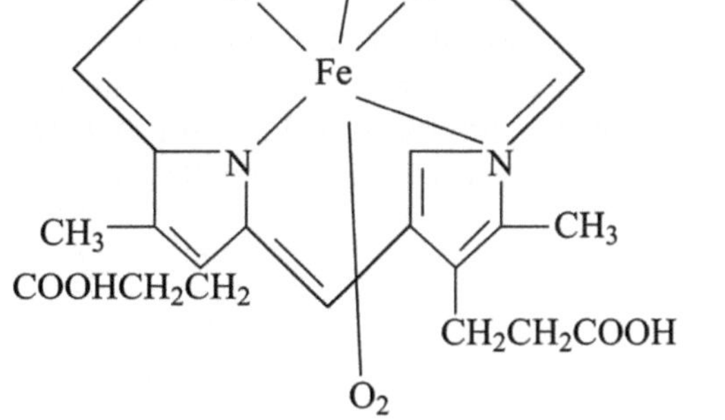

Sulfur Dioxide

Sulfur dioxide is a colorless pungent gas, and almost 30% of SO_2 pollution in atmosphere is due to burning of coal in industries. The remaining 70% is due to the gases evolved from volcano eruption. Sulfur dioxide causes headache and vomiting and may lead to respiratory problems. The combustion of coal or oil leads to the formation of SO_2 due to sulfur present in them. The SO_2 converts to

$$S + O_2 \longrightarrow SO_2$$

$$2SO_2 + O_2 \longrightarrow 2SO_3$$

sulfur trioxide in excess of oxygen. When sulfur trioxide comes in contact with water vapors of atmosphere, it is converted to H_2SO_4.

$$SO_3 + H_2O \longrightarrow H_2SO_4$$

The formation of H_2SO_4 is also possible by partial oxidation of SO_2 in air in catalytic process involving hydrocarbons having three oxygen and nitrogen oxides. Thus,

$$SO_2 + RO_3 \longrightarrow SO_3 + RO_2$$

$$SO_3 + H_2O \longrightarrow H_2SO_4$$

The pollution of SO_2 by industries can be reduced to some extent by using chemical reactions of SO_2. For example, if the exhaust of industries is passed through a slurry of $CaCO_3$, the SO_2 will react with it to form solid $CaSO_4$, and thus SO_2 can be removed. A solution of citrate salt can also be used effectively for removal of SO_2:

$$2CaCO_3 + 2SO_2 + O_2 \longrightarrow 2CaSO_4 + CO_2$$

In this process the gases evolved from the chimney are cooled to ~50 °C and then passed through an absorption tower containing solution of citrate. Finally, it is passed through a closed chamber in which H_2 S gas is bubbled. Sulfur is precipitated in the process and SO_2 is completely removed from the chimney gases. The chemical reactions involved in this process are as follows:

$$SO_2 + H_2O \longrightarrow H^+ + HSO_3^-$$

$$HSO_3^- + H_2cit^- \longrightarrow (HSO_3 . H_2cit)^{2-}$$

$$(HSO_3 . H_2cit)^{2-} \xrightarrow{H_2/H_2S} 2S + H_2cit^- + 3H_2O$$

where citrate is

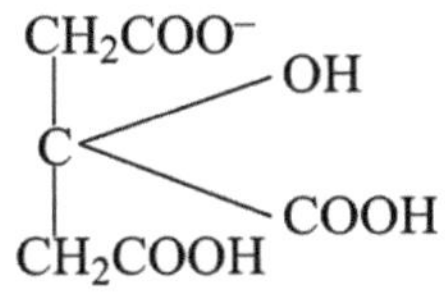

6.8 Hydrocarbons

One of the important hydrocarbons which enters in the atmosphere through natural sources is methane (CH_4), which is produced during anaerobic decomposition of organic waste by bacteria in water, soil, and sediments. The simple reaction is

$$[CH_2O] \longrightarrow \frac{1}{2}CO_2 + \frac{1}{2}CH_4$$

Other hydrocarbons such as ethane, ethylene, propane, *n*-butane, isopentane, toluene, etc. have also been detected in the atmosphere; however, their amount is relatively low and is of the order of hundred parts per billion. Hydrocarbons are removed by chemical or photochemical oxidation reactions leading to CO_2 as the ultimate major product together with water-soluble products. *It must be remembered that hydrocarbons themselves are not as harmful as their products of photochemical decomposition in the atmosphere.*

6.9 SMOG

The combination of smoke and fog is called as smog. It was realized long ago due to the formation of a mixture of coal, smoke, and fog. Basically smog is of two types:

- Photochemical smog
- Reducing smog

6.9.1 Photochemical SMOG

Photochemical smog consists of brown hazy fumes and reduces visibility and causes irritation in the eyes, whereas reducing smog is characterized by irritation in sinus tract and becomes worst in sunlight.

In general, the main constituents of smog are unsaturated hydrocarbons, nitrogen oxides, and some sulfur compounds; however, photochemical smog also possesses organic peroxides and hydroperoxides, peracetic acid, peroxyacyl nitrate (PAN), and similar compounds. The formation of these compounds occurs due to reaction of industrial exhaust and automobile exhaust with sunlight. The formation of ozone also occurs due to the reaction of SO_2, aldehyde, and nitrogen oxides in the presence of sunrays, which helps in the formation of photochemical smog. The presence of ozone in the smog causes irritation in the eyes and throat and difficulty in breathing, whereas polynuclear hydrocarbons present in smog are carcinogenic in nature. The constituents of smog vary with the time of day, depending upon the level of pollutants released in atmosphere by factories and automobiles. Some common reactions involved in smog formation of photochemical type are as follows.

Oxides of Nitrogen

NO, NO_2, and N_2O are the major primary pollutants. NO plays an important role in formation of photochemical smog. These nitrogen oxides are produced from automobile exhausts and from burning of fuels.

Nitrous oxide N_2O dissociated in U.V region to develop NO pollution:

$$N_2O \xrightarrow[\text{UV}]{hv} NO + O$$

The atomic oxygen combines with O_2 to produce O_3 which further oxidizes NO to NO_2

$$O_2 + O \longrightarrow O_3$$

$$O_3 + NO \longrightarrow NO_2 + O_2$$

NO is also obtained by oxidation of NH_3 and nitrogen:

$$2NH_3 + \frac{5}{2}O_2 \longrightarrow 2NO + 3H_2O$$

$$N_2 \xrightarrow{hv} N + N$$

$$N + O_2 \longrightarrow NO + O$$

$$NO + O_3 \longrightarrow NO_2 + O_2$$

NO_2 thus produced combines with water to form HNO_3 which reduces the pH of water causing acid rain.

Toxicity of NO and NO_2

At lower concentration NO and NO_2 absorbed by plants do not cause any damage. The uptake rate of NO_2 is very high in comparison with NO. Concentration of NO_2 in excess of 2mg/kg of air causes leaf damage of sensitive plants. At higher concentrations, it causes irritation in the eyes and damages the respiratory system.

Reaction of hydrocarbons with nascent oxygen or ozone to give free radicals

$$RH + O \longrightarrow R^{\bullet} + - - - - - -$$

$$RH + O_3 \longrightarrow R^{\bullet} + - - - - - -$$

The free radicals so produced react with oxygen to form new free radicals which on reaction with O_2, nitrogen oxides leads to the formation of several products as shown below.

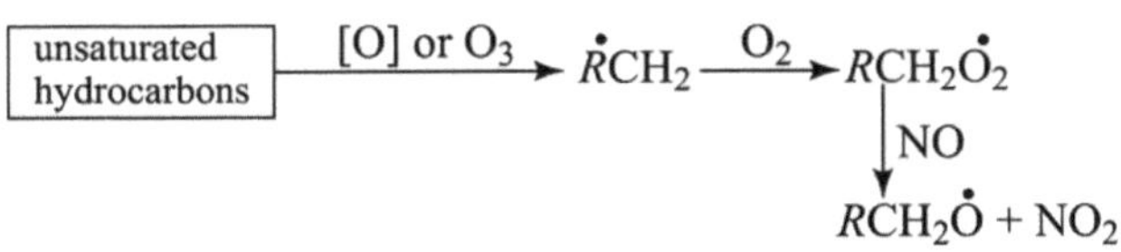

NO_2 is regenerated in the process and a large number of new free radicals are produced. $RCH_2O^{\bullet}$ can give aldehyde on reacting with oxygen which can react with $OH^{\bullet}$ free radical to give peroxyacyl nitrate (*PAN*) in the smog. This is also known as photochemical smog or Los Angeles smog:

$$RCH_2O^{\bullet} + O_2 \longrightarrow RCHO + HO_2^{\bullet}$$

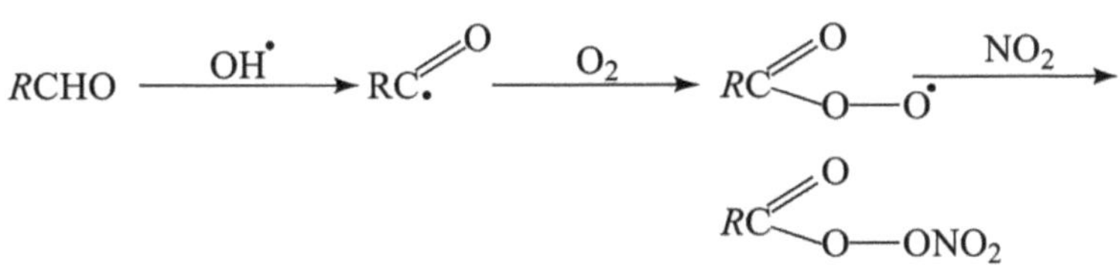

Toxicity of PAN

The maximum concentration of PAN in smog has been found to be about 0.04 mg/kg of air. Sensitive plants damage at 0.01mg/kg of PAN concentration. PAN causes headache, eye irritation, soreness, respiratory irritation, etc. Presence of O_3 also decolorizes paintings and damages textiles and rubber articles.

It must be realized that as soon as a free radical is formed, chain reactions start to give different new free radicals which lead to a variety of products. These reactions continue till the free radicals are deactivated by some reactions. To reduce photochemical smog, it is necessary to control the emission of hydrocarbons in automobile exhaust and the reactions that lead to the formation of ozone in troposphere region.

6.9.2 *Reducing SMOG*

Reducing smog also called as London smog mainly consists of fog, smoke, and SO_2 gas. As this type of smog was very common in London in the early 1950s, it was named as London smog. The presence of SO_2 in the smog causes respiratory problems and may lead to death. In 1952, the concentration of SO_2 at 1.3 ppm in smog prevailed for nearly 5 days and caused death of more than 4000 people. SO_2 is the main constituent of this type of smog together with particulate and water.

Coal used as a reducing mixture in industries possesses small amount of sulfur. The combustion of coal gives SO_2 and SO_3 gases. The formation of SO_2 is common, whereas SO_3 is formed in presence of

$$S + O_2 \longrightarrow SO_2$$

$$2\,S + 3O_2 \longrightarrow 2SO_3$$

sunlight or particulate present in atmosphere. The SO_2 then reacts with H_2O to give sulfurous acid which is oxidized to sulfuric acid. Another reaction of SO_2 is also common in which it reacts with NH_3:

$$SO_2 + H_2O \longrightarrow H_2SO_3$$
$$H_2SO_3 + 1/2O_2 \longrightarrow H_2SO_4$$

in moisture to give ammonium sulfate.

$$H_2SO_4 + 2NH_3 \longrightarrow (NH_4)_2\,SO_4$$

This type of smog is more dangerous and is common in the areas where sulfur-rich coal is used in industries. In India industrial cities such as Mumbai, Kolkata, Delhi, Chennai, etc. need to be monitored continuously for pollution levels to avoid the formation of smog.

6.10 Acid Rain

It is expected that the natural rainwater should have a pH close to 7.0; however, the carbon dioxide present in the atmosphere dissolves in it and forms carbonic acid (H_2CO_3), and hence its pH becomes acidic. If we measure the pH of rainwater during normal days, it will be between 5.5 and 6.5. Thus, the pH is slightly acidic due to the formation of carbonic acid. If the rainwater possesses excess of acids, its pH can go down to 2.5 and such rainwater is called as acid rain.

As mentioned earlier nitrogen oxides, SO_2, SO_3, etc., enter in the atmosphere from the automobile exhaust, combustion of fossil fuels, or industrial waste gases and convert into HNO_3 and H_2SO_4 in the atmosphere. A number of reactions are involved in the conversion of these gases into corresponding acids. Some of these reactions have been described earlier in this chapter. Other photochemical reactions involved in the formation of acids are as follows:

$$SO_2 + h\nu \longrightarrow SO_2^{\bullet} \text{ (excited state)}$$
$$SO_2^{\bullet} + O_2 \longrightarrow SO_3 + O$$
$$SO_2^{\bullet} + SO_2 \longrightarrow SO_3 + SO$$
$$SO + SO_2 \longrightarrow SO_3 + S$$
$$H_2O + SO_3 \longrightarrow H_2SO_4$$

Similarly for nitrogen oxides, we can write:

$$NO + O_3 \longrightarrow NO_2 + O_2$$
$$NO_2 + O_3 \longrightarrow NO_3 + O_2$$
$$NO_2 + NO_3 \longrightarrow N_2O_5 \quad \text{(nitrogen pentaoxide)}$$
$$N_2O_5 + H_2O \longrightarrow 2HNO_3$$

Apart from photochemical oxidation, catalytic oxidation of SO_2 in the presence of Fe(II), Cu(II), Mn (II) ions in water also gives sulfuric acid:

$$SO_2 + \frac{1}{2}O_2 + H_2O \xrightarrow{\text{metal oxides}} H_2SO_4$$

Now the question comes: is an acid rain harmful: The answer is Yes. In fact acid rain is a major pollution problem these days because it causes several adverse effects. Some important adverse effects are:

- It causes damage to buildings, monuments of marble/limestone, etc., which reacts with acids to form soluble sulfates:

$$CaCO_3 + H_2SO_4 \longrightarrow CaSO_4 + H_2O + CO_2$$

You must have read about the problems of air pollution and acid rains, the historical monument Taj Mahal at Agra is facing. The white-colored stone is slowly turning to yellow due to SO_2, nitrogen oxides, and hydrocarbons in the atmosphere. Acid rain further aggravates the problem and is of much concern these days.

- Due to acidification of soil, many bacteria are killed. Fishes in lakes and pools are killed and thus ecological balance is disturbed.
- In many areas the acid rain has destroyed crops in the past and thus productivity is decreased. North America, Canada, and Europe have seen acid rains, which reduced the growth of potatoes, peas, spinach, and many more vegetables.
- Acid rain may wash away important minerals from the soil, which are necessary for the plants' growth. Acid deposition also leads to weakling of big trees and hence causes decrease of trees in forests
- Acid rain contaminates the potable water and thus can cause digestive, respiratory, neurological, and other diseases.

It is highly desirable that acid rain should be controlled by decreasing the atmospheric pollution of SO_2, nitrogen oxides, etc. Use of diesel fuel should also be minimized because its emission contains a large amount of nitrogen oxides.

6.10.1 Methods for Pollution Control

The pollution can be controlled or minimized by reducing the formation of pollutants. We can modify the process design of industrial plants to minimize pollutant production. Control of the pollutants at the source can be accomplished in several ways through raw material changes, modification, or change of equipment. Some suggestions for control of air pollution are given here:

1. Modification of industrial instruments using smokeless fuel, precipitator on chimneys, smokeless chulhas, solar cooker, and biogas.
2. SO_2 pollution can be prevented by treating high sulfur fuel or by using natural gas and nonconventional energy sources.
3. Smoke can be reduced during combustion of fuel by synchronized combustion of fuel.
4. Automobile vehicles should be fitted with antipollution devices.
5. Sanitation condition of communities should be improved by using modern methods.
6. Awareness about pollution hazards should be distributed to the general public.
7. Deforestation should be strictly banned and fast growing plants should be planted.

Sink for Pollutant Gases

In atmosphere the polluted gases spread through the whole troposphere. These pollutant gases are not retained in atmosphere for a long time but sink in oceans and vegetation when these gases combine with long-lived pollutants. These gases diffuse on the surface of ocean, cross the gas liquid interface, and finally go into the bulk of ocean, e.g., SO_2 is transferred from atmosphere to ocean as

$$SO_2 + H_2O \longrightarrow H_2SO_3$$

$$H_2SO_3 + H_2O \rightleftharpoons H_3O^+ + HSO_3^-$$

$$HSO_3^- + H_2O \rightleftharpoons H_3O^+ + SO_3^{--}$$

Vegetation are also capable to take up the atmospheric gases as CO_2. The limestone serves as sink for atmospheric acids:

$$CaCO_3 + H_2SO_4 \longrightarrow CaSO_4 + H_2O + CO_2$$

Exercises

1. Explain the term "biotic."
2. Give the details of part of atmosphere having O_3 in it.
3. How is ozone produced in the environment?
4. Why radio waves do not move through atmosphere?
5. What is the toxic effect of CO pollution on human system?
6. Explain why a cigarette smoker normally has respiratory problems.
7. Explain the reason of submerging of coastal cities.
8. Describe the various segments of the environment.
9. Discuss the characteristics of the major regions of the atmosphere.
10. What gases are mainly responsible for causing greenhouse effect?
11. How is ozone formed and decomposed in nature? Explain the reason and consequences of the depletion of O_3 layer.
12. How pollution is responsible for causing an acid rain?
13. Write a short note on:
    (i) Acid rain
    (ii) Photochemical smog
    (iii) Greenhouse effect
    (iv) Ozone hole
    (v) Radiation balance of earth.
14. Explain how does automobile exhaust contribute to air pollution.
15. Give one practical example of pollution due to SO_2, NO_2, and CO_2.
16. How is air pollution classified? What are the major effects of CO, SO_2, NO, and hydrocarbons on human health?
17. O_2 has an important role in troposphere, while O_3 in stratosphere. Explain.
18. How can internal combustion engine be modified to make auto-exhausts free from pollution?

Chapter 7
Corrosion

Corrosion refers to the degradation of metals through redox processes, where electron transfer reactions lead to material deterioration. This phenomenon significantly impacts metal-based structures, including those made from iron, copper, or brass, causing substantial economic losses. For instance, rusting, a prevalent form of iron corrosion, transforms the metal into a reddish-brown compound, primarily hydrated iron (III) oxide ($Fe_2O_3 \cdot nH_2O$), along with other by-products. In the United States, approximately one-quarter of steel production is dedicated to replacing materials lost to corrosion, highlighting its widespread impact.

In nature, metals are typically found as oxides or similar compounds, which are energetically more stable than their pure metallic forms. Consequently, most metals, excluding noble metals like gold and platinum, readily react with atmospheric oxygen to form oxide layers. For example,

$$4Fe + 3O_2 \longrightarrow 2Fe_2O_3$$

$$4Cu + O_2 \longrightarrow 2Cu_2O$$

$$4Al + 3O_2 \longrightarrow 2Al_2O_3$$

These oxide layers often create a protective barrier on the metal surface, reducing further oxidation. For metals like aluminum and copper, this layer is dense and robust, offering significant protection against continued corrosion.

Corrosion is a surface-based electrochemical process requiring distinct anodic and cathodic regions on the metal. At anodic sites, the metal undergoes oxidation, releasing metal ions and electrons:

$$M \longrightarrow M^{n+} + ne^-$$

At cathodic sites, these electrons participate in reduction reactions, such as

© The Author(s), under exclusive license to Springer Nature Switzerland AG 2026

R. N. Goyal, H. Goel, *Textbook of Engineering Chemistry*,
https://doi.org/10.1007/978-3-032-22068-4_7

$$2H^+ + 2e^- \longrightarrow H_2$$

$$H_2O + 2e^- \longrightarrow 2OH^- + H_2$$

$$O_2 + 2H_2O + 4e^- \longrightarrow 4OH^-$$

Anodic regions often develop at surface imperfections, such as cracks in the oxide coating, grain boundaries, or areas near impurities. Cathodic regions typically form around metallic impurities, such as sulfides, embedded within the metal. In aqueous environments, iron corrosion leads to rust formation, as illustrated in a typical diagram of the process.

The rates of anodic and cathodic reactions determine the overall corrosion process. The anodic reaction, involving metal oxidation, typically proceeds faster than the cathodic reaction, which involves electron consumption. However, the slower cathodic step governs the overall corrosion rate, acting as the rate-limiting factor. Understanding these electrochemical steps is crucial for both analyzing corrosion mechanisms and developing strategies to mitigate it.

Unlike iron, metals like aluminum and copper form dense, impermeable oxide layers during corrosion, which shield the underlying metal from further degradation. In contrast, the rust layer on iron is porous and flaky, offering little protection and allowing continued corrosion of the underlying metal. This distinction explains why iron structures are more susceptible to ongoing deterioration compared to aluminum or copper.

7.1 Iron Corrosion Under Different Environmental Conditions

The corrosion of iron consistently involves an anodic reaction where iron oxidizes to form Fe^{2+} ions and releases electrons. However, the cathodic reaction varies depending on environmental factors, influencing the overall corrosion process and its products.

In conditions with limited or no oxygen, such as iron pipes embedded in moist soil, the cathodic reaction involves the reduction of water (Fig. 7.1):

$$Fe \longrightarrow Fe^{2+} + 2e^- \qquad \text{Anodic reaction}$$

$$2H_2O + 2e^- \longrightarrow 2OH^- + H_2 \qquad \text{Cathodic reaction}$$

$$Fe + 2H_2O \longrightarrow Fe(OH)_2 + H_2 \qquad \text{Net reaction}$$

The resulting iron(II) hydroxide ($Fe(OH)_2$) has low solubility and precipitates on the iron surface, forming a layer that partially inhibits further oxidation. Consequently, corrosion in oxygen-deficient environments proceeds slowly. Over time, any available oxygen may further oxidize $Fe(OH)_2$ into magnetite (Fe_3O_4), a mixed oxide of FeO and Fe_2O_3:

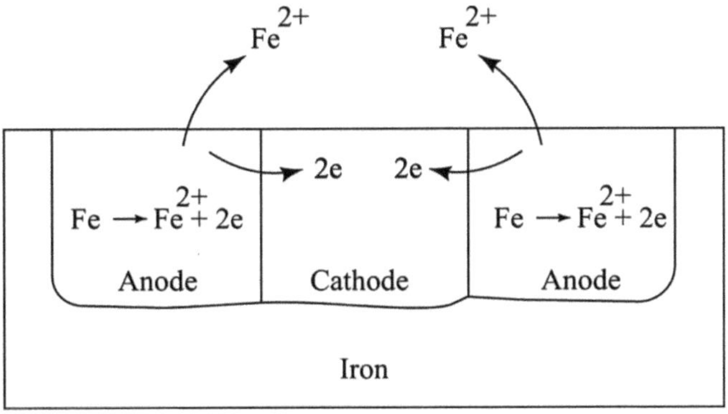

Fig. 7.1 Reactions involved in corrosion of iron

$$6Fe(OH)_2 + O_2 \longrightarrow 2Fe_3O_4 \cdot H_2O + 4H_2O$$

$$Fe_3O_4 \cdot H_2O \longrightarrow Fe_3O_4 + H_2O$$

This black magnetite layer forms on iron surfaces in moist, low-oxygen settings, contributing to the corrosion observed in buried iron structures.

When both oxygen and water are abundant, the cathodic reaction shifts, significantly accelerating corrosion—up to 100 times faster than in low-oxygen conditions. The reactions are

$$Fe \longrightarrow Fe^{2+} + 2e^- \qquad \text{Anodic reaction}$$

$$O_2 + 2H_2O + 4e^- \longrightarrow 4OH^- \qquad \text{Cathodic reaction}$$

$$2Fe + O_2 + 2H_2O \longrightarrow 2Fe(OH)_2 \qquad \text{Net reaction}$$

In the presence of ample oxygen, $Fe(OH)_2$ further reacts to form hydrated iron(III) oxide ($Fe_2O_3{\cdot}nH_2O$), commonly known as rust:

$$4Fe(OH)_2 + O_2 \longrightarrow 2Fe_2O_3 \cdot H_2O + 2H_2O$$

This reddish-brown rust is visible on iron surfaces, such as vehicles, structural components, or even in tap water after periods of stagnation. Rust is porous and does not effectively shield the underlying iron, allowing continued corrosion.

The corrosion process is exacerbated by chloride ions, often present in coastal environments due to sea air. Chloride ions penetrate the oxide layer on iron, forming soluble metal chlorides. These chlorides dissolve and migrate to the surface, exposing fresh iron to further attack by oxygen and water, thus accelerating rust formation.

7.2 Fundamentals of Chemical Corrosion

Corrosion involves chemical reactions where metal atoms lose their loosely bound electrons, typically through interactions with nonmetals like oxygen and sulfur. These nonmetals, possessing incomplete valence shells, readily form compounds with metals. The resulting compounds, such as oxides or sulfides, have lower free energy than the pure metal, making these reactions nearly spontaneous. This process occurs at the metal's surface, forming a layer that influences further corrosion. The degree of protection provided by this layer depends on its characteristics.

The nature of the surface layer formed during corrosion significantly affects the metal's resistance to further degradation. The following outlines the different types of layers and their impact:

- **Stable and Adherent Layers:** When the surface layer is robust and firmly adheres to the metal, it acts as a barrier, preventing oxygen or other corrosive agents from reaching the underlying metal. For instance, oxides formed on aluminum, copper, tin, and platinum are dense and impermeable, effectively shielding the metal from further corrosion.
- **Unstable Layers:** In some cases, the surface layer is unstable and may decompose back into the metal and the nonmetal (e.g., oxygen). Such instability prevents sustained corrosion, as the layer does not persist to facilitate ongoing reactions.
- **Volatile Layers:** Certain metals, like molybdenum, form oxide layers that are volatile and evaporate upon formation. This exposes fresh metal surfaces to continuous attack, leading to rapid corrosion rates due to the absence of a protective barrier.
- **Porous or Cracked Layers:** Many oxide layers contain cracks or pores, allowing atmospheric oxygen to penetrate and interact with the underlying metal. Such porous layers fail to halt corrosion, enabling it to continue until the metal is fully degraded.

7.3 Factors Influencing Corrosion

The corrosion of metals and their alloys is governed by multiple factors that influence the rate and extent of degradation. Below are key factors affecting corrosion:

- **Position in the Galvanic Series:** The tendency of a metal or alloy to corrode depends on its position in the galvanic series, which ranks metals by their electrochemical activity. More active (less noble) metals are more prone to corrosion. The corrosion rate increases with a larger potential difference between two metals in contact within the series.

- **Ratio of Cathodic to Anodic Areas:** Corrosion, being an oxidative process, accelerates when the anodic area (where oxidation occurs) is smaller relative to the cathodic area (where reduction occurs). A smaller anodic area must supply electrons to a larger cathodic area, leading to faster corrosion of the anode.
- **Protective Oxide Layers:** Exposure to air results in the formation of thin oxide layers (a few angstroms thick) on most metals, which can slow corrosion. The effectiveness of this layer depends on the specific volume ratio, defined as the volume of the metal oxide relative to the metal. Metals like aluminum, chromium, and nickel form dense, protective oxide layers that resist further corrosion. For example, stainless steel's corrosion resistance stems from chromium, which forms a self-healing oxide layer under oxidizing conditions.
- **Presence of Impurities:** Pure metals exhibit greater corrosion resistance, but impurities create localized electrochemical cells, initiating corrosion at anodic sites. For instance, trace amounts of lead or iron in zinc act as cathodic sites, promoting zinc corrosion. The corrosion rate correlates with the level of impurities present.
- **Removal of Corrosion Products:** Corrosion accelerates if corrosion products are removed from the metal surface, exposing fresh metal to further attack. This occurs when products dissolve in the surrounding medium or volatilize. For example, molybdenum forms volatile molybdenum oxide, which evaporates, leaving the metal surface vulnerable.
- **Environmental Conditions:** The surrounding environment significantly impacts corrosion rates:

  - **Temperature:** Higher temperatures increase diffusion rates, accelerating corrosion.
  - **Humidity:** High relative humidity facilitates the formation of electrochemical cells by providing a conductive medium, enhancing corrosion.
  - **Pollutants:** Atmospheric pollutants, such as sulfur dioxide (SO_2) from coal combustion, exacerbate corrosion. SO_2 is highly soluble in water (approximately 1000 times more than oxygen), forming sulfuric acid and contributing to acid rain. The corrosion of iron in the presence of SO_2 proceeds as follows:

  $$Fe + SO_2 + O_2 \longrightarrow FeSO_4$$

  $$4FeSO_4 + O_2 + 6H_2O \longrightarrow 2FeO_3 \cdot H_2O + 4H_2SO_4$$

  $$4H_2SO_4 + 4Fe + 2O_2 \longrightarrow 4FeSO_4 + 4H_2O$$

  The sulfuric acid produced is persistent, making corrosion difficult to halt even after cleaning, as residual sulfates continue to promote degradation.

- **Electrolyte Conductivity:** In environments like seawater, sodium and chloride ions increase the conductivity of the solution, enhancing the corrosion rate by facilitating electron transfer across the metal surface.
- **Solution pH:** The pH of the environment alters the corrosion process. Lower pH (more acidic conditions) reduces hydrogen gas evolution and increases

oxygen ionization at the cathode, accelerating corrosion. Conversely, metals prone to acid corrosion can be protected in higher pH (alkaline) environments, where the corrosion mechanism is less favorable.

7.4 Corrosion Inhibitors

Corrosion inhibitors are chemical substances that, when added in small amounts, significantly reduce the rate of metal corrosion. They work by either preventing the onset of corrosion or slowing the rate of corrosive attack. Since corrosion is an electrochemical process, inhibitors primarily function by decreasing the rate of electrode reactions at the metal surface.

Corrosion inhibitors can be categorized based on the electrode reaction they target (electrochemical mechanism):

- **Anodic Inhibitors:** These reduce the anodic reaction (metal oxidation). Examples include phosphates, vanadates, and benzoates.
- **Cathodic Inhibitors:** These suppress the cathodic reaction (reduction process). Common examples are salts of magnesium, calcium, or zinc.
- **Mixed Inhibitors:** These affect both anodic and cathodic reactions. Examples include organic compounds containing nitrogen or sulfur, as well as substances like gelatin and agar-agar.

Inhibitors can also be classified based on how they mitigate corrosion (mode of action):

- **Type A Inhibitors:** These form a protective layer on the metal surface, acting as a barrier against corrosive agents. Examples include phosphates, chromates, and silicates.
- **Type B Inhibitors:** These modify the surrounding environment to reduce corrosivity, such as dehumidifying agents (e.g., calcium chloride, silica gel) or oxygen scavengers (e.g., hydrazine, sulfites).
- **Dual-Action Inhibitors:** These combine the effects of Type A and Type B by both forming a protective film and altering the environment. For instance, adding sodium hydroxide or calcium hydroxide to water increases the pH and forms a thin oxide layer on steel, enhancing corrosion resistance.

Another approach categorizes inhibitors based on their chemical properties:

- **Solubility:** Soluble or insoluble in the medium.
- **Volatility:** Volatile or nonvolatile.
- **Acidity:** Acidic, basic, or neutral.
- **Composition:** Organic or inorganic.

However, these classifications are not always definitive, as a single inhibitor may fit into multiple categories, making clear distinctions challenging.

7.5 Mechanisms of Corrosion Inhibition

Several theories explain how corrosion inhibitors mitigate metal degradation. These include the adsorption theory, hydrogen overvoltage theory, and quantum chemical approach. Each is discussed below.

7.5.1 Adsorption Theory

This theory posits that inhibitors form a protective layer on the metal surface through adsorption, which can occur via two mechanisms:

- **Chemisorption:** Inhibitors chemically bond to the metal surface through electron transfer, often forming coordinate bonds. Transition metals with vacant, low-energy d-orbitals readily accept electron pairs from inhibitors, such as organic molecules with lone-pair electrons.
- **Physical Adsorption:** Inhibitors adhere to the metal surface through weaker, nonchemical interactions. This adsorbed layer acts as a barrier, shielding the metal from corrosive agents.

7.5.2 Hydrogen Overvoltage Theory

According to this theory, inhibitors adsorb onto the metal surface and impede either the anodic (oxidation) or cathodic (reduction) reactions, or both. This adsorption causes rapid polarization of the anodic or cathodic sites, reducing the corrosion rate. Most inhibitors act as cathodic polarizers, suppressing hydrogen evolution at the cathode. This, in turn, limits the anodic reaction, effectively slowing corrosion.

7.5.3 Quantum Chemical Approach

This theory suggests that the outermost electrons of organic inhibitors interact directly with the metal surface, leading to chemisorption and corrosion inhibition. The interaction can be modeled as a linear combination of the molecular orbitals of the inhibitor and the metal surface atoms. The strength of this bond correlates with the energy difference between the inhibitor's highest occupied molecular orbital (HOMO) and lowest unoccupied molecular orbital (LUMO) relative to those of the metal. Higher inhibition efficiency is often associated with a favorable HOMO energy level, facilitating stronger interactions with the metal surface.

7.6 Passivity in Corrosion

Passivity refers to a specific phenomenon in corrosion where a metal or alloy, typically active in the electrochemical series, exhibits enhanced resistance to corrosion. Unlike general corrosion resistance, passivity involves a selective suppression of the anodic reaction (metal dissolution), resulting in a positive shift in the electrode potential and reduced corrosion rates. Several theories explain this passive state, including the film theory, adsorption theory, and electron configuration theory, each described below.

7.6.1 *Film Theory*

The film theory suggests that passivity arises from the formation of a thin, three-dimensional layer on the metal surface, typically composed of metal oxides or other reaction products with the environment. This layer, often only a few angstroms thick and invisible to the naked eye, isolates the metal from corrosive agents, slowing corrosion. Evidence for this theory includes the isolation of a thin oxide film from passive iron, which was shown to protect the metal. The protective mechanism of the film includes:

- Complete separation of the metal surface from the corrosive environment.
- Inhibition of the cathodic process by limiting oxygen diffusion to the metal.
- Reduction of anodic activity, promoting passivation.

7.6.2 *Adsorption Theory*

The adsorption theory proposes that passivity results from a chemisorbed monolayer of oxygen or oxygen-containing anions on the metal surface. This layer displaces water molecules, reducing the rate of anodic dissolution. Over time, the adsorbed layer may thicken and transform into a stable oxide film. Studies on passive films reveal:

- For noble metals like platinum, oxygen forms a purely adsorbed layer. For other metals, adsorption of oxygen leads to the formation of an adsorbed compound, followed by an oxide layer.
- Effective passivation occurs when the adsorbed layer exhibits:
  - Low ionic conductivity.
  - Low dissolution rate.
  - High electron conductivity.
  - Strong adhesion to the metal surface.
  - Thermodynamic stability over a wide range of potentials.

7.6.3 *Electron Configuration Theory*

The electron configuration theory attributes passivity to the electronic structure of transition metals, particularly those with vacant d-orbitals. These empty d-orbitals accept electrons from environmental species, such as oxygen, forming strong bonds that stabilize the metal surface. This theory explains the passivity of alloys like stainless steel, which contains transition metals (iron, chromium, nickel) with unfilled d-orbitals. For example, chromium atoms with vacant 3d orbitals become passive upon electron filling. However, this theory has limitations, as it does not account for the passivity of metals like magnesium, aluminum, and beryllium, which lack empty d-orbitals.

7.7 Types of Corrosion

Corrosion can be classified based on the visual appearance of the corroded metal, broadly categorized into two types: uniform corrosion and nonuniform (localized) corrosion.

7.7.1 *Uniform Corrosion*

Uniform corrosion is the most prevalent form, contributing to significant material degradation. It involves a consistent chemical or electrochemical reaction across the entire exposed metal surface. For instance, when a zinc plate is immersed in dilute sulfuric acid, dissolution occurs evenly across the submerged area, resulting in a uniformly corroded surface, as shown in Fig. 7.2.

7.7.2 *Nonuniform (Localized) Corrosion*

Nonuniform corrosion, also referred to as localized corrosion, affects specific areas of the metal surface. It is subdivided into the following types:

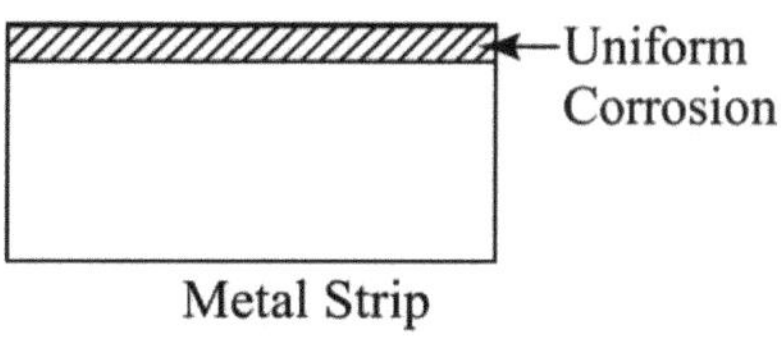

Fig. 7.2 Uniformly corroded surface of metal after dipping in sulfuric acid

(a) Galvanic corrosion
(b) Crevice corrosion
(c) Filiform corrosion
(d) Pitting corrosion
(e) Erosion corrosion
(f) Intergranular corrosion
(g) Stress corrosion
(h) Corrosion fatigue

While a detailed discussion of these types is beyond the scope of this text, a brief overview is provided below.

Galvanic Corrosion

Galvanic corrosion occurs when two dissimilar metals are in contact within a conductive electrolyte, such as a saline solution. The potential difference between the metals drives electron flow, with the less corrosion-resistant metal acting as the anode, undergoing oxidation and dissolution, while the more resistant metal serves as the cathode, accepting electrons and undergoing reduction.

Crevice Corrosion

Crevice corrosion is a localized attack within narrow gaps where a metal contacts nonmetallic materials like wood, plastic, and rubber. Oxygen-rich solutions outside the crevice act as the cathode, while the oxygen-depleted solution within the crevice becomes the anode. Factors such as pH changes, chloride ions, or stagnant solutions within the crevice accelerate this corrosion.

Filiform Corrosion

Filiform corrosion manifests as irregular, threadlike patterns under coatings like paint and rubber on metals such as aluminum, steel, and zinc, as shown in Fig. 7.3. It begins with a blue-green discoloration, followed by brown filaments. As a surface phenomenon, it does not compromise structural integrity but affects appearance. High relative humidity, particularly above 90%, significantly promotes filiform corrosion.

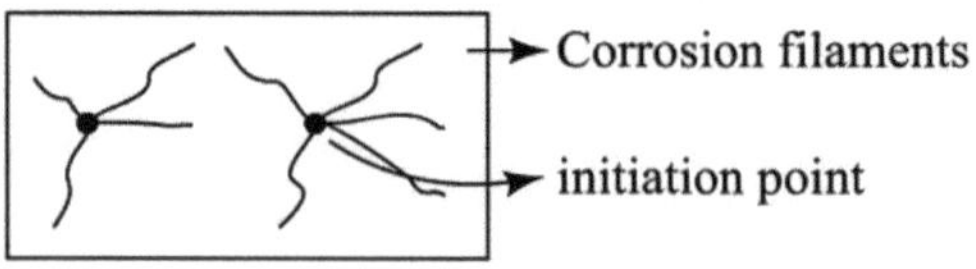

Fig. 7.3 Filiform corrosion in metals

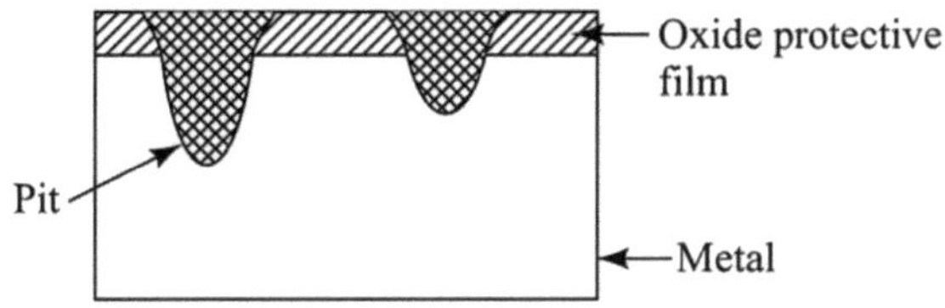

Fig. 7.4 Formation of pits in pitting corrosion

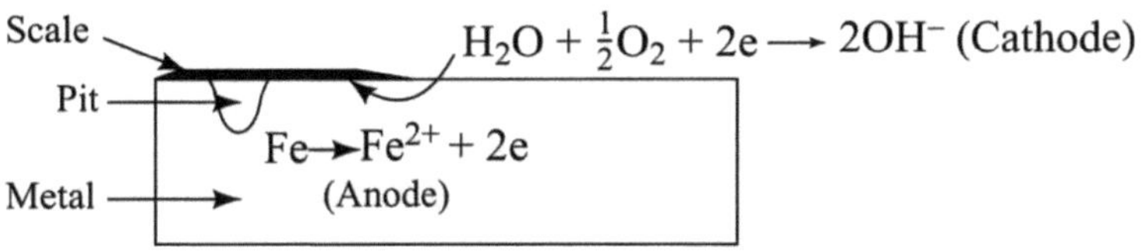

Fig. 7.5 Reactions normally occur in pitting corrosion

Pitting Corrosion

Pitting corrosion involves localized damage where small cavities or pits form on the metal surface, while the rest remains unaffected, as illustrated in Fig. 7.4. Pits typically have similar width and depth and form due to breaches in the protective film. These cavities are often hard to detect due to their small size or coverage by corrosion products, but they can grow rapidly, leading to perforations and fluid leaks. Pitting corrosion resembles crevice corrosion, as both involve a small anode and large cathode generating a corrosion current. Surface impurities, such as scales, can also initiate pitting by creating anodic sites under the impurity, as shown in Fig. 7.5.

Erosion Corrosion

Erosion corrosion results from the combined effects of mechanical wear and chemical or electrochemical attack. High-velocity or turbulent liquid flow removes the protective oxide layer or metal surface, exposing fresh metal to corrosion, as depicted in Fig. 7.6. The nature of the metal surface, flow rate, and turbulence influence this corrosion type. Metals like copper and lead are particularly susceptible, while stainless steel and aluminum, with protective oxide films, are more resistant. Pits formed in erosion corrosion align with the direction of fluid or gas flow.

Intergranular Corrosion

Intergranular corrosion occurs at grain boundaries due to structural defects or compositional differences in the metal. Grain boundaries, where grains meet at low atomic density planes, become active anodic sites under certain conditions, leading to localized attack, as shown in Fig. 7.7. This corrosion reduces mechanical

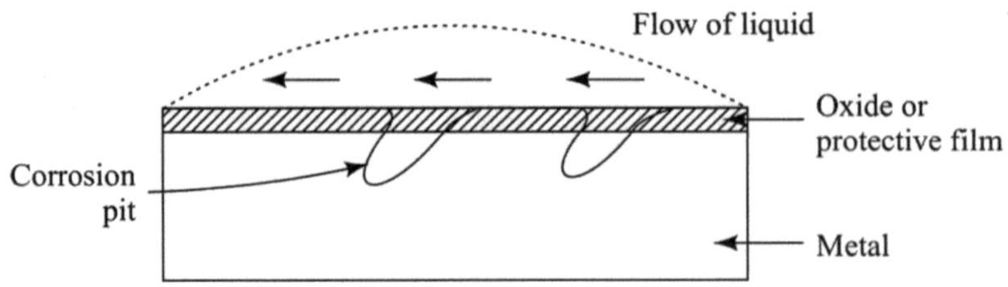

Fig. 7.6 The pit formation in erosion corrosion

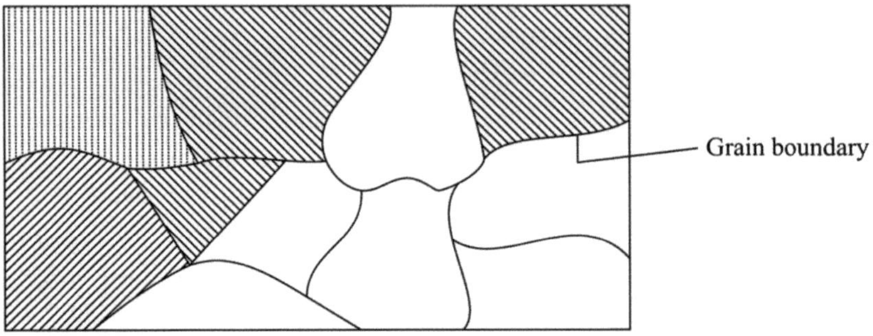

Fig. 7.7 Separation of kinks in intergranular corrosion

Fig. 7.8 Fine crack development in stress corrosion

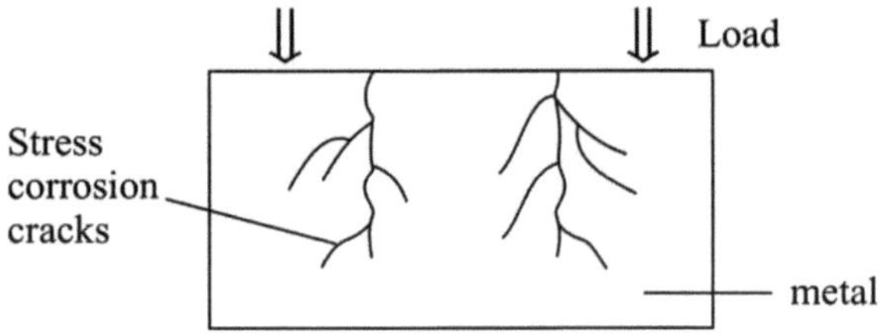

strength and, in severe cases, causes the metal to disintegrate into powder. Heating metals to 800–1100 K or slow cooling through this temperature range sensitizes them to intergranular corrosion. For example, carbon diffuses to grain boundaries at high temperatures, forming chromium-rich carbide precipitates in alloys, creating chromium-depleted anodic zones prone to corrosion.

Stress Corrosion

Stress corrosion, often termed stress corrosion cracking (SCC), results from the combined effects of mechanical stress and a corrosive environment, leading to crack formation, as illustrated in Fig. 7.8. Cracks may be:

- **Intergranular:** Propagating along grain boundaries.
- **Transgranular:** Crossing through individual grains.

Fig. 7.9 Comparison of steel in air and corrosive medium for cyclic stress

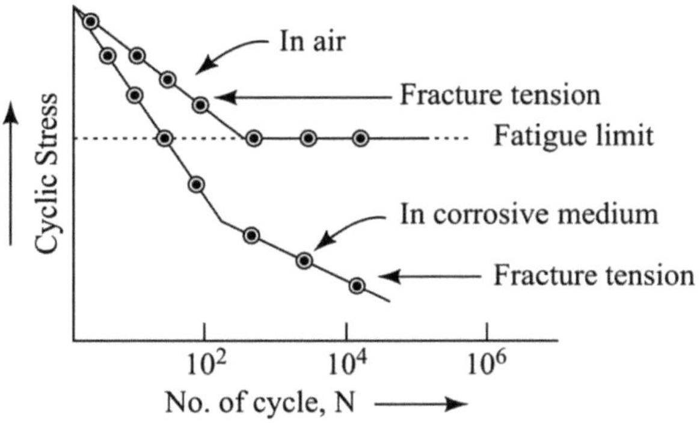

The metal surface remains largely intact, but fine, branching cracks develop, particularly in specific corrosive environments where corrosion products dissolve without reforming protective layers. For instance, carbon steel cracks in strong alkaline solutions, and brass cracks in nitrate solutions.

Corrosion Fatigue

Corrosion fatigue refers to the reduced fatigue resistance of a metal under repeated cyclic stress in a corrosive environment, as shown in Fig. 7.9. Fatigue is the fracture of a metal under cyclic loading below its stress limit. In a noncorrosive environment, a metal stressed below its fatigue limit can endure infinite cycles without failure. However, in a corrosive medium, failure occurs after fewer cycles due to decreased fatigue resistance. For example, high-strength steel exhibits a shorter fatigue life in moist air compared to dry air, and aluminum and bronze show reduced fatigue resistance in seawater. Environmental factors like temperature, pH, humidity, and aeration, along with the metal's properties, influence corrosion fatigue. Cracks initiate at surface defects like scratches or corrosion pits, where the crack base acts as an anode with a more negative potential, accelerating crack propagation through combined stress and corrosion.

7.8 Protection from Corrosion

Corrosion in metals and alloys can be assessed through two primary methods: (i) measuring weight changes in a specific environment and (ii) monitoring alterations in physical, electrical, or electrochemical properties over time. Given the destructive nature of corrosion, implementing preventive measures is essential. Common strategies to protect metals from corrosion include:

- Selection of appropriate materials
- Modification of metal composition
- Application of surface coatings
- Alteration of environmental conditions
- Adjustment of electrode potential

These methods are discussed in detail below.

7.8.1 Selection of Materials

Choosing the right metal or alloy for a specific environment is a cost- effective approach to corrosion prevention. In dry conditions, metals like aluminum and stainless steel can be used without additional protection due to their inherent corrosion resistance. In wet environments, economical materials such as mild steel with protective coatings are preferable. For applications involving high temperatures and pressures, enhancing the intrinsic corrosion resistance of the metal is more effective than relying on coatings.

7.8.2 Modification of Metal

The corrosion resistance of a material can be improved by altering its alloy composition or processing conditions. For instance, adding chromium, nickel, or molybdenum to stainless steel enhances its passivity. Additionally, modifying the metal's microstructure through techniques like annealing and redesigning the alloy's structure can increase its resistance to corrosion.

7.8.3 Barrier Protection or Surface Coatings

Applying protective coatings, such as paints, varnishes, enamels, or lacquers, creates a barrier that shields the metal from corrosive agents. Proper surface preparation is critical to ensure adhesion, involving the removal of grease, dust, sand, or scale using phosphate or chromate solutions to form corrosion-resistant films. Organic coatings offer two key benefits:

- They prevent air and moisture from contacting the metal surface.
- Pigments in the coatings can act as corrosion inhibitors.

The effectiveness of these coatings depends on meticulous application, as errors can lead to cracks and coating failure. The coating process typically involves:

- Surface preparation
- Priming
- Filling
- Sanding
- Finishing

The final coating must be continuous, free of pores, impervious to gases and water, and chemically stable to ensure effective protection.

7.8.4 Alteration of Environmental Conditions

Modifying the environment can significantly reduce corrosion rates. For example, lowering the temperature slows corrosion, as it reduces the rate of chemical reactions. In some cases, boiling fresh or saltwater decreases oxygen solubility, thereby reducing corrosion. Atmospheric pollutants like CO_2, SO_2, and NO_2 dissolve in water to form acids, accelerating corrosion. Adjusting the pH of the environment also influences corrosion; lower pH values favor hydrogen evolution over oxygen ionization at the cathode, increasing corrosion rates. Adding small amounts of corrosion inhibitors or retardants can further mitigate corrosion.

7.9 Cathodic Protection

Corrosion in aqueous environments is an electrochemical process, making cathodic or anodic protection an effective control method. Cathodic protection, widely used since the 1836 application of zinc coating on iron sheets for warships, is more common. In this method, two metals in a corrosive medium are electrically connected, with the more active metal acting as the anode, undergoing oxidation:

$$\mathrm{M} \longrightarrow \mathrm{M}^{n+} + ne^- \qquad \text{(anodic process)}$$

$$n\mathrm{H}^+ + ne^- \longrightarrow \frac{n}{2}\mathrm{H}_2$$

$$\frac{n}{4}\mathrm{O}_2 + \frac{n}{2}\mathrm{H}_2\mathrm{O} + ne^- \longrightarrow n\mathrm{OH}^- \qquad \text{(cathodic process)}$$

When disconnected, both metals corrode, but when connected, the noble metal is protected at the expense of the active metal. By supplying electrons to the active metal, it remains cathodic, preventing metal ion dissolution. Two approaches to cathodic protection are:

(a) Galvanic Cathodic Protection In this method, a more electronegative auxiliary anode is connected to the metal to be protected. Electrons flow from the anode to

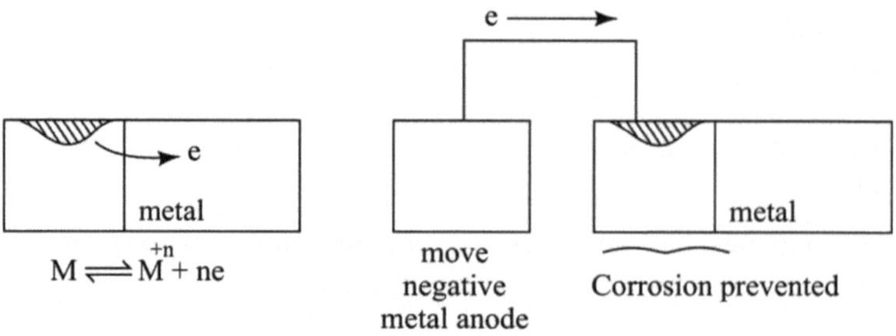

Fig. 7.10 Prevention of corrosion by galvanic protection

the metal, preventing metal ion dissolution and corrosion. Figure 7.10 illustrates electron transfer from anodic to cathodic areas of a metal, and with an auxiliary anode, metal dissolution is halted, protecting the metal. This method, often called sacrificial protection, is effective for underground structures and ships.

(b) Electrolytic Cathodic Protection This method involves constructing an electrolytic cell with the metal as the cathode and materials like nickel, aluminum, zinc, and graphite as the anode. Unlike galvanic protection, the anode need not be more electronegative, as an external power source (e.g., battery or DC generator) supplies the current. This approach is suitable for protecting large metal surfaces where power is available, offering low maintenance and eliminating the need for anode replacement.

7.10 Anodic Protection

Anodic protection involves maintaining a metal in a passive state by applying an external electric current to form a protective film on its surface. Metals like iron, nickel, chromium, titanium, and their alloys undergo rapid active-to-passive transitions when current is applied. Effective anodic protection requires careful control of:

- Critical current density for passivation
- Passivation current
- Potential needed for complete passivation

This method is effective for imparting corrosion resistance to carbon steel, stainless steel, and other passivatable metals.

Given the significant economic losses caused by corrosion, ongoing efforts to develop and refine prevention methods remain critical.

Exercises

1. What is rust? Explain its formation with suitable equations.
2. Explain the reason for faster corrosion of impure metal than pure metal.
3. Why does corrosion occur in steel pipes connected to Cu plumbing?
4. Fe corrodes faster than Al, although the electrode potential of Al is greater than Fe. Explain.
   [*Hint:* Al forms a protective layer of Al_2O_3 to stop corrosion.]
5. What is galvanic corrosion? Why does a steel screw in brass corrode? Explain.
6. Explain the reason for fast rusting of iron in saline water than in ordinary water.
7. Calculate how much rust will be produced from 100 g iron.
   [*Hint:* 112 g of iron gives—214 g rust ($Fe_2O_3 \cdot 3H_2O$)]

$$100g = \frac{214 \times 100}{112} = 191 \text{ g}$$

8. Explain how you can protect a surface from corrosion.
9. What type of corrosion can be successful for ships? Describe it in detail.
10. What is the basic difference between uniform and nonuniform corrosion. Explain any three types of nonuniform corrosion?
11. What do you understand by corrosion inhibitors? Describe quantum chemical approach for mechanism of inhibition.
12. Write the equations to explain corrosion of iron objects in the presence of SO_2 in atmosphere.
13. Write True or False for the following:

    (i) Relative humidity significantly affects the rate of corrosion.
    (ii) If a metal rapidly reacts with H^+, it can be protected by keeping it at higher pH environment.
    (iii) In the presence of air, most metal form oxide.

14. Explain: (i) Pitting corrosion. (ii) Stress corrosion
15. Explain why stainless steel utensils are corrosion resistance.

Chapter 8
Lubricants

Lubricants are defined as substances capable of reducing the friction between two rubbing surfaces. But before we discuss them in detail, let us see why do we need lubricants. The reason is that whenever two surfaces come in contact with each other, the loss of material occurs due to rubbing. Such loss of material is harmful because the debris formed can either act as abrasive and will remove more material or act as adhesive to fill the depression on the metallic surface. However, adhesive action is possible only in the initial phase of sliding.

If we look at the metallic surface on a molecular scale, even the best polished surface has irregularities as shown in Fig. 8.1. Thus, the surface shows roughness and several asperities and valleys are noticed. When two solid surfaces come in contact with each other, a real contact between them occurs only at a limited number of asperities, and, thus, the real contact area is always a very small fraction of the apparent contact area as shown in Fig. 8.2. The height and depth of asperities and valleys differ, and hence whenever two metals are in contact, they bear high local pressure on asperities and cause cold welding (adhesion). The wear of metal actually depends on the nature of two metals opposing each other.

If one of the two metals is hard and the other one is soft, it is obvious that the asperities of the hard metal will plow furrows in the softer metal. Many times, the asperities of the two metals weld so strongly that nothing happens at the junction of the two asperities, whereas grain displacement creates minute pits in the softer metal. The situation becomes complicated when the two metallic surface move over one another. The friction becomes now operative and opposes motion irrespective of the shape, size, or nature of the material in contact. You must realize that wheel rolling over the road or train moving on the track , etc. also has friction and the frictional force is overcome by the applied force to cause movement. Also remember that friction is necessary; otherwise, it will be difficult even to walk on the road. When one walks on ice, where friction is low, the chances of getting slipped increase tremendously. The amount of friction developed depends on the following factors:

© The Author(s), under exclusive license to Springer Nature Switzerland AG 2026
R. N. Goyal, H. Goel, *Textbook of Engineering Chemistry*,
https://doi.org/10.1007/978-3-032-22068-4_8

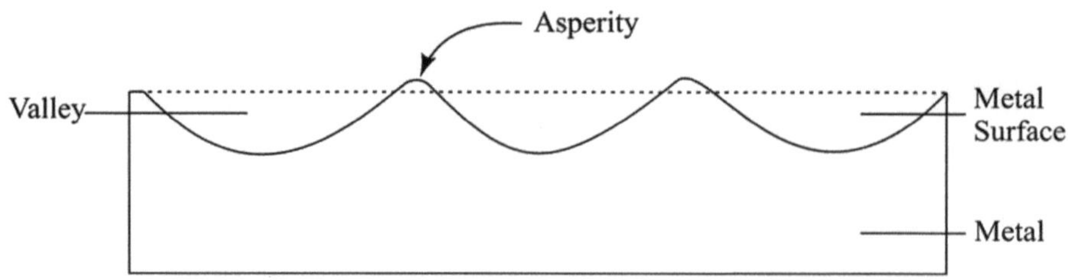

Fig. 8.1 An enlarged look of metallic surface

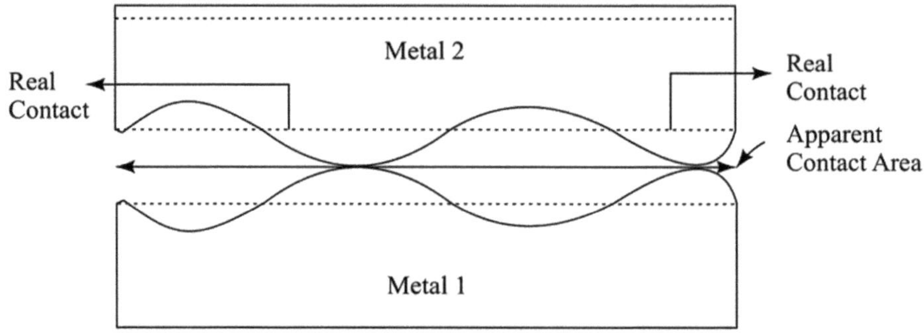

Fig. 8.2 The apparent and real contact area in the case of two metals in contact with each other

- Nature of the rubbing surfaces
- Condition of the rubbing surfaces and
- Pressure applied and the extent to which friction developed can be indicated by
- Heat generated
- Wear of surface
- Force required to initiate the motion and maintain it. The friction is of several types. Some common types of friction are the following.

8.1 Sliding Friction

It is produced when two solid bodies, clean and dry, slide over each other. This type of friction is of highest magnitude because the contact area of the two solids is very large. Sliding of metallic surfaces of iron on iron, iron on brass, steel on Babbitt (alloy of tin, copper, and antimony) are some common examples of sliding friction (Fig. 8.3).

When two surfaces slide over each other, the asperity of one metal tries to block the movement of other metal. If pressure is large, the temperature will also rise. In many cases the temperature reaches very high and causes metal softening, leading

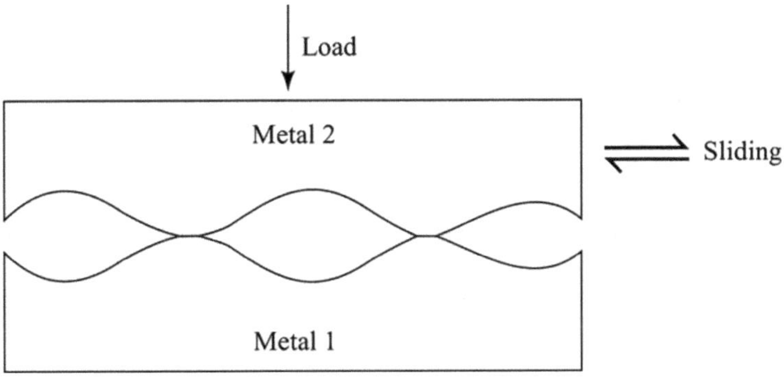

Fig. 8.3 Sliding friction

Fig. 8.4 Rolling of wheel on a flat surface

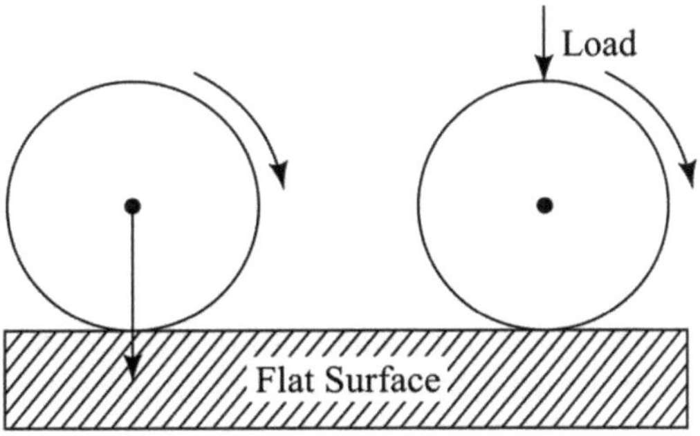

to permanent welding. The sliding friction is also dependent on the load applied and is approximately proportional to the pressure between the sliding surfaces.

It must be realized that the force required to initiate motion of a body at a rest is always higher than the force required to continue its motion. The force required to initiate motion is called as *static friction*, whereas the force required to continue motion is called as *kinetic friction*.

8.2 Rolling Friction

Rolling friction is developed when a ball or wheel rolls over a hard flat surface. The coefficient of rolling friction is very low as compared to sliding friction. Under pure rolling condition, the friction will be small, whereas if load is applied, the contact area is increased and friction increases as shown in Fig. 8.4. During rolling friction some sliding action also occurs in an effort of the rolling wheel or ball to lift over the obstructions. Rolling friction like sliding friction is also proportional to the load and is nearly independent of speed. It decreases with hardness of surface and varies inversely with the radius of the rolling member.

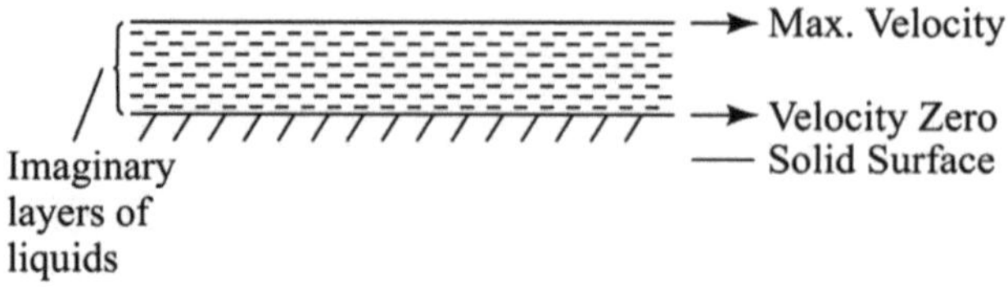

Fig. 8.5 Flow of liquid on a solid surface

8.3 Fluid Friction

The force that resists the movement of numerous imaginary molecular layers of the liquid/fluid is called as fluid friction. The fluid friction has been found high for the liquids that do not flow easily such as castor oil, glycerin, mobile oil, etc., whereas for easily flowing liquids such as water, alcohol, acetone, etc., fluid friction is small. Let us consider a liquid layer flowing over solid surface as shown in Fig. 8.5. The velocity of top layer of the liquid is higher, whereas at bottom, the velocity is practically zero. Thus, from moving bottom to top layer of liquid, the velocity of flow of liquid increases. It can be assumed that layers of liquid slide over each other and this action is called as *shear*.

The speed with which these layers slide over each other is called as rate of shear, which within a given fluid mass controls its *fluidity*. The fluidity of a liquid is expressed in terms of its viscosity and is inversely proportional to viscosity:

$$\text{viscosity} \propto \frac{1}{\text{fluidity}}$$

Hence, increase in viscosity of a liquid would cause a decrease in its fluidity.

8.4 Boundary Friction

The force that resists the relative movement of two solid bodies, whose surfaces are barely separated by a lubricant film, is called as boundary friction. Usually one or more layers of the lubricant are enough for the development of boundary friction.

Thus, the friction in machines causes wear and tear of various parts and requires their replacement. Also the heat generated due to friction causes loss in efficiency of the machines. To minimize the bad effect of friction, chemical substances are added between the two solid bodies, which form a thin layer between them. *A lubricant thus can be defined as any substance introduced between two surfaces in a view to reduce friction between them.*

In industries the damage to the surfaces of machines occur due to the following important reasons:

- Corrosive environment
- Lack of lubricant
- Improper lubricant
- Improper use of metals or alloys in machine parts, etc.
- Overloading
- Overspeeding
- Sudden shock or vibration

We will discuss the first two aspects, i.e., lubricant and its proper use, in this chapter. The use of lubricant minimizes the damage in various ways:

(i) As lubricant film separates the moving or sliding parts of the two bodies, it reduces wear and tear of machine parts.
(ii) It reduces loss of energy as heat and acts as a coolant.
(iii) It also prevents welding of machine parts under extreme pressure conditions. Thus, the maintenance cost of machines is reduced.
(iv) Also it introduces corrosion inhibitors between the two bodies and thus protects them against corrosion.
(v) In many cases lubricant also acts as a seal and prevents the leakage of gases at high pressure, for example, in internal combustion engines.
(vi) As high temperature is prevented by the lubricant, the possibility of expansion of metals due to local frictional heat is reduced.

8.5 Classification of Lubricants

Lubricants can be classified according to their physical state. For example,

- **Solid Lubricants**—Graphite, mica, molybdenum disulfide, lead iodide, calcium iodide, wax, etc.
- **Liquid Lubricants**—A variety of liquid lubricants are in use and can be subdivided into four groups:

  (i) *Mineral oils*—Straight or unadulterated oils obtained from petroleum crudes or after mixing with additives are widely used as lubricants.
  (ii) *Fixed oils*—Animal oils (lard oil, tallow oil, etc.), vegetable oils (palm oil, castor oil), etc.
  (iii) *Synthetic fluids*—Silicones, polyglycols, silicate esters, etc.
  (iv) *Soluble oils*—Mineral oil or synthetic fluids compounded with emulsifying agents.

- **Gaseous lubricants**—CO_2, air, helium, etc.
- **Plastic lubricants**—mineral oil or synthetic, fluids thickened with soaps, semi-fluid grease, synthetic grease, etc.

Let us now discuss some general properties of these lubricants.

Table 8.1 Pretreatment recommended for some metals before using lubricants

Metals	Pretreatment suggested
Stainless steel	Chemical etching or vapor blasting
Copper	Bright dipping
Aluminum	Anodizing
Steel	Phosphating

8.5.1 Solid Lubricants

Solid lubricants are used in extreme conditions of temperature, load, vacuum, etc., under which oils and grease do not work satisfactorily. For example, at very high temperature, oxidation of oils occur and volatile constituents evaporate, whereas at low temperatures, oil lubricants or greases are solidified and thus moving parts of the machines are jammed. Similarly high vacuum also causes evaporation of oil lubricants.

Solid lubricants such as graphite and molybdenum disulfide are used in dry state or after mixing in oil or grease or combined with binder so that they can stay at the target surface. Phenolics, epoxy, polyamide resins, and fused ceramics are some of the common binders used. Solid lubricants are usually applied by spraying, brushing, or dipping, and a pretreatment of the metal surfaces is recommended before applying solid lubricants. The common pretreatments suggested for some metals are presented in Table 8.1.

Graphite and molybdenum disulfide are two largely used solid lubricants that work well at high-temperature environment. Molybdenum disulfide also performs satisfactorily in lubricating control rod mechanisms of nuclear reactor that are directly exposed to radiations. With the presence of moisture, molybdenum disulfide causes an increase in kinetic friction, whereas graphite requires an adsorbed film of moisture for performing satisfactory lubrication.

Graphite

In elementary form, carbon exists in two allotropic forms, diamond and graphite. Graphite is a crystalline form of element carbon and is a product of hard coal or coke. The sp^2 hybridized carbon atom forms a planar network of six membered rings in graphite as shown in Fig. 8.6. Many resonance structures can be written for the structure of graphite due to delocalization of the π-electrons over the entire sheet. The free π-electrons are responsible for making graphite conducting to electricity. The availability of free π-electrons also imparts black color to graphite, because for a photon of almost any energy, there is an electron available to absorb it. The covalent bonds of graphite are quite strong; however, the different sheets

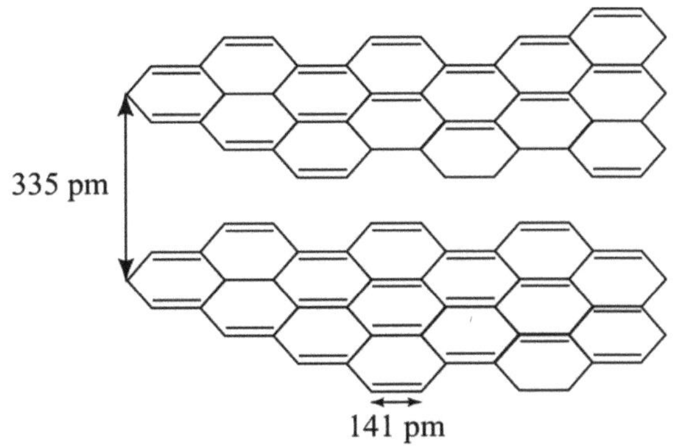

Fig. 8.6 Layered structure of carbons in graphite

are bonded to each other by weak dispersion forces, due to which the sheets have a tendency to readily slide over each other and impart lubricating property to graphite.

Graphite has a melting point of 3570 °C and is used as a lubricant in a powdered or colloidal form. It can be used alone or after mixing with oil or grease. It is not soluble in water and has high chemical stability. For its lubricating action, graphite requires an adsorbed film of moisture on its surface, and if such a film is not present, graphite acts as abrasive. Graphite is resistant to heavy temperature and pressure and hence is successfully used under extremely severe operating conditions.

Dispersion of graphite in oil, commonly known as *oil dag*, is used in Internal combustion engines as a lubricant between the piston rings and cylinder to provide a tight fit contact. It is also used in lathe machines, cast iron bearings, etc.

Molybdenum Disulfide

Like graphite, molybdenum disulfide has also a layer structure in which alternate layers of molybdenum and sulfur atoms exist. The distance between these layers is ~3.1 Å and is nearly similar to distance between S—S and Mo—Mo atoms as shown in Fig. 8.7.

Molybdenum disulfide can be used in powder form or after mixing with oil or grease. In powder form it is sprinkled on metallic surfaces sliding over each other at high velocities. The main advantage of using molybdenum disulfide is that it can be used in a large range of temperature range (−72 °C to 400 °C). At temperature >400 °C, oxidation of molybdenum disulfide occurs. It has a very low coefficient of friction (0.032) and is blue-gray or black in color. It has affinity for metal surfaces and is noncorrosive in nature. The thin film of disulfide is resistant to water and chemicals and can be used in extreme operating conditions.

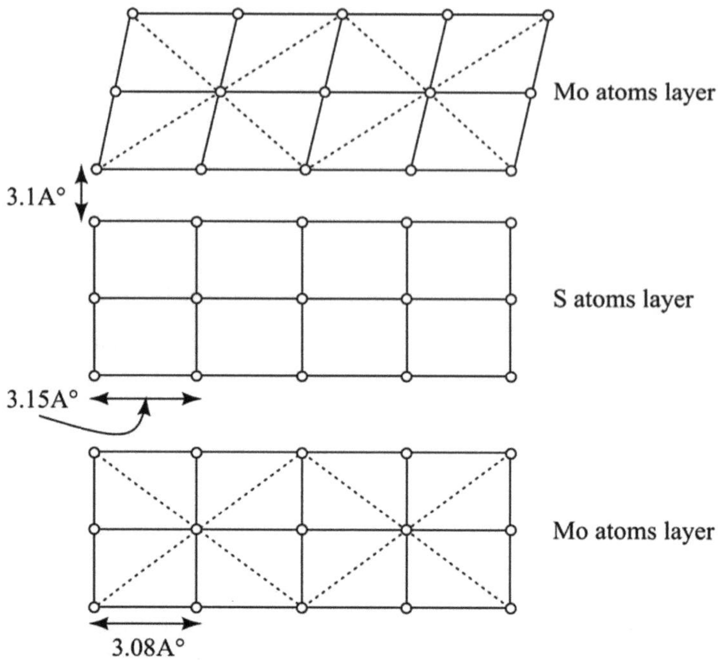

Fig. 8.7 Layer structure of molybdenum disulfide

8.5.2 Liquid Lubricants

Liquid lubricants are the most important group of lubricants, and besides reducing friction between two moving surfaces, they also provide cooling effect. A liquid lubricant must possess the following qualities:

- high boiling point
- low freezing point
- adequate viscosity and
- stable at operating temperature.

Liquid lubricants are further classified as the following.

(i) Mineral Oils

Mineral oils are produced from petroleum crudes by distillation and other refining methods. These are organic compounds having carbon and hydrogen as the major elements. Nitrogen, sulfur, and oxygen may also be present in small amounts depending upon the source of the crude petroleum. Mineral oils are cheap as they are available in abundance and quite stable under normal working conditions. However, they possess many undesirable properties like poor oiliness and impurities such as wax, asphalt, etc. It is therefore essential that mineral oils should be purified before they put to use. *Remember good lubricants are always made, not found.*

To remove impurities present in the mineral oils, generally the following steps are used:

- **Dewaxing:** Mineral oil is mixed with solvents like trichloroethylene and then cooled to about −5 to −10 °C. The precipitated wax is removed through a continuous filter or using centrifugation.
- **Refining:** The dewaxed oil is further purified by using acid or solvent refining. In acid refining, oil is treated with concentrated H_2SO_4, and impurities dissolve in acid and form sludge which is removed by filtration, whereas in solvent refining, solvents like benzene, nitrobenzene, cresol, furfural, etc. are used to dissolve impurities and the oily layer is separated. The traces of solvents are removed by distillation to get pure oil.

(ii) Fixed Oils

The oils obtained from vegetables, animals, or fish are called as fixed oils and used as lubricants because of their good oiliness. They are called as fixed oils because they do not volatilize without decomposition. However, these oils have some limitations like the following: (i) they are costly in comparison with mineral oils; (ii) they combine with oxygen of air to form gummy product. This process is known as drying and (iii) they have tendency to hydrolyze in contact with moist air.

Fixed oils are composed of fatty acids and alcohols. Some of these acids combine with alcohol to form fatty esters. It is possible to determine free and combined fatty acids in a fixed oil by determining its saponification number, which is defined as *the number of milligrams of KOH used by 1 g of fixed oil under test conditions*. Higher values of saponification number indicate that the oils possess higher amount of base-consuming constituent. Another property of these oils is their drying capacity which is determined by iodine number. *The iodine number is defined as the amount of iodine absorbed (in percentage) by an oil under test conditions*. Some properties of the fixed oils are summarized in Table 8.2.

Some fixed oils are used as an additive for mineral oils to improve their oiliness or increase their emulsibility, whereas some of them are used in making greases. The important uses of some fixed oils are described in Table 8.3.

Synthetic Fluids

In many industries special lubricants are needed in the view of their specific requirements such as extreme temperatures, high vacuum, chemical reducing environment, nuclear radiations, etc. Petroleum oils should not be used where ever any fire hazard exists. For extreme conditions, synthetic fluids such as silicones,

Table 8.2 Common properties of some fixed oils

Fixed oil	Iodine	Saponification number	sp gravity number	Fatty acid present
Coconut	–	–	0.96	Acetic acid $C_6H_{12}O_2$—$C_{14}H_{28}O_2$
Castor	85	180	0.96	Ricinoleic acid, stearic acid
Lard	70	194	0.91	Oleic acid
Linseed	180	150	0.93	Linoleic acid
Olive	85	192	0.91	Oleic acid, palmitic acid
Palm	55	20	0.92	Palmitic acid
Peanut	101	191	–	Oleic acid, linoleic acid
Tallow	65	194	0.95	Stearic acid

Table 8.3 Uses of some fixed oils

	Fixed oil	Uses
1.	Castor oil	Blended with mineral oils to produce caster machine oil
2.	Lard oil	Used as such or after blending with mineral oils for lubrication of ordinary metal parts of machine
3.	Olive oil	Used for lubrication of machine parts in textile finishing and machine parts subjected to low pressure
4.	Tallow oil	Used as an emulsifying agent for steam cylinder oils
5.	Fish oils	Used in lubricating greases
6.	Neat foot oil	Used for lubrication of guns, clocks, watches, etc.

fluorocarbons, polyglycol esters, etc. are used with fairly good thermal stability. The general structure of silicone oils is presented below.

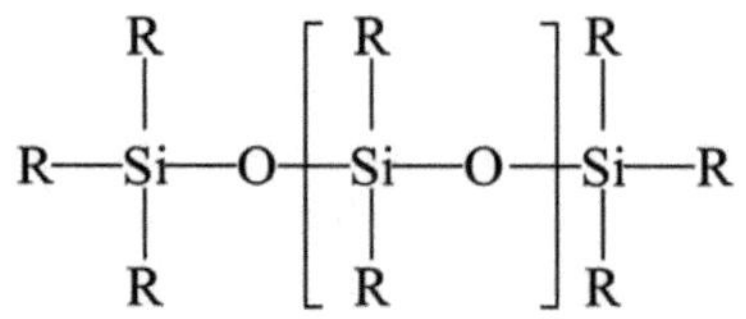

where *R* is methyl or phenyl group.

The viscosity of these oils increases with increase in chemical units. They have good thermal stability and resist the action of water. On exposing them to oxidation conditions, they maintain their lubricating properties for a long time. Some characteristics of synthetic fluids are given in Table 8.4.

Table 8.4 Useful characteristics of synthetic fluids

Fluids	Lubricant property	Oxidation resistance	Thermal stability
Silicones	Poor	Good	Very good
Polyglycol esters	Good	Poor	Fair
Fluorocarbons	Good	Excellent	Excellent

Fluorocarbons are widely used these days as lubricants for machines that handle reactive elements and compounds, like turbopump bearings of aircraft, oxygen compressor cylinders, piston rods, etc. They have general structure where R and R' are the terminal groups.

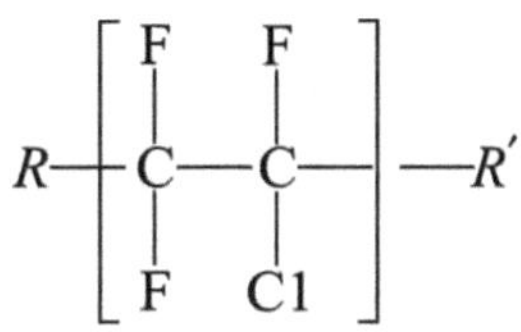

The addition of fluorene in such molecules makes them highly resistant to acids, alkalies, and oxidizing compounds. In electrical equipments they are used as spark-repressing and spark-insulating fluids. They can be safely used in the temperature range of −40 °C to 200 °C.

8.5.3 *Use of Water as a Lubricant*

In spite of the fact that water does not possess lubricating properties, it is used to dissipate heat generated in grinding or cutting machines. In nuclear reactors water is used as a coolant and also to lubricate its pressurizing equipment. In high-speed cutting machines, like glass- or stone-cutting machines, water is used as a coolant. Emulsion of oils and water, which is a two-phase system consisting of two immiscible liquids, is also used as lubricant. In emulsions one liquid is dispersed as fine droplets in other, which is called as dispersing phase. The size of dispersed droplets is very fine and varies in the range of 1–6 micron.

An emulsion is generally unstable; however, its stability can be increased by adding some agents commonly known as *emulsifiers* or *emulsifying agents* such as sodium palmitate, $C_{15}H_{31}COONa$. Emulsifiers have a long hydrocarbon chain and a polar end. The hydrocarbon end is called as *hydrophobic end* and polar end is called as *hydrophilic end*. The hydrophobic end is liked by oil, whereas hydrophilic end is liked by water.

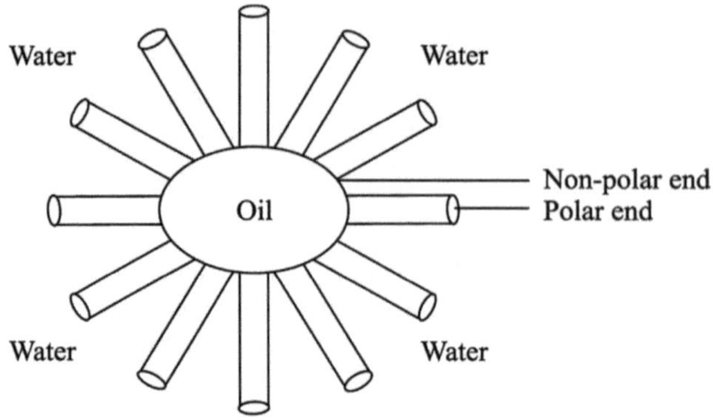

Micelle
Oil in water emulsion

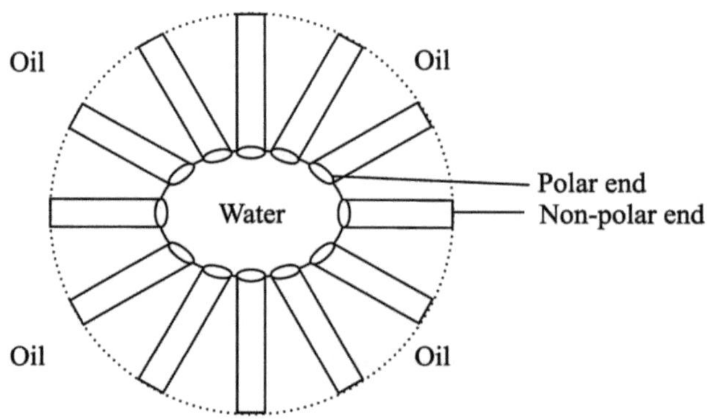

Reverse micelle
Water in oil emulsion

Due to adsorption of these emulsifiers at the interface of two immiscible liquids, a protective film is formed around the dispersed droplets, and the emulsion becomes stable.

Depending on the percentage of oil in an emulsion, two types of emulsions can be formed:

(i) Less than 26% oil—oil in water emulsion
(ii) More than 74% oil—water in oil emulsion

If the percentage of oil is in between 26 and 74%, both types of emulsions can be formed. In oil and water emulsion, oil droplets are surrounded by emulsifiers in such a way so that their hydrophobic end is toward oil. Such structure formed in an emulsion is called as *micelle*. Oil in water emulsion is used as cooling as well as lubricating liquid for cutting tools. It is also used in heavy sliding components of various machine parts such as pistons of diesel engines. This type of emulsion also acts as rust-protecting agent.

In contrast to oil in water emulsion, water in oil emulsion has water droplets surrounded by emulsifiers in a reverse fashion. The polar end of the emulsifiers now faces toward water droplets and *reverse micelle* is formed. Such emulsions have high viscosity than oil in water emulsion or that of oil from which they are prepared. Emulsions having 40% water are largely used to lubricate compressors.

8.5.4 Plastic Lubricants

Plastic lubricants such as grease are widely used in the machines where operating conditions do not permit the use of solid or liquid lubricants. Grease, a semisolid material at room temperature, acts as plastic under operating conditions. A large number of chemical variations are possible in the ingredients of grease and their processing methods, and hence a large variety of greases are available these days in the market.

Greases are made by the saponification of fat with a metal base and adding hot lubricating oil while agitating the mixture under controlled temperature conditions. The metal bases commonly used in the preparation of greases are hydroxides of calcium, sodium, aluminum, lithium, and lead and named as corresponding greases, viz., calcium soap, soda soap, etc. Hydroxides of barium and strontium are also used in many cases. The physical properties of grease depend primarily on the base used in its preparation. The greater soap content in grease imparts consistency to it, whereas to improve its heat resistance nature, inorganic solid-thickening agents such as clay, carbon black, bentonite, etc. are added. Typical uses of some common greases are presented in Table 8.5.

Graphite grease prepared by adding up to 20% powdered graphite in calcium soap is widely used for wet conditions like pumps, pump plungers, hydraulic rams, wet plain beamings, etc. The use of grease is normally required in the following: (i) sudden or intermittent operation is required due to heavy loads, for example, rail axle boxes, (ii) dripping of oil is undesirable, for example, in machines preparing paper, textile, or edible items, and (iii) bearings need to be protected against dust.

Table 8.5 Some common properties and uses of different types of greases

Base used	Solubility in water	Maximum operating temperature	Use
Calcium	Insoluble	80 °C	In bearings of moderate speed and temperature
Sodium	Soluble	175 °C	In bearings of high speed. Used in dry conditions
Lithium	Slightly soluble	160 °C	General-purpose grease in auto chassis, flexible, etc.
	Insoluble	90 °C	

8.5.5 *Gaseous Lubricants*

Gases like air, helium, hydrogen, nitrogen, etc. are used as lubricants mostly for lighter bearings, thrust bearings of some vertical turbines, and the rotating mount of large telescopes. The advantages of using gas as a lubricant are that its viscosity does not change with temperature, has non-drip property, and does not cause contamination. In nuclear environment, gases are useful hydrostatic lubricants. Similarly under extreme temperature conditions like space exploration and cryogenics, gases are extremely useful as lubricants.

8.6 Mechanism of Lubrication

As stated earlier, during lubrication a film of lubricant is interposed between the two metallic parts to reduce friction between them. The mechanism of lubrication is of three types.

8.6.1 *Fluid Film, Thick-Film, or Hydrodynamic Lubrication*

A thick film of lubricant ($\sim$1 Å thick) is used to separate moving or sliding surfaces. The thick film produced prevents direct surface-to-surface contact and consequently reduces wear as shown in Fig. 8.8. The resistance in the movement of two moving or sliding surfaces is due to internal resistance of the lubricant particles moving over each other. The essential requirement of this type of lubrication is that the lubricant should remain in place and separate the two surfaces. Thus, its viscosity should be minimum. Another important factor associated with the development of such a film is adequate and frequent supply of the lubricant particularly during motion of the surfaces.

The thickness of the film produced controls the load-carrying capacity of the hydrodynamic area and is a function of:

(i) load per unit area
(ii) viscosity of the lubricant and
(iii) relative velocity of two moving surfaces

The coefficient of friction in this mechanism is low ($\sim$0.01). Mineral oils are usually considered suitable to provide hydrodynamic lubrication.

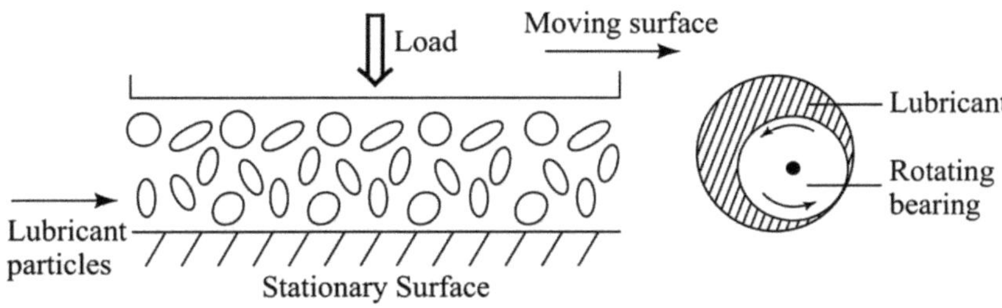

Fig. 8.8 Thick-film lubrication during moving surface

8.6.2 Boundary Lubrication or Thin-Film Lubrication

Another type of mechanism of lubrication is called as *boundary lubrication* and is required when a film of lubricant cannot persist due to the following reasons:

(i) Load applied is very high
(ii) Speed is low and
(iii) Surface starts to move from the rest position.

In all these cases, direct metal-to-metal contact of the moving surfaces is possible, and hence a lubricant capable of adsorbing at the surfaces of moving or sliding parts is used. The adsorption occurs due to physical or chemical forces or both. The adsorbed layer is sufficiently stable and does not remove under operating conditions. Thus, the adsorbed layer of lubricant avoids direct metal-to-metal contact. Vegetable oils, animal oils, and their soaps possess property of adsorption. Graphite alone or its suspension in oil also provides boundary lubrication.

8.6.3 Extreme Pressure Lubrication

When the moving surfaces are under high pressure and move with high speed, then the lubricant may decompose. For these conditions, some additives are added to lubricants. These additives are called as "extreme pressure additives." The additives are reactive compounds of phosphorus, sulfur, etc. At a very high temperature, they react with metal and convert as sulfide or phosphides. These lubricants have very high melting points and act under extreme pressure lubrication conditions.

To determine the suitability of a lubricant for a particular use, its properties are assessed. Some common properties are the following.

(i) Viscosity

Viscosity of a liquid is defined as its property by virtue of which it offers resistance to its own flow. The details of such resistance by different layers of liquid are presented in fluid friction described earlier in this chapter. Let us assume that two layers of a liquid separated by a distance, d, are moving with a relative velocity difference v. The top layer of the liquid always moves faster than the layer below it. To maintain this velocity difference, force F required per unit area is given by an expression:

$$F = \frac{\eta v}{d}$$

where

F = force per unit area
η = coefficient of viscosity and is constant for a liquid
v = velocity difference between the two layers
d = distance between the two layers.

If velocity difference between the two layers is unity and distance is also unity, then

$$F = \eta$$

Hence, we can define *coefficient of viscosity as force per unit area required to maintain a unit velocity gradient between two moving layers of a liquid, 1 unit apart*. Say if the two parallel layers are 1 cm apart and to maintain a velocity gradient of 1 cm/s between them, a force of 1 dyne is required. In such conditions, the coefficient of viscosity of the liquid is 1 poise. The smaller unit of viscosity is centipoise which is equal to 1/100 of poise.

Viscosity is an important property of liquids and provides an idea whether the liquid will be able to make thin film on metallic surfaces or not. If viscosity of a liquid is too low, a liquid film cannot be maintained, whereas if viscosity is too high, an excessive friction will result. Thus, the viscosity of a lubricant should not be at two extremes.

The viscosity of a liquid is also dependent on temperature and generally decreases with increase in temperature. Such a decrease in viscosity at higher temperature affects the lubricating property as the oil becomes thinner. For a good lubricant, the viscosity should not change significantly with increase in temperature. To determine the change of viscosity of an oil with temperature, an arbitrary scale is used which is known as viscosity index. The lubricating oil for which viscosity falls sharply with increase in temperature is termed as low viscosity index oil, whereas if viscosity does not change significantly with increase in temperature, the oil is called as high viscosity index oil.

Viscosity Index

The viscosity index of a lubricant is calculated by determining the viscosity at two temperatures (37 and 99 °C) and comparing these values with the standard tables of viscosities available for standard oils. The viscosities of two types of standard oils, viz., paraffinic base Pennsylvanian oils (V.I = 100) and napthenic base gulf oils (V.I. = 0), have been determined at 37 and 99 °C and available in tabular form in the literature.

For calculating the viscosity index, the following steps are followed:

- The viscosity of the test oil is determined at 37 and 99 °C. As stated earlier the difference between the two viscosities should be minimum for a good lubricating oil. Let us assume that the values of these viscosities are a and b, respectively.
- Now from the standard tables of the paraffin base Pennsylvanian oil (V.I. = 100) having viscosity similar to the oil under test at 99 °C is selected, and its viscosity at 37 °C is read from the table. Similarly from the standard tables of the napthenic base gulf oils (V.I. = 0), an oil having viscosity similar to the one for the oil under test is selected, and its viscosity at 37 °C is read from the table. Let these values be c and d respectively.

The viscosity index is then calculated using a formula:

$$\text{V.I.} = \frac{d - a}{d - c} \times 100$$

where

d = viscosity of low viscosity standard oil at 37 °C, having the same viscosity at 99 °C as oil under test.
a = viscosity of oil under test at 37 °C.
c = viscosity of high viscosity standard oil at 37 °C, having the same viscosity at 99 °C as oil under test.

The viscosity of an oil can be increased by the addition of certain organic polymers. These polymers are slightly soluble in lubricating oil at low temperature and more soluble at high temperature. Thus, at higher temperatures the viscosity of oils will increase. By adding correct amount of such polymers, a lubricating oil can be produced having very slight effect of temperature on its viscosity.

Example 8.1 A student determined the viscosity of a lubricating oil using Redwood viscometer and found it to be 60 s at 99 °C and 528 s at 37 °C. He compared the viscosity at 99 °C with tables of Pennsylvanian oils and Gulf oils, and the corresponding values of viscosities at 37 °C were 480 s and 760 s for these oils. Calculate the viscosity index of the lubricating oil.

Solution

$$\textit{Viscosity Index} = \frac{d-a}{d-c} \times 100$$

$$d = 760\,\text{s}$$

$$a = 528\,\text{s}$$

$$c = 480\,\text{s}$$

$$V.I. = \frac{760-528}{760-480} \times 100$$

$$= \frac{232}{280} \times 100$$

$$= 82.8$$

Determination of Viscosity

The viscosity of a lubricating oil is determined by allowing a fixed volume of it to flow through a capillary under gravity force and the time is recorded in seconds. The time of flow so obtained is directly proportional to the viscosity of the oil. Two viscometers are used for the determination of viscosity; they are as follows: (i) Redwood viscometer and (ii) Saybolt viscometer. The time of flow recorded in seconds from these viscometers is represented as Redwood seconds or Saybolt seconds. In India, Redwood viscometer is mostly used in the laboratories and is simple to operate.

The Redwood viscometer has two parts:

- **Oil cylinder:** A brass cylinder having a capillary attached at its bottom is fixed with a big metallic cup as shown in Fig. 8.9. The wall of the cup makes the viscometer a two-compartment instrument. The diameter and the length of the capillary are 1.82 mm and 10 mm, respectively.
- **The outer compartment:** Is used for filling water and thus controls the temperature at which viscosity is determined. The brass cylinder has an arrangement of ball at the mouth of the capillary by which the mouth of the capillary is blocked. The oil starts flowing through the capillary only when the ball is pulled with the attracted string.

For determining the viscosity of the desired oil sample, the brass cylinder is filled up to the pointer level with the oil. Now the temperature at which viscosity is to be determined is maintained using water heating, and when the temperature of the oil is reached to the desired level, the ball is pulled and the time to flow the oil through the capillary is recorded and expressed in Redwood seconds. If viscosity of the oil is less, it takes more time to flow through the capillary.

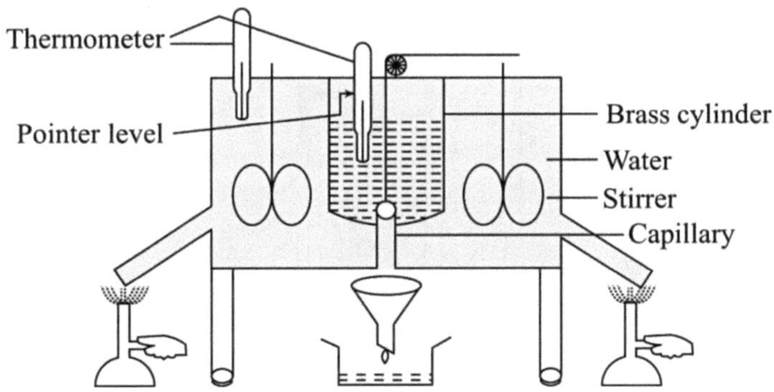

Fig. 8.9 Redwood viscometer used for determining viscosity

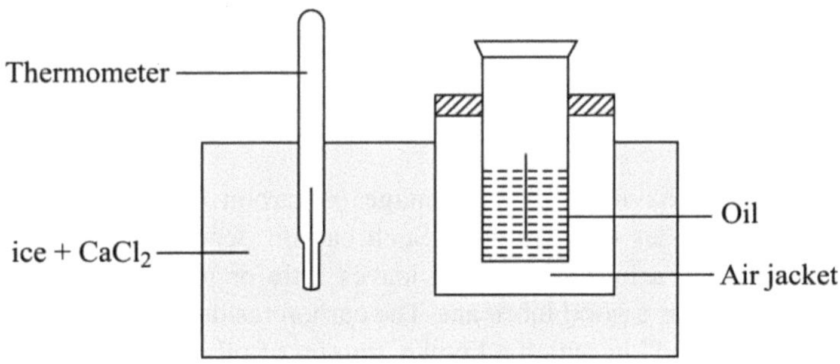

Fig. 8.10 A setup used for determining pour point of a lubricant

(ii) Pour Point

The determination of pour point of a lubricating oil is useful to provide an idea about the temperature up to which it can be safely used. At temperature lower than pour point, the oil could solidify and hence will cause jamming of the machine parts.

If a lubricating oil is cooled slowly, it becomes cloudy at a particular temperature. Further cooling of the oil leads to a stage where the oil ceases to flow. The temperature at which oil becomes cloudy or hazy is called as *cloud point*, and the temperature at which oil ceases to flow is called as *pour point*. Cloud points and pour points can be determined by cooling the desired oil slowly in a double-walled tube freezing mixture (ice + + $CaCl_2$) as shown in Fig. 8.10.

The oil is taken in a double-walled tube. The outer layer provides an air jacket for cooling the oil. For determining cloud point, the tube is taken out for every 3–4 °C fall in temperature. The temperature at which the oil becomes cloudy is noticed and called as cloud point. The cooling is further continued. The temperature at which oil does not flow in the tube even after keeping it horizontally is called as pour point.

It is essential that lubricating oils used in cold climates should have pour point well below the climatic temperatures; otherwise, they will solidify and will cause damage to the machines.

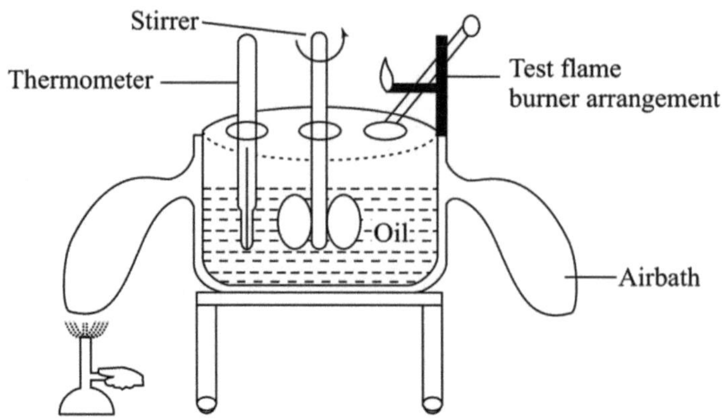

Fig. 8.11 Pensky-Marten's apparatus used for determining flash and fire point

(iii) Carbon Residue

The lubricating oils having high percentage of carbon leave carbon residue on decomposition at higher temperatures. Such carbon deposition is undesirable in machines, and hence a lubricant which leaves little or practically nil amount of carbon is considered as a good lubricant. The carbon residue of an oil is determined by "Conradson method" in which a known amount of oil is taken in silica crucible and is burnt in chimney-shaped iron hood. Initially the temperature is raised slowly and then strong heating is carried till all volatile matters are removed. The crucible is then allowed to cool and the weight of the crucible is taken. The residue left is determined and expressed as the percentage of oil weight taken.

(iv) Flash Point

Flash point of an oil is the temperature at which it gives enough vapors that ignite for a moment when these vapors are exposed to a tiny flame. The information about the flash point is essential as a lubricating oil should be used up to the temperature well below its flash point.

If temperature of the oil is further increased, it gives enough vapors that ignite for at least 5 seconds. The temperature at which vapors ignite for at least 5 seconds is called as *fire point* of the oil. The fire point of an oil is usually 5° to 20 °C higher than the flash point. Flash point and fire point are determined by using Pensky-Martens apparatus, which has a thick double-walled metallic cup of about 5 cm diameter and 5–6 cm deep as shown in Fig. 8.11. On the top of the cup, there is an arrangement of pilot burner with flame exposure device. The cup is covered with a lid having openings for thermometer and stirrer.

The oil is filled in the cup up to the marked level and the lid is covered. Slow heating is done by heating the air bath, and stirrer is rotated at the rate of

2–3 revolutions per second for uniform heating. Initially for every 5 °C rise in temperature, the flame is introduced with pilot burner. The temperature at which vapors burn for a second with a weak sound (a distinct flash appears) is recorded. The experiment is then repeated, and when temperature reaches to 5° below the recorded temperature, the flame is introduced for every 1 °C rise of temperature, and exact flash point is determined.

The temperature is further increased, and test flame is introduced as previously for every 1 °C rise of the temperature. The temperature at which vapors of the oil continuously burn for 5 s or more is recorded as fire point of the lubricant oil. It is essential that a lubricant oil should be used at temperatures well below its flash and fire point; otherwise, it may catch fire.

(v) Oiliness

The property of a lubricant to stick on the surface at which it is applied is known as oiliness. It is an important property of the lubricant. Many lubricants do not stick at the metallic surfaces and squeeze out under heavy load. Thus, it is desirable that the lubricant should remain attached with the surface to provide lubricating action. Oiliness of a lubricant is its property for which no tests are available for its measurement.

(vi) Specific Gravity

Specific gravity of a lubricating oil is defined as the ratio of weight per unit volume of oil to the weight of an equal volume of water when both are taken at 15.5 °C or 60 °F:

$$\text{Specific gravity} = \frac{\text{Weight per unit volume of oil}}{\text{Weight of equal volume of water}}$$

The oil industries use another scale to represent specific gravity which is based on the method suggested by the American Petroleum Institute.

According to this method, a hydrometer having scale from 0 to 100° is used, and the value of specific gravity is expressed as API gravity. The relation between API gravity and specific gravity is given by the expression:

$$\text{API gravity} = \frac{14.5}{\text{Specific gravity}} - 131.5$$

The gravity of a lubricating oil provides an idea about the type of crude from which oil is produced. Most of the lubricating oils are lighter than water, and hence their API gravity usually lies in the range of 10–30 as shown in Table 8.6.

Table 8.6 API gravity for some selected lubricating oils

Oil	API
Napthenic base oils	19–23
Asphaltic base oils	24–27
Paraffinic base oils	29–32
Mixed base oil	27–30

Example 8.2 A lubricating oil has a specific gravity of 0.86; calculate its API gravity:

$$\begin{aligned}\text{API graivty} &= \frac{141.5}{\text{specific gravity}} - 131.5\\ &= \frac{141.5}{0.86} - 131.5\\ &= 164.5 - 131.5\\ &= 33\end{aligned}$$

(vii) Neutralization Number

The neutralization number of an oil is described as the number of milligrams of KOH required to produce a pink color in 1 g of oil during titration using phenolphthalein as an indicator. The neutralization number indicates all the acidic constituents present in the oil. A rise in neutralization number does not always indicate that the lubricating oil should be changed. The more important consideration is that how much time it takes to reach the neutralization number to that level. In general, rise in neutralization number with increase in viscosity of the oil suggests that a serious condition has been reached and the lubricating oil needs to be examined.

(viii) Emulsification

According to old belief, water and oil do not mix. However, such a belief is true only under extraordinary conditions. Emulsification is the property of oil to mix with water forming an emulsion. In most of the machines, emulsification is undesirable because it causes:

- rust and corrosion in machine parts
- lubricant detonation with time
- replacement of lubricant from the frictional areas causing mechanical failure
- foaming in some cases.

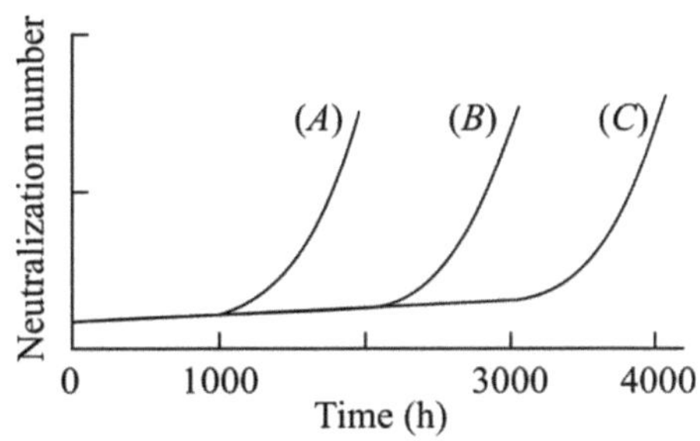

Fig. 8.12 Variation of neutralization number of oils (A to C) with time

The emulsification tendency of a lubricating oil can be determined by a test known as steam emulsion test. In this test 20 ml of oil is taken in a test tube and steam is passed through it for 4–6.5 seconds. The tube is then kept at 90 °C, and the time required for the oil to separate from water is monitored in seconds and called as SE number. For example, 20 ml of oil does not separate in 10 min (10 $\times$ 60 seconds); its SE number is reported as $10 \times 60 = 600$ plus. The SE number of an oil can be increased by adding certain types of additives.

(ix) Oxidation Stability Test

The oxidation stability test of a lubricating oil provides an idea about the stability of oil in the presence of air or other oxidizing conditions. The test is carried out by bubbling oxygen through a 300-ml sample of oil maintained at 90 °C. The rate at which neutralization number changes is monitored with time. The time (number of hours) between the start of the test and sharp rise in neutralization number indicates the oxidation stability of the test. The higher time required to reach sharp increase in neutralization number indicates more stable oil. A comparison of neutralization number of three oils with time at 90 °C is presented in Fig. 8.12, and obviously oil C is more stable than oils A and B.

(x) Friction and Wear Tests

A variety of friction and wear tests are available to determine the capacity of lubricants to resist friction and wear. A common test employed to determine wear is called as precision shell four-ball test. In this test relation of friction to wear is determined at different conditions. The test is carried out in an instrument having four balls of 1/2" diameter arranged in such a way so that an equilateral tetrahedron is formed as shown in Fig. 8.13.

The bottom three balls are fixed in a pot, whereas the top ball is rotated at different speed. About 10 ml of the lubricant oil is put in the pot, and at fixed temperature and load, the ball is allowed to rotate. The load is then increased and experiment is carried out for 1 minute. The ball is withdrawn and examined for

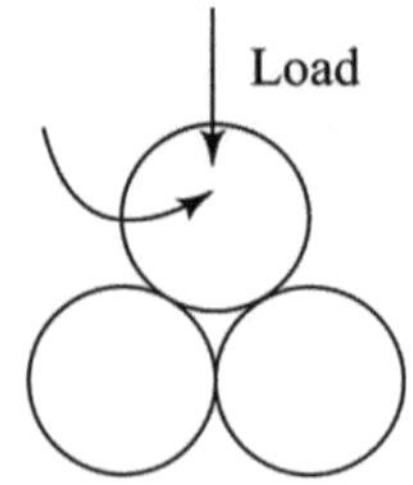

Fig. 8.13 Four-ball test used for frictional test

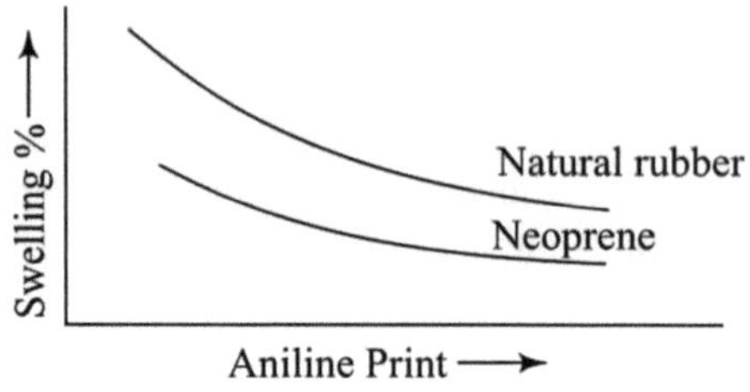

Fig. 8.14 Dependence of swelling on aniline point for natural rubber and neoprene

scratches, scale formation, etc. The test is capable of determining weld loads and seizure loads by calculating measured wear.

(xi) Aniline Point

The aniline point of a lubricant is defined as the lowest temperature at which equal volumes of a lubricating oil and aniline are completely miscible in each other. The aniline point is an important parameter of a lubricating oil as it provides an idea about the suitability of a lubricant in contact with rubber sealings or other rubber products.

Aromatic and napthenic oils possess a tendency to dissolve rubber material more in comparison with paraffinic base oils. Hence, the low aromatic or napthenic content in a lubricant will make it more useful. The higher aniline point of an oil indicates that it has less effect on rubber packings or other rubber materials.

Aniline point of a lubricant is determined by mixing equal volumes of aniline and oil sample in a tube. The mixture is then slowly heated till a homogenous solution is obtained. The tube is then allowed to cool slowly, and the temperature at which two phases separate out is recorded and called as aniline point of the lubricating oil. It has been found that increase in aniline point of a lubricant causes less effect on swelling of natural rubber or synthetic rubber material as shown in Fig. 8.14.

In spite of all the characteristics of a lubricant mentioned above, it is necessary that a proper lubricant is selected for a given requirement. The method of application, care of the lubricant, and care of the moving or sliding parts of the machine make a good lubrication.

Exercises

1. What is the difference between fluid film and boundary lubrication?
2. What is a lubricant? Explain its mechanism.
3. What is lubricant emulsion?
4. Write the type of lubricant normally used for:

   (a) Road roller
   (b) Lift
   (c) Door joints
   (d) Auto engine
   (e) Watch
   (f) Tire
   (g) Sewing machine
   (h) Transformers

5. What is the oiliness of lubricant oil?
6. Why graphite is used as lubricants on Moon?
   [*Hint:* Temperature on moon is 110 °C during day and −150 °C at night.]
7. Describe the method for determining the viscosity of a lubricant using Redwood viscometer.
8. Explain how pour point of a lubricating oil is different than cloud point.
9. What do you understand by the term "aniline point" of a lubricant? Why is it useful to know the aniline point?
10. Classify the following as solid, liquid, or gas lubricants:
    Graphite, wax, castor oil, silicones, lead iodide
11. Explain why a pretreatment is usually needed for metal surfaces before applying solid lubricants?
12. Can water act as a lubricant? If yes write two processes in which it is still used as a lubricant.
13. Describe:

    (i) Micelle
    (ii) Reverse Micelle

14. What are plastic lubricants?
15. Explain thin-film lubrication. How is it different from thick-film lubrication?

Chapter 9
Water Chemistry

Water is a vital and abundantly available substance. It forms an essential component of all living organisms, both animals and plants. It exists in the atmosphere as water vapor and can be sourced from various natural reservoirs such as rivers, lakes, seas, and wells. Among all forms, rainwater is considered the purest, as it results from the evaporation of surface water, leaving most impurities behind. However, with the rise in industrial activity, rainwater tends to absorb several gases from the atmosphere before falling to the ground. Consequently, the first showers of rain often contain high concentrations of dissolved gases like sulfur dioxide (SO_2), nitrogen dioxide (NO_2), and carbon dioxide (CO_2). With ongoing industrialization and changing societal practices, water pollution has significantly increased, making its prevention a critical concern. A prominent example of the adverse effects of pollution is the recent ruling by the Honorable Delhi High Court, which mandated the relocation of polluting industries away from residential areas to safeguard public health. In this chapter, we will explore various properties of water and the analytical techniques used to assess its purity.

9.1 General Properties of Water

Water plays a crucial role in regulating the Earth's climate and sustaining life. Chemically, water is represented by the formula H_2O, where the oxygen atom is located at the center of a tetrahedral arrangement, with two hydrogen atoms and two lone pairs of electrons occupying the four corners. Each point of this tetrahedron engages in hydrogen bonding with adjacent water molecules, giving rise to a complex structural network.

In its solid form, ice exhibits a well-organized and aesthetically pleasing structure. The oxygen atoms in ice are positioned at the vertices of hexagons, with each edge composed of a typical O—H bond and a hydrogen bond. The

© The Author(s), under exclusive license to Springer Nature Switzerland AG 2026
R. N. Goyal, H. Goel, *Textbook of Engineering Chemistry*,
https://doi.org/10.1007/978-3-032-22068-4_9

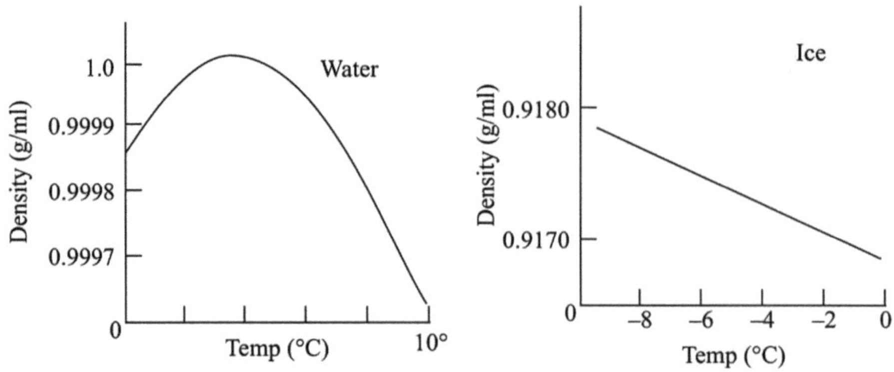

Fig. 9.1 Dependence of density on temperature for ice and water

structure resembles a repeating hexagonal pattern. The normal O—H bond length is approximately 99 pm, while the hydrogen bonds are about 170 pm long. Since hydrogen bonds are longer, they are inherently weaker than covalent bonds.

When ice melts into liquid water, the organized structure partially collapses but retains about 75–80% of its hydrogen bonds. This structural transformation accounts for many of the distinctive properties of water, particularly its behavior in terms of density. Water reaches its maximum density of 1.00 g/mL at 4 °C. As the temperature increases to 10 °C, the density slightly decreases to 0.9997 g/mL. Likewise, when the temperature drops below 4 °C, the density becomes less than 1.0000 g/mL, as illustrated in Fig. 9.1. Ice at 0 °C has a density of 0.9170 g/mL, and this density increases as the temperature continues to fall.

This unique property is highly beneficial, especially in colder regions. During winter, lakes and rivers cool down, causing denser water to sink and less dense water to rise. This circulation continues until the entire body of water stabilizes at 4 °C. Below this temperature, the colder water remains at the surface and begins to freeze, since water at 4 °C is denser than colder water.

In its pure state, water is a colorless, odorless, and tasteless liquid. It melts at 0 °C and boils at 100 °C under standard atmospheric pressure (760 mm Hg). Due to its polar nature, water has a high dielectric constant (around 80), which makes it an exceptional solvent.

9.2 Chemical Properties

1. **Stability and Decomposition:**
   Water is quite stable at atmospheric pressure (1 atm) and undergoes minimal decomposition—about 0.5%—into hydrogen and oxygen gases, following the reaction:

$$2H_2O \longrightarrow 2H_2 + O_2$$

2. **Reaction with Alkali Metals:**
   Water reacts readily with alkali metals even at room temperature, forming metal hydroxides and releasing hydrogen gas. For example, when a small piece of sodium is introduced into water, it reacts intensely:

$$2Na + 2H_2O \longrightarrow 2NaOH + H_2$$

   Lithium reacts more mildly, while potassium, rubidium, and cesium react with extreme violence.
3. **Reaction with Alkaline Earth Metals:**
   Alkaline earth metals typically react with water upon heating. Freshly cleaned beryllium or magnesium, when heated, reacts slowly and releases hydrogen gas:

$$Mg + 2H_2O \longrightarrow Mg(OH)_2 + H_2$$

   The other alkaline earth metals are highly reactive and must be stored carefully away from air and moisture.
4. **Reaction with Transition Elements:**
   Noble metals such as silver, gold, mercury, and members of the platinum group (except osmium) do not react with water. However, certain metals like zinc and iron react with steam to form metal oxides, as shown:

$$3Fe + 4H_2O \longrightarrow Fe_3O_4 + 4H_2$$

   Lead and copper can decompose water but only when subjected to extremely high temperatures.
5. **Water of Crystallization:**
   Water readily combines with many salts during the process of crystallization, forming compounds known as hydrates. Examples of common hydrates include copper(II) sulfate pentahydrate ($CuSO_4 \cdot 5H_2O$) and sodium carbonate decahydrate ($Na_2CO_3 \cdot 10H_2O$).

9.3 Hard Water and Soft Water

Water that produces lather easily with soap is referred to as *soft water*, whereas water that does not lather readily is called *hard water*. The hardness of water arises due to the presence of dissolved calcium and magnesium salts, which interfere with the action of soap.

Soap, such as sodium stearate ($C_{17}H_{35}COONa$), is the sodium salt of long-chain fatty acids. These sodium salts are soluble in water. However, in hard water, calcium or magnesium ions react with soap to form insoluble salts like calcium stearate, causing precipitation and preventing lather formation:

$$2C_{17}H_{35}COONa + CaCl_2 \longrightarrow (C_{17}H_{35}COO)_2Ca + 2NaCl$$

The precipitate of calcium stearate does not dissolve in water, and until all calcium ions are removed, soap will not form a lather. Water hardness is generally classified into two types: *temporary* and *permanent*.

Temporary hardness is caused by dissolved bicarbonates of calcium or magnesium. It can be removed by *boiling*, which converts the soluble bicarbonates into insoluble carbonates that can be filtered out:

$$Ca(HCO_3)_2 \longrightarrow CaCO_3 + CO_2 + H_2O$$
$$Mg(HCO_3)_2 \longrightarrow MgCO_3 + CO_2 + H_2O$$

Another effective method for removing temporary hardness is by adding *milk of lime* ($Ca(OH)_2$). This reacts with bicarbonates to produce insoluble carbonates:

$$Ca(HCO_3)_2 + Ca(OH)_2 \longrightarrow 2CaCO_3 + 2H_2O$$
$$Mg(HCO_3)_2 + Ca(OH)_2 \longrightarrow MgCO_3 + CaCO_3 + 2H_2O$$

Permanent hardness results from the presence of soluble calcium and magnesium salts like chlorides and sulfates. The following methods are commonly used to treat permanent hardness:

(a) **Washing Soda Method:** Adding sodium carbonate (Na_2CO_3), also known as washing soda, to hard water converts calcium and magnesium salts into insoluble carbonates:

$$Ca(HCO_3)_2 + Na_2CO_3 \longrightarrow CaCO_3 + 2NaHCO_3$$
$$MgCl_2 + Na_2CO_3 \longrightarrow MgCO_3 + 2NaCl$$
$$CaCl_2 + Na_2CO_3 \longrightarrow CaCO_3 + 2NaCl$$

(b) **Calgon Method:** Sodium hexametaphosphate, known as *Calgon* ($(NaPO_3)_6$), effectively removes calcium and magnesium ions. It works by exchanging these ions with sodium ions, which do not contribute to hardness:

$$(NaPO_3)_6 \longrightarrow 2Na^+ + (Na_4P_6O_{18})^{2-}$$
$$2Ca^{2+} + (Na_4P_6O_{18})^{2-} \longrightarrow (Ca_2P_6O_{18})^{2-} + 4Na^+$$

This reaction results in the formation of a soluble complex and the removal of hardness-causing ions (Fig. 9.2).

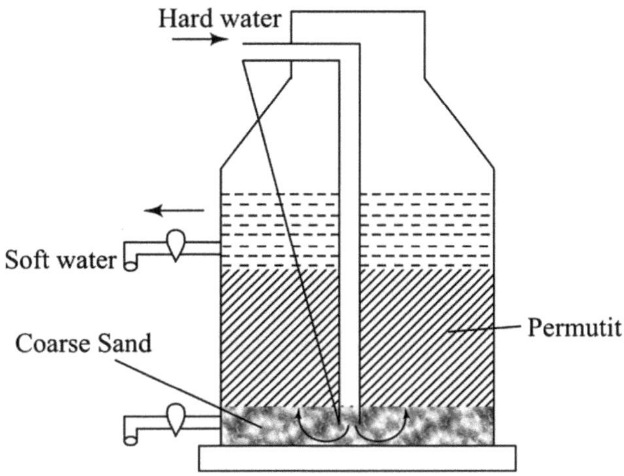

Fig. 9.2 Permutit process used for softening the water

(c) **Permutit Process (Ion Exchange Method):** *Permutit*, or sodium aluminum orthosilicate ($Na_2Al_2Si_2O_8 \cdot xH_2O$), is produced by fusing sodium carbonate, alumina, and silica. It is used in water softening by packing it in a treatment vessel. As hard water passes through, calcium ions are exchanged for sodium ions:

$$CaCl_3 + Na_2Al_2Si_2O_8 \longrightarrow Ca(Al_2Si_2O_8) + 2NaCl$$

The resulting product, *calcium zeolite*, reduces water hardness. However, over time, the permutit becomes saturated with calcium and loses effectiveness. To regenerate it, a sodium chloride (NaCl) solution is passed through the vessel, restoring the sodium ions:

$$Ca(Al_2Si_2O_8) + 2NaCl \longrightarrow Na_2(Al_2Si_2O_8) + CaCl_2$$

9.4 Degree of Hardness of Water

The *degree of hardness* refers to the concentration of calcium and magnesium salts present in a specific quantity of water. This is commonly expressed in terms of an equivalent amount of *calcium carbonate ($CaCO_3$)*. $CaCO_3$ is chosen as the reference substance because it is among the least soluble salts in water and has a molecular mass of 100 g/mol. Accordingly, the concentrations of various salts like bicarbonates, chlorides, and sulfates are converted into their $CaCO_3$ equivalents.

Hardness is typically reported in *parts per million (ppm)*, which indicates the number of milligrams of $CaCO_3$ per liter of water (i.e., mg/L) or, equivalently, parts of $CaCO_3$ per one million parts of water.

Example 9.1 A water sample contains 60 mg of $MgSO_4$ per liter. Calculate its hardness.

Solution Molecular weights: Mg = 24, S = 32

$$O_4 = 64 \longrightarrow MgSO_4 = 120 \ \text{g/mol}$$

$$CaCO_3 = 100 \ \text{g/mol}$$

So, 120 g of $MgSO_4 \equiv 100$ g of $CaCO_3$. Therefore, 60 mg of

$$MgSO_4 = (100/120) \times 60 = 50 \text{ mg}$$

Thus, the *hardness of the water sample is 50 ppm.*

The *total hardness* of water (accounting for both calcium and magnesium ions) can be determined by a *complexometric titration* using *EDTA (ethylenediaminetetraacetic acid)* as the titrant.

In this titration, the disodium salt of EDTA first forms a complex with Ca^{2+} ions, and then with Mg^{2+} ions. The titration employs *Eriochrome Black T* as the indicator. The end point is observed by a color change from *wine red to blue.*

If EDTA is represented by its disodium salt formula, $\mathbf{Na_2H_2X}$, the following complexation reactions occur:

- $Ca^{2+} + H_2X^{2-} \longrightarrow CaX^{2-} + 2H^+$
- $Mg^{2+} + H_2X^{2-} \longrightarrow MgX^{2-} + 2H^+$

The *volume of EDTA used* in the titration corresponds to the total hardness of the water. When using *0.01 M EDTA*, each *1 mL* of this solution is equivalent to *1.0 mg of* $CaCO_3$.

To determine *temporary hardness*, the water sample should be *boiled for 15–20 minutes*, then *cooled, and filtered.* A titration of this treated sample gives the *permanent hardness*. Hence:

$$\text{Temporary hardness (ppm } CaCO_3) = \text{Total hardness} - \text{Permanent hardness}$$

9.4.1 *Disadvantages of Hard Water*

The use of *hard water* in domestic life and industrial operations leads to several drawbacks. During bathing or laundering, hard water does not produce sufficient lather with soap, instead forming sticky precipitates of *calcium and magnesium salts*

Fig. 9.3 A section of boiler with $CaCO_3$ scales

of fatty acids. These insoluble compounds reduce the cleaning efficiency of soap and result in wastage.

When consumed for *drinking or cooking*, hard water may have harmful effects. The *calcium and magnesium ions* can accumulate in the form of *oxalates* in the kidneys, urinary tract, and other internal organs. For this reason, *hard water is not recommended for drinking*. Additionally, these dissolved salts raise the *boiling point* of water, meaning more *heat and energy* is needed for cooking food in hard water.

Several *industries* are adversely affected by hard water as well. In *paper manufacturing*, hard water interferes with the finishing process, reacting with certain chemicals and preventing the production of smooth, glossy paper. A far more critical issue arises in *industrial boilers* used for generating steam or hot water. As water continues to evaporate inside these boilers, *calcium and magnesium salts* get deposited along the interior surfaces.

Initially, salts that are *more soluble in hot water* but *less soluble in cold water* form loose, floating solids known as *sludge*. Over time, *hard, stubborn deposits* known as *scale* form and adhere tightly to the boiler walls (see Fig. 9.3).

Common salts responsible for *sludge* include $CaCl_2$, $MgCl_2$, and $MgSO_4$, while scales typically result from the precipitation of $CaCO_3$ or $CaSO_4$. The formation of $Mg(OH)_2$ can also contribute to scale buildup in the presence of Mg^{2+} ions. The reaction below illustrates scale formation due to decomposition of calcium bicarbonate:

$$Ca(HCO_2)_2 \longrightarrow CaCO_3 + H_2O + CO_2$$

Scales from $CaSO_4$ form because its *solubility decreases with temperature*—as the water heats up in boilers, $CaSO_4$ becomes less soluble and precipitates.

These scales drastically reduce the *thermal conductivity* of boiler walls, requiring greater fuel input to maintain the desired steam output. For example, a *scale layer of 0.25 mm* increases fuel consumption by *10%*, while a *1.0 mm thickness* demands *40% more fuel*. Further buildup can block valves and pipes, risking *equipment failure or explosion*. Therefore, *preventing scale formation* is critical for efficient fuel use and the safe operation of boilers.

9.5 Removal of Scales

Because scale formation inside boiler walls causes serious problems, it is essential either to *use soft water* or to *periodically remove the scales* that develop. One basic method of removing scales involves *mechanical cleaning* using tools like *sandpaper* and *wire brushes.*

However, when the scales are primarily made of *calcium sulfate* ($CaSO_4$), they tend to be extremely *hard and resistant*, requiring *chemical treatment.* One effective agent is *EDTA*, which works by forming *soluble complexes with calcium ions*, thereby dissolving the $CaSO_4$ deposits. In contrast, *calcium carbonate* ($CaCO_3$) scales can be removed by treating them with a *10% hydrochloric acid (HCl)* solution.

In certain cases, a *thermal shock method* is employed—by first *heating the boiler* and then *rapidly cooling* it with water. This sudden change in temperature causes *cracks in the scale layer* due to uneven expansion and contraction, making it easier to remove the deposits.

Since *prevention is more effective than removal*, various strategies are commonly recommended to *prevent scale formation.* The next section discusses several such preventive measures.

9.5.1 *Prevention of Formation of Scales*

To prevent scale formation in boilers, various methods are used that involve *chemical reactions* to convert relatively *insoluble compounds* like $CaCO_3$ and $CaSO_4$ into *loose precipitates* that do not stick to boiler surfaces.

(i) **Use of Sodium Carbonate:**
Adding *sodium carbonate* (Na_2CO_3) to boiler water helps prevent scale by reacting with *calcium sulfate* ($CaSO_4$). This reaction produces *calcium carbonate* ($CaCO_3$) as a soft, loose precipitate, which can be removed using *air-blowing techniques.*

$$CaSO_4 + Na_2CO_3 \longrightarrow CaCO_3 + Na_2SO_4$$

(ii) **Use of Calgon:**
Calgon, chemically known as *sodium hexametaphosphate* $(NaPO_3)_6$, is another effective scale-prevention agent. When added to boiler water, it forms *soluble complex compounds* with calcium sulfate, thus preventing it from precipitating. The reaction is

$$(NaPO_3)_6 \longrightarrow 2Na^+ + (Na_4P_6O_{18})^{2-}$$

$$2CaSO_4 + (Na_4P_6O_{18})^{2-} \longrightarrow 4Na^+ + (Ca_2P_6O_{18})^{2-}$$

This mechanism is similar to the method used for *removing hardness* from water.

(iii) **Use of Sodium Phosphates:**
Various *sodium phosphate compounds*, such as *disodium hydrogen phosphate* (Na_2HPO_4), *sodium dihydrogen phosphate* (NaH_2PO_4), and *trisodium phosphate* (Na_3PO_4), react with *calcium and magnesium salts* to produce *calcium or magnesium phosphate*. These form soft sludges that can be easily removed from the boiler:

$$3CaCl_2 + 2Na_3PO_4 \longrightarrow Ca_3(PO_4)_2 + 6NaCl \qquad \text{Calcium phosphate}$$

(iv) **Use of Organic Substances:**
At times, *organic additives* such as *kerosene*, *agar-agar*, and *tannin* are introduced into boiler water. These substances form a *coating around the precipitates*, making them *nonsticky*. As a result, the precipitates do not adhere to the inner surfaces and can be *periodically removed* from the boiler system.

9.6 Treatment of Hard Water

In industrial settings, the water fed into boilers is chemically treated to eliminate hardness, thus preventing the formation of scales. Two common methods employed for this purpose are:

(i) The Lime-Soda Process
(ii) The Zeolite Process

9.6.1 *Lime-Soda Process*

This method involves treating hard water with *lime [*$Ca(OH)_2$*]* and *soda ash [*Na_2CO_3*]*, which react with dissolved calcium and magnesium salts to produce *insoluble compounds*. Since lime and soda are used for softening, this method is referred to as the *lime-soda process* or simply *L–S process*.

The quantities of lime and soda required are determined using stoichiometric equations that describe the conversion of hardness-causing salts into precipitates. For example, in the presence of *calcium and magnesium bicarbonates*, the reactions are

$$Ca(HCO_3)_2 + Ca(OH)_2 \longrightarrow 2CaCO_3 + 2H_2O$$

$$Mg(HCO_3)_2 + 2Ca(OH)_2 \longrightarrow CaCO_3 + Mg(OH)_2 + 2H_2O$$

Hence, lime required

$$= (\text{Temporary } Ca^{2+} + 2 \times \text{Temporary } Mg^{2+})$$

and to express this in $CaCO_3$ equivalents:

$$= (74/100) \times (\text{Temporary } Ca^{2+} + 2 \times \text{Temporary } Mg^{2+})$$

because

$$74 \text{ g}Ca(OH)_2 \equiv 100 \text{ g } CaCO_3$$

In cases of *permanent hardness* due to salts like $CaCl_2, MgCl_2, CaSO_4, and\, MgSO_4$, the reactions with soda and lime are

$$Ca^{2+} + Na_2CO_3 \longrightarrow CaCO_3 + 2Na^+$$

$$Mg^{2+} + Ca(OH)_2 \longrightarrow Mg(OH)_2 + Ca^{2+}$$

So, total soda and lime needed

$$(\text{Permanent } Ca^{2+} + \text{ Permanent } Mg^{2+})$$

When converted to $CaCO_3$ equivalents:

$$= (74/100) \times \text{Permanent } Ca^{2+} + (106/100) \times \text{Permanent } Mg^{2+}$$

The same logic applies for any other impurities in water that contribute to hardness, and their chemical removal must be accounted for accordingly.

9.6.2 Types of Lime-Soda Processes

Cold Lime-Soda Process

In this method, water is treated at *room temperature* with pre-calculated amounts of lime and soda before entering the boiler. The treatment produces *fine precipitates* of calcium carbonate and magnesium hydroxide. However, due to their *very fine nature*, these precipitates do not settle easily. To aid in their removal, small amounts of *coagulants* like *sodium aluminate* ($NaAlO_2$) and *alum* are added. These substances undergo *hydrolysis* to produce a gelatinous precipitate of *aluminum hydroxide* [$Al(OH)_3$], which helps trap the fine particles:

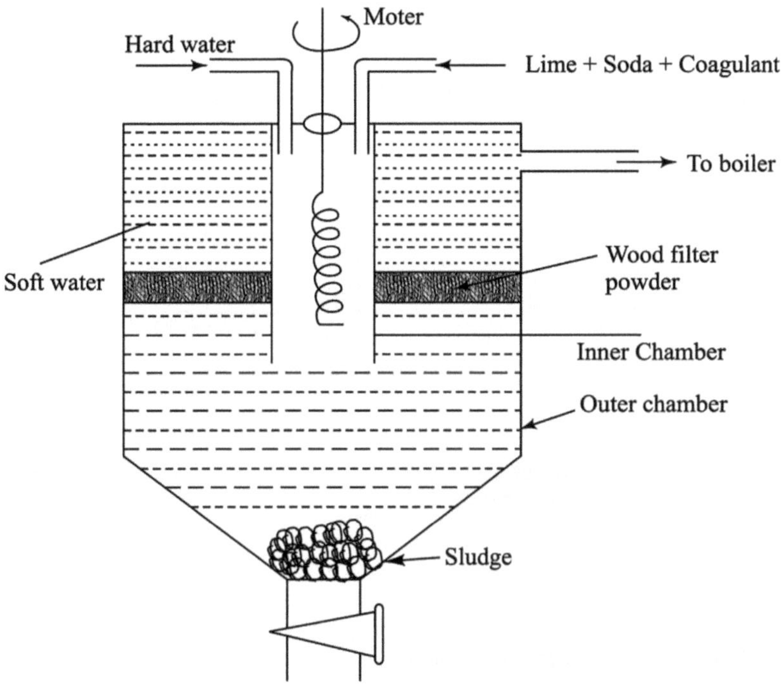

Fig. 9.4 A plant used for cold lime-soda process

$$NaAlO_2 + 2H_2O \longrightarrow NaOH + Al(OH)_3$$
(Sodium aluminate) (Aluminum hydroxide)

Figure 9.4 illustrates an industrial setup for the *cold lime-soda process*. The apparatus consists of *two chambers*—a small inner chamber for mixing and a larger outer one where settling occurs. A rotor in the smaller chamber ensures thorough mixing of lime, soda, and coagulant with the incoming water. As the reactions proceed, the *sludge* settles to the bottom and is periodically removed. The softened water passes through a *coarse filtration layer* to eliminate residual sludge and is then directed from the top outlet into the boiler.

Hot Lime-Soda Process

In the *hot lime-soda process*, hard water is treated with lime and soda at *elevated temperatures* (usually between *80 and 100 °C*). Operating at a higher temperature offers several benefits:

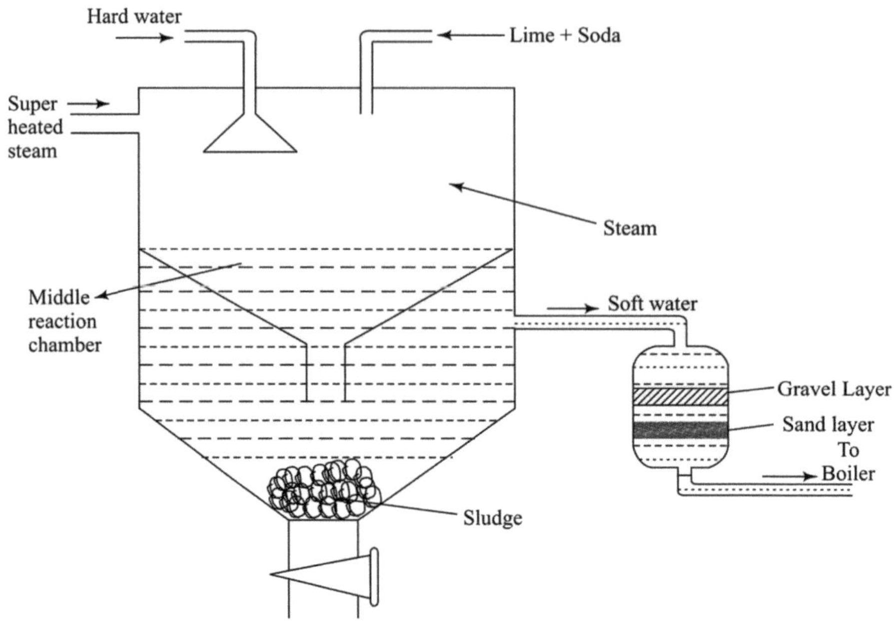

Fig. 9.5 Basic unit used for the hot lime-soda process

- The chemical reactions involved in softening occur more rapidly.
- The precipitates of calcium and magnesium compounds settle more readily, eliminating the need for coagulants.
- The overall *softening efficiency improves*, reducing hardness to approximately *15–20 ppm*.
- It becomes easier to *filter* the treated water.

The typical setup for this process, shown in Fig. 9.5, features *two chambers*. The first chamber is where hard water is mixed with lime and soda in the presence of *superheated steam*. Unlike the cold lime-soda method—where reactions occur in the inner chamber—here the chemical reactions take place in the *middle section* of the first chamber, which is separated from its base. The resulting soft water flows into a *second chamber* containing *sand and gravel layers* for fine filtration. The *sludge* that forms settles at the bottom and is periodically removed.

This method is *cost-effective* and also increases the *pH of water* due to the formation of hydroxides during treatment. The elevated pH minimizes *corrosion in pipelines* and reduces the number of *pathogenic microbes*. Nevertheless, the process has a few drawbacks:

- It generates a large amount of *sludge*, which presents a *disposal challenge*.
- It demands *careful monitoring* to ensure efficiency and safety.

Despite its benefits, it can only reduce hardness down to *15 ppm*, which is acceptable for boilers but still *higher than ideal* for long-term boiler performance.

9.6.3 Zeolite Process

Zeolites are *sodium aluminosilicates* with a *porous structure*, consisting of cavities and tunnels of various sizes (typically *280–1000 picometers* in diameter). This porous nature allows small molecules like H_2O and NH_3 to fit inside, making zeolites highly effective *drying agents*.

The structural formula of zeolites can be written as

$$Na_2O \cdot Al_2O_3 \cdot xSiO_2 \cdot yH_2O$$

The framework mainly contains SiO_4 *tetrahedra*, occasionally substituted with Al^{3+} in place of Si^{4+}, which introduces a negative charge. This charge is balanced by Na^+ *ions* embedded in the lattice.

Zeolites are used for water softening because they can *exchange sodium ions* (Na^+) with *calcium* (Ca^{2+}) and *magnesium* (Mg^{2+}) ions found in hard water. A common synthetic zeolite used industrially is *Linde A*, with the chemical formula:

$$Na_{12}(Al_{12}Si_{12}O_{48}) \cdot 27H_2O$$

The softening process involves the following reactions:

$$CaSO_4 + Na_2Ze \longrightarrow CaZe + Na_2SO_4$$

$$MgSO_4 + Na_2Ze \longrightarrow MgZe + Na_2SO_4$$

$$CaCl_2 + Na_2Ze \longrightarrow CaZe + NaCl$$

$$Ca(HCO_3)_2 + \underset{\text{(Sodium Zeolite)}}{Na_2Ze} \longrightarrow CaZe + 2NaHCO_3$$

Thus, *both temporary and permanent hardness* are efficiently removed using the *zeolite process* (Fig. 9.6).

The industrial setup is straightforward. It contains a *zeolite-filled column* layered over *gravel* (illustrated in Fig. 9.7). *Hard water is sprayed* onto the top, and as it passes through the zeolite, the Ca^{2+} *and* Mg^{2+} *ions are replaced by* Na^+ *ions*, effectively softening the water.

However, with prolonged use, the zeolite's *exchange capacity decreases*, as sodium ions are gradually replaced by calcium and magnesium. When this happens, the zeolite is *regenerated* using a *10% NaCl solution*. This process converts calcium or magnesium zeolite back into *sodium zeolite*:

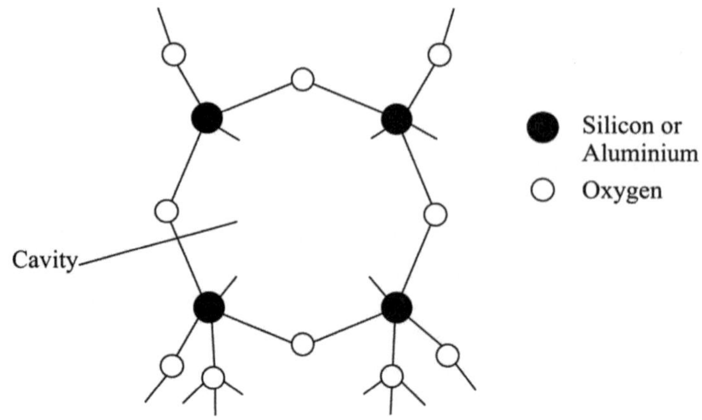

Fig. 9.6 Structure of a zeolite

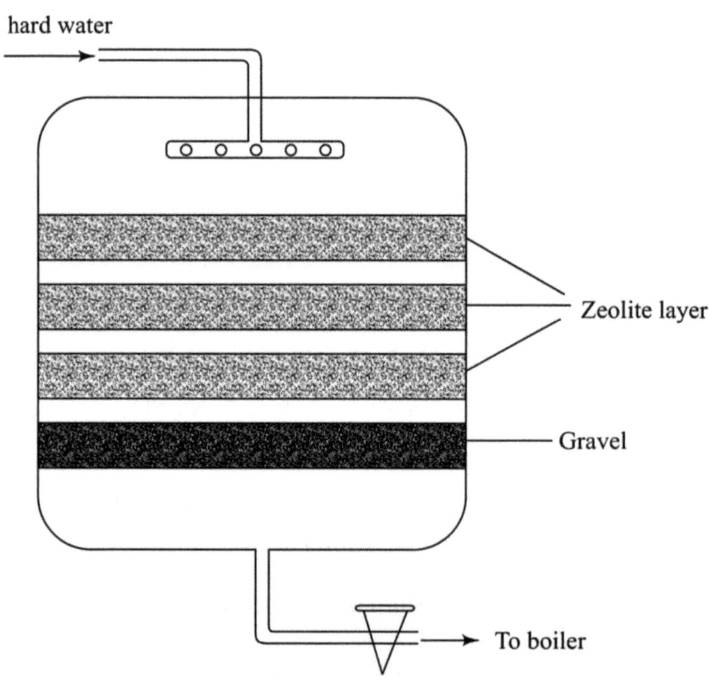

Fig. 9.7 Plant used in industries for zeolite process

$$CaZe + 2NaCl \longrightarrow Na_2Ze + CaCl_2$$

This regeneration allows the zeolite to be reused for further softening. The water produced typically has a *residual hardness of 10–15 ppm*. Despite its effectiveness, the *zeolite method is relatively expensive*, and the softened water contains a *high concentration of* Na^+ *ions*. Moreover, *acidic water* cannot be treated with this method, as it would *damage the zeolite structure.*

9.7 Ion Exchange Method

In various industrial processes, *deionized water* is essential. This type of water must be free from *both cations and anions*. One effective and straightforward way to achieve this is through the *ion exchange method.* Today, a wide range of *synthetic ion exchange resins* are commercially available. These resins are made of *long-chain organic polymers* that are cross-linked to form a *microporous network.*

Cation exchange resins contain acidic functional groups such as —*COOH* (carboxylic acid) or —SO_3H (sulfonic acid), which enable them to exchange H^+ *ions* for *metal cations* like Ca^{2+} and Mg^{2+} present in hard water. A commonly used resin in industries is *styrene-divinylbenzene copolymer*. When sulfonated or carboxylated, this resin becomes capable of replacing its H^+ ions with the calcium and magnesium ions in the water.

On the other hand, *anion exchange resins* have *basic groups*, such as *ammonium, amino*, and *quaternary ammonium*. After being treated with *NaOH*, these resins incorporate OH^- *ions* into their structure and can then exchange them with *anions* like Cl^-, SO_4^{2-}, and NO_3^- in hard water. A commonly used anion exchange resin is also a *styrene-divinylbenzene copolymer*, functionalized to contain these basic groups.

To produce *deionized water*, hard water is passed *first through a cation exchange resin chamber*, followed by an *anion exchange resin chamber*. The cation resin removes positively charged ions, and the anion resin removes negatively charged ones, yielding water that is virtually free of ions—*deionized or ion-free water*. A schematic of this system is shown in *Fig. 9.8*.

Over time, both resins become *exhausted*—the cation resin loses its H^+ ions and the anion resin loses its OH^- ions. To *regenerate* the cation resin, *dilute HCl or* H_2SO_4 is passed through the resin bed, replacing Ca^{2+} and Mg^{2+} ions with fresh

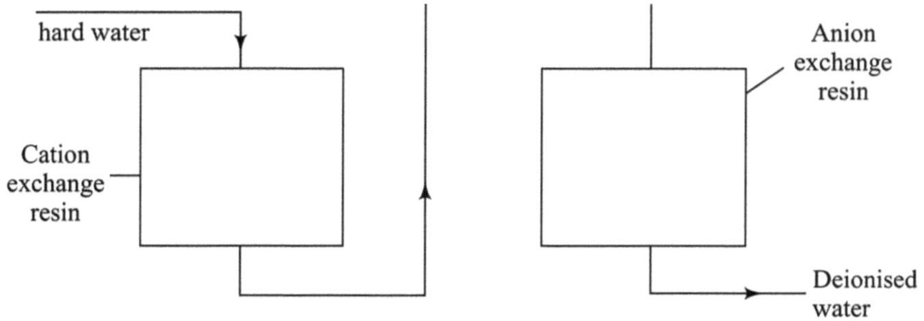

Fig. 9.8 A block diagram for preparing deionized water

H^+ ions. For the anion resin, *dilute NaOH* is used to restore OH^- ions. After regeneration, both resins are *washed with distilled water* and are ready to be reused for further deionization.

Ion exchange resins are not only important for producing deionized water but also play a key role in *removing toxic metal ions* and *hazardous anions* from water. Thus, they contribute significantly to *water purification* and *pollution control*, which will be explored further in the section on environmental concerns.

9.8 Water Treatment for Industrial Purposes: Boiler Feed Water

Water used in industries must meet specific quality standards, as it often contains the following impurities:

(a) Suspended or dissolved substances such as clay, silica, and salts
(b) Dissolved gases like CO_2, NO_2, SO_2, and O_2
(c) Oils, greases, and emulsions

These contaminants can lead to *operational issues*, particularly *scaling*, where metal salts deposit inside tubes and drums, reducing efficiency. Therefore, *water must be treated* before it is used in industrial systems.

9.8.1 *Issues Arising from Heating Water*

In many industrial settings, water is converted into *steam* by boiling it in boilers. This process can lead to several challenges:

(1) **Sludge and Scale Formation**
When hard water is boiled, *bicarbonates* decompose to produce *insoluble carbonates*:

$$Ca(HCO_3)_2 \longrightarrow CaCO_3 + CO_2 + H_2O$$

Additionally, evaporation causes even *soluble salts* to precipitate, leading to the formation of oxides like Fe_2O_3, Al_2O_3, MgO, and CaO:

$$Fe + 3H_2O \longrightarrow Fe_2O_3 + 6H$$
$$Ca^{2+} + SO_4^{2-} \longrightarrow CaSO_4$$

These compounds contribute to *scale*, which is hard and adherent.

Disadvantages of Scale

(1) Scale has *poor thermal conductivity*, reducing boiler efficiency and increasing fuel consumption.
(2) It prevents *direct contact* between water and heating surfaces, delaying boiling.
(3) Scale can also cause *corrosion* of heating elements.

Scale-Prevention Methods

(1) Adding *colloidal agents* like kerosene to coat particles and inhibit aggregation.
(2) Using *lime-soda (L–S) softening*, which causes precipitation of Fe, Al, and Mg hydroxides:

$$Al^{3+} + 3OH^- \longrightarrow Al(OH)_3$$

$$HCO_3^- + 2OH^- \longrightarrow CO_3^{2-} + H_2O$$

(3) Applying *hot L–S softening* at $\sim$50 °C enhances precipitation and also removes silica.
(4) Employing *ion exchange softening*, which removes all cations, thereby preventing scale.

(2) **Corrosion**
Corrosion in boilers occurs due to:
(a) Hydrolysis of salt-forming acids
(b) Hydrolysis of fatty oils
(c) Formation of galvanic cells
(d) Dissolved gases like CO_2 and O_2

Corrosion Prevention

- Neutralizing acids with bases
- Using *oxygen scavengers* such as Na_2SO_3 and NH_2NH_2:

$$2Na_2SO_3 + O_2 \longrightarrow 2Na_2SO_4$$

$$NH_2NH_2 + O_2 \longrightarrow N_2 + 2H_2O$$

- CO_2 is removed using ammonia:

$$CO_2 + 2NH_3 + H_2O \longrightarrow (NH_4)_2CO_3$$

(3) **Priming and Foaming**

- *Priming* is the carryover of *water droplets with steam*, often caused by high concentrations of *alkali sulfates or chlorides*.
- These droplets can trap solid matter, leading to *foaming*.

Prevention Measures

(a) Provide sufficient *steam space*
(b) Remove *oils and soaps* with alum and soda ash
(c) Eliminate *saline content* through blow-off
(d) Prevent accumulation of *suspended solids*

(4) **Caustic Embrittlement**
In L–S softening, Na_2CO_3 is often added. At high temperatures, it *hydrolyzes*:

$$Na_2CO_3 + H_2O \longrightarrow 2NaOH + CO_2$$

The resulting *NaOH* can seep into microscopic cracks in boiler walls. As water evaporates, NaOH *concentrates* and reacts with the boiler's iron, forming *sodium ferrate*. This causes *stress corrosion* and *crack formation*—a phenomenon called *caustic embrittlement.*
This typically occurs near *riveted joints*, leading to *intergranular cracks.*

Prevention Add inhibitors like:

- $NaOH : NaNO_3$ in *1:4 ratio*
- $Na_2SO_4 : NaOH$ in *1:1 ratio*

(5) **Silica Removal**
At high pressure, *silica* can vaporize and be carried with steam, contributing to *priming*. It also reacts with Ca or Al to form *hard silicates* that deposit on surfaces.

Silica can be removed by:

(i) *Hot L–S softening* with added *magnesia*, which either *adsorbs* silica or forms *magnesium silicate*
(ii) Using $Fe_2(SO_4)_3$ with $Ca(OH)_2$ at *pH 9* to form $Fe(OH)_3$, which coagulates suspended material:

$$Fe_2(SO_4)_3 + 3Ca(OH)_2 \longrightarrow 2Fe(OH)_3 + 3CaSO_4$$

(iii) **Fluosilicate Process**:

- Convert silica to *fluosilicic acid* using *NaF* and a *cation exchanger*:

$$SiO_2 + NaF + 3H_2Re \longrightarrow Na_2Re + H_2SiF_6 + 2H_2O$$

- Then pass the water through a *weak base anion exchanger* to remove fluosilicic acid.

9.9 Water Quality

Water intended for *drinking* must be adequately purified to ensure safety and acceptability. The standard *requirements* for potable water are as follows:

- It should be *colorless and odorless*
- The *pH* value should lie between *7.0 and 8.0*
- It should be *reasonably soft*, with total hardness *not exceeding 500 ppm*
- It must be *free from suspended impurities*
- It should be *free from pathogenic microorganisms*

In urban areas, *municipal corporations* are responsible for water supply. They typically treat water to remove *suspended solids* and *harmful microbes*, but they *do not address hardness*. If hard water is supplied, users should take additional steps to make it drinkable. A common and simple method is to *boil the water for 5–10 minutes* and then allow it to *cool* before consumption. This process helps in reducing hardness, making the water suitable for drinking.

Municipal water purification generally involves two major steps.

9.9.1 Removal of Suspended Impurities

Water is stored in *large settling tanks* (approximately 10 × 10 × 5 meters) and allowed to remain undisturbed for *4–6 hours*. Under the influence of gravity, heavier suspended solids settle at the bottom. If the particles are *colloidal in nature*, they remain dispersed and do not settle. In such cases, *coagulants* like *alum and* $FeSO_4$ are added. These neutralize the *negative charge* on colloidal particles, causing them to aggregate and settle. The clear water above (supernatant) is then transferred to another tank for *disinfection*.

9.9.2 Removal of Microorganisms

Three major methods are used to eliminate microbes from water:

(a) **Using Bleaching Powder**

Bleaching powder ($CaOCl_2$) reacts with water to release *chlorine gas*, which subsequently forms *hypochlorous acid* (*HClO*)—a strong disinfectant:

$$CaOCl_2 + H_2O \longrightarrow Ca(OH)_2 + Cl_2$$

$$Cl_2 + H_2O \longrightarrow HCl + HClO \qquad \text{(Hypochlorous acid)}$$

The standard dosage is *1 kg per 10^6 liters* of water. Despite its effectiveness, bleaching powder has *two key drawbacks*:

(1) It contributes *Ca^{2+} ions*, thereby *increasing water hardness*
(2) Over time, *stored bleaching powder loses chlorine*, diminishing its disinfecting ability. Hence, it is essential to *test for available chlorine* before use to ensure effectiveness.

(b) **Using Chlorine Gas**
To overcome the limitations of bleaching powder, *chlorine gas* or *liquid chlorine* is more commonly used. When dissolved in water, chlorine also forms *hypochlorous acid*, which efficiently kills bacteria:

$$Cl_2 + H_2O \longrightarrow HCl + HClO$$

Even *small doses* of chlorine are sufficient, making this method both *cost-effective* and *efficient*. As a result, it is the *preferred disinfection method* adopted by most *municipal corporations.*

(c) **Using Ozone**
Ozone (O_3), a blue-colored allotrope of oxygen, is another disinfectant used for water treatment. It is generated by passing *oxygen gas through an electric discharge*, following the reaction:

$$3O_2 \longrightarrow 2O_3$$

Though ozone is *unstable* and decomposes back to oxygen, this breakdown releases *nascent oxygen [O]*, a *highly reactive oxidizing agent* that rapidly kills bacteria:

$$O_3 \longrightarrow O_2 + [O] \qquad \text{(Nascent oxygen)}$$

For disinfection, *2–5 ppm* of ozone is sufficient, with a *contact time of 10–15 minutes*. A major *advantage* of using ozone is that *excess ozone does not leave harmful residues*, as it simply reverts to oxygen. *European countries* commonly use this method for purifying drinking water due to its safety and efficiency.

9.10 Water Pollution and Testing

With rising consumption of water for *domestic* and *industrial* uses, *water pollution* has emerged as a serious issue. *Domestic sewage* generally contains *nontoxic organic matter*, whereas *industrial effluents* may include a broad range of *hazardous chemical pollutants*. Therefore, it is essential to *test industrial wastewater* before it is released into the environment. Several parameters are used to assess the *quality of wastewater*, and depending on the level and type of contaminants, appropriate *treatment methods* are applied.

The *World Health Organization (WHO)* has set *permissible limits* for various toxic elements commonly found in water, as shown below.

Toxic elements	Maximum allowed (ppm)
Lead	0.05
Chromium	0.05
Arsenic	0.05
Cadmium	0.01
Copper	1.50
Iron	5.00
Zinc	1.50

Additionally, the *permissible limits* for *cyanide* and *fluoride* ions are *0.20 ppm* and *1.50 ppm*, respectively. Two other important indicators of water pollution are the *biochemical oxygen demand (BOD)* and *chemical oxygen demand (COD)*. Their maximum permissible levels are *6 ppm* for BOD and *10 ppm* for COD.

9.11 Biochemical Oxygen Demand (BOD)

The *BOD* of a water sample refers to the *amount of oxygen* (in mg or ppm) needed to *biologically degrade* the organic matter present in *one liter* of that water. Organic contaminants require oxygen for their *oxidation*, and a higher oxygen demand reflects *greater pollution*:

$$\text{Organic material} + O_2 \longrightarrow CO_2$$

To calculate the BOD of wastewater, one must first determine the *dissolved oxygen (DO)* content. The procedure for measuring DO is outlined below.

9.11.1 Determination of Dissolved Oxygen (DO)

The *dissolved oxygen* content in water varies based on its *source*. *Industrial wastewaters* usually have *very low DO* levels. The process for measuring DO involves the following steps:

- Collect the water sample in *special DO bottles* (sealed to prevent air contact).
- Add *1.0 ml of M/50* $MnSO_4$ solution and *1.0 ml of 1 M alkaline KI* solution.
- Immediately *stopper* and shake the bottle to mix the contents.

- A *precipitate* will form; dissolve it by adding *1.0–2.0 ml of concentrated* H_2SO_4.
- The resulting solution is titrated with *M/40 sodium thiosulfate* ($Na_2S_2O_3$).
- Convert the result to ppm for one liter of sample:

$$1 \text{ ml of M/40 } Na_2S_2O_3 \equiv 0.2\text{ppm } O_2$$

For example, if the titration requires *50 ml* of thiosulfate, then:

$$DO = 50 \times 0.2 = 10\text{ppm}$$

9.11.2 Reactions Involved in DO Determination

(1) **Initial Dissociation:**

$$MnSO_4 \longrightarrow Mn^{2+} + SO_4^{2-}$$

$$NaOH + KI \longrightarrow K^+ + I^- + OH^- + Na^+$$

(2) **Formation of Precipitates:**

$$Mn^{2+} + 2OH^- \longrightarrow Mn(OH)_2$$

(white precipitate in the absence of O_2)

$$Mn^{2+} + 2OH^- + \frac{1}{2}O_2 \longrightarrow MnO_2 + H_2O$$

(brown precipitate in the presence of O_2)

(3) **Oxidation of Iodide Ions:**

$$MnO_2 + 2I^- + 4H^+ \longrightarrow I_2 + Mn^{2+} + 2H_2O$$

(4) **Titration with Sodium Thiosulfate:**

$$I_2 + 2Na_2S_2O_3 \longrightarrow 2NaI + Na_2S_4O_6$$

(Starch is used as an indicator in the titration.)

Note: A *higher DO value* indicates *less polluted water* and better *aerobic conditions*, making it more suitable for supporting aquatic life.

9.11.3 Determination of BOD of Water

The *biochemical oxygen demand (BOD)* of a wastewater sample represents the *amount of oxygen required* to biologically oxidize the *organic pollutants* present. This oxidation process is carried out by *microorganisms* over a period of *5 days at a constant temperature*. Therefore, the *selection and addition of microorganisms* is a crucial step. In samples that naturally lack these microbes, a process known as *seeding* is carried out by adding suitable microorganisms. However, in the case of *domestic sewage or surface water*, natural populations of microbes are typically sufficient, and seeding is unnecessary.

For samples that require seeding, *12 ml* of *settled domestic wastewater* (which has been stored for at least *24 hours at 20 °C*) serves effectively as a microbial seed. The procedure for determining BOD involves the following steps:

(1) **Preparation of Dilution Water**
To dilute various wastewater samples appropriately, a *dilution water* is first prepared by mixing the following:

- 1.0 ml of *phosphate buffer (pH 7.2)*
- 1.0 ml of *magnesium sulfate solution (22.5 g/liter)*
- 1.0 ml of *calcium chloride solution (27.5 g/liter)*
- 1.0 ml of *ferric chloride solution (0.25 g/liter)*

This mixture is made up to *1 liter* with distilled water. Then, *2.0 ml of settled domestic wastewater* is added as a seed, and the mixture is *aerated for 5–10 minutes* by bubbling air through it.
The *wastewater sample* is now diluted using this prepared dilution water. The *degree of dilution* depends on how polluted the sample is:

- For *industrial wastewater*, a typical dilution is *1.0 ml sample to 100 ml.*
- For *polluted river water*, between *25 and 50 ml* of sample is diluted to *100 ml.*

(2) **Measurement of DO in the Undiluted Sample**
Determine the *dissolved oxygen (DO)* in the *original (undiluted)* wastewater sample using the method described earlier. Denote this as S_1.

(3) **Incubation and DO Measurement**
Now, place the *diluted wastewater* and a sample of the *dilution water alone* in *incubation* at *20 °C for 5 days*. After incubation, measure the DO values:

- S_2 = DO of *diluted wastewater* after incubation
- B_1 = DO of *dilution water* (without sample) after incubation

9.11.4 BOD Calculation

The *BOD* (in ppm or mg/liter) is calculated using the formula:

$$BOD = [(B_1 - S_2) \times 100\%] - (B_1 - S_1)$$

where:

- B_1 = DO of dilution water after incubation
- S_1 = DO of undiluted wastewater
- S_2 = DO of diluted wastewater after incubation

For instance, if the BOD of a sample is *50 ppm*, it means that *50 mg of oxygen* is required to completely *biodegrade* the organic content present in *one liter* of that wastewater sample.

9.12 Chemical Oxygen Demand (COD)

Chemical oxygen demand (COD) is a commonly used parameter to assess the *pollution level* in a wastewater sample. COD is often preferred over *biochemical oxygen demand (BOD)* because it allows for the *complete oxidation* of all organic substances to *carbon dioxide and water* using *strong chemical oxidants*. The COD test employs *potassium dichromate* in the presence of *concentrated sulfuric acid* to chemically oxidize the organic matter. The COD determination process consists of two main steps.

9.12.1 Step 1: Preparation of Reagents

Prepare the following standard solutions:

- 0.25 N Potassium dichromate ($K_2Cr_2O_7$)
- 0.25 N Ferrous Ammonium Sulfate ($FeSO_4 \cdot (NH_4)_2SO_4 \cdot 6H_2O$)

The *ferrous ammonium sulfate* solution is *standardized* using *ferroin indicator*, and the end point of the titration is marked by the appearance of a *red color*.

9.12.2 Step 2: Oxidation and Titration

Take a *known volume* (5–50 ml) of the wastewater sample in a *round-bottom flask* and *dilute to 50 ml* with distilled water. Then, add *1 g each* of $HgSO_4$ and Ag_2SO_4, followed by *5 ml of concentrated* H_2SO_4. Allow the mixture to cool.

Next, slowly add *25 ml of 0.25 N* $K_2Cr_2O_7$ and *70 ml of concentrated* H_2SO_4 to the mixture. Attach a *reflux condenser* and *reflux* the mixture for *2–3 hours* (as illustrated in Fig. 9.9). After cooling, *titrate* the *excess dichromate* with *standard ferrous ammonium sulfate* using ferroin indicator.

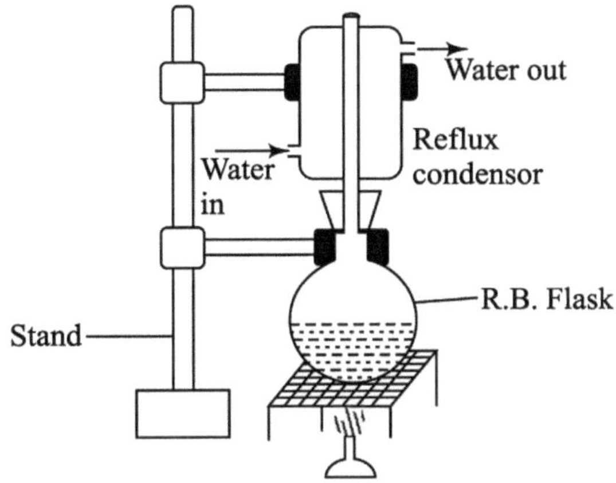

Fig. 9.9 Experimental setup used for determining COD

A *blank experiment* is also conducted using *distilled water*, and the reading from this is used for comparison.

9.12.3 COD Calculation

The COD (in mg/L or ppm) is calculated using the formula:

$$\text{COD} = \frac{(B - S) \times N \times 8 \times 1000}{\text{Volume of sample (ml)}}$$

where:

- B = Volume of ferrous ammonium sulfate used for blank (in ml)
- S = Volume used for the sample (in ml)
- N = Normality of ferrous ammonium sulfate

9.12.4 Determination of Toxic Elements in Water and Wastewater Samples

A commonly employed technique for identifying *toxic metal ions* is *atomic absorption spectroscopy (AAS)*. This method is *accurate*, detects *ppm-level concentrations*, and is suitable for a wide variety of metals. However, the equipment is *expensive* and requires *specialized maintenance* (Fig. 9.10).

(i) **Polarographic Method**
The *polarographic technique* is a *cost-effective* alternative that can also detect metal ions at *ppm levels*. This method relies on the *electrochemical reduction*

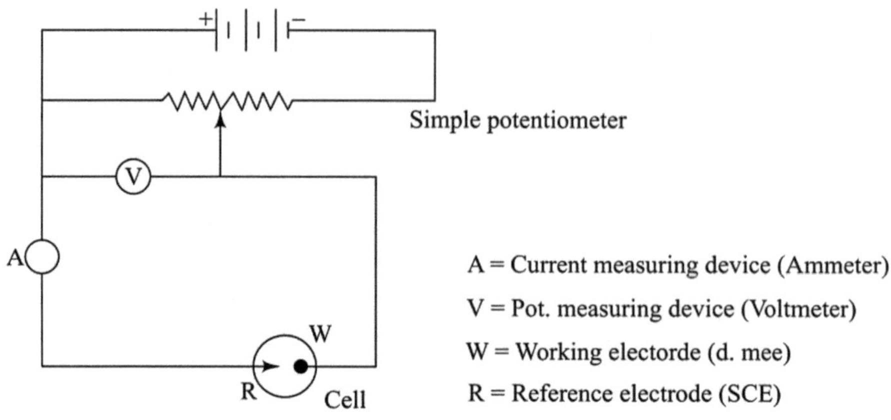

Fig. 9.10 A simple circuit of manual polarograph

of metal ions by applying a potential. For example, the presence of Cd^{2+} *ions* in wastewater leads to their reduction as follows:

$$Cd^{2+} + 2e^{-} \rightarrow Cd$$

Since each metal ion has a *unique reduction potential*, they can be *identified* through this technique. The setup includes:

- A *dropping mercury electrode (DME)* as the *working electrode*
- A *saturated calomel electrode (SCE)* as the *reference electrode*

The *water sample* is placed in the electrochemical cell, along with a few ml of *1 M KCl* as the supporting electrolyte. The *potential is gradually varied*, and the *resulting current* is recorded. This gives a *polarogram*, an *S-shaped curve* that is characterized by two main parameters:

- $E_{1/2}$ = Half-wave potential (specific to the metal ion)
- i_d = Diffusion current (proportional to metal ion concentration)

The *diffusion current* (i_d) is described by the *Ilkovic equation*:

$$i_d = 607nD^{\frac{1}{2}}Cm^{\frac{2}{3}}t^{\frac{1}{6}}$$

where:

- i_d = diffusion current (in μA)
- n = number of electrons involved
- D = diffusion coefficient (cm^2/s)
- C = concentration of metal ion (mol/L)
- m = rate of mercury flow (mg/s)
- t = drop time (s)

A *calibration curve* is prepared by plotting i_d values against known concentrations of metal ions. The concentration in unknown samples is determined by comparing the measured i_d to this plot.
Metal ions like Pb^{2+}, Zn^{2+}, Cd^{2+}, Cu^{2+}, Fe^{3+}, and others can be detected this way. However, a *limitation* is that the method cannot *distinguish* between metal ions with $E_{1/2}$ *values that differ by less than 100 mV.*
Here is a reference table showing $E_{1/2}$ values for some metal ions in *0.1 M KCl.*

Metal Ion	$-E_{1/2}, mV$
Cd^{2+}	600
Co^{2+}	1200
Cr^{2+}	340
Fe^{2+}	1300
Ni^{2+}	1100
Pb^{2+}	400
Zn^{2+}	995

(ii) **Spectroscopic Method**
Spectrophotometry is widely applied to detect *metal ions* that form *colored complexes* with specific reagents:

- For instance, Fe^{2+} ions form a *reddish-orange complex* with *1,10-phenanthroline*, which absorbs strongly at *510 nm.*
- Similarly, Cu^{2+} forms a *yellow complex* with *2,9-dimethylphen anthroline*, which is soluble in a *chloroform-methanol mixture* and can be measured at *457 nm.*

9.13 Wastewater Treatment

This section addresses the critical issue of *wastewater treatment* in the context of *pollution control.* Although the treatment method depends on the *source and composition* of the wastewater, it generally falls into two categories:

(i) Domestic waste water
(ii) Industrial waste water

The *degree of contamination* from *metal ions* and *organic pollutants* in *industrial effluents* varies significantly depending on the type of industry. For instance, *textile industries* discharge large amounts of *dyes and toxic chemicals*, whereas *paper mills* release *phenols*, *chlorinated phenols*, and other *organic compounds.*

9.13.1 Domestic Waste Water

Domestic wastewater usually contains *nonhazardous organic matter* and typically has a *BOD* (biochemical oxygen demand) value of *less than 200 ppm*. It is generally treated through a *two-stage process*.

Stage 1: Physical Separation

This step focuses on *removing solid and suspended impurities*. The wastewater is passed through a *fine mesh screen*, and the *settled sludge* is collected in a *settling chamber*.

Stage 2: Biological Decomposition

In this stage, the wastewater is *exposed to activated sludge* that contains *microorganisms* such as *Zoogloea*, *Pseudomonas*, etc., in the presence of *ample oxygen*. These microbes *decompose* the organic material, resulting in the formation of *solid particles* that *settle at the bottom* and are removed.

The treated water after this step has a *much lower BOD value* of around *25–30 ppm*, making it *safe for discharge* into rivers. In certain situations, an *additional treatment* using *calcium hydroxide* may be applied to *eliminate phosphate ions* before final discharge.

9.13.2 Industrial Waste Water

Most *industrial wastewaters* contain *toxic* and *nonbiodegradable* substances, making it essential to *treat them properly* before discharging into *rivers*. One of the most commonly used techniques for eliminating *toxic organic compounds* is the application of *activated charcoal*, illustrated in Fig. 9.11. In this process, the effluent is passed through a *column* filled with *activated charcoal* possessing a *large surface*

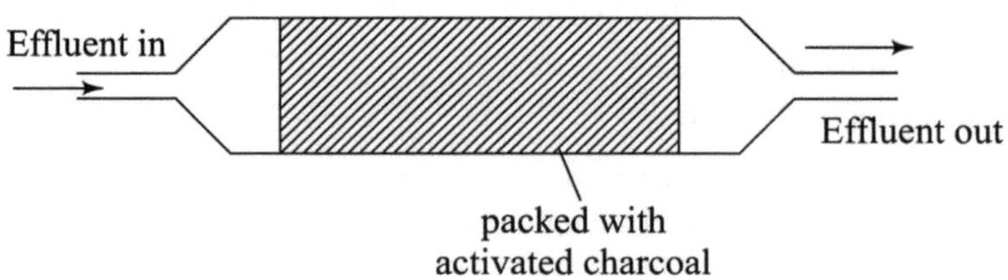

Fig. 9.11 Use of activated charcoal packed column in removing toxic substances

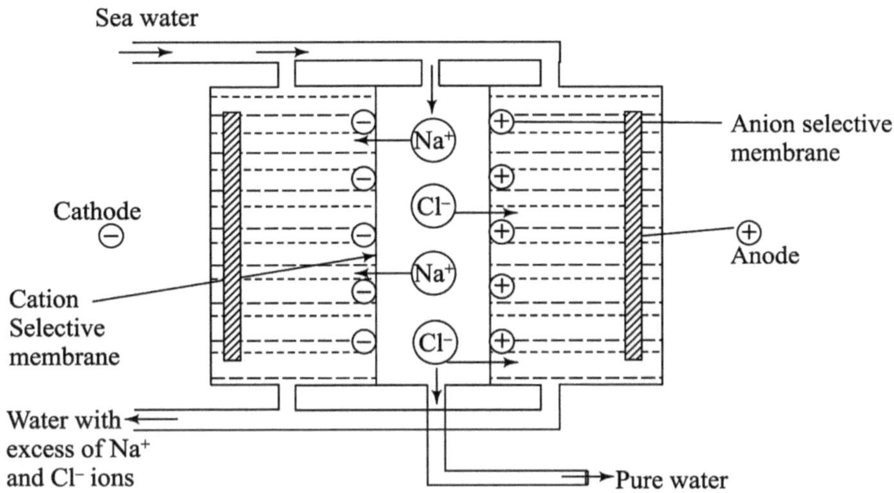

Fig. 9.12 Electrodialysis used for preparation of fresh water from seawater

area. Activated charcoal has a *high capacity for adsorption* of a wide range of organic substances and therefore effectively removes the *major nonbiodegradable toxic compounds* from the effluent.

It has been found that *pesticides* such as *DDT*, *aldrin*, and *dieldrin* can be efficiently removed by passing the effluent through a *column of activated charcoal.*

If the industrial effluent contains a *high concentration of metal ions*, it is passed through *ion exchangers* composed of *synthetic resins*. Commonly used *cation exchange resins* are based on *styrene-divinylbenzene copolymers*, while *anion exchange resins* contain OH^- *groups*. These resins function effectively, and their working principle and structure have already been discussed in the section on *deionized water*.

Another approach to *metal ion removal* in industrial setups is *electrodialysis*, where the effluent flows through an *electric field*. In this setup, *anions* move toward the *anode* and *cations* toward the *cathode*, allowing both types of ions to be extracted.

Electrodialysis is also applied for the *desalination of seawater* in *coastal regions*. Seawater, which contains high concentrations of Na^+ and Cl^- along with other ions, is passed through large chambers fitted with *cation-selective and anion-selective membranes* (see Fig. 9.12). When an *electric current* is applied and seawater is allowed to pass through, *anions and cations* are separated, resulting in the production of *fresh water*. The *side chambers*, receiving Na^+ and Cl^- ions from the central chamber, become more saline. These can be *recirculated* through the system to recover more *pure water*.

As the *demand for clean water* continues to grow, it is important to *curb water pollution* and ensure the *judicious use of water resources*. The *Ganga pollution control initiative* is a notable government effort, undertaken not only for *environmental reasons* but also due to the river's *religious significance*.

9.14 Useful Hints for Numerical on Water Chemistry

9.14.1 Units of Hardness

(1) *1 mg/liter* is equivalent to *1 part per million (ppm)*.
(2) **Degree Clark** (°**Cl**): It denotes the number of *grains of* $CaCO_3$ present in *one gallon of water*.
(3) **Part per million (ppm)**:

$$1 \text{ ppm} = 1 \text{ part of } \mathrm{CaCO_3} \text{ in } 10^6 \text{ parts of water}$$

(4) Degree French (°**Fr**): Represents *1 part of* $CaCO_3$ equivalent hardness in 10^6 *parts of water*.

Total Hardness

This is the *sum of temporary and permanent hardness* in water. It is typically determined by performing a *complexometric titration* of the hard water sample using *EDTA as titrant* and *Eriochrome Black T as the indicator*. The hardness, expressed in terms of $CaCO_3$, can be calculated using the formula:

$$\text{hardness in ppm} = \frac{\text{ml of EDTA used } \times 100 \times M}{\text{ml of sample taken}} \times 1000$$

where M is the *molarity of EDTA* solution.

Alkalinity

Alkalinity in water is attributed to the presence of *bicarbonates*, *carbonates*, and *hydroxides* of alkali and alkaline earth metals. These are quantified by titration with a *standard acid*, using *phenolphthalein* and *methyl orange* indicators.

- *Phenolphthalein* gives a *pink color* in the presence of OH^- and CO_3^{2-} salts.
- *Methyl orange* produces a *yellow-orange color* in the presence of OH^-, CO_3^{2-}, and HCO_3^- ions.

The values obtained from phenolphthalein and methyl orange titrations are referred to as *HpH alkalinity* and *MeOH alkalinity*, respectively.

9.14.2 Determination of Phenolphthalein (HpH) and Methyl Orange (MeOH) Alkalinity

To determine alkalinity, take *100 ml of water* in a *conical flask*, add a few drops of *phenolphthalein indicator*, and titrate the solution with *N/50* H_2SO_4 until the *pink color disappears*. The *HpH alkalinity*, expressed in ppm, is calculated as

$$\{P\} = \frac{\text{Volume of acid used} \times \text{Normality of acid} \times 10^6}{1000 \times \text{ml Volume of sample taken}}$$

Next, to the same flask, add a few drops of *methyl orange indicator* and continue titrating with the same standard acid until the end point is reached. The *total alkalinity or MeOH alkalinity*, also in ppm, is calculated as

$$\{M\} = \frac{\text{Total amount of acid used for MeOH} \times \text{Normality of acid} \times 10^6}{1000 \times \text{Volume of water sample}}$$

Based on the values of *P* and *M*, different forms of alkalinity can be determined using the following table.

Titration results (HpH and MeOH)	OH^- ppm	CO_3^{2-} ppm	HCO_3^- **ppm**
$P = 0$	NIL	NIL	M
$P = M$	*P* or *M*	NIL	NIL
$P = 1/2M$	NIL	$2P$	NIL
$P > 1/2M$	$2(P\text{-}M)$	$2(M\text{-}P)$	NIL
$P < 1/2M$	NIL	$2P$	$M\text{-}2P$

Note: The simultaneous presence of OH^- and HCO_3^- is not possible because OH^- *converts* HCO_3^- *into* CO_3^{2-} through the following reaction:

$$OH^- + HCO_3^- \longrightarrow CO_3^{2-} + H_2O$$

Additionally, it is not possible for *all three species (*OH^-*,* CO_3^{2-}*, and* HCO_3^-*)* to coexist in a solution simultaneously.

9.14.3 Determination of Chlorides

After completing the methyl orange (MeOH) alkalinity titration, add a few drops of potassium chromate (K_2CrO_4) indicator to the solution. Then titrate with a standard silver nitrate ($AgNO_3$) solution until the color changes from yellow to reddish brown. The chloride ion concentration (Cl^-) in ppm can be calculated using the following formula:

$$Cl^- \text{in ppm} = \frac{\text{Volume of } AgNO_3 \times \text{ Normality of } AgNO_3 \times 35.5 \times 10^6}{1000 \times \text{Volume of water sample}}$$

9.14.4 Determination of Lime and Soda Requirement

Lime ($Ca(OH)_2$) is required for the following purposes:

1. **To neutralize free acid**:

$$2H^+ + 3Ca(OH)_2 \longrightarrow 3Ca^{2+} + 2H_2O$$

2. **To coagulate alum and iron(II) sulfate:**

$$Al_2(SO_4)_3 + 3Ca(OH)_2 \longrightarrow 2Al(OH)_3 + 3CaSO_4$$
$$FeSO_4 + Ca(OH)_2 \longrightarrow CaSO_4 + Fe(OH)_2$$

To react with CO_2 and bicarbonate ions:

$$CO_2 + Ca(OH)_2 \longrightarrow CaCO_3 + H_2O$$
$$2HCO_3^- + Ca(OH)_2 \longrightarrow CaCO_3 + CO_3^{2-} + H_2O$$

3. **To precipitate magnesium ions:**

$$Mg^{2+} + Ca(OH)_2 \longrightarrow Mg(OH)_2 + Ca^{2+}$$

Soda ash (Na_2CO_3) is required:

4. **To remove calcium ions by precipitation:**

$$Ca^{2+} + Na_2CO_3 \longrightarrow CaCO_3 + 2Na^+$$

9.14.5 Hints for Solving Numericals

(1) Units of impurities should be converted into *ppm*, *grains/gallon*, or *degree Clark* for consistency.

(2) **Substances that do not contribute to hardness** include:

$$KCl,\ NaCl,\ Na_2SO_4,\ SiO_2,\ Fe_2O_3,\ K_2SO_4,\ etc.$$

(3) **Convert all substances to $CaCO_3$ equivalent** using the formula:

$$\text{Substance (as } CaCO_3 \text{ equivalent)} = \left(\frac{\text{Weight of substance}}{\text{Molecular weight}}\right) \times 100$$

9.14.6 *Formula for Lime Requirement*

Lime requirement (as $CaCO_3$) is calculated using

$$\begin{aligned}&\text{Lime (mg/l)}\\&= \frac{74}{100}\Big[\text{Temp. } Ca^{2+} + 2 \times \text{Temp. } Mg^{2+}\\&\qquad + \text{Perm. } \left(Mg^{2+} + Fe^{2+} + Al^{3+}\right) + CO_2 + HCl + H_2SO_4\\&\qquad + HCO_3^- - NaAlO_2 + OH^-\Big]\end{aligned}$$

All terms must be *expressed in $CaCO_3$ equivalents.*

9.14.7 *Formula for Soda Requirement*

Soda requirement (as $CaCO_3$) is calculated using

$$\begin{aligned}&\text{Soda (mg/l)}\\&= \frac{106}{100}\Big[\text{Perm. } Ca^{2+} + \text{Perm. } Mg^{2+} + Al^{3+} + Fe^{2+} + HCl + H_2SO_4\\&\qquad - HCO_3^- - NaAlO_2 - O_3^{2-} + OH^-\Big]\end{aligned}$$

9.14.8 *Conversion Units*

- 1 ppm = 1 mg/L
- 1 ppm = 0.07° Clark
- 1 ppm = 0.1° French

Solved Problems

1. Calculate temporary hardness and total hardness of a sample of water containing $Ca\,(HCO_3)_2 = 16.2$ mg/l; $MgHCO_3 = 7.3$ mg/l; $CaSO_4 = 13.6$ m/l and $MgCl_2 = 9.5$ mg/l.

Salts	Quantity	Multiplication factor	In terms of $CaCO_3$
$Ca\,(HCO_3)_2$	16.2 mg/l	100/162	10 mg/l
$Mg\,(HCO_3)$	7.3 mg/l	100/146	5 mg/l
$CaSO_4$	13.6 mg/l	100/136	10 mg/l
$MgCl_2$	9.5 mg/l	100/95	10 mg/l

Solution Temporary Hardness $= 10 + 5 = 15$ mg/l or ppm
Permanent Hardness $= 10 + 10 = 20$ mg/l or ppm
Total Hardness $= 15 + 20 = 35$ mg/l or ppm.

2. How many grams of $FeSO_4$ are dissolved per liter to give 210.5 ppm of hardness? (Atomic weight: Fe = 56, S = 32, O = 16, Ca = 40, C = 12)

Solution $FeSO_4$ in terms of $CaCO_3$

$$100 \text{ ppm of } CaCO_3 \equiv 136 \text{ ppm of } FeSO_4$$

$$210.5 \text{ ppm} \equiv \frac{136 \times 210.5}{100}$$

$$= 286.3 \text{ ppm of } FeSO_4$$

3. Calculate the temporary and permanent hardness of water sample having the following analysis:

$Mg\,(HCO_3)_2 = 73$ mg; $Ca\,(HCO_3)_2 = 162$ mg/l;

$CaSO_4 = 136$ mg/l; $MgCl_2 = 95$ mg/l;

$NaCl = 100$ mg/l; $CaCl_2 = 111$ mg/l

S. no.	Salts	Quantity	M.F.	In terms of $CaCO_3$
1	$Mg\,(HCO_3)_2$	73 mg/l	100/146	50 mg/l
2	$Ca\,(HCO_3)_2$	162 mg/l	100/162	100 mg/l
3	$CaSO_4$	136 mg/l	100/136	100 mg/l
4	$MgCl_2$	95 mg/l	100/95	100 mg/l
5	$CaCl_2$	111 mg/l	100/111	100 mg/l
6	NaCl	100 mg/l	×	×

Solution

$$\text{Temporary hardness: } 1 + 2 = 50 + 100 = 150 \text{ mg/l or ppm}$$

$$\begin{aligned}\text{Permanent hardness} &= 3 + 4 + 5 = 100 + 100 + 100\\ &= 300 \text{ mg/l or ppm.}\end{aligned}$$

Note: NaCl does not contribute to hardness.

4. 100 ml of water sample has a hardness due to 12.5 ml 0.08 $NMgSO_4$. Calculate the hardness in ppm.

Solution

$$\begin{aligned}12.5 \text{ ml of } 0.08 \text{ N } MgSO_4 &\cong 12.5 \times 0.08 \text{ ml N } MgSO_4\\ &\equiv 1 \text{ mlNMgSO}_4\\ \text{eq. wt of } CaCO_3 &= 50\end{aligned}$$

therefore,

$$1 \text{ ml N } MgSO_4 \cong 1 \text{ ml} \times 50 \text{ mg } CaCO_3 \text{ eq.} \cong 50 \text{ mg } CaCO_3,$$

i.e., in 100 ml of water sample hardness of $MgSO_4$ = 50 mg in terms of $CaCO_3$. For

$$\begin{aligned}1000 \text{ ml} &= \frac{50 \times 1000}{100} = 500 \text{ mg/l } CaCO_3\\ &= 500 \text{ ppm of } CaCO_3\end{aligned}$$

5. Calculate the quantity of lime and soda for softening 50,000 liters of water sample having $MgCO_3$ = 144 ppm; $CaCO_3$ = 25 ppm, $MgCl_2$ = 95 ppm, $CaCl_2$ = 111 ppm; Fe_2O_3 = 25 ppm; Na_2SO_4 = 15 ppm.

S. no.	Salts	Quantity	M.F.	In terms of $CaCO_3$
1	$MgCO_3$	144 ppm	100/84	171.4 ppm
2	$CaCO_3$	25 ppm	100/100	25.0 ppm
3	$MgCl_2$	95 ppm	100/95	100.0 ppm
4	$CaCl_2$	111 ppm	100/111	100.0 ppm

Solution Requirement of lime

$$= \frac{74}{100}\left\{2 \times MgCO_3 + CaCO_3 + MgCl_2\right\} \times 50{,}000$$
$$= \frac{74}{100}\{2 \times 171.4 + 25 + 100\} \times 50{,}000$$
$$= 17{,}309{,}000 \text{ mg} = 17.3 \text{ kg}.$$

Requirement of soda

$$= \frac{106}{100}\{171.4 + 100 + 100\} \times 50{,}000$$
$$= \frac{106}{100}\{371.4\} \times 50{,}000$$
$$= 19{,}684{,}200 \text{ mg}$$
$$= 19.7 \text{ kg}$$

6. Calculate the amount of lime and soda required for softening of 50,000 liters of water having the following analysis results.
$Mg^{2+} = 12$ ppm; $Ca^{2+} = 40$ ppm; $HCO_3^- = 164.7$ ppm; $CO_2 = 30.8$ ppm

Ions	Quantity	M.F.	In terms of $CaCO_3$
Mg^{2+}	12 ppm	100/24	50
Ca^{2+}	40 ppm	100/40	100
HCO_3^-	164.7 ppm	100/61	135
CO_2	30.8 ppm	100/44	70

Solution Lime requirement

$$= \frac{74}{100}\left\{HCO_3^- + Mg^{2+} + CO_2\right\} \times 50{,}000$$
$$= \frac{74}{100}\{135 + 50 + 70\} \times 50{,}000$$
$$= 9{,}435{,}000 \text{ mg}$$
$$= 9.43 \text{ kg}$$

7. Calculate the quantity of lime-soda requirement for 200,000 liters of water. The water is coagulated with 16.4 ppm of $NaAlO_2$. The data of analysis of water is Raw water:

$$Ca^{2+} = 160 \text{ ppm; } Mg^{2+} = 72 \text{ ppm; } HCO_3^- = 732 \text{ ppm; } CO_2 = 44 \text{ ppm}$$

Softened water:

$$CO_3^{2-} = 30 \text{ ppm; } OH^- = 17 \text{ ppm}$$

Ion/salts	Quantity	M.F.	For lime requirement	For soda requirement
Ca^{2+}	160	100/40	–	400
Mg^{2+}	72	100/24	300	300
HCO_3^-	732	100/(61 × 2)	600	−600
CO_2	44	100/44	100	–
$NaAlO_2$	16.4	100/(82 × 2)	−10	−10
CO_3^{2-}	30	100/60	–	−50
OH^-	17	100/(17 × 2)	50	50
			Total = 1040	90

Solution Lime requirement

$$= \frac{74}{100} \times 1040 \times \frac{200{,}000}{10^6} \text{ kg} = 153.92 \text{ kg}$$

Soda requirement

$$= \frac{106}{100} \times 90 \times \frac{200{,}000}{10^6} \text{ kg} = 19.08 \text{ kg}$$

8. 10,000 liters of hard water was made soft with zeolite. The recharging of zeolite required 8 liters of NaCl solution having 150 g/l of NaCl. Calculate the hardness of water.

Solution

$$\text{Amount of NaCl in 8 liters of solution}$$
$$= 8 \text{ liter} \times 150 \text{ g/l}$$
$$= 1200 \text{ gms. in 8 liters}$$

i.e., for 10,000 liters of hard water, NaCl required = 1200 gms.

$$\text{In terms of } CaCO_3 = \frac{1200 \times 100}{58.5 \times 2} \text{ g/10,0001}$$

$$\text{for 1 liter the hardness} = \frac{1200 \times 100}{58.5 \times 2 \times 10{,}000} \text{ g/l}$$
$$= \frac{6}{58.5} \text{ g/l}$$
$$= 0.1025 \text{ g/l}$$

or 102.5 ppm.

9. A water reservoir supplies 1 m^3/s of water. The analysis of water reservoir gave the following results:
   $Mg(HCO_3)_2$ = 219 ppm;
   Mg^{2+} = 36 ppm;
   HCO_3^- = 18.3 ppm and H^+ = 1.5 ppm.
   Calculate the cost of treatment of water per day if lime (90% pure) and soda (95% pure) cost Rs. 500 per ton and Rs. 7,000 per ton, respectively.

Salts	Quantity	M.F.	In terms of $CaCO_3$
$Mg(HCO_3)_2$	219 ppm	$\frac{100}{146}$	150 ppm
Mg^{2+}	36 ppm	$\frac{100}{24}$	150 ppm
HCO_3^-	18.3 ppm	$\frac{100}{122}$	15 ppm
H^+	1.5 ppm	$\frac{100}{2}$	75 ppm

Solution Lime required:

$$\frac{74}{100}[2 \times 150 + 150 + 15 + 75] = \frac{74}{100} \times 540 = 399.60 \text{ mg/l}$$

$$\begin{aligned} \text{in 1 day} &= 1\left(\text{m}^3/\text{s}\right) \times 60 \times 60 \times 24 \times 399.60 \\ &= 3.45 \times 10^7 \text{ m}^3/\text{day} \quad (\text{as } 1 \text{ m}^3 = 10^3 \text{ liter}) \\ &= 3.45 \times 10^{10} \text{ liter/day} \\ &= 34.5 \text{ tonnes.} \end{aligned}$$

Soda required:

$$\frac{106}{100}[150 + 76 - 15] = 222.6 \text{ mg/l}$$

$$\begin{aligned} \text{In 1 day} &= 1(\text{m}^3/\text{s}) \times 60 \times 60 \times 24 \times 222.6 \\ &= 1.92 \times 10^7 \text{ m}^3/\text{day} \\ &= 1.92 \times 10^{10} \text{ liter/day} \\ &= 19.2 \text{ ton} \end{aligned}$$

Cost of treatment 1 day

$$\begin{aligned} &= 34.5 \times \frac{100}{90} \times 500 + 19.2 \times \frac{100}{95} \times 7000 \\ &= 19{,}166 + 141{,}473 \\ &= \text{Rs. } 160{,}640.00 \end{aligned}$$

10. A sample of water was found to contain the following species:

$$\begin{aligned} &Ca^{2+} = 40 \text{ mg/l}, \quad Mg^{2+} = 24 \text{ mg/l}, \\ &Na^{+} = 8.05 \text{ mg/l}, \quad HCO_3^{-} = 18.3 \text{ mg/l} \\ &SO_4^{2-} = 55.68 \text{ mg/l and } Cl^{-} = 6.74 \text{ mg/l} \end{aligned}$$

Solution Calculate alkalinity and hardness and amount of salts present

Ions	Quantity	M.F.	In terms of $CaCO_3$
Ca^{2+}	40	$\frac{100}{40}$	100 ppm
Mg^{2+}	24	$\frac{100}{24}$	100 ppm
Na^+	8.05	$\frac{100}{23 \times 2}$	17.5 ppm
HCO_3^-	18.3	$\frac{100}{61 \times 2}$	150 ppm
SO_4^{2-}	55.68	$\frac{100}{96}$	58 ppm
Cl^-	6.74	$\frac{100}{35.5 \times 2}$	9.5 ppm

does not take part in hardness

Calcium alkalinity	100 ppm
Magnesium alkalinity	50 ppm
Total alkalinity	150 ppm
Total hardness	200 ppm
Ca temporary hardness	100 ppm
Mg temporary hardness	50 ppm
Mg permanent	50 ppm
Salts are	$Ca\ (HCO_3)_2 = 100$ ppm
	$Mg\ (HCO_3)_2 = 50$ ppm
	$MgSO_4 = 50$ ppm
	$NaCl = 9.5$ ppm

11. A water sample has no phenolphthalein alkalinity. The 100-ml sample requires 16.9 ml of 0.02 N (N/50) HCl with methyl orange. Calculate the type and amount of alkalinity present.

Solution $P = 0$ and $M = 16.9$, so the alkalinity is due to HCO_3^- ions:

$$1 \text{ ml of } 1 \text{ N HCl} \equiv 50 \text{ mg } CaCO_3$$

$$16.9 \text{ ml of } \frac{N}{50} \text{ HCl} \equiv 50 \times 16.9 \frac{N}{50} = 16.9 \text{ mg } CaCO_3$$

So in 100 ml of water sample, the alkalinity = 16.9 mg/s

$$\text{for 1 liter of water sample the alkalinity} = \frac{16.9 \times 1000}{100} = 169 \text{ mg/l}$$

12. 100 ml of a water sample required 4.7 ml of N/50 HCl using HpH indicator. When a few drops of MeOH are added in the same solution and titrated, the change from yellow color to red required 10.5 ml of N/50 HCl. What are the type and amount of alkalinity present in sample?

Solution

$$P = 4.7 \text{ ml} \quad M = 15.2 \text{ ml}$$

as $P < \frac{1}{2}M$ so sample contain CO_3^{2-} and HCO_3^- alkalinity.

$\Rightarrow$ Volume of $\frac{N}{50}$ HCl for CO_3^{2-} in 100 ml of water sample = $2P$

$$= 2 \times 4.7 = 9.4 \text{ ml}$$

Since 1 ml of 1 NHCl $\equiv$ 50 mg $CaCO_3$

$$9.4 \text{ ml} \frac{N}{50} \text{ HCl} = 50 \times 9.4 \times \frac{1}{50} = 9.4 \text{ mg } CaCO_3$$

in 1 liter = 94 ppm

$\Rightarrow$ Volume of $\frac{N}{50}$ HCl for HCO_3^- in 100 ml of water sample

$$= M - 2P = 5.8 \text{ ml}$$

1 ml of 1N HCl = 50 mg $CaCO_3$

$$5.8 \text{ of } \frac{N}{50} \text{ HCl} = 50 \times \frac{N}{50} = 5.8 \text{ mg}$$

in one liter = 58 ppm

Total alkalinity = 94 + 58 = 152 mg/l or ppm.

13. A water sample, using $FeSO_4 \cdot 7H_2O$ as a coagulant in the rate of 139 ppm, gave the following results. Ca^{2+} = 160 ppm, CO_2 = 88 ppm, Mg^{2+} = 72 ppm, HCO_3^- = 488 ppm
Calculate lime and soda required to soften 100,000 liters of water.

Ions	Quantity	M.F.	In terms of $CaCO_3$
Ca^{2+}	160 ppm	$\frac{100}{40}$	400 ppm
Mg^{2+}	88 ppm	$\frac{100}{24}$	300 ppm
CO_2	72 ppm	$\frac{100}{44}$	200 ppm
HCO_3^-	488 ppm	$\frac{100}{61 \times 2}$	400 ppm
$FeSO_4 \cdot 7H_2O$	139 ppm	$\frac{100}{278}$	50 ppm

Solution Lime required

$$= \frac{74}{100} \begin{matrix} Mg^{2+} \; CO_2 \; HCO_3^- \; FeSO_4 \\ (300 + 200 + 400 + 50) \end{matrix} \frac{74}{200} \{950\} = 703 \text{ ppm}$$

For

$$100{,}000 \text{ liters} = \frac{703 \times 100{,}000}{10^6} = 70.3 \text{ kg}$$

Soda required

$$= \frac{106}{100} \begin{matrix} Ca^{2+} \; Mg^{2+} \; FeSO_4 \; HCO_3^- \\ (300 + 400 + 50 - 400) \end{matrix} \frac{106}{100} \{350\} = 371 \text{ ppm}$$

For

$$100{,}000 \text{ liters} = \frac{371 \times 100{,}000}{10^6} = 37.1 \text{ kg}$$

14. A zeolite softener was 90% exhausted, when 10,000 liters of hard water was passed through it. The softener required 200 liters of NaCl solution having 50 g/l NaCl. What is the hardness of water?

Solution Amount of NaCl used for 10,000 liters of water = 200 × 50 = 10,000 g

$$\text{in terms of } CaCO_3 = \frac{10{,}000 \times 100}{58.5 \times 2}$$
$$= 8547 \text{ g } CaCO_3$$

$$\text{For 1 liter of hard water} = \frac{8547}{10{,}000} = 0.8457\ \text{g/l}$$
$$= 845.7\ \text{mg/l}\ \text{ or ppm}$$

15. 0.28 g of $CaCO_3$ was dissolved in 1 liter of water. 100 ml of this solution on titration required 28 ml of EDTA. 100 ml of unknown hard water sample required 33 ml of same EDTA. After boiling and cooling, 100 ml of this unknown sample required 10 ml EDTA. Calculate the temporary and permanent hardness.

Solution

$$\text{1-liter sample has} = \frac{0.28 \times 100}{100} CaCO_3 = 0.28\ \text{g in terms of}\ CaCO_3$$
$$1\ \text{-ml sample has} = \frac{0.28}{1000} = 0.28\ \text{mg}\ CaCO_3\ \text{eg.}$$
$$28\ \text{ml EDTA} \equiv 100\ \text{ml of}\ 0.28\ \text{mg}\ CaCO_3\ \text{eg.}$$
$$= 28\ \text{mg}\ CaCO_3$$
$$1\ \text{ml EDTA} = 1\ \text{mg}\ CaCO_3 \tag{9.1}$$

For Unknown Solution:

$$\text{Total hardness 100 ml of unknown water} = 33\ \text{ml EDTA}$$
$$= 33\ \text{mg}\ CaCO_3 \quad \text{(from Eq. 9.1)}$$
$$\text{1000 ml of unknown} = \frac{33 \times 1000}{100}$$
$$= 330\ \text{mg}\ CaCO_3\ \text{Eq.}$$

Permanent Hardness:

$$\text{100 ml of boiled unknown water} = 10\ \text{ml EDTA}$$
$$= 10\ \text{mg}\ CaCO_3\ \text{eq.}$$
$$\text{1000 ml of boiled unknown} = \frac{10 \times 1000}{100}$$
$$= 100\ \text{mg}\ CaCO_3\ \text{eq.}$$

Temporary Hardness:

$$= 330 - 100 = 230\ \text{mg} \cdot CaCO_3$$

Exercises

1. Calculate the temporary and permanent hardness of a water sample, having the following analysis: $Mg(HCO_3)_2 = 73$ mg, $Ca(HCO_3)_2 = 162$ mg, $CaSO_4 =$ 136 mg, $MgCl_2 = 95$ mg, $CaCl_2 = 111$ mg, $NaCl = 100$ mg.
2. 100 ml of a raw water sample on titration with $N/50$ H_2SO_4 required 12.4ml of the acid to HpH end point and 15.2 ml of the acid to MeOH end point. Determine the type and extent of alkalinity present in the water sample.
3. A water sample is not alkaline with HpH. However, 100 ml of the sample, on titration with $N/50$ HCl, required 16.9 ml to obtain the end point, using MeOH as indicator. What are the types and amount of alkalinity present in the sample?
4. A water sample is alkaline with both HpH and MeOH.100 ml of the sample, on titration with $N/50$ HCl, required 4.7 ml to obtain the end point, using HpH as an indicator and 10.5 ml for MeOH. What are the types and amount of alkalinity in the sample?
5. A water sample is alkaline with both HpH and MeOH.100 ml of the sample, on titration with $N/50$ HCl, required 5.8 ml to obtain the end point, using HpH as indicator and 11.6 ml using MeOH indicator. What are the types and amount of alkalinity in the sample?
6. A water sample is alkaline with both HpH and MeOH.100 ml of the sample, on titration with N/50 HCl, required 5.0 ml to obtain the end point, using HpH as indicator and 5.0 ml for MeOH indicator. Determine the type and amount of alkalinity in the sample.
7. Calculate the amount of lime (84%) and soda (92%) required for treatment of 20,000 liters of water whose analysis is as follows:
   $Mg(HCO_3)_2 = 36.5$ ppm, $Ca(HCO_3)_2 = 40.5$ ppm,
   $CaSO_4 = 34.0$ ppm, $MgSO_4 = 30$ ppm,
   $CaCl_2 = 27.75$ ppm, $NaCl = 10$ ppm.
   Also calculate the temporary and permanent hardness of the sample.
8. A water sample, on analysis, gave the following data:
   $MgCO_3 = 84$ ppm, $CaCO_3 = 40$ ppm,
   $CaCl_2 = 5.5$ ppm, $Mg(NO_3)_2 = 37$ ppm, $KCl = 20$ ppm.
   Calculate the amount of lime (86% pure) and soda (83% pure) required for the treatment of 80,000 liters of water.
9. A water sample, on analysis, gave the following data:
   $MgSO_4 = 120$ ppm, $CaSO_4 = 272$ ppm,
   $MgCl_2 = 95$ ppm, $H_2SO_4 = 49$ ppm, $SiO_2 = 4$ ppm.
   Calculate the amount of lime (95% pure) and soda (97% pure) needed for the treatment of 100,000 liters of water. If the costs of lime and soda are Rs. 40 and Rs. 2000 per 100 kg, respectively, calculate the total cost of chemicals required for treatment of 100,000 liters of water.
10. A water sample, on analysis, gave the following data in grains per gallon:
    $MgCl_2 = 9.5$, $CaCO_3 = 5.0$, $CaSO_4 = 3.4$,
    $Mg(HCO_3)_2 = 7.3$, $MgSO_4 = 6.0$, $SiO_2 = 2.4$.

Calculate the amount of lime (95% pure) and soda (90% pure) needed for the treatment of 20,000 gallons of water. The costs for 100 kg of lime and soda are Rs. 28 and Rs. 96, respectively.

11. A water sample using $FeSO_4 \cdot 7H_2O$ (139 ppm) as coagulating agent on analysis gave the following data:
$Mg^{+2} = 72$ ppm, $Ca^{+2} = 160$ ppm,
$CO_2 = 88$ ppm, $HCO_3 = 488$ ppm.
Calculate the amount of lime and soda needed for the treatment of 100,000 liters of water.
12. A water sample that required lime and soda for cold softening of 200,000 liters of water, using sodium aluminate (16.4 ppm) as coagulating agent, on analysis, gave the following data:
$Mg^{+2} = 72$ ppm, $Ca^{+2} = 160$ ppm,
$CO_3^{2-} = 30$ ppm, $HCO_3^- = 732$ ppm,
$OH = 17$ ppm, $CO_2 = 44$ ppm.
Calculate the amount of lime and soda required for the above quantity of water.
13. An exhausted zeolite softener was regenerated by passing 150 liters of NaCl solution, having strength of 150 g/l of NaCl. How many liters of hard water sample, having hardness of 600 ppm, can be softened using this softener?
14. Calculate the amount of lime and soda required, for the treatment of 100,000 liters of water, whose analysis is as follows:
$Mg\,(HCO_3)_2 = 73$ ppm, $Ca\,(HCO_3)_2 = 243$ ppm,
$CaSO_4 = 102$ ppm, $MgCl = 95$ ppm, $NaCl = 500$ ppm.
Also calculate the temporary and permanent hardness of the sample.
15. 50 ml of standard hard water containing 1.5 mg of pure $CaCO_3$ per ml consumed 44.0 ml EDTA. 40 ml of the same water sample needed 20 ml of EDTA using Eriochrome Black T indicator. Calculate the total hardness of water sample in ppm.
16. 50 ml of standard hard water containing 1 mg of pure $CaCO_3$ per ml consumed 25 ml EDTA. 50 ml of water sample needed 25 ml of EDTA using EBT indicator. Calculate the total hardness of water sample in ppm.
17. Calculate the amount of lime and soda required to soften 25,000 liters of water having the following analysis:
$Ca\,(HCO_3)_2 = 4.86$ ppm, $Mg\,(HCO_3)_2 = 7.3$ ppm,
$CaSO_4 = 6.8$ ppm, $MgCl_2 = 5.7$ ppm,
$MgSO_4 = 9.0$ ppm, $SiO_2 = 3.5$ ppm, $NaCl = 5.85$ ppm.
18. Calculate the amount of lime (84% pure) and soda (92% pure) required for treatment of 20,000 liters of water, whose analysis is as follows:
$Ca\,(HCO_3)_2 = 40.5$ ppm, $Mg\,(HCO_3)_2 = 36.5$ ppm, $MgSO_4 = 30.0$ ppm,
$CaSO_4 = 34.0$ ppm, $CaCl_2 = 27.75$ ppm, $NaCl = 10.00$ ppm.
Also calculate the temporary and permanent hardness of the water sample.
19. Explain with equations and calculate the quantity of quick lime and soda ash required to soften 10,000 liters of water containing:

    (a) 219 ppm of magnesium bicarbonate and 234 ppm of sodium chloride.

(b) 36 ppm of Mg^{2+} and 18.3 ppm of HCO_3^-.
(c) 1.5 ppm of the free acids, 144 ppm of sulfate ions and 71 ppm of chloride ions.

20. A precipitate of 0.110 g of $CaC_2O_4 \cdot H_2O$ (calcium oxalate) was obtained from 250 ml of a water sample. Express the calcium content in ppm.
21. The hardness of 50,000 liters of a sample of water was removed by passing it through a zeolite softener. The softener then required 200 liters of sodium chloride solution containing 125 g/liter of NaCl for regeneration. Calculate the hardness of sample of water.
22. Calculate the quantities of lime (CaO) and soda (anhyd Na_2CO_3) required for cold softening of 125,000 L of water with the following analysis, using 10 ppm of sodium aluminate as coagulant. Analysis of raw water gave the following data:
$CO^{2+} = 95$ ppm, $Mg^{2+} = 36$ ppm,
$CO_2 = 66$ ppm, $HCO_3^- = 264$ ppm, $H^+ = 2$ ppm.
Analysis of treated water gave the data.
$CO_3^{2-} = 45$ ppm, $OH^- = 34$ ppm.
Write the chemical reactions involved.
23. An exhausted zeolite softener was regenerated by passing 150 liters of NaCl solution, having a strength of 150 g/l of NaCl. If the hardness of water is 600 ppm, calculate the total volume of water that is softened by this softener.
24. Calculate the amount of lime and soda required to soften 10,000 liters of water containing the following ions per liter:
$Mg^{2+} = 4.8$, $Ca^{2+} = 16.9$ mg, $HCO_3^- = 73.2$ mg.
25. 20 ml of water sample was treated with excess of 10% KI solution and then titrated against N/100 hypo solution using starch as an indicator. 2.5 mL hypo was used for starch end point. Calculate the amount of chlorine in water.
26. Calculate the amount of lime and soda required for softening 10,000 liters of water containing the following salts per liter:
$Ca(HCO_3)_2 = 162$ mg, $CaSO_4 = 136$ mg,
$MgSO_4 = 95$ mg and NaCl = 56.1 mg.
Assume that the purity of lime is 93% and that of soda is 99%.
27. Calculate the amount of lime and soda required to soften a million liters of water having the following composition:
Free $CO_2 = 20$ ppm, $Ca(HCO_3)_2$ as $CaCO_3 = 2000$ ppm,
$Mg(HCO_3)_2$ as $MgCO_3 = 5$ ppm, $CaSO_4$ as Ca^{2+} ions = 30 ppm
$MgSO_4$ as Mg^{2+} ions = 10 ppm.
28. A water sample was found to contain the following impurities:
$Mg(HCO_3)_2 = 73$ mg/l, $CaCl_2 = 222$ mg/l,
$MgSO_4 = 120$ mg/l, $Ca(NO_3)_2 = 82$ mg/l.
Calculate the quantity of lime (84% pure) and soda (95% pure) needed for softening 5, 000 L of water.
29. A sample of water contains the following dissolved salts:
$Mg(HCO_3)_2 = 22$ mg/liter, $CaCl_2 = 85$ mg/liter,

$MgCl_2 = 30$ mg/liter, $CaSO_4 = 28$ mg/liter.
Calculate the temporary and permanent hardness.

30. Explain why NH_3—NH_4Cl buffer solution is added during determination of hardness of water by EDTA titration.
31. Write the Ilkovic equation and explain how polarography can be used to determine concentration of metal ions.

Chapter 10
Fuels and Combustion

The word fuel primarily means a material that is used to provide heat/chemical energy. It also indicates that the material under consideration is fossil fuel, that is, those fuels which occur naturally in the Earth's crust. Coal, crude petroleum, and natural gas are some of the common examples of fossil fuels. It is believed that the decay of organic matter supplemented by water vapors and relatively very small amounts of vegetable waste materials lead to the formation of fossil fuels in thousands of years. Fuel being a combustible substance gives large amount of heat energy when burnt properly. The heat energy generated can be economically used for industrial or domestic purposes. However, we must realize that the chemical energy, a usual energy source of fuel, can be rapidly converted to thermal, mechanical, or electrical energy. Human system converts chemical energy in the form of food within the body to thermal and mechanical energies.

From the point of view of chemistry, most of the fuels release energy due to a chemical reaction in which outer electrons of the atoms are rearranged. Thus, the nucleus of the atom remains unaffected. However, there are many fuels in which nucleus of the atom can be disturbed by the bombardment with high-energy subatomic particles. Such fuels are called as nuclear fuels. An isotope of uranium $\left({}^{235}U\right)$ is a common nuclear fuel in which energy conversion takes place due to reactions in nucleus. As we all know, the nucleus of an atom is tightly packed with protons and neutrons having forces thousands of times greater than ordinary chemical forces; it is expected that the amount of energy released due to the nuclear fission will be very high. It has been found that the energy released by 5×10^{11} g of coal combustion is equal to the energy released by the fission of all the nuclei in 1 kg of ${}^{235}U$. A common nuclear fission reaction between ${}^{235}U$ and a neutron can be represented as

$${}^{235}_{92}U + n \longrightarrow {}^{94}_{37}Rb + {}^{140}_{55}Cs + 2n + 3.2 \times 10^{-11}\ J$$

© The Author(s), under exclusive license to Springer Nature Switzerland AG 2026

R. N. Goyal, H. Goel, *Textbook of Engineering Chemistry*,
https://doi.org/10.1007/978-3-032-22068-4_10

The neutrons produced in the reaction either escape or collide with other nuclei to cause more fissions, and hence a chain reaction is initiated.

As stated earlier, fuel is combustible substance which means it combines with oxygen in an exothermic reaction. Thus, the process in which heat energy from the fuel is liberated is termed as *combustion*. As most of the fuels possess carbon and hydrogen, they are converted to CO_2 and H_2O, and these new products formed have comparatively lesser energy than the reactants, and hence heat is liberated. The reaction of combustion generally takes place at high temperature and at a high speed. The incomplete combustion of a fuel can be avoided by maintaining proper air supply and avoiding cooling of combustion gases before the reaction is complete.

10.1 Classification of Fuels

The non-nuclear fuels can be classified into two groups:

- Primary fuels
- Secondary fuels

10.1.1 Primary Fuels

Occur naturally (in the earth's crust) and no chemical processing is done before their utilization. Coal, wood and peat, crude petroleum, and natural gas are some of the examples of primary fuels.

10.1.2 Secondary Fuels

These are the fuels which are obtained from primary fuels and involve some chemical treatment or change in their composition or structure is made at the processing stage. Charcoal, coke, coal gas, diesel, kerosene, etc. are some of the examples of secondary fuels.

Primary and secondary fuels are further divided into three types on the basis of their state of aggregation. Thus, the three subtypes are:

(i) Solid
(ii) Liquid and
(iii) Gas in both types of fuel.

10.2 Some Solid Fuels

10.2.1 Wood

Wood being one of the earliest of all the fuels is most generally used. Well, this is due to the fact that it has been available to man in plenty throughout man's existence, and its products are obtained quite easily. Though wood grows at a rapid rate as compared to all other fuels, it is of importance as a fuel only in the regions where other fuels are not readily available.

Lignin is the cellulose which is present in the combustible portion of wood. It contains a high percentage of carbon and oxygen, the exact composition being C = 49.4; H = 6.2; O = 43.4, and ash = 1%. One can see that the composition of hydrogen toward the calorific value is negligible and it is largely dependent on the carbon. The average calorific value of cellulose is 4150 calories per gm.

It has been observed that resinous (substance passed through the pores by certain plants) woods, such as fir, pine, etc., have a higher calorific valve, for example, a pine knot wood give about 6005 calories.

Dry wood, therefore, is a poor fuel in terms of thermal valve. Normally, dry wood contains 10 to 15 percent of moisture. This moisture content is much higher when freshly cut, ranging from 25% to 50%.

10.2.2 Wood Charcoal

Wood charcoal is obtained by the destructive distillation of wood carried out with an initial temperature of 200°C and final temperatures of 330–340°C. Only about 20% of the dry wood is obtained as charcoal, but as this charcoal has lost high composition of oxygen, its calorific valve is quite high, of about 8000–9000 kcal/kg. The other products available due to destructive distillation of wood are turpentine, light oils, and heavy oils, and in addition to these liquid products, some of the gases are also obtained as shown below:

CO_2	30%
C_nH_m	8%
CO	35%
H_2	11%
CH_4	12%
N_2	4%

Charcoal is a fibrous black solid with crumbled character. Its capacity for absorbing gases is quite high, and this is because of the fact that it has a very large surface area. It is used as a fuel in the form in which it is produced. It is used as a purifier after some changes.

10.2.3 Coal

The decay of plant materials deposited in the earth's crust millions of years ago resulted in the formation of many layers which was further altered by heat and pressure. It is here in these layers that coal is found. Thus, coal is a stored fuel. It mainly contains C, H, and O which are also called the true coal substance. Apart from these elements, N, S, As, P, and many gases are also present in coal in small amount.

The plant remains accumulated have been explained according to two theories:

(a) **In Situ Theory**: This theory holds that the area where coal is found in the layers is the area where original matter grew and accumulated.
(b) **The Drift Theory**: This theory holds that the accumulation of plants is not the original place where it used to be. The material was transported by rivers to lakes or other depressions thereby forming many layers one after the other. As a result, under various conditions of temperature, pressure, etc., being acted upon for a long time, the wood got decomposed. It could well be assumed that during the transportation of wood, plants, etc., huge quantities of mud moved along layers. The occurrence of very thick seams is satisfactory explained by this theory.

Classification of Coal

Peat Peat is not a coal truly speaking, but it is the first stage of coal series. Dry peat contains 62 to 64% carbon and around 30% of oxygen. The dry peat also contains a moisture content of 0.33%, hydrogen 5.7 to 6.3%, nitrogen 1.3 to 2.7%, and sulfur 1%.

As labor is cheap in country districts and also because of local market, peat is still obtained by hand-cutting. But for full-scale production, a long series of processes are involved. The calorific valve of peat (dry) varies between 5000 and 5500 kcal/kg. It is mainly used as a fuel for domestic purposes. It is a good source of wax of the Montan type.

Lignite The composition of lignites varies according to the metamorphosis, which the lignin of the plant has undergone. They have been formed more recently than those of true coal formations.

Lignites may be black or brown and earthy having a woody structure. They readily break or disintegrate when air-dried. The dried material contains 60 to 75% of carbon and 20 to 25% of oxygen with moisture content of 15 to 20%. Their calorific valve ranges widely from 6000 to about 7000 kcal/kg. Lignite is used in the production of town gas and as a pulverized fuel in steam locomotives.

Bituminous Coal The bituminous coal is formed from lignitous coal, but the process is not clearly defined. They contain 24 to 40% of volatile matter and 84 to 91.2% of carbon.

They are black and banded in appearance. Since the bituminous coal has a large number of properties, it is therefore used with a prefix like caking, medium, strongly, etc., to clearly define the type of bituminous coal. As stated earlier, it has a variety of uses and is a general purpose coal of house hold and industrial purposes. Its calorific valve is about 8000 to 8500 kcal/kg.

They are of two types:

(a) **Sub-bituminous coal:** The carbon content in them is 75 to 83% and 10 to 20% of oxygen. They do not have any caking power but contain high moisture and volatile matter. They are black in color and have a calorific valve of about 7000 kcal/kg.
(b) **Semi-bituminous:** The carbon content is 90 to 93%, oxygen from 2 to 4%, and volatile matter from 5 to 10% in this type of coal. They are used in steam raising, in domestic stoves, and in the production of metallurgical coke.

Anthracite It is the highest rank coal, containing over 93% carbon and less than 8% of volatile matter, with no caking power. It is one of the most valuable coal and least widely distributed. Anthracite is hard and lustrous and does not soil the fingers. It is a valuable fuel for use in stones because of its high density, and as a result of low content of volatile matter, it burns almost without smoke and does not soften. They have a calorific value of about 8700 kcal/kg. The chief industrial uses of anthracite are steam raising, domestic stoves, and central heating and malt and hop drying.

Analysis of Coal

Coal can be analyzed in two different ways depending upon the details of the analysis and data required. These analysis are called as:

- Ultimate analysis.
- Proximate analysis.

Ultimate Analysis The ultimate analysis of the coal deals with the chemical analysis of the material. Elements like carbon, hydrogen, nitrogen, and sulfur are determined by the usual procedures. The amounts of C and H are determined by burning a known amount of coal in oxygen. The amount of CO_2 and H_2O formed is determined from which percentage of carbon and hydrogen are calculated. The percentage of nitrogen is determined by Kjeldahl's method and that of sulfur is calculated by precipitating it as barium sulfate. As the details of carbon, hydrogen, nitrogen, and sulfur determination in a sample are taught to the students in 10 + 2 class, these methods are not discussed here. The various well-known formula used in these calculations are as follows:

$$\% \text{ of C} = \frac{12}{44} \times \frac{\text{Weight of CO}_2}{\text{Weight of coal}} \times 100$$

$$\% \text{ of H} = \frac{2}{18} \times \frac{\text{Weight of H}_2\text{O}}{\text{Weight of coal}} \times 100$$

$$\% \text{ of S} = \frac{32}{233} \times \frac{\text{Weight of BaSO}_4}{\text{Weight of coal}} \times 100$$

$$\% \text{ of N} = \frac{1.4 \times \text{N} \times \text{V}}{\text{Weight of coal}} \times 100 \qquad \text{(Kjeldahl's method)}$$

where N = normality of the acid used and V = volume of the acid used.

However, the method used for the determination of phosphorus is presented. If the percentage of phosphorus and chlorine is expected to be significant, their analysis is also carried out. The oxygen present in the material is determined by subtracting the sum of elements and moisture from 100. The results of these percentages of various elements provide an idea about the quality of coal and are used in its combustion calculations. Remember coal is also used during preliminary roasting in the metallurgy of iron; hence, the amount of sulfur and phosphorus present in the coal has a significant effect on the metallurgy of iron and steel.

Several methods are available for the determination of phosphorus in coal samples. A common method employs oxidation of phosphorus to phosphoric acid which is then treated with solution of magnesium carbonate (magnesia), and phosphorus is precipitated as magnesium ammonium phosphate $(MgNH_4\ PO_4)$. The magnesium ammonium phosphate is then ignited, and magnesium pyrophosphate is obtained.

$$2MgNH_4PO_4 \xrightarrow{\text{ignite}} \underset{\text{Magnesium pyrophosphate}}{Mg_2P_2O_7} + 2NH_3 + H_2O$$

The pyrophosphate so obtained is cooled and weighed. The percentage of phosphorus is calculated using the formula where the term 62/222 represents the presence of 62 g of phosphorous in 222 g of $Mg_2P_2O_7$.

$$\% \text{ of P} = \frac{62}{222} \times \frac{\text{Weight of Mg}_2\text{P}_2\text{O}_7}{\text{Weight of coal sample}} \times 100$$

If the coal sample is found to possesses high sulfur content, it should be used carefully because on combustion sulfur forms dioxide and trioxide. These oxides of sulfur can either form corrosive layer of sulfuric acid on the cooler part of the plants or can lead to atmospheric pollution. The presence of sulfur oxides in atmosphere has been shown as a major hazard to the human health. The role of SO_2 in smog

Table 10.1 Percentage of various elements and other matter present in a normal coal sample

	As sampled	Dry, ash free
C	80.3	87.3
H	5.0	5.4
N	1.5	1.6
S	0.8	0.9
O	4.3	4.7
Moisture	2.2	–
Ash	5.8	–

formation has also been discussed in the earlier chapter of this book. Similarly it is recommended that the coke used in blast furnace should not contain more than 0.01% phosphorus.

Proximate Analysis Proximate analysis of the coal deals with the determination of percentage of the moisture, coke, ash, volatile matter, and fixed carbon in a given material. The results of these analysis depend on the procedure adopted for the analysis like temperature of heating, duration of heating, the material of the crucible used, etc., and hence the word proximate is used. This term should not be confused with the word approximate, as a high accuracy is desired in the analysis.

The results of the analysis can be expressed as either sampled or on a dry, ash-free basis. The analysis results for a typical coal sample are as shown in Table 10.1.

The information about the moisture, ash, and volatile matter helps in deciding about the quality of coal. For example the higher percentage of moisture in a coal sample indicates it to be a lower grade as high moisture reduces the gross calorific value of a fuel.

Similarly ash is the incombustible material which remains when coal is burnt. It affects

- the calorific value of the coal, and
- as ash covers the fire bars, the flow of air is blocked and hence it affects the burning process.

Determination of Moisture and Volatile Matter The moisture content of a coal sample is determined by heating the known amount of coal powder in a silica or porcelain crucible at 105°–110°C for about 1 h in an oven. The crucible is then cooled in a desiccator and weighed. The loss of weight on heating tells about the moisture content present in the coal sample. Similarly, the volatile matter present in the coal can be calculated by heating the known amount of sample in a platinum crucible at 950°C in a furnace. The crucible should have a proper cover so that no air is allowed to come in contact with the sample. Heating is done at 950°C for exactly 7 min and then the crucible is taken out and cooled. The loss in weight of the sample is determined, from which the percentage of volatile matter present is calculated.

Example 10.1 1.508 g of a Durham coal sample was heated at 110°C for 1 h. On cooling the weight of the sample was found as 1.478 g. Strong heating of the sample at 950°C for 7 min was then carried out in a closed platinum crucible. The sample on cooling exhibited weight as 1.068 g. Calculate the percentage of moisture and volatile matter present in the sample.

Solution Moisture present in the sample

$$= 1.508 - 1.478$$

$$= 0.030 \text{ g}$$

$$\% \text{ of moisture} = \frac{0.030}{1.508} \times 100$$

$$= 1.98\%$$

Similarly, volatile matter present in the sample

$$= 1.478 - 1.068$$

$$= 0.410 \text{ g}$$

$$\% \text{ of volatile matter} = \frac{0.410}{1.508} \times 100$$

$$= 27.19\%$$

Determination of Ash Content and Fixed Carbon The material left in the crucible after determining volatile matter is again heated in a muffle furnace at 700°C for about 30 min. For complete combustion, the sample is stirred with a platinum wire. The crucible is then taken out from the furnace and cooled. The procedure of heating and cooling is continued till the weight of the crucible becomes constant. The percentage of ash will be equal to

$$= \frac{\text{weight of ash}}{\text{weight of coal sample taken}} \times 100 \qquad (10.1)$$

The fixed carbon of the coal sample can be determined by subtracting the percentage of moisture, volatile matter, and ash from 100. The percentage of fixed carbon of the coal tells us about the calorific value. Higher percentage of fixed carbon in a sample indicates higher calorific value and, thus, is useful in determining the quality of the coal.

Storage of Coal

The storage of coal is a costly procedure as it requires large space; however, storage is essential in view of its increased demand in winter. It must be realized that the storage of coal causes deterioration of its quality and the possibility of serious

Table 10.2 Loss of calorific value of coal due to storage

	Loss in calorific value (%)	
Size of coal (cm)	1 year	2 years
5–8	0.11	0.24
0.4–1	0.50	1.00
< 0.4	1.00–1.8	2.00–3.00

damage by spontaneous combustion. The nature of deterioration may be due to any one or more than one of the following:

- Physical breakdown of the size of coal.
- Reduction in calorific value.
- Loss of caking power.
- Increase in crumble nature.

All of these deteriorations during storage have suggested that most serious deterioration takes place during the early stages of storage. Hence, short-term storage of coal should be avoided. Table 10.2 presents the effect of storage of coal on the calorific value.

In addition to deterioration of quality of coal during storage, a first precaution needed is that surface ventilation is necessary. In fact when coal is freshly mined, it liberates absorbed hydrocarbons, particularly methane, which varies in amount from 0.04 Cu ft/kg for bituminous coal to 0.03 Cu ft/kg in anthracite coal. Hence, it is necessary to have proper ventilation at the place of storage of coal. After the absorbed hydrocarbons are escaped, the coal is liable to oxidation, which can occur in three stages:

(i) coal absorbs as well as adsorbs oxygen with liberation of heat.
(ii) if the heat is not dissipated properly, it leads to rise in temperature and rate of oxidation is accelerated.
(iii) the rise of temperature may cause breakdown of sorbed phases.

It is recommended that a concrete floor is ideal for the storage of coal and should be stored in rectangular bays. The temperature at frequent intervals of the stored coal should be monitored by inserting iron pipes which allows use of thermometer. If the temperature recorded is more than 70°C, the stored coal should be spread to dissipate the heat generated.

10.3 Liquid Fuels

Crude petroleum is the main primary fuel which occurs naturally in the liquid form and is usually associated with natural gas. The crude petroleum obtained from the earth is not used directly as a fuel; however, various fractions obtained from its distillation are used as fuel.

The most important question one would ask is as to how crude petroleum is formed. To explain the formation of petroleum, many theories have been proposed; however, none of them is able to satisfactorily explain the formation. It is, however, generally accepted that petroleum was formed from vegetable, animal, or mineral origins by anaerobic decay of marine plants and animal life. Depending upon the source of crude petroleum, its color, composition, and appearance changes widely. Many crudes are very dark in color, whereas others are light colored. The compounds present in the crude petroleum can be divided in four groups:

- Hydrocarbons
- Oxygen compounds
- Sulfur compounds
- Inorganic compounds

As mentioned earlier, the primary process used for the separation of various components of crude petroleum is distillation. Fractional distillation is carried out to separate various components on the basis of their boiling points. Many components of the crude petroleum have close boiling points, and hence large fractionating columns are used. In the event of two liquids having close boiling points, a secondary distillation is carried out. The important fractions obtained during fractional distillation of crude petroleum are shown in Fig. 10.1.

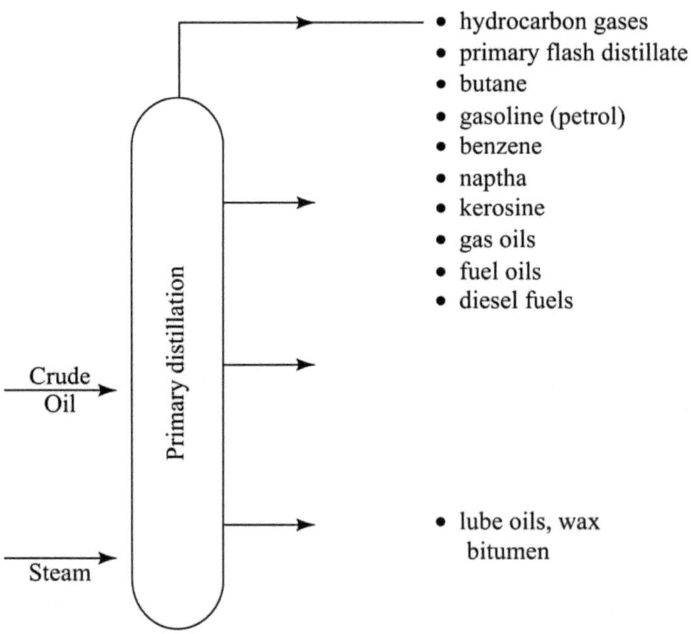

Fig. 10.1 Important fractions obtained from the primary fractional distillation of the crude petroleum

Each of these fractions has uses in domestic or industrial applications as well as in manufacturing of wide range of chemicals used in medicines.

10.4 Natural Gas

Natural gas is an important fuel associated with crude petroleum and is produced in large volumes from oil wells. In many cases the gas migrates through porous rocks away from the oil. The actual composition of natural gas varies from one source to another; however, the basic composition is presented in Table 10.3, which includes the major components as methane, ethane, propane, and butane. Natural gas may be used as a fuel in both domestic and industrial applications; however, special burners are necessary. Also as natural gas is odorless, it is advisable to add an odorant (organic sulfides) to enable the leaks to be detectable. Carbon black is usually prepared by burning the natural gas under fuel-rich conditions. The smoky flame produced is used for collecting the soot.

10.4.1 Liquified Petroleum Gases

Liquified petroleum gases commonly known as LPG are considered as primary fuel these days and is mainly produced from natural gas or crude oil. The main hydrocarbons present in LPG are propane and butane. As both these hydrocarbons possess low vapor pressure, they can be easily liquified. The amount of air required

Table 10.3 Basic composition of natural gas

Hydrocarbons	
CH_4	70–90%
C_2H_6, C_3H_8, C_4H_{10}	0 – 20%
Other gases	
CO_2	0–7%
H_2	0–1%
N_2	0–5%
H_2S	0–5%
O_2	0–0.2%

Table 10.4 A comparison of physical properties of aviation and regular gasoline

Property	Aviation gasoline	Regular gasoline
Boiling point range	40–170°C	37–185°C
Specific heat cal/g/°C	0.53	0.53
Latent heat cal/g	70	70
gravity	0.724	0.733
C (%)	85.1	85.5
H (%)	14.9	14.4
S (%)	~ 0.01	~ 0.1
Value (net) cal/g	10,510	10,440

for combustion of these gases is much higher (25 : 1) than needed for coal gas (6 : 1) or natural gas (10 : 1). Hence, special burners are required to fulfill the air requirement of these gases.

Another important feature of LPG is that on evaporation, one volume of propane or butane produces nearly 270 volumes of gas. This feature makes it possible to storage large amount of heat energy in a small volume. In India, a standard LPG cylinder holds nearly 0.036 m^3 liquid, which is approximately equal to 14 kg of butane. The calorific value of butane is 60400 kJ/kg, and hence a standard LPG cylinder possesses 8,45,600 kJ of stored energy. Due to this property, LPG has been considered suitable for providing fuel in remote locations. Special precautions are, however, needed to use LPG due to its inflammable nature.

10.4.2 Gasolines

Gasolines consist a complex mixture of hydrocarbons in the range C_4 to C_8 and are very similar in their physical properties. Various grades of gasolines are marketed these days which include aviation grade, premium grade, and regular grade gasoline. The basic difference in these grades of gasoline is of their knock rating.

The anti-knock nature of a liquid fuel is defined in terms of its octane number. Iso-octane (2, 2, 4-trimethylpentane) has been assigned an octane number 100 and *n*-heptane as zero. The octane number is thus *defined as the percentage of iso-octane in an iso-octane/n-heptane mixture by volume, which matches the petrol in its detonation property during the engine test.* Thus, if a fuel has an octane number 70, it means that the fuel matches with 70 : 30 iso-octane-heptane mixture.

Aviation gasoline has hydrocarbons in low boiling range (40° – 170°C) which shows high octane number. Its latent heat of vaporization is about 70 cal/g. A comparison of some physical properties of aviation gasoline and regular grade gasoline is presented in Table 10.4.

During "take-off," an aircraft engine develops increased power which is provided by an increase of fuel supply to give a rich mixture of air and fuel. Usually air-fuel ratio required for complete combustion is 15 : 1; however, air is provided in excess at the time of "take-off." It has been found that by using higher octane number fuel in aircraft engines increases the power to take-off. A change in fuel from 90 octane number to 100 octane number may lead to 10–30% increase in power available for "take-off," due to increase in compression ratio. The efficiency of engine cycle is defined by an expression where

$$E = 1 - \left(\frac{1}{CR}\right)^{0.258} \tag{10.2}$$

CR is the compression ratio of the fuel. Let us consider an example.

Example 10.2 The octane number of an aviation fuel was increased from 90 to 100, and the compression ratio changed from 7.8 to 9.5. Calculate the increase in efficiency of the engine.

Solution Efficiency of the engine in the first case will be

$$\begin{aligned} E &= 1 - \left(\frac{1}{7.8}\right)^{0.258} \\ &= 1 - 0.589 \\ &= 0.411 \end{aligned}$$

Efficiency of the engine in the second case will be equal to

$$\begin{aligned} E &= 1 - \left(\frac{1}{9.5}\right)^{0.258} \\ &= 1 - 0.559 \\ &= 0.441 \\ \text{Increase in efficiency} &= 0.441 - 0.411 = 0.030 \\ \%\ \text{increase} &= \frac{0.030}{0.441} \times 100 \\ &= 7.29\% \end{aligned}$$

10.4.3 Kerosene

Kerosene is a fraction obtained during fractional distillation of crude petroleum between the temperature range 140° and 280°C. It is used as a fuel in stoves and also used as an illuminant. The physical properties of kerosene depend upon the amount of paraffins, naphthenes, and aromatic hydrocarbons present in it. Some typical properties of kerosene oil are summarized in Table 10.5.

Kerosene with high percentage of aromatic hydrocarbons burns with smoky flame. In domestic appliances, kerosene is always vaporized before combustion.

10.4.4 Diesel Fuel

Diesel fuel is an important fraction obtained from crude petroleum. It has a boiling range 170–370°C and possesses saturated hydrocarbons with usually higher percentage of sulfur. It is largely used in diesel engines and for furnace heating.

In diesel engines, heat of compression is used to ignite the fuel. Thus, the fuel is injected at the end of compression stroke. During this stroke, the air for combustion is compressed to about 500 lb per square inch, and this compression raises the air temperature to about 600°C. This temperature is sufficient to ignite the fuel almost spontaneously. The time interval between the start of injection and ignition of fuel is called as " **ignition delay**." If the ignition delay is short, the fuel is considered as a high-quality fuel or a fuel having high cetane number. However, it is necessary that the ignition delay must be sufficient to permit compression stroke to be completed to achieve maximum power output and minimum fuel consumption. Excess of air is necessary to carry out complete combustion otherwise some of the hydrocarbons will be only partially oxidized.

Cetane number of a fuel is used to evaluate its ignition quality on the basis of its compression ignition in an engine. Cetane is a saturated hydrocarbon, $C_{16}H_{34}$, and has been assigned a value of 100. On the other hand, α-methylnaphthalene ($C_{11}H_{10}$) is designated a value of zero. The actual cetane number of a fuel is the ratio by volume of cetane in a mixture of two hydrocarbons, having the same ignition delay after injection of the fuel under test. A diesel fuel having cetane number between 50 and 55 is considered satisfactory for use and has a relative density of 0.84.

The calculation of cetane number can also be carried out using the following formula: Cetane number $= 175 \times 4 \log$ (mid boiling point °F) $+ 1.98$ (°API) $- 496$, where ° API is the scale of relative density as proposed by American Petroleum Industry and is equal to

Table 10.5 Typical properties of kerosene oil

Specific gravity	0.79–0.82
Flash point °C	40–60
Initial boiling point °C	140
50% boils at °C	210
Sulfur %	0.05–0.08
Smoke point, mm	20–25

$$^\circ\text{API} = \left(\frac{141.5}{\text{relative density at 288 K}}\right) - 131.5 \quad (10.3)$$

As oils expand on heating, it is necessary to know its relative density at 288 K (10°C). It also provides an idea about the volume of given mass of oil for the purpose of transport or storage.

10.5 Essential Properties of Fuels

The choice of a fuel in industries largely depend on the availability in a given area and to the source of supply. Thus, for example, you will find steel industries in the area close to coal mines. In the blast furnace, coke is used as a fuel and at the same time also used for reducing the iron ore to the metal. The following are some essential properties of fuels.

The first requirement of a fuel is that it should have a high calorific value. Let us now discuss what do we mean by calorific value and how it can be determined, but before that, it is necessary to understand the process of combustion.

10.5.1 Combustion

Almost all chemical reactions absorb or produce energy. Energy is produced when a change leads to a more stable product, i.e., when less stable molecules are converted to more stable molecules or in other words when product of a reaction is more stable than the reactant. On the other hand, energy is absorbed, if the change leads to a less stable state. In a chemical reaction, heat is the form of energy that is most commonly absorbed or released.

The combustion of hydrogen in oxygen is one of the common chemical reaction which releases energy.

$$2H_2(g) + O_2(g) \longrightarrow 2H_2O(l) + \text{ heat}$$

The heat generated by the combustion process is transferred from the system to the surroundings. Any process that gives off heat is called as *exothermic process*.

The decomposition of mercury (ii) oxide at high temperature occurs in the reaction.

$$2HgO(s) + \text{Heat} \longrightarrow 2Hg(l) + O_2$$

and is an example of *endothermic process*.

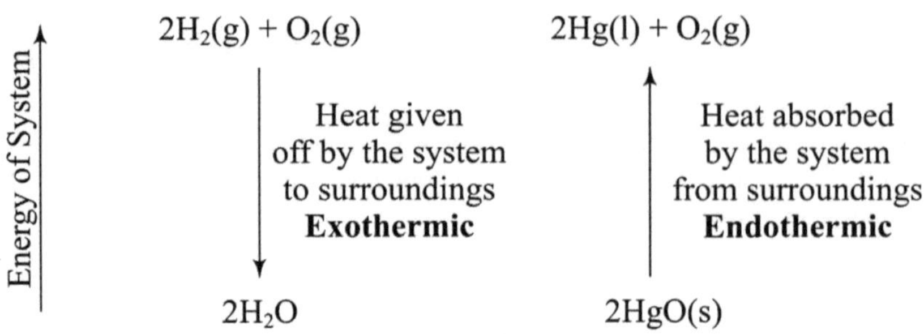

Most of the physical and chemical processes take place at constant pressure conditions (one atmospheric pressure). Chemists use a quantity called as heat content or enthalpy, represented by the symbol H. The change in enthalpy (ΔH) of a system during a reaction, under a constant pressure, is equal to the heat given off or absorbed by the system. The ΔH is defined as

$$\Delta H = H \text{ (products)} \; - \; H \text{ (reactants)}$$

For an exothermic process, ΔH is negative, whereas for endothermic process, this value is positive.

Whenever a solid, liquid, or gaseous fuel is combusted, the atoms of carbon or gases combine with oxygen of air to form products with simultaneous liberation of heat. However, for the combustion to occur, the fuel must be brought to its ignition temperature and continues to burn without further requirement of heat. The common reactions involved are

$$C(s) + O_2(g) \longrightarrow \text{Products } + \text{Heat}$$

or

$$\text{Fuel } + O_2(g) \longrightarrow \text{Products } + \text{Heat}$$

$$CH_4(g) + 2O_2(g) \longrightarrow CO_2(g) + 2H_2O(g) + \text{Heat}$$

The requirement of oxygen for burning the fuels is fulfilled by using air. The composition of air (in a broad sense) can be considered as shown below in percentage by weight and by volume. Though it depends on several factors, percentage of oxygen may vary depending upon the pollutants present in the air.

Air	By weight	By volume
O_2	23%	21%
N_2	77%	79%

Thus, if for a combustion, 1 kg of oxygen is needed, that means we will require $(1 \times 100)/23$ kg of air.

10.5.2 Combustion of Carbon

Combustion of carbon depends on the availability of oxygen. Thus, two possible reactions are

$$2C + O_2 \longrightarrow 2CO \tag{10.4}$$

$$C + O_2 \longrightarrow CO_2 \tag{10.5}$$

In terms of weights, we can say that for Eq. (10.4), 24 parts of carbon need 32 parts of

$$\underset{C}{2 \times 12} + \underset{O_2}{2 \times 16} \longrightarrow \underset{CO}{2 \times 28}$$

oxygen to give 56 parts of carbon monoxide. In other words, it can be concluded that 1 g of C reacts with $\frac{32}{24}$ g of O_2 to give $\frac{56}{24}$ g of CO.

Similarly for Eq. (10.5), which is complete combustion of carbon, it can be represented as

	C	+	O_2	$\longrightarrow$	CO_2
wt. (g)	12		32		44
and	1		2.667		3.667

In terms of volume, it can be written as

$$C(s) + O_2(g) \longrightarrow CO_2(g)$$

$$\begin{array}{lll} 12\text{ g} & 22.4\text{ lit} & 22.4\text{ lit} \\ 1\text{ g} & 1.86\text{ lit} & 1.86\text{ lit} \end{array}$$

The volumes of O_2 and CO_2 are in liter at 0°C and 1 at pressure (760 mm). As combustion normally occurs in air and not pure oxygen, we can rewrite Eq. (10.5) as

$$\text{Carbon} + \text{Air} \longrightarrow \text{Products}$$

$$C + O_2 + N_2 \longrightarrow CO_2 + N_2$$

and

$$\underbrace{12\text{ g}}_{1\text{ g}} + \underbrace{22.4\text{ lit } + 84.3\text{ lit}}_{8.87\text{ lit}} \longrightarrow \underbrace{22.4\text{ lit } + 84.3\text{ lit}}_{8.87\text{ lit}}$$

Thus, 1 g of carbon will react with 8.87 lit of air.

10.5.3 Combustion of Hydrogen and Sulfur

$$2H_2 + O_2 \longrightarrow 2H_2O$$
$$4\text{ g} + 32\text{ g} \longrightarrow 2 \times 18\text{ g}$$

or

$$1\text{ g} + 8\text{ g} \longrightarrow 9\text{ g}$$

It is clear from this equation that 1 g of hydrogen combines with 8 g of oxygen to give 9 g of water. We can also say that 2 volumes of hydrogen combine with 1 volume of oxygen to produce H_2O.

Combustion of sulfur normally leads to SO_2

$$S + O_2 \longrightarrow SO_2$$
$$32\text{ g} + 32\text{ g} \longrightarrow 64\text{ g}$$

or

$$1\text{ g} + 1\text{ g} \longrightarrow 2\text{ g}$$

or 1 g of sulfur needs 1 g of oxygen to give 2 g of SO_2.

Example 10.3

1. Calculate the weight and volume of air requirement for combustion of 3 kg of carbon.

   Solution Weight of air required for combustion of carbon.

$$= 3\text{ kg} \times \frac{32}{12} \times \frac{100}{23}$$
$$= 34.7\text{ kg}$$

   Volume of air required

$$= \text{mol of air } \times 22.4$$

where (1 mol. = 22.4l at NTP) and (Mol. wt. of air = 28.92)

$$= \frac{34.7 \times 10^3\ \text{g}}{28.92} \times 22.4$$

$$= 26.8 \times 10^3\ \text{litre} = 26.7\ \text{m}^3\ \text{air.}$$

2. Calculate the volume of air required for complete combustion of 1 m^3 of gaseous fuel having CO = 46%; H_2 = 40%; CH_4 = 10%; C_2H_2 = 20%; N_2 = 1.0% and rest is CO_2.

Solution For 1 m^3 of gaseous fuel
CO = 0.46 m^3; H_2 = 0.40 m^3; CH_4 = 0.1 m^3;
C_2H_2 = 0.02 m^3 and N_2 = 0.01 m^3
The gases which will take part in combustion are

	O_2 required.
$CO + \frac{1}{2}O_2 = CO_2$	$0.46 \times 0.5 = 0.23\ m^3$
$CH_4 + 2O_2 = CO_2 + H_2O$	$0.10 \times 2.0 = 0.20\ m^3$
$H_2 + \frac{1}{2}O_2 = H_2O$	$0.40 \times 0.5 = 0.20\ m^3$
$C_2H_2 + \frac{5}{2}O_2 = 2CO_2 + H_2O$	$0.02 \times 2.5 = 0.05\ m^3$
	Total = $0.68\ m^3$

∴ volume of air required

$$= 0.68 \times \frac{100}{21} = 3.238\ \text{m}^3$$

3. A coal containing 62.4%C; H = 4.1%; O = 6.9%; N = 1.2%; S = 0.8% and moisture = 15.1%; ash = 9.71% was burnt in such a way that dry fuel gas contained CO_2 = 12.9%; CO = 0.2%; O_2 = 6.1% and N_2 = 80.8%. Calculate

(i) weight of air theoretically required to burn 1 kg of coal
(ii) weight of dry fuel gas/kg of fuel
(iii) weight of actually supplied air/kg coal burnt.

Solution

(i) In 1 kg of coal
$C = 624\ \text{g}, H = 41\ \text{g}, O = 69\ \text{g},\ N = 12\ \text{g and } S = 8\ \text{g}.$

$$\text{Air required} = \left[624 \times \frac{32}{12} + 41 \times \frac{16}{2} + 8\left(\frac{32}{32}\right) - 69\right] \times \frac{100}{23}$$

$$= 8.396\ \text{kg}.$$

(ii) Weight of fuel gas/kg of coal.

$$= \frac{\text{Weight of C in coal}}{\text{Weight of } C \text{ in fuel gas}}$$

Wt. of C in fuel gas/kg=Wteight of $CO_2 \times \frac{12}{44}$ + Wt. of CO $\times \frac{12}{28}$

$$= 129 \times \frac{12}{44} + 2 \times \frac{12}{28} = 36.04\ \text{g}$$

Weight of fuel gas/kg of coal $= \frac{624}{36.04} = 17.314$ kg.

(iii) O_2 is needed to convert CO to CO_2

$$= 2 \times \frac{16}{28} = 1.4\ \text{g}$$

$$\begin{array}{lll} CO+ & \frac{1}{2} O_2 \longrightarrow & CO_2 \\ 28 & 16 & \end{array}$$

Remaining weight of O_2/kg of fuel gas

$$= 61 - 1.4 = 59.6\ \text{kg} = 0.059\ \text{kg}.$$

Weight of O_2 obtained by burning 1 kg coal

$$= \text{Wt. of fuel gas } \times \text{Wt. of } O_2/\text{kg of fuel gas}$$

$$= 17.314 \times 0.059$$

$$= 1.0319\ \text{kg}.$$

Air required

$$= 1.0319 \times \frac{100}{23} = 4.487\ \text{kg}$$

Weight of air actually supplied/kg coal burnt

$$= 8.396 + 4.487 = 12.885\ \text{kg air}.$$

10.6 Calorific Value

The heat liberated by complete combustion of a unit mass of the fuel with oxygen and the condensation of the products to the temperature of determination is known as calorific value of the fuel. For solid and liquid fuels, the quantity of fuel present is sufficiently defined by the mass present, but in the case of gaseous fuels, both the pressure and temperature must be specified in order to define the quantity.

For the efficient use of a fuel in combustion and furnace plant, the main factor required is the calorific value.

The various units in which calorific value is measured are:

(a) **Calorie:** It is the amount of heat required to raise the temperature of one gram of water from 14.5°C to 15.5°C, i.e., through one degree centigrade. This unit is used as the international standard in calorimetry.

(b) **Kilo Calories:** The amount of heat which can raise the temperature of one kilogram of water through one degree centigrade is known as kilo calorie. It is basically equal to 1000 calories.

(c) **British Thermal Unit:** This is a British unit and is defined as the amount of heat which increases the temperature of one pound of water (454 g) through one degree Fahrenheit.

One mean British Thermal Unit (B. Th. U)

$$= 251.996 \text{ calories}$$

$$\approx 252 \text{ calorie}$$

$$= 0.252 \text{ kcal}$$

(d) **Pound Degree Centigrade Unit:** Being the amount of heat required to raise 1 lb (pound) of water through 1°C. This unit is frequently employed in stating the calorific value of solid and liquid fuels. 1 kilogram degree centigrade = 2.205 pound degree centigrade.

$$1 \text{ pound degree centigrade} = 1 \times 78 \text{ B. Th. U}$$

10.6.1 Gross Calorific Value or High Calorific Value

To determine the calorific value of a fuel, it is completely burnt and the combustion products are cooled as stated earlier. Now, if a fuel containing hydrogen is burnt, water is produced. As this is the product, it is condensed thereby giving up latent heat as steam. This latent heat together with the heat liberated on cooling from its condensation point to the temperature of the calorimeter constitutes the gross or higher calorific value. Thus, the gross calorific value of the fuel is the number of heat units liberated where unit weight of fuel (or volume, in the case of gases) is burnt and the products of combustion are all cooled to 288 K. It is represented as GCV or HCV (high calorific value).

10.6.2 Net Calorific Value

If the water formed by the combustion of hydrogen containing fuel is not condensed and is allowed to escape remaining in the vapor state, then it will not be available for conversion into work. Thus, it will not be able to play any role in raising the flame temperature of the burning gases. As a result, the heat liberated per unit mass of fuel is less. Thus, the net calorific value is defined as the heat liberated by burning unit mass of fuel completely and the products of combustion, i.e., the water remains as vapor and is allowed to escape.

One can see that the net calorific value is similarly defined to the gross value except that the products of combustion are condensed to 288 K in the case of gross value. They are allowed to escape, or in other words, water vapor does not condense in the case of net calorific value. Thus, we can say that

$$\begin{pmatrix}\text{Net calorific}\\ \text{value}\end{pmatrix} = \begin{pmatrix}\text{Gross calorific}\\ \text{value}\end{pmatrix} - \begin{pmatrix}\text{Latent heat of water}\\ \text{vapors formed}\end{pmatrix}$$

It is represented as LCV, i.e., low calorific value.

10.7 Bomb Calorimeter

The modern bomb calorimeter is a development of the original calorimeter of Berthelot.

The modern bomb calorimeter is made of corrosion resisting steel in which the combustion of fuel takes place. For the bomb to act like a gas-light jar, a lid is provided which can be screwed as desired. This screw is a rubber ring which is not subjected to corrosion troubles. The lid of the bomb is made of stainless steel and has three holes in it, two for electrodes and one for an oxygen inlet. Another ring is attached to one of the electrodes which supports a nickel crucible. A piece of the fuse wire is tied across the inner terminals, and a piece of sewing cotton is attached to it with the other end in contact with the fuel. To prevent the heat loss, the bomb is immersed in a known mass of water. The heat released due to combustion is transferred to the water, and the change in temperature of water is measured accurately using Beckmann thermometer. A simple sketch of bomb calorimeter is shown in Fig. 10.2.

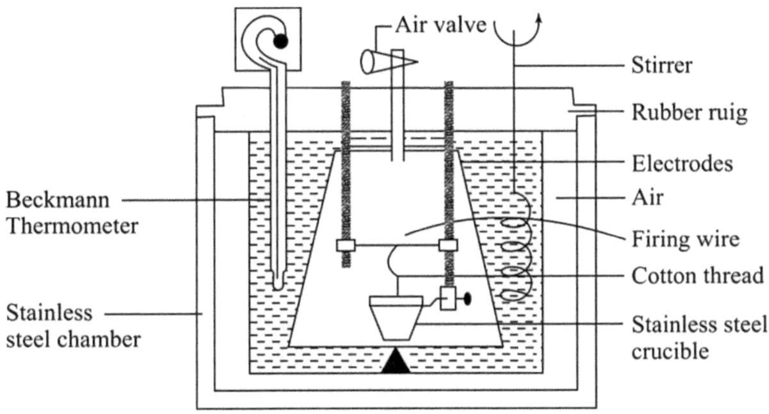

Fig. 10.2 A typical bomb calorimeter

Procedure About 1 gram of the given fuel, i.e., coal or coke (air-dried), is weighed into the crucible. In the case of solid fuels, the moisture content must be determined first, so that any correction of the result to a bias of dry fuel shall be accurate. Approximately, 2 ml of water is placed in the bottom and the cover properly assembled. Oxygen at a pressure of 25 atm should be charged in the bomb.

The calorimeter should also be charged with a weighed quantity of water that should be sufficient to cover the whole apparatus. The water must be at room temperature so that correction is minimized. The electrodes are then connected to a 6 V battery, and spark is produced.

After about 5 minutes, take the first reading of temperature and continue the readings at intervals of 1 minute. Water should be stirred continuously so as to take the maximum temperature. The calculation of the calorific value is based on simple calorimetry.

Let us assume that x gm of a fuel is burnt in calorimeter, and the following data is obtained:

(i) Mass of water taken in the calorimeter $= w$ g
(ii) Water equivalent of calorimeter, stirrer, thermometer etc. $=$ Wg
(iii) Initial temperature of water $= t_1$°C
(iv) Final temperature of water $= t_2$°C
(v) Latent heat for condensation of steam $= 587$ cal/g

If net calorific value of fuel is Q, we can write that the heat produced by burning the fuel is $= xQ$ and heat absorbed by the water is[1]

[1] Heat given or taken is expressed as mst where m is the mass, s is specific gravity, and t is the difference in temperature.

$$= w \times l \times (t_2 - t_1) + W \times l \times (t_2 - t_1)$$
$$= (w + W)(t_2 - t_1)$$

Thus, heat produced should be equal to heat absorbed, if there is no loss

$$xQ = (w + W)(t_2 - t_1)$$
$$Q = \frac{(w + W)(t_2 - t_1)}{x} \text{ cal/g} \tag{10.6}$$

The water equivalent of a calorimeter can be determined by taking a fuel of known calorific value. A well-known compound used for this purpose is benzoic acid whose gross calorific value is reported as 6.325 kcal/kg. Thus, all other parameters in the above equation are known, and hence value of W can be determined.

The value of Q obtained is the value of GCV or HCV. If latent heat of water vapor produced is subtracted from it, the net calorific value can be obtained.

$$\text{Net calorific value (LCV)} = [\text{GCV} - 0.09\text{H} \times 587] \text{ cal/g}$$

To get an idea about the theoretical calorific value of a fuel, Dulong proposed a formula, which is based on the percentages of carbon, hydrogen, and oxygen present in the fuel. The formula is

$$\text{Gross Calorific Value (GCV)} = \frac{1}{100}[8080\text{C} + 34,400(H - O)/8] \text{ cal per g.}$$

and

$$\text{LCV} = \left[\text{GCV} - \frac{9\text{H}}{100} \times 587\right] \tag{10.7}$$

The term 587 is appearing as latent heat for condensation of steam.

This formula was modified by several workers, and percentages of nitrogen and sulfur were also included. Thus, modified Dulong's formula is used for gross calorific value.

If only sulfur is considered as

$$= \frac{1}{100}\left[8,080\text{C} + 34,500\left(H - \frac{O}{8}\right) + 2,240\text{ S}\right]$$

and if sulfur and nitrogen both are considered

$$= \frac{1}{100}[8080\text{C} + 34,500(H - (O + N - l/8) + 2240\text{ S})] \text{ cal per g.} \tag{10.8}$$

The theoretical values calculated were found to match with experimental calorific values only in few cases. Many corrections were than suggested; however, nowadays, it is much simpler to experimentally determine the calorific value of a fuel.

Precautions in Determining Calorific Value Some precautions are needed to determine the correct calorific value of fuel due to various elements presents in it. For example:

- The nitrogen and/or sulfur present in a fuel can be converted to corresponding acids during combustion.

$$S + 2H + 2O_2 \longrightarrow H_2SO_4 + \text{Heat}$$

$$2N + 2H + 3O_2 \longrightarrow 2HNO_3 + \text{Heat}$$

  As these reactions are exothermic, the heat liberated will also be used to raise the temperature of water. Thus, a correction is needed in such cases. The amount of H_2SO_4 or HNO_3 formed can be determined by volumetric or gravimetric method from the washings of calorimeter and following corrections should be made corresponding to sulfur of H_2SO_4 and nitrogen of HNO_3.
  For 1mg of sulfur 2.25 cals
  For 1 ml of $N/10$ HNO_3 1.43 cals.
- The firing fuse wire used for ignition of fuel also gives some heat, which requires a correction.
- In spite of all insulation in calorimeter, some heat is lost. The amount of heat lost can be corrected by using *cooling correction*. For determining the value of cooling correction, the time required to cool the water in calorimeter from maximum temperature to the room temperature is noted, and then rate of cooling is calculated in degree per minute. The cooling correction is then determined by multiplying the rate of cooling with total time needed for bringing the higher temperature to room temperature. For example, if the total time required in an experiment to bring the higher temperature to room temperature is T min and rate of cooling is $r°/\text{min}$, the cooling correction is equal to rT. The gross calorific value is then calculated using the expression
  Gross Calorific Value

$$= \frac{(W + w)\,(t_2 - t_1 + \text{cooling correction}) - (\text{Acid} + \text{fuse correction})}{\text{Mass of fuel}} \tag{10.9}$$

The cooling correction simply indicates that the higher temperature would have reached to the room temperature in time T. Thus, the actual difference in the temperature $(t_2 - t_1)$ should be higher by minus the value rT.

Table 10.6 provides an idea about the percentage of various elements present in some solid fuels and their calorific values.

To make the formulae more clear, let us consider some examples for the calculation of calorific value.

Table 10.6 Properties of some useful solid fuels

Fuel	Analysis (%)	Calorific value kcal/kg	
		Gross	Net
Wood	$C = 42.5$ $H = 5.1$ $N + S = 0.9$ $O = 36.5$ Moisture ~ 15	3770	3430
Charcoal	$C = 90.2$ $H = 2.4$ $N + S = 1.5$ $O = 2.9$ Moisture ~ 2	8080	7930
Coke	$C =\sim 80.0$ $H = 0.4$ $N + S = 1.7$ $O = 0.9$ Moisture =~ 8.0	6840	6760

Example 10.4 In an experiment, 0.680 g of fuel was burnt in bomb calorimeter in presence of excess of oxygen. The rise in temperature due to heat generated was 3.61°C for 1080 g of water in the calorimeter. If water equivalent of calorimeter is 150 g, calculate the net calorific value of the fuel.

Solution Using the formula for net calorific value, we can write that

$$\text{LCV} = Q = \frac{(W + w)\,(t_2 - t_1)}{x}\ \text{cal/g}$$

on substituting the values

$x = 0.681$ g—mass of fuel
$W = 150$ g—water equivalent of calorimeter
$w = 1080$ g—mass of water in calorimeter
$t_2 - t_1 = 3.61$°C—rise in temperature

$$\begin{aligned} Q &= \frac{(150 + 1080)(3.61)}{0.681} \\ &= \frac{1230 \times 3.61}{0.681} = 6520.2\text{cal/g} \\ &= 6.520\ \text{kcal/g} \end{aligned}$$

Example 10.5 When the above experiment was repeated for acid correction, cooling correction, and fuse wire correction, the following data was noted:

(i) Acid correction = 50.0 cals

(ii) Fuse wire correction = 5.0 cals
(iii) Cooling correction = 0.05°C

Calculate the value of gross calorific value for the fuel.

Solution Using the expression for GCV i.e., Eq. (10.9), we can write

$$\text{GCV} = \frac{(W + w)\,(t_2 - t_1 + \text{Cooling Correction}) - (\text{Acid} + \text{fuel correction})}{\text{Mass of fuel}}$$

on substituting the values

$$\begin{aligned}
\text{GCV} &= \frac{(150 + 1080)(3.61 + 0.05) - (50 + 5)}{0.681} \\
&= \frac{(1230)(3.66) - (55)}{0.681} \\
&= \frac{4501.8 - 55}{0.681} \\
&= 6529.80\ \text{cal/g} \\
&= 6.529\ \text{kcal/g}
\end{aligned}$$

Similarly Dulong's formula can also be used to calculate the gross and net calorific value of a fuel, if the percentage of various elements present in it are known.

Example 10.6 On analysis, a coal sample gave following results C = 90.2%; H = 2.4% and O = 2.9%. Calculate the net and gross calorific value of the sample.

Solution Using Dulong's formula

$$\begin{aligned}
\text{GCV} &= \frac{1}{100}\left[8080\text{C} + 34,400\left(H - \frac{O}{8}\right)\right]\text{cal/g} \\
&= \frac{1}{100}\left[8080 \times 90.2 + 34,400\left(2.4 - \frac{2.9}{8}\right)\right] \\
&= \frac{1}{100}[728816 + 34,400(2.04)] \\
&= \frac{1}{100}[728816 + 70176] \\
&= \frac{798992}{100} = 7989.92\ \text{cal/g} \\
&= 7.98\ \text{kcal/g} \\
&= 7.98 \times 10^3\ \text{cal/g}
\end{aligned}$$

The students must also remember that SI units for energy is Joules. Hence, we can also convert calories into Joules using the well-known relations

$$\begin{aligned} 1\ \text{Joule} &= 1\ \text{kg m}^2/\text{s}^2 \\ &= 1\ \text{Nm} \\ 1\text{cal} &= 4.184\ \text{Joule} \end{aligned}$$

or

$$1\ \text{Joule} = 0.2390\ \text{cal}.$$

Some other properties of fuels are:

1. It is desirable that the solid fuels should possess low content of moisture, ash, mineral matter, etc., which are inert in nature. Their increased percentage reduces heating value of the fuel. However, this does not apply to the liquid or gaseous fuels.
2. Combustion of the fuel should be complete in presence of theoretical amount of oxygen. The incomplete combustion or addition of more than required amount of air causes loss of heat in the waste gases. The incomplete combustion of coal can be seen by the formation of smoke or formation of carbon monoxide. In the case of liquid fuels, smoke formation usually does not occur, and their incomplete combustion can be detected by release of carbon monoxide gas.
3. The temperature obtained from the combustion of fuel should be sufficiently high to maintain the flame; otherwise the flame is extinguished. In solid fuels, the temperature of the flame produced largely depends on the ratio of volatile combustible material and the carbon residue present in it. It is also necessary that for the active combustion, the temperature of the fuel should be raised to its *ignition temperature*. The ignition temperature of a fuel depends on the conditions of heating. For example, finely divided coal ignites near its ignition temperature, whereas its lump requires higher temperature for ignition.
4. **Calorific Intensity:** Whenever a fuel is burnt, the actual temperature produced by its combustion not only depends on the calorific value of fuel but on several other factors such as:

   - Weight of the products of combustion
   - Weight of the excess air, if any
   - Specific heat of the products, and
   - Heat losses.

   If the entire heat produced is used to raise the temperature, that is, the loss is minimized, it is possible to determine the theoretical temperature that should be attained. However, it is very difficult to attain theoretical temperature due to inevitable loss of heat due to radiation. *Calorific intensity* is therefore described

as the maximum temperature attained when the fuel is completely burnt in the theoretical amount of air required.
Let us consider a case of a fuel which is burning, and a constant temperature is obtained from it. In this condition, the heat generated and heat losses have attained an equilibrium. If the amount of fuel added is increased at this stage, two things will happen:

(i) the rate of combustion will increase; however, the amount of heat produced per kg of fuel will remain same, and
(ii) the heat produced will be increased in comparison to heat loss, and the overall effect will be the increase in calorific intensity of the fuel.

5. Sulfur and phosphorus content in the fuel should be as low as possible because the gases produced from their combustion are corrosive in nature and cause environmental pollution.

In the case of liquid fuels, several other properties like viscosity, flash point, fire point, carbon residue, etc., are determined by the methods reported in Chapter of Lubricants, and their quality is determined.

10.8 Calorific Value of Gaseous Fuel by Boy's Gas Calorimeter

This calorimeter is used for gaseous or easily vaporizable fuels. It consists of suitable gas burner in which a known volume of gas is burnt at known pressure at constant rate. Around burner, there is a chimney or combustion chamber having cooling tube inside, and water at constant rate is passed from outer coil to inner coil. The temperature t_1 and t_2 are measured for incoming and outgoing water, respectively. The circulation of water and burning of gas is continued for about 15 minutes to warm the calorimeter. After that the rate of the fuel and water is so adjusted so that outcoming water comes out at atmospheric pressure. The heat produced is transferred to water and steam formed (if any) condense separately. When steady conditions are reached then following observation are made (Fig. 10.3):

1. Volume of gas burnt at a given temperature and pressure in a certain time at NTP.
2. Quantity of water passed through the coil in same time.
3. Temperature t_1 and t_2.
4. Weight of condensed water.

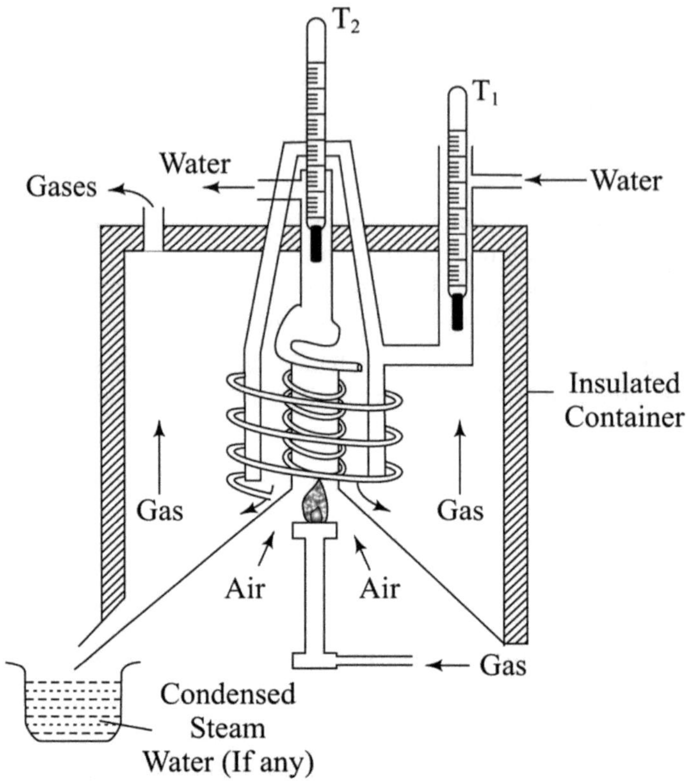

Fig. 10.3 Boy's Gas Calorimeter

10.8.1 Observation

Volume of gas burnt at NTP in time $T = V$
Weight of water passed in time $T = W$
Temperature of incoming water $= t_1$
Temperature of outgoing water $= t_2$
Weight of condensed steam $= w$

Thus,

$$\text{Heat absorbed by water} = W\,(t_2 - t_1)$$

$$\text{Heat produced by fuel} = V \times C$$

As

$$\text{Heat produced} = \text{Heat absorbed}$$

$$V \times C = W\,(t_2 - t_1)$$

$$(\text{Gross C.V.})\ C = \frac{W\,(t_2 - t_1)}{V} \tag{10.10}$$

Net Calorific Value

$$\text{Weight of condensed water} = \frac{W}{V}$$

From unit V of gas

$$\text{Latent heat of steam}/V \text{ of gas} = \frac{W}{V} \times 587$$

$$\begin{aligned}
\text{Net C.V.} &= \text{Gross C.V.} - \text{Latent heat} / V \text{ of gas} \\
&= C - \frac{W \times 587}{V} \\
&= \frac{W\,(t_2 - t_1)}{V} - \frac{W \times 587}{V} \\
\text{Net C.V.} &= \frac{(W\,(t_2 - t_1) - W(587))}{V}
\end{aligned}$$

10.9 Nonconventional Sources of Energy

The various fuels discussed so far are considered as conventional sources of energy; however, they are nonrenewable sources which implies that since they are natural fuels, once exhausted they cannot be replaced. Coal, petroleum, etc. are such fuels, and hence it is desirable that they should be used wisely. Prior to 1970s, coal and petroleum were the main source of energy used for domestic and industrial purposes; however, a sudden rise in the cost of crude oil by oil-producing countries initiated search for alternatives of coal and petroleum. Several nonconventional sources of energy such as solar, wind, water ocean, tidal, and geothermal energy were discovered. The possibility of use of these energy sources were considered advantageous because of their renewable nature. The use of nonconventional energy sources is required because of several reasons:

- The use of coal, petroleum, oil etc. causes environmental problems. The incomplete combustion of these fuels lead to the formation of toxic gas carbon monoxide. The CO_2 formed from the combustion is largely responsible for the greenhouse effect and global warming.

- The energy produced from the thermal power plants is also accompanied with the oxides of carbon, nitrogen, and sulfur and thus cause pollution of air, water, and soil.
- The nuclear power plants also pose a threat as the radioactive waste when buried in soil lead to soil pollution as well as underground water pollution. If the waste is dumped in rivers, it causes poisoning of aquatic life.

Let us now discuss some nonconventional sources of energy.

10.9.1 Wind Energy

The air in motion is called as wind and its power can be calculated on the basis that power is defined as the energy per unit time. The kinetic energy for a particle is equal to

$$= \frac{1}{2} mv^2 \tag{10.11}$$

where m is the mass of the particle and v is its velocity.

If we consider wind moving with velocity v through an area A, the volume per unit time will be Av and if density of air is considered as d, the mass of the air will be $d \times Av$. On substituting the value of m in Eq. (10.11)

$$\begin{aligned} \text{Density} &= \frac{\text{Mass}}{\text{volume}} \text{ or Mass} \\ &= \text{Density} \times \text{volume} \\ &= d \times Av \end{aligned}$$

we get power of the wind

$$\begin{aligned} P &= \frac{1}{2}(dAv)v^2 \\ &= \frac{1}{2}dAv^3 \text{ Watt} \end{aligned}$$

Under normal conditions, the density of wind is reported as $d = 1.2\,\text{kg m}^{-3}$, and if we consider unit area ($A = 1$), the maximum power that can be obtained theoretically will be

$$P = 0.6\,\text{v}^3\,\text{Wm}^{-2} \tag{10.12}$$

Thus, the power of wind is proportional to the cube of its velocity, and therefore wind energy can be utilized in the areas where sufficient wind is available. The meteorological data on wind speed is required for a sufficient period of time to calculate mean speed throughout the year. The average annual wind speeds have been found to vary in the range 0.5 to 18 m/s, whereas for the economic use of wind, a speed in the range 6.0 to 11.0 m/s is needed. Several experiments on wind mill efficiency were carried out, and it was found that the theoretical maximum efficiency of a windmill is 59 to 60%. Thus, the maximum power that can be obtained from the wind energy is 60% of Eq. (10.12). Hence, the normal maximum power that can be derived from the wind is

$$P = 0.36v^3 \text{ Wm}^{-2}$$

The use of wind will require several precautions such as:

- To avoid damage due to wind, a wind mill should not be used in winds having a velocity greater than 20.6 m/s, which is the upper limit of Beaufort force.
- As load on a windmill increases with the square of the wind velocity, the support should be capable of taking enough load.

The largest windmill of the United States having twin bladed propeller of diameter 53.5 m and having an output of about 1.2 MW collapsed in 1941 as it was unable to withstand high winds.

It must also be realized that a large windmill might be capable of producing power equal to 1 MW; however, looking into the requirements, this amount is very small and may be of use for providing electricity to small towns.

10.9.2 Solar Energy

Solar energy is one of the important sources of energy available on earth, and it is received in the form of radiations from the sun. At the earth's surface solar energy can be visualized in a number of ways such as evaporation of the water by sun rays leads to gain in potential energy and is the source of power for hydroelectricity. Similarly, green plants use sunlight to convert CO_2 and H_2O into carbohydrates and thus provide food for animals. An enormous amount of energy is provided by sun by series of thermonuclear reactions and its order has been reported as 3.5×10^{27} W. Out of the total solar radiations produced, a very small fraction falls on the surface of earth. If we assume this small fraction as x, nearly 30% of x is reflected as short wave radiations, whereas 45% is directly converted to heat. The remaining 25% is used in the photosynthesis, evaporation of water, and so on. A detailed data on use of solar energy in earth's climatic system is presented in Fig. 10.4.

Under the most favorable conditions of the sun light an energy of 1000 W/m^2 is available in desert areas, whereas this amount in considerably reduced over the 24 h of the day due to variation in climatic conditions. However, the value of 1000 W/m^2

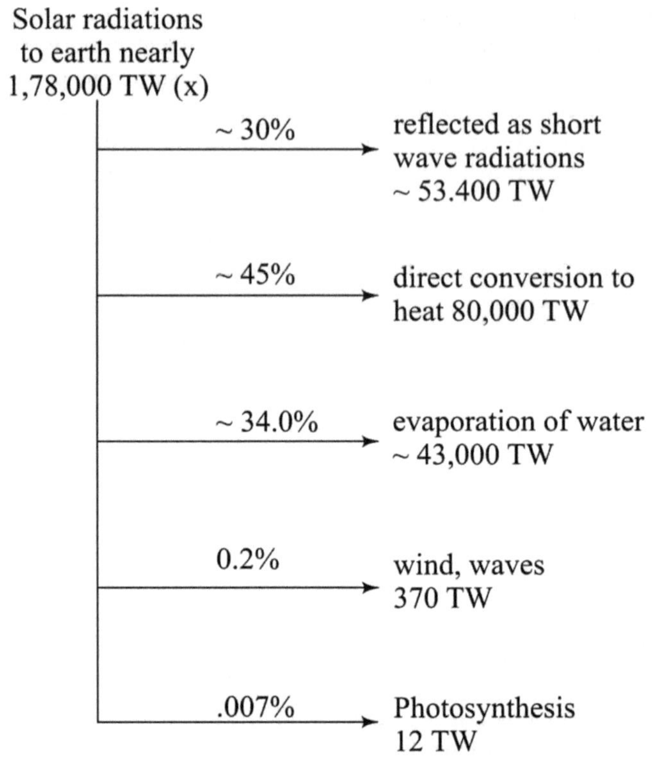

Fig. 10.4 Use of solar energy in the earth's climate system (unit 1 TW = 10^{12} W)

is usually the amount considered as renewable energy source from the sun. To get an idea about this energy into practical terms, one can say that energy of nearly 208 W/m^2 is equivalent to 5 kwh day^{-1} m^{-2}. Thus, a solar energy collector of collecting solar energy from about 4 m^2 area should be capable of supplying enough energy for a family. This requirement of area may be less in summer and more in winter.

Basics of Solar Energy Collection

To use solar energy in a convenient form, say, in generating electricity or for providing hot water, its collection is necessary. For a heat engine operating between two temperatures T_H and T_L (high and low temperature in kelvin respectively), the Carnot efficiency is given by the expression

$$\eta = \frac{T_H - T_L}{T_H}$$

where η is Carnot efficiency. From this expression, it is obvious that to achieve high efficiency, the value of T_H must be as high as possible. Similarly, the temperature T which can be reached for an incoming radiation of power density P can be given by the relation

$$\alpha P = \sigma e T^4 \tag{10.13}$$

where

σ is the Stefan-Boltzmann constant
e is the emissivity of the absorbing material
α is the absorptivity.

If a body re-emits all the radiations falls on it, has been assigned the value of e as unity, and is known as a block body, if no radiations are emitted by a body, its emissivity has been considered as zero. Similarly black body also absorbs all radiant energy falls on it and hence its absorptivity α is considered as unity. In practice, however, not more than 80 to 90% energy falls on a body is absorbed by it and hence the value of α varies between 0.8 to 0.9 for such bodies.

Let us consider an example when a solar energy collector is placed so that the available day time energy for collection is 800 $\mathrm{W/m^2}$. If the collector has an absorption capacity 80% and emissivity of 0.5, the maximum temperature that can be achieved can be calculated using Eq. (10.13).

$$\alpha P = \sigma e T^4$$

on substituting the values

$$0.8 \times 800 = 5.67 \times 10^{-8} \times 0.5T^4$$

where

$$\sigma = 5.67 \times 10^{-8}\ \mathrm{Wm^{-2}\ K^{-4}}$$

on solving the equation, we get

$$T = 115°\mathrm{C} \text{ or } 388\ \mathrm{K}$$

Thus, a temperature of 115°C can be achieved using the solar collector under normal conditions. For providing hot water in homes, such a temperature is more than enough for routine uses. However, if we use this temperature in a perfect heat engine, its efficiency at room temperature ($T_L = 20°\mathrm{C}$) will be equal to

$$\eta = \left[\frac{388 - 293}{388}\right] \times 100$$
$$= 24.5\%$$

one of the possible ways to increase the temperature is to use collector material having a low value of e and high value of α.

Example 10.7 A solar collector material having absorptivity 0.9 and emissivity 0.04 is used for collecting solar energy. If the sun light energy available is 750 W/m^2, calculate the maximum temperature that can be achieved.

Solution Using Eq. (10.3), we can write

$$\alpha P = \sigma e T^4$$

as

$$\alpha = 0.9$$
$$P = 750$$
$$\sigma = 5.67 \times 10^{-8}$$
$$e = 0.04$$

The equation becomes

$$0.9 \times 750 = 5.67 \times 10^{-8} \times 0.04 \times T^4$$
$$T^4 = \frac{0.9 \times 750}{5.67 \times 10^{-8} \times 0.04}$$
$$= 2976.2 \times 10^8$$
$$T = 738 \text{ K}$$

A common application of solar energy in producing distilled water can be seen on the roofs of many laboratories. The sheet of glass is placed over a black colored painted vessel of approximate size 1 m × 1 m. The glass sheet placed normally makes an angle of 45 to 60% as shown in Fig. 10.5. The vessel has a provision of collecting the distilled water. Ordinary tap water is filled in the vessel. The sunshine heats the inside of vessel and water is vaporized and condenses on the inner surface of glass. As the glass sheet is kept at an angle of 45 to 60°C, the condensed water droplets moves easily and are collected in a bottle kept outside. It is an economical and convenient method for preparing distilled water.

Another important area where solar energy can be effectively used is photoelectric effect. Photovoltaic cells are used for the direct generation of electricity when the sun shines for a long time. In such cells a silicon of high-grade purity is used and

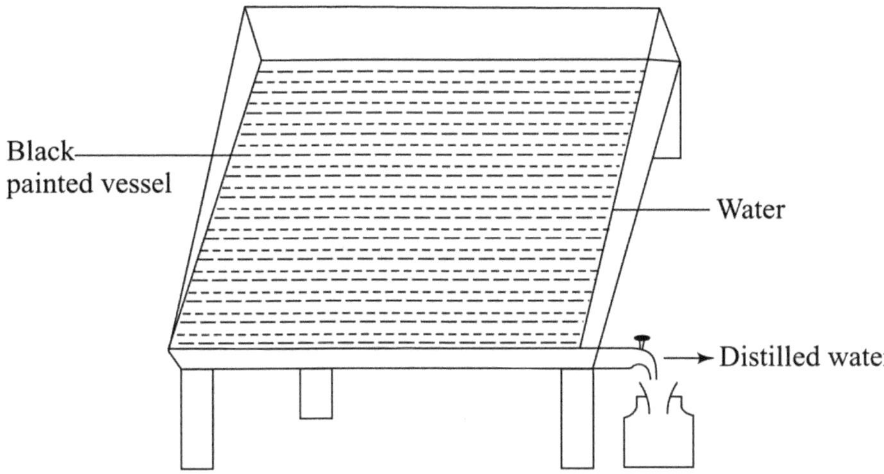

Fig. 10.5 Solar energy device for producing distilled water

called as solar grade silicon. A single crystal, polycrystals, amorphous silicon, etc. are used in cells, and these cells are free from pollution. The only difficulty realized with these cells is that the conversion efficiency of these cells is only 10–15%, and hence their cost of production of electricity is very high ranging 80 to 100 times that of a conventional power station. The basic principle involved in photovoltaic cells is the excitation of electrons from the valence band to the conduction band with the help of energy received from sun rays.

10.10 Biogas and Biomass

Biogas is a mixture of several gases such as methane, hydrogen, hydrogen sulfide, and small amount of CO_2. The major amount of gas present in biogas is methane which constitutes about 60 to 70% of the total gases. Biogas is obtained by the anaerobic degradation of biomass, which basically includes cattle dung, plant wastes, agricultural wastes, and so on. The degradation of biomass occurs in presence of water, and biogas produced can be used as an excellent fuel. In villages and other rural areas, the use of biogas can save consumption of kerosene or other conventional fuels.

The basic unit for biogas production consists of a tank having inlet and outlast pipes of nearly 10 to 15 cm diameter as shown in Fig. 10.6. The animal *dung* and other biomass material are allowed to enter in the tank after proper mixing with water. Under anaerobic conditions, microorganisms cause a type of fermentation, and biogas produced is collected in gas holder steel tank. A pipe line from the steel tank then supplies gas to the place of its use. The biogas produced can be used

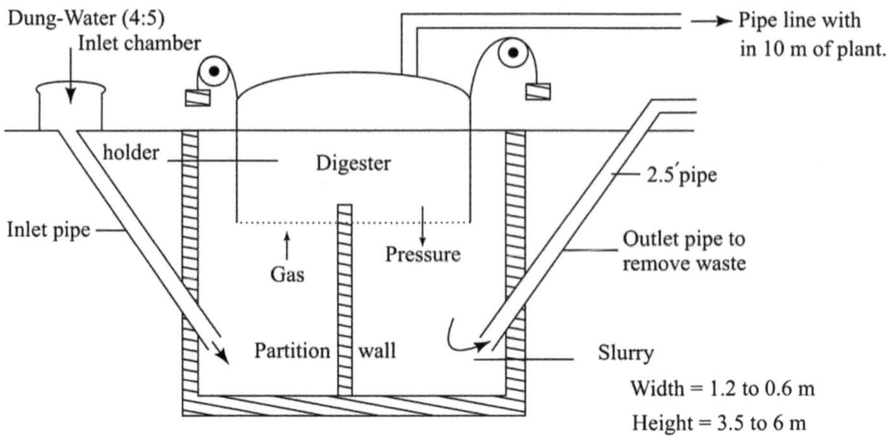

Fig. 10.6 A conventional biogas plant

not only for burning in place of kerosene but can also be used for cooking purposes. The use of biogas is capable of saving conventional fuel energy to a large extent. For example, an average size biogas plant provides energy in a year which is equal to the energy produced by four tones of wood. Thus, it is desirable that biogas production should be increased by utilizing the biomass. However, the limitations associated with biogas production are its:

(i) dependence on the animal dung and other waste material. Normally, a cattle or buffalo produces 10 to 12 kg of dung per day; however, the amount of gas produced is nearly 65 m^3 per ton of wet dung. Thus, to get a proper supply of biogas, there must be three to four animals to provide enough dung.
(ii) The anaerobic degradation requires specific temperature to produce biogas. The rate of this degradation becomes slow in rainy season as well as in winter season, and hence, the supply of gas becomes irregular.
(iii) Due to the formation of H_2S, the stove should be kept at least 10 to 12 meters away from the biogas plant to avoid foul smell.

In view of increasing use of conventional fuels in the entire globe, it has become necessary to use these fuels efficiently and explore more sources of alternative or nonconventional energy sources.

Solved Problems

1. Complete combustion of 1.050 g of a fuel was carried out in a calorimeter in presence of excess of oxygen. The temperature of 1500 g of water of calorimeter raised from 25°C to 27.8°C. If water equivalent of the calorimeter is 125 g, calculate the calorific value of the fuel.

Solution Using Eq. (10.6) we can write

$$Q = \frac{(w + W)\,(t_2 - t_1)}{x}$$

On substituting the values

$$Q = \frac{(1500 + 125)(27.8 - 25)}{1.050}$$

$$\begin{aligned} Q &= \frac{1625 \times 2.8}{1.050} \\ &= 4,333.3\ \text{cal/g} \\ &= 4.3\ \text{kcal/g} \end{aligned}$$

Remember: If only calorific value is asked, it means high calorific value (HCV) or gross calorific value (GCV).

2. A coal sample on analysis gave the following results: C = 90%; O = 3.0%, S = 0.5%, N = 0.5% and ash 2.5%. The law calorific value of the coal was found to be 8500 kcal/kg. Calculate the higher calorific value of coal as well as percentage of Hydrogen in the coal sample.

Solution Using Eqs. (10.7) and (10.8) we can write

$$\begin{aligned} \text{GCV} &= \text{LCV} + 9H/100 \times 587 \\ &= \text{LCV} + 0.09H \times 587 \\ &= [8500 + 52.8H]\text{kcal/kg} \\ \text{GCV} &= \frac{1}{100}\left[8080\text{C} + 34{,}500\left(H - \frac{0}{8}\right) + 2240\ \text{S}\right] \\ &= \frac{1}{100}\left[8080 \times 90 + 34{,}500\left(H - \frac{3.0}{8}\right) + 2240 \times 0.5\right] \\ &= 7272 + 345H - 129.4 + 11.2 \\ &= [7153.8 + 345H]\ \text{kcal/kg} \end{aligned}$$

The two GCV values should be equal, thus

$$\begin{aligned} 8500 + 52.8H &= 7153.8 + 354H \\ 292.2H &= 1346.2 \\ H &= 4.60\% \end{aligned}$$

and

$$\begin{aligned} \text{GCV} &= [8500 + 52.8 \times 4.6] \\ &= [8500 + 242.9] \\ &= 8742.9 \text{ kcal/kg} \end{aligned}$$

3. 1.000 g of a coal sample obtained from Bihar on analysis gave C = 93% and H = 6%. The ash content in the sample was 1%. The coal sample was burnt in bomb calorimeter and following results were obtained

   (i) Wt. of coal taken = 1.000 g
   (ii) Weight of water taken = 500 g
   (iii) Initial temperature of water = 24°C
   (iv) Final temperature of water = 26.2°C
   (v) Fuse wire correction = 10 cal
   (vi) Acid correction = 50 cal
   (vii) Water equivalent of calorimeter = 2000 g

   If latent heat of condensation of steam is 580 cal/g, calculate gross and net calorific values of the sample.

   Solution Using Eq. (10.9), we can write

$$\text{GCV} = \frac{(W + W)\,(t_2 - t_1 + \text{cooling correction})}{\text{mass of fuel}} - (\text{Acid} + \text{fuse correction})$$

On substituting the values

$$= \frac{(2000 + 500)(26.2 - 24) - (50 + 10)}{1.000}$$

Note: As cooling correction is not given, it is considered as zero.

$$= \frac{(2500)(2.2) - (60)}{1.000}$$

$$\text{GCV} = 5540 \text{ cal/g}$$

and net calorific value

$$\begin{aligned}\text{LCV} &= [\text{GCV} - 9H/100 \times 580] \\ &= [5540 - 9 \times 6/100 \times 580] \\ &= [5540 - 313.2] \\ &= 5226.8\ \text{cal/g}\end{aligned}$$

4. Analysis of a coal sample was carried out for moisture, ash, and volatile matter. 1.5642 g of the sample was weighed in a platinum crucible and heated at 110°C for 1 h. On cooling in a desiccator for 1 h, the crucible was weighed again and found as 1.5022 g. Then crucible was covered with a vented lid and strongly heated at 950°C for exactly 7 min. The residue on cooling weighed 0.7628 g. The heating was then done without lid for nearly 1 hour, and a constant weight of the residue was then obtained as 0.2140 g. Calculate the percentage of fixed carbon.

Solution

(i) Weight of coal sample = 1.5642 g
(ii) Weight of sample after

$$\begin{aligned}\text{heating at } 110^\circ\text{C} &= 1.5022\ \text{g} \\ \text{Moisture} &= 1.5642 - 1.5022 \\ &= 0.0620\ \text{g} \\ \%\ \text{of moisture} &= \frac{0.062 \times 100}{1.5642} \\ &= 3.96\%\end{aligned}$$

(iii) Weight of sample after

$$\begin{aligned}\text{heating at } 950^\circ\text{C} &= 0.7628 \\ \text{Volatile matter} &= 1.5022 - 0.7628 \\ &= 0.7394\ \text{g} \\ \%\ \text{of volatile matter} &= \frac{0.7394 \times 100}{1.5642} \\ &= 47.27\%\end{aligned}$$

(iv)

$$\begin{aligned}\text{Ash content left in the last} &= 0.2140\ \text{g} \\ \%\ \text{of ash} &= \frac{0.214 \times 100}{1.5462} \\ &= 13.68\%\end{aligned}$$

Therefore percentage of fixed carbon can also be calculated as

$$
\begin{aligned}
\%\ \text{fixed carbon} &= [100 - (\text{moisture} + \text{volatile matter} - \text{ash})] \\
&= 100 - (3.96 + 47.27 - 13.68)] \\
&= [100 - 37.55] \\
&= 62.45\%
\end{aligned}
$$

5. A sample of coal (0.5248 g) was strongly heated at 950°C in a muffle furnace for 1 h. The ash formed was extracted with HCl and made up to 100ml. The solution was heated and barium nitrate solution was added slowly. The precipitate of $BaSO_4$ was filtered, dried, and weighed. The weight of the precipitate was found as 0.0482 g. Calculate the percentage of sulfur in the coal sample.

   Solution

$$
\begin{aligned}
\%\ \text{of sulfur} &= \frac{32}{233} \times \frac{\text{Wt. of BaSO}_4}{\text{Wt. of sample}} \times 100 \\
&= \frac{32}{233} \times \frac{0.0482}{0.5248} \times 100 \\
&= \frac{154.24}{122.27} \\
&= 1.26\%
\end{aligned}
$$

6. In a Boy's gas calorimeter, a gaseous fuel was burnt, and the following data were obtained:

   Volume of gas used = 0.1 m^3 at STP
   Weight of water heated = 10 kg
   Temperature of inlet water = 22°C
   Temperature of outlet water = 30°C
   Weight of steam condensed = 0.025 kg

   Calculate the gross and net calorific values of fuel. Assume the latent heat of water vapors formed as 580 kcal/kg.

   Solution Using Eq. (10.10)

$$
\begin{aligned}
\text{GCV} &= \frac{W\ (t_2 - t_1)}{V} \\
&= \frac{10(30 - 22)}{0.1\ \text{m}^3} \\
&= 800\ \text{kcal/m}^3
\end{aligned}
$$

and net calorific value or low calorific value

$$
\begin{aligned}
&= \text{GCV} - (m/V) \times 580 \\
&= 800 - \left(\frac{0.025}{0.1}\right) \times 580 \\
&= 800 - 145 \\
&= 655\text{kcal/m}^3
\end{aligned}
$$

7. Analysis of a fuel gave C = 90%; H = 3.5%; O = 3.0%; S = 0.5; H_2O = 0.1%; N = 0.5% and ash = 2.4% Calculate the minimum weight of air required for complete combustion of 1 kg of fuel.

Solution If 1 kg of fuel is taken, the percentages can be converted as C = 900 g; H = 35 g; O = 30 g and S = 5 g.

The complete combustion would occur according to the reactions

$$
\begin{aligned}
C + O_2 &\longrightarrow CO_2 \\
2H + 1/2O_2 &\longrightarrow H_2O \\
S + O_2 &\longrightarrow SO_2
\end{aligned}
$$

Thus, 12 g of carbon will need 32 g of oxygen
2 g of hydrogen will need 16 g of oxygen
32 g of sulfur will need 32 g of oxygen.
Amount of oxygen required is

$$
\begin{aligned}
&= \left[900 \times \frac{32}{12} + 35 \times \frac{16}{2} + 5 \times \frac{32}{32}\right] \\
&= 2400 + 280 + 5 \\
&= 2685 \text{ g.}
\end{aligned}
$$

As 30 g oxygen is already available, the amount of oxygen required = 2685 − 30 = 2655 g. The weight of air can be calculated assuming 23% oxygen in air by weight

$$
\begin{aligned}
\text{Amount of air needed} &= \frac{2655 \times 100}{23} \\
&= 11{,}543 \text{ g} \\
&= 11.543 \text{ kg}
\end{aligned}
$$

8. A gaseous fuel was found to have following gases on analysis: CH_4 = 14%; H_2 = 32%; N_2 = 40% and O_2 = 14%. If 25% excess air is used for its complete combustion, calculate the volume of air in litre actually supplied per m^3 of this gaseous fuel.

Solution The combustion reactions of various gases present in the fuel are

$$CH_4 + 2O_2 \longrightarrow CO_2 + 2H_2O$$

$$H_2 + \frac{1}{2} O_2 \longrightarrow H_2O$$

In 1 m^3 of gaseous fuel, the volume of gases will be

$$CH_4 = 0.14\ m^3$$

$$H_2 = 0.32\ m^3$$

$$N_2 = 0.40\ m^3$$

$$O_2 = 0.14\ m^3$$

Volumes of oxygen required for methane and hydrogen as per combustion reactions are

$$CH_4 = 0.14\ m^3 \times 2 = 0.28\ m^3$$

$$H_2 = 0.32\ m^3 \times 0.5 = 0.16\ m^3$$

$$\text{Total oxygen needed} = 0.28 + 0.16$$

$$= 0.44\ m^3$$

$$\text{Oxygen already present} = 0.14\ m^3$$

$$\text{Net } O_2 \text{ needed} = 0.44 - 0.14$$

$$= 0.30\ m^3$$

$$= 300\ L$$

Volume of air required assuming 21% oxygen in air by volume

$$= 300 \times (100/21) \times (125/100)$$

$$= 1785.7\ L$$

Note: The term 125/100 is appearing because 25% excess air is passed.

9. A coal sample obtained from a mine was analyzed for various elements and exhibited the following results: C = 75%; H = 5.2%; N = 3.2%; O = 12.1% and ash 4.5%.
If 1 kg of this coal sample is burnt, how much minimum air is needed in kg. Also calculate the GCV and net calorific value of the coal sample.

Solution For 1 kg of coal sample, the weight of different elements will be

$$\begin{aligned} C &= 750\ \text{g} \\ H &= 52\ \text{g} \\ N &= 32\ \text{g} \\ O &= 121\ \text{g} \\ \text{Ash} &= 45\ \text{g} \end{aligned}$$

Minimum weight of oxygen needed will be

$$\begin{aligned} &= [C \times (32/12) + H \times (16/2) - O] \\ &= [750 \times (32/12) + 52 \times (16/2) - 12] \\ &= [2000 + 416 - 121] \\ &= [2295]\text{g} \end{aligned}$$

Assuming 23% oxygen in air by weight, minimum amount of air will be equal to

$$\begin{aligned} &= \frac{2295 \times 100}{23} = 9978.3\ \text{g} \\ &= 9.978\ \text{kg} \end{aligned}$$

As only carbon and hydrogen percentage is given

$$\begin{aligned} \text{GCV or HCV} &= \frac{1}{100}\left[8080 \times C + 34500\left(H - \frac{O}{8}\right)\right] \text{kcal/kg} \\ &= \frac{1}{100}\left[8080 \times 75 + 34500\left(5.2 - \frac{12.1}{8}\right)\right] \\ &= \frac{1}{100}[606{,}000 + 127{,}218.7] \\ &= 7332.18\ \text{kcal/kg} \end{aligned}$$

and net calorific value will be

$$= (\text{GCV} - 0.09\text{H} \times 587)\ \text{kcal/kg}$$
$$= (7332 - 0.09 \times 5.2 \times 587)$$
$$= 7057\ \text{kcal/kg}$$

10. Analysis of a coal sample gave C = 81%; O = 8%, S = 1%, H = 5%; N = 1% and remaining were ash contents. The 1 kg of sample was burnt in air. Calculate the minimum amount of air by weight for the complete combustion of fuel. Also determine the percentage composition of the dry products of combustion formed.

    Solution For 1 kg of fuel, the amount of various elements will be equal to C = 810 g; O = 80 g, S = 10 g; H = 50 g and N = 10 g.

    The complete combustion of C, S, and H would occur through the reactions

$$C + O_2 \longrightarrow CO_2$$
$$S + O_2 \longrightarrow SO_2$$
$$2H + 1/2O_2 \longrightarrow H_2O$$

and weight of oxygen needed will be

$$\text{for C} = 810 \times 32/12 = 2160\ \text{g}$$
$$\text{S} = 10 \times 32/32 = 10\ \text{g}$$
$$\text{H} = 50 \times 16/2 = 400\ \text{g}$$
$$\text{Total oxygen needed} = 2570\ \text{g}$$
$$\text{Oxygen already available in fuel} = 80\ \text{g}$$
$$\text{Net } O_2 \text{ needed} = 2490\ \text{g}$$

Assuming 23% oxygen in air by weight, the amount of air needed

$$= 2490 \times \frac{100}{23} = 10{,}826\ \text{g}$$
$$= 10.826\ \text{kg}$$

The products of combustion are CO_2 and SO_2.
From the equations written above, 44 g CO_2 is obtained from 12 g of carbon, hence, weight of CO_2 obtained from 810 g of carbon = $810 \times 44/12 = 2970$ g.
Similarly, Wt. of SO_2 obtained from 10 g of sulfur

$$= 10 \times 64/32 = 20\ \text{g}$$

Nitrogen would also be present in the products and its weight will be equal to

$$= 10 \text{ g (present in the fuel)} + 10{,}826 \times 77/100 \text{ (from air used)}$$
$$= 10 + 8336 = 8346 \text{ g}$$

$$\text{Total weight of dry products} = 2970 + 20 + 8346$$
$$= 11{,}336 \text{ g}$$
$$\text{Percentage composition of } CO_2 = \frac{2970}{11{,}336} \times 100$$
$$= 26.19\%$$
$$\text{of } SO_2 = \frac{20}{11{,}336} \times 100$$
$$= 0.176\%$$
$$\text{of } N_2 = \frac{8346}{11{,}336} \times 100$$
$$= 73.62\%$$

11. The percentage analysis of a producer gas was found as $H_2 = 19.4\%$, $CO = 20.5\%$, $N_2 = 50.1\%$, $CH_4 = 4.2\%$, $CO_2 = 6.0\%$.
If 30% excess air is supplied for the complete combustion of the gas, calculate percentage of dry products formed.

Solution In 1 m^3 of the producer gas, the volume of contents will be $H_2 = 0.194\ m^3$; $CO = 0.205\ m^3$; $N = 0.501\ m^3$; $CH_4 = 0.042\ m^3$ and $CO_2 = 0.060\ m^3$
Combustion reactions involved will be

$$H_2 + \frac{1}{2} O_2 \longrightarrow H_2O$$
$$CO + \frac{1}{2} O_2 \longrightarrow CO_2$$
$$CH_4 + 2O_2 \longrightarrow CO_2 + H_2O$$

and corresponding volume of oxygen required will be 0.5, 0.5, and 2.0 times.

Thus,

$$\text{for } H_2 = 0.194 \times 0.5 = 0.097 \text{ m}^3$$
$$CO = 0.205 \times 0.5 = 0.102 \text{ m}^3$$
$$CH_4 = 0.042 \times 2.0 = 0.084 \text{ m}^3$$
$$\text{Total } = 0.283 \text{ m}^3$$

Taking 21% oxygen by volume in air and 30% air in excess, the volume of air will be

$$= 0.283 \times \frac{100}{21} \times \frac{130}{100}$$
$$= 1.752 \text{ m}^3$$

Volume of dry products formed

$$CO_2 = 0.06 \text{ m}^3 \text{ (available in fuel)}$$
$$0.205 \times 1 \text{ m}^3 \text{(from CO)}$$
$$0.042 \times 1 \text{ m}^3 \text{ (from } CH_4)$$
$$\text{Total } 0.307 \text{ m}^3$$

$$N_2 = 0.501 \text{ m}^3 \text{ (available in fuel)}$$
$$1.752 \times \frac{79}{100} \text{ m}^3 \text{ (from air used)}$$
$$\text{Total } 1.885 \text{ m}^3$$

$$O_2 = 1.755 \times 21/100 \text{ m}^3 \text{ (from air used)}$$
$$= 0.368 \text{ m}^3$$

Total volume of dry products

$$= 0.307 + 1.885 + 0.368 = 2.56 \text{ m}^3$$

Percentage of dry products

$$CO_2 = \frac{0.307 \times 100}{2.56}; \quad N_2 = \frac{1.885 \times 100}{2.56}; \quad O_2 = \frac{0.368 \times 100}{2.56}$$
$$CO_2 = 11.99\%, \quad N_2 = 73.63\%, \quad O_2 = 14.37\%$$

12. A sample of coal was analyzed for the elements present, and the percentage of various elements was found as follows:
$C = 78\%, H = 4.1\%, O = 12.0\%,\ S = 1.2\%,$
$N = 2.1\%$ and remaining was ash.
Calculate the minimum amount of air necessary for complete combustion of 1 kg of coal.

Solution In 1 kg of coal, the weight of different elements will be C = 0.78 kg, H = 0.041 kg, O = 0.12 kg, S = 0.012 kg, and N = 0.021 kg.

Minimum amount of oxygen required for combustion

$$= \left[0.78 \times \left(\frac{32}{12}\right) + 0.041 \times \left(\frac{16}{2}\right) + 0.012 \times \left(\frac{32}{32}\right) - 0.12\right] \text{kg}$$

$$= [2.08 + 0.328 + 0.012 - 0.12]$$

$$= 2.30 \text{ kg}$$

Minimum amount of air required assuming 23% oxygen by weight

$$= 2.30 \times \frac{100}{23} = 10 \text{ kg}$$

13. 1.5 kg of a carbon sample was burnt in air. Calculate the weight and volume of air requirement for the complete combustion of carbon.

Solution For complete combustion of carbon

$$C + O_2 \longrightarrow CO_2$$

The weight of O_2 needed is

$$= \frac{32}{12} \times 1.5 = 4.00 \text{ kg}$$

Weight of air required will be

$$= 4.00 \times \frac{100}{23} = 17.39 \text{ kg}$$

As 32 g of oxygen occupies 22.4 lit at NTP

$$\text{4000 g of oxygen occupies } = \frac{22.4}{32} \times 4000$$

$$= 2800 \text{ lit}$$

Hence volume of air (having 21% oxygen by volume) is

$$= \frac{100}{21} \times 2800$$
$$= 13{,}333 \text{ lit}$$
$$= 13.33 \text{ m}^3.$$

Exercises

1. Calculate the GCV and NCV of a coal sample having the following composition: C = 85%, H = 8%, S = 1%, N = 2%, and ash 4%. Latent heat of steam 587 cal/g.
2. A coal sample has the following composition: C = 90%, O = 3%, S = 0.5%, N = 0.5%, and ash 2.5%. If latent heat of steam is 587 cal/g and NCV = 8490.5 kcal/kg, calculate GCV and percentage of hydrogen.
3. 0.72 g of a fuel containing 80% carbon when burnt in bomb calorimeter, increased the temperature of water from 27.3 to 29.1°C. If the calorimeter contains 250 g of water and its water equivalent is 150 g, calculate the HCV in kJ/kg.
4. A coal sample has the following composition: C = 93%, H = 6% and ash = 1%. Latent heat of steam = 580cal/g. The following data were obtained when the above coal was tested in bomb calorimeter

   Weight of coal = 0.92 g, Weight of water = 550 g

   Water equivalent of calorimeter = 2200 g, Rise in temperature = 2.42°C

   Fuse wire correction = 10.0cal, Acid correction = 50.0 cal

   Calculate the values of GCV and NCV.
5. 0.5 g of a sample of coal was used in bomb calorimeter to get GCV = 8600 cal/g. The ash obtained is extracted with acid and fluxed with barium nitrate to get the ppt of barium sulfate. After drying, the ppt was weighed and found to be 0.05 g. Calculate the % of S in coal and NCV if H in the sample is 2%.
6. 1.0 g of a sample of coal was used for nitrogen estimation by Kjeldahl's method. The ammonia obtained was absorbed in 25 ml $N/10H_2SO_4$. To neutralize the remaining acid, 15 ml of 0.1 N NaOH was required. Determine the % of N in coal.
7. A sample of coal (2.5 g) was weighed in a silica crucible. After heating for one hour at 110°C, the residue weighed 2.415 g. The crucible is next covered with

lid and strongly heated at 950°C and then heated without lid to get a constant weight of 0.245 g. Calculate the % results of the above analysis.

8. Calculate the volume of gas required for complete combustion of 1 m^3 of fuel having composition: $CO = 46\%, CH_4 = 10\%, H_2 = 4\%, C_2H_2 = 2\%,\ N_2 = 1\%$ and remaining CO_2.
9. Calculate the volume of gas required for complete combustion of 1 m^3 of fuel having composition $O_2 = 14\%, CH_4 = 14\%, H_2 = 32\%,\ N_2 = 40\%$.
10. A coal sample containing $C = 62.4\%, H = 4.1\%, O = 1.2\%,\ S = 0.8\%,\ N = 1.2\%$, moisture $= 15.1\%$ and ash 9.7% was burnt. Calculate the weight of air theoretically required for complete burning of per kg of coal.

Chapter 11
Instrumental Techniques

A variety of organic/inorganic compounds are synthesized these days for their use as drugs, dyes, and many more. To establish their structure, instrumental techniques are used. In this chapter, we will deal with some of these common techniques; however, before that let us have an idea about electromagnetic radiations and the common terms associated with it.

11.1 Electromagnetic Radiations

It is well-known that all forms of radiation can be described as oscillating or wave-like electric and magnetic field in space as shown below. The electric and magnetic fields are described by sine waves and at right angles with each other. Radiations, such as light, microwaves, television, x-rays, etc., have been called as electromagnetic radiation and provide an idea as to how electrons behave in atoms.

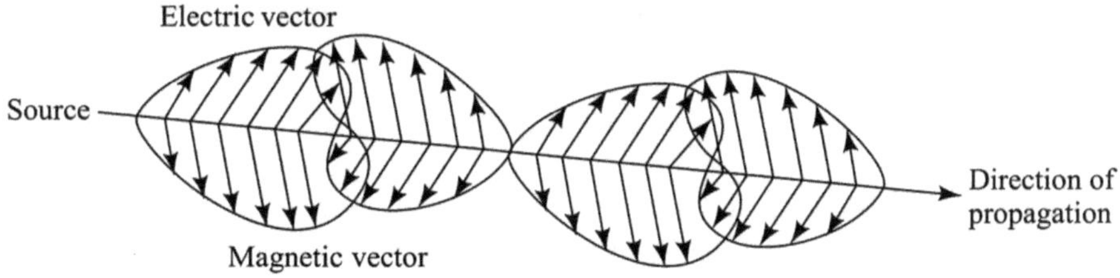

Let us consider two sine waves as shown in Fig. 11.1. The essential difference between them is that one wave completes two cycles in the time other completes one, assuming that both are traveling at the same speed. The two common terms used for the propagation of these waves are wavelength and wave amplitude. The wavelength of a wave is the distance between two successive crests (high points) or troughs (low points) and can be measured in meters, nanometers, or any other

© The Author(s), under exclusive license to Springer Nature Switzerland AG 2026
R. N. Goyal, H. Goel, *Textbook of Engineering Chemistry*,
https://doi.org/10.1007/978-3-032-22068-4_11

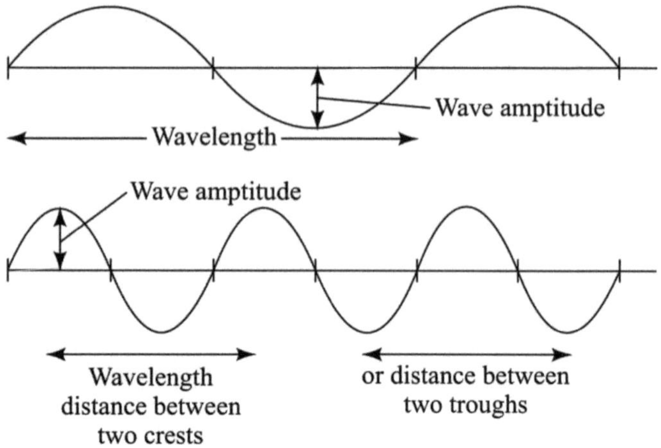

Fig. 11.1 Representation of sine waves

convenient unit. The wavelength is represented by the Greek letter λ (lambda). The maximum height of the wave, measured from the axis of propagation, is called as its *amplitude*. The points of zero amplitudes are called as *nodes*. Another term used in connection with waves is called as *frequency*, represented by Greek letter v (nu). The frequency is described as the number of waves passing through a point in some given amount of time. It is a number represented in per second or commonly called nowadays as hertz. If above characteristics of a wave is known, its velocity can be determined by expression

$$\text{Velocity} = \text{wavelength} \times \text{frequency}$$

and if wavelength is measured in meter and frequency in s^{-1}, the unit of velocity will be (m/s), that is, meters per second. For visible light and other electromagnetic radiations, the velocity is well-known and is given as

$$c = 2.9979 \times 10^8 \text{ m/s}$$

and the symbol c is commonly used for the velocity of light.

We all know that visible light consist of a spectrum of colors ranging from red light to violet light. The wavelength of red light is long whereas of red light is short. Remember, visible light is only a very small portion of the total electromagnetic radiations as shown in Fig. 11.2. When the waves have wavelength shorter than visible light, we get ultraviolet radiations which cause sun burn and are injurious to eyes. X-rays and γ-rays have wavelength further lower than ultraviolet rays.

On the other hand, at longer wavelength than ultra violet rays, infrared radiations are obtained. The micro wave radiation have still longer wavelength as shown in Fig. 11.2.

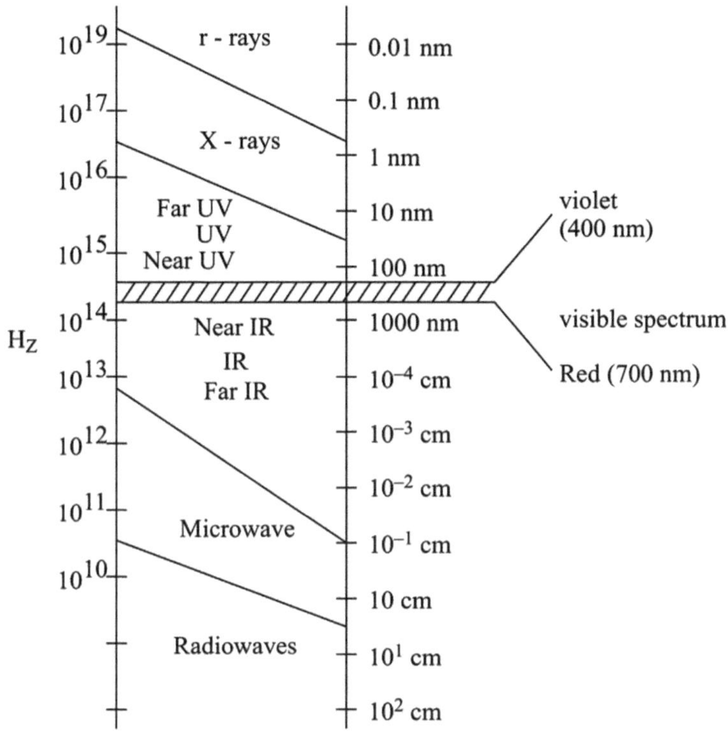

Fig. 11.2 Total electromagnetic spectrum

11.2 Ultraviolet and Visible Spectroscopy

The absorption of ultraviolet or visible radiation transitions of a molecule are observed in the electronic energy levels, and hence it is also referred as electronic spectroscopy. Electronic spectroscopy measures the extent of multiple bond or aromatic conjugation within molecules. The non-bonding electrons present on oxygen, nitrogen, and sulfur also enter the conjugation of multiple bond systems. Electronic spectroscopy also helps to differentiate between conjugated and non-conjugated dienes, conjugated dienes and trienes, and $\alpha - \beta$ unsaturated ketones from their $\beta - \gamma$ analogs. It is also successfully used to study the effect of substituent in conjugated dienes and carbonyl compounds.

11.3 Chromophore Concept

Compounds that absorb light in the range of 100 to 800 nm appear colored to human eye. However, the precise color is the function of which wavelength of light the compound subtracts from white light.

Chromophore was used to define systems responsible for imparting color to the compound. However, it is now used to define any compound that absorbs electromagnetic radiation, whether or not it imparts color, e.g., carbonyl group is a chromophore.

Similarly, auxochrome was defined as any compound that enhances the color imparting properties of a chromophore, e.g., $-OR$, $-NH_2$, $-NR_2$ etc. Actually, an auxochrome extends the conjugation of a chromophore, as its shares its non-bonding electrons with it. Thus, it becomes part of a new conjugated chromophore.

11.4 Theory

When ultra violet/visible light is absorbed, only one electron is excited to a high energy level, and other electrons remain unaffected. This excited state is produced in a very short time $\left(10^{-15}\ \text{s}\right)$. Probably, the excitation of electron takes place from the highest occupied molecular orbital to the lowest available unfilled orbital. But in some cases, many transitions are observed and hence show several bands in the spectrum as shown in Fig. 11.3. Some transitions are forbidden; thus, intensity of the associated band is low.

In alkanes, the transition involved is $\sigma \longrightarrow \sigma^*$, i.e., from the σ orbital to high energy σ^* antibonding orbital. This being high energy process involves short wavelength ultraviolet light (around 150 nm).

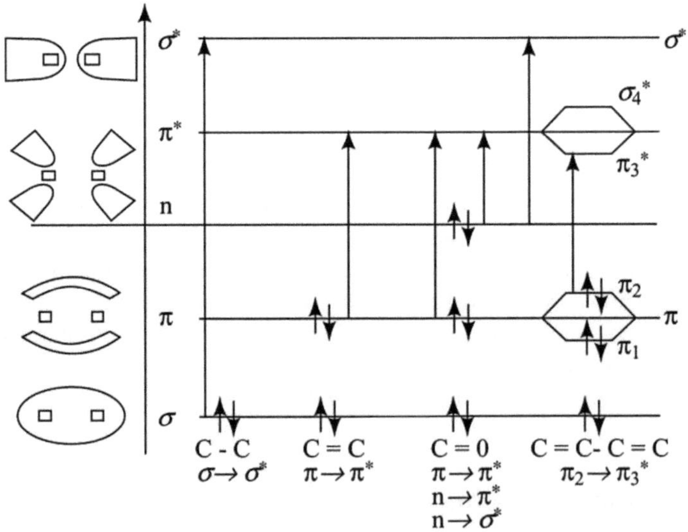

Fig. 11.3 Possible transitions in UV spectroscopy

In alkenes, $\pi \longrightarrow \pi^*$ transition is the most important, showing absorption band around 170–190 nm. In saturated aliphatic ketones, the lowest transition involves the non-bonding electrons on oxygen. The $n \longrightarrow \pi^*$ transition is forbidden, hence its intensity is low. The wavelength being long (around 280 nm) allowed transition are $n \longrightarrow \sigma^*$ and $\pi \longrightarrow \sigma^*$. Energy being higher, wavelength is shorter (around 185 for $n \longrightarrow \sigma^*$ and 160 for $\pi \longrightarrow \pi^*$).

In conjugated dienes, π orbitals of separate alkene groups combine to form two bonding (π_1 and π_2) and two antibonding (π_3^* and π_4^*) orbitals. The transition $\pi_2 \longrightarrow \pi_3^*$ is most possible. Conjugated dienes show absorption at longer wavelength (typically around 217 nm).

11.5 Beer's and Lambert's Laws

These laws govern light absorption by molecules. Beer's law relates absorption to the concentration of absorbing solute, and Lambert's law relates total absorption to optical path length.

$$A = \log \frac{I_0}{I} = \varepsilon C l$$

or

$$\varepsilon = \frac{A}{Cl}$$

where

I_0 is the intensity of incident light
I is the intensity of light transmitted through sample
$\log \frac{I_0}{I}$ is the intensity of absorbance (A) of the solution
C is the intensity of concentration of solute
l is the intensity of path length of sample
ε is the intensity of molar absorptivity

The molar absorptivity, ε, is constant for a compound at a particular wavelength.

Absorption spectra are plotted with increasing $\in$ on the ordinate and wavelength on abscissa, increasing from left to right. Presence of substituent may shift spectra to longer or shorter wavelength. Bathochromic shift (or red shift) is shift to longer wavelength. On the other hand, hypsochromic shift (or blue shift) is shift to shorter wavelength.

11.6 Sample Handling

Ultraviolet spectra of compounds are recorded in vapor phase or in solutions. A variety of quartz cells are used for determinations. Cells used for determination of spectra in solution vary in path length from 1 to 10 cm, however, 1 cm^2 quartz cells are commonly used.

For recording the spectrum of a small amount of solution microcells are available these days. For preparing the solution, the sample is weighed and dissolved in required amount of solvent in a volumetric flask. Further dilution, if required, can be carried out. It is necessary to use clean cells for recording the spectrum, and hence cells should be washed either with detergent or dilute solution of nitric acid followed by water. The cells should then be rinsed with the solvent used in preparing the solution. Usually methanol, ethanol, cyclohexane, 1, 4-dioxane etc. are used as a solvent. In case the compound is expected to be unstable in the solution, its spectrum should be recorded as quickly as possible after preparing the solution.

A simple UV/vis is spectrophotometer can be represented by a block diagram as shown in Fig. 11.4. The source on the left side produces radiations. In the UV region, tungsten lamp and in visible region sodium lamp is used. The radiations are passed through a monochromator which selects the required wavelength. The beam than passes through sample and a plot of absorbance against wavelength is recorded using a detecting device. Typical UV spectra of potassium permanganate and uric acid are presented in Fig. 11.5.

11.6.1 Applications

1. Electronic spectroscopy is applied to conjugated dienes, trienes, and polyenes whose increased conjugation impart color. Red color of tomatoes and carrots arise due to these conjugated molecules.
2. Plants contain many conjugated poly-ynes and eneynes whose structures have been elucidated with electronic spectroscopy.
3. The UV spectrum helps to distinguish between $\alpha - \beta$ unsaturated carbonyl compounds.

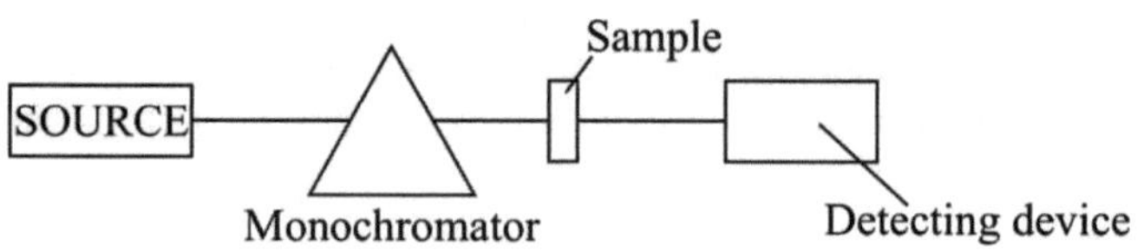

Fig. 11.4 A block diagram of UV spectrophotometer

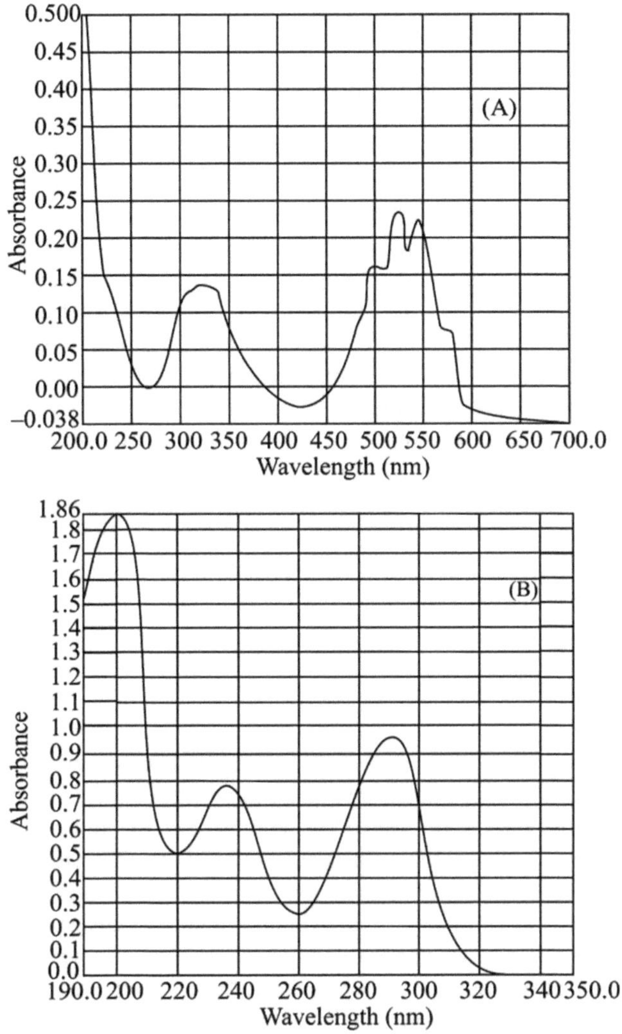

Fig. 11.5 UV spectra of 0.1 mM solution of potassium permanganate (A) and uric acid (B) in water

4. Substituted benzenes are generally studied as alkyl substituents and shift absorption slightly to lower wavelength while non-bonding pair substituents (—OH, —OR, —NH_2 etc.) shift the absorption substantially to longer wavelength.
5. Compounds having benzene rings fused together give rise to very complex spectra in ultraviolet/visible spectroscopy.

11.7 Infrared Spectrometry (IR)

Infrared region is a part of electromagnetic spectrum between the visible and microwave regions. It is widely used for the detection of functional groups in various compounds. An organic chemist matches the IR of an unknown compound with that of an authentic sample. It is, however, found that the IR of two compounds are never totally identical. The limited portion for study of IR is between 4000 and 400 cm^{-1}. The IR region is divided into three, of which the mid-infrared region (650–4000 cm^{-1}) is of most interest also known as the *finger-print region.* Frequencies lower than 650 cm^{-1} lie in the far infrared region and that of frequencies higher than 4000 cm^{-1} in the near infrared region. In an infrared spectrophotometer, where infrared radiation is passed through the sample, some of the frequencies are absorbed, while others are transmitted. Hence, an infrared spectrum is the representation of absorbance or transmittance against frequency. Infrared absorption is measured in frequency units by wave number ($\bar{v}$) measured in reciprocal centimeters $\left(cm^{-1}\right)$ or by wavelength λ, measured in micrometers (in older literature):

$$cm^{-1} = 10^4/\mu m$$

11.8 Molecular Vibrations

At room temperature, atoms in a molecules constantly keep on vibrating, each bond in it having its characteristic stretching and bending frequency as it absorbs light of particular frequency. Two atoms joined by a chemical bond can be compared as two balls connected by a spring. Due to molecular vibrations, there is change in dipole moment of the molecule, and this gives rise to absorption bands in IR. Otherwise molecule is referred to as infrared inactive, and hence no absorption in IR region occurs. Molecular absorptions are of two types: stretching and bending.

The stretching absorptions always appear at higher frequencies in the infrared spectrum than the bending absorptions. A stretching vibration involves movement along the bond axis such that the inter atomic distance either increases or decreases while a bending one involves change in bond angle of a common atom or the movement of a part of the molecule with respect to the whole molecule such that no movement among the part is observed with respect to each other, e.g., twisting, rocking, and torsional vibrations. The different stretching modes of water are shown in Fig. 9.6.

The degrees of freedom of a molecule is the sum of degrees of freedom of individual atoms. For a non-linear molecule with atoms, the number of vibrational modes is $(3n - 6)$ whereas for linear molecules it is $(3n - 5)$. However, it is necessary that a particular vibration must cause a change in the dipole moment of the molecule so that absorption of Infrared energy results. Thus, water molecule being non-linear possesses three fundamental vibrations as shown in Fig. 11.6. Consider a

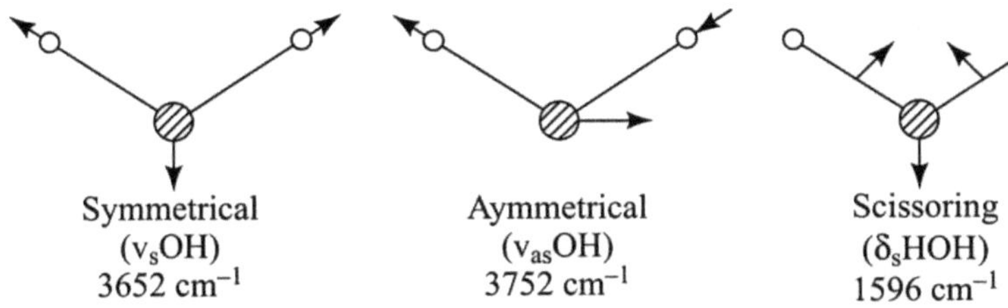

Fig. 11.6 Different stretching modes of water molecule

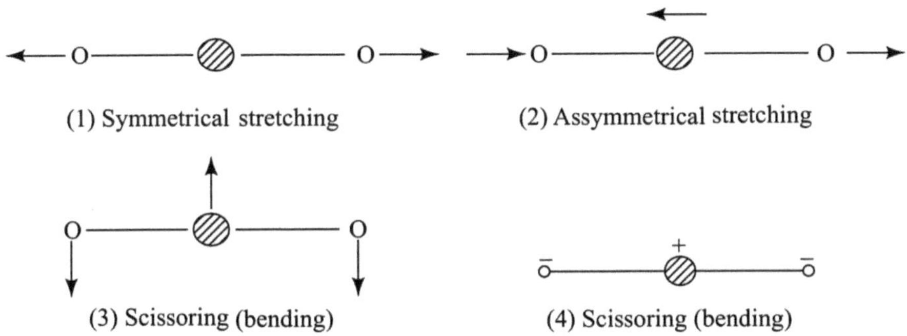

Fig. 11.7 Fundamental vibrations for nonlinear carbon dioxide molecule

CO_2 molecule, and being a linear molecule, it contains four fundamental vibrations as shown in Fig. 11.7.

The stretching frequencies for important bonds are given below:

Bond	Frequency Range (cm^{-1})
C—C, C—O, C—N	1300 – 800
C = C, C = O, C = N, N = O	1900 – 1500
C ≡ C, C ≡ N	2300 – 2000
C—H, O—H, N—H	3800 – 2700

In case of aromatic rings, C—H bending gives absorption in the region 650–900 cm^{-1}.

11.8.1 Calculations of Vibrational Frequencies

The equation used for calculation of vibrational frequencies is Hooke's law which relates frequency with bond strength and mass as follows:

$$v \alpha \sqrt{\frac{\text{bond strength}}{\text{mass}}}; \quad \text{thus } v = \frac{1}{2\pi c}\left(\frac{k}{m_1 m_2/m_1 + m_2}\right)^{1/2}$$

where

v = frequency
c = velocity of light
k = constant related to the strength of the spring and is called as bond force constant.
m_1, m_2 = masses of atoms

The terms $\frac{m_1 m_2}{m_1+m_2}$ is commonly expressed as μ, the reduced mass of the system.

11.8.2 Factors Affecting Vibrational Frequencies

As in the case of ball and spring system when energy is gained, only change in amplitude is observed, and there is no change in the frequency. Similarly in case of chemical bonds, quantum theory also imposes some restrictions. Hence, vibrational energy only increases due to quantum jumps, and thus energy difference between successive vibrational levels is

$$\Delta E = hv \qquad \text{or} \qquad \Delta E = N_A hv$$

where N_A is Avogadro constant ($= 6.02 \times 10^{23}$ mol^{-1}).

When in excited state, molecule can give its energy in the form of rotational, translational, or collision processes. However, it is found that interpretation of infrared spectrum is not easy as overtones (twice the frequency of fundamental) are also found. Vibration frequencies are also affected by nature of the neighboring substituent, conjugation, and strains felt by them. Hydrogen bonding also affects these frequencies.

11.8.3 Instrumentation

IR spectrophotometers operate on simple principle. A block diagram of double beam IR spectrophotometer is shown in Fig. 11.8.

Infrared radiations are produced by electrically heating rods of Nernst filament (fabricated from oxides of Zr, Th, Ce, Y, Er, etc.) and Globar (silicon carbide) in the range 1000–1800°C. The radiations coming from the source are divided into two, a sample beam and reference beam, both being focused into the respective regions. The reference beam and sample beam combine together into a single beam and gives a switching frequency at the detector. When the two beams are of equal intensity, the instrument is at an optical null, and 100% transmittance is observed. The attenuator

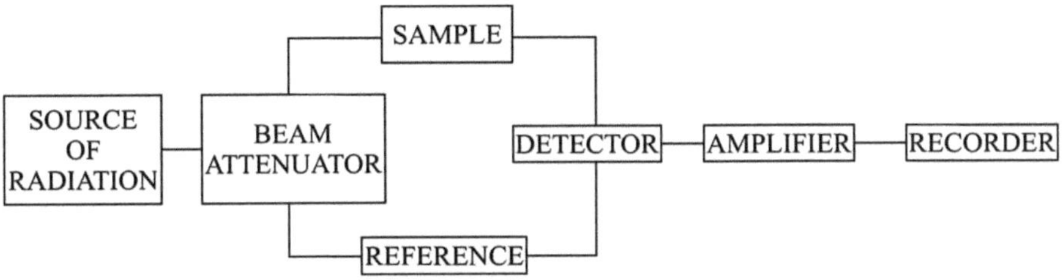

Fig. 11.8 Block diagram of an IR spectrophotometer

moves in and out of the reference beam in response to the signal at detector, due to absorbance by the sample beam. When absorbance is observed by the sample beam, the attenuator drives into the reference beam so as to match its intensity with the sample beam. Detector measures the signals by means of heating effect. There are two types of detectors used in IR spectroscopy. They are thermocouple type and bolometer type.

In thermocouple, one of the two bimetallic junctions is heated due to radiant energy, and an electromotive force is produced. In bolometer, however, change in resistance is recorded. Detector sees the reference and sample beam alternately, and any change due to absorption in the intensity of radiations is recorded. The amplified signal of the detector is used to position attenuator so that both the beams have equal intensity. The amount of attenuation is the absorption by the sample, which is recorded by the pen.

Fourier transform infrared spectroscopy is an advancement over the ordinary IR technique and is extensively used these days. In this technique, the entire radiation range is passed through the sample in less than a second, and its resolving power is very high. Further, FT-IR technique can be combined with other techniques such as liquid chromatography or high-performance liquid chromatography (HPLC) to get the IR spectrum of various fractions obtained in it.

11.8.4 Handling of Sample

IR spectra can be recorded for solids, liquids, or gases. Gas samples cells are available in lengths from few cm to even meters. Liquids are examined while neat or when solution is made. Neat samples always show higher absorbance. For recording the IR of liquid sample, NaCl plates are used. Silver chloride plates may be used for samples that dissolve sodium chloride plates (particularly aqueous solution). For recording spectra of solid samples KBr disks, mulls and deposited films are used. KBr disks are prepared by mixing thoroughly sample with KBr and then grinding well in a smooth agate mortar. The mixture is than pressed with special dies to give a transparent disk. Mulls are prepared by grinding the sample with oil. Mull is examined as a thin film between flat salt plates. Nujol is commonly used for preparing mull. When hydrocarbon bands interfere, fluorotube mulls are used. Solid films can be prepared by depositing solution of sample in a volatile solvent drop by drop on NaCl plates. Deposited film technique is commonly used to study resin and plastics.

11.8.5 Applications of IR Spectroscopy

IR spectra of complex compounds give absorptions in certain regions which are characteristic of it and hence is referred to as the fingerprint region which lies roughly from 900 to 1400 cm^{-1}. When an IR of a substance is recorded, it is compared with that of an authentic sample. If two samples have same IR, they are referred to as identical. Thus, samples can be identified by comparing the IR of an authentic sample with the observed IR.

Moreover, IR can be used to identify the functional groups of various substances. Interpretation of IR first recommends study of carbon skeleton, i.e., alkane, alkene, alkyne, and aromatic residues. Some common functional groups are than seen such as $C = O$, if present, whether associated with $N - H$, $C-$ H str, etc. Presence of $O - H$ str or $C \equiv N$ str is then examined. Also sulfur and phosphorous compounds are studied with IR. However, study of only IR is not always sufficient. The use of other techniques such as mass, NMR etc. helps in identifying a compound well. Characteristic infrared absorption frequencies of some important groups is given in Table 11.1.

A typical IR spectrum of *o*-chlorobenzoic acid is presented in Fig. 11.9.

11.9 Nuclear Magnetic Resonance Spectroscopy (NMR)

Nuclear magnetic resonance (NMR) is another form of absorption spectrometry, much more superior compared to IR and UV spectroscopy. NMR helps to record the change in magnetic properties of atomic nuclei and also their positions within the molecule. When a sample is placed in a magnetic field under certain conditions, it absorbs radiations in the radio-frequency regions characteristic of the sample. Hence, a plot of frequencies of the absorption peaks versus peak intensities constitutes an NMR spectrum. NMR thus involves study of radio-frequency radiations. It was first developed by E.M. Purcell and Felix Bloch in 1946.

11.9.1 Theory

The nuclei of certain atoms have been considered to spin, and such spinning of these charged particles produces a magnetic moment. The magnetic moment so produced

Table 11.1 IR frequencies of some important groups

Bond	Compound type	Frequency range (cm^{-1})
C—H	Alkanes	2800–2950 1340–1480
C—H	Alkenes	3010–3080 670–980
C—H	Aromatic rings	3000–3150 670–880
C—H	Alkynes	3310
C = C	Alkenes	1650–1680
C ≡ C	Alkynes	2000–2250
C—C	Aromatic rings	1500
C—O	Alcohols, ethers, carboxylic acids	1050–1300
C = O	Aldehydes, ketones, carboxylic acid, esters	1650–1750
O—H	Monomeric alcohols, phenols Hydrogen bonded alcohols, phenols carboxylic acid	3600–3640 3200–3600 2400–3000
N—H	Amines	3310–3500
C—N	Amines	1150–1350
C ≡ N	Nitriles	2200–2260
—NO_2	Nitro compounds	1510–1565 1340–1385

is along the axis of spin, and hence these nuclei act like tiny bar magnet. A common nucleus is of hydrogen, i.e., ^{1}H. Thus, if a proton is placed in external magnetic moment can be aligned with or against the external field. Naturally alignment with field will be stable, and energy must be absorbed to change over to less stable alignment. The amount of energy needed to flip the proton to less stable alignment depends on the strength of external field.

Consider a top, as well as its spinning motion; the top also performs a slower motion in which the spinning motion is around a particular axis. This is called as the precessional motion. Similarly, a proton behaves as a spinning top as it precesses around the axis of the external magnetic field B_0 and either aligns in the direction of field or opposite to field as shown in Fig. 11.10.

The spinning frequency of a proton does not change, but its precessional frequency does. Hence, precessional frequency is directly proportional to external field: (B_o)

$$\upsilon \alpha B_o$$

A precessing proton, thus, absorbs energy according to the source of radio frequency. If the precessing frequency is similar to the frequency of radio frequency, the two are said to be in resonance; hence, the term nuclear magnetic resonance was given.

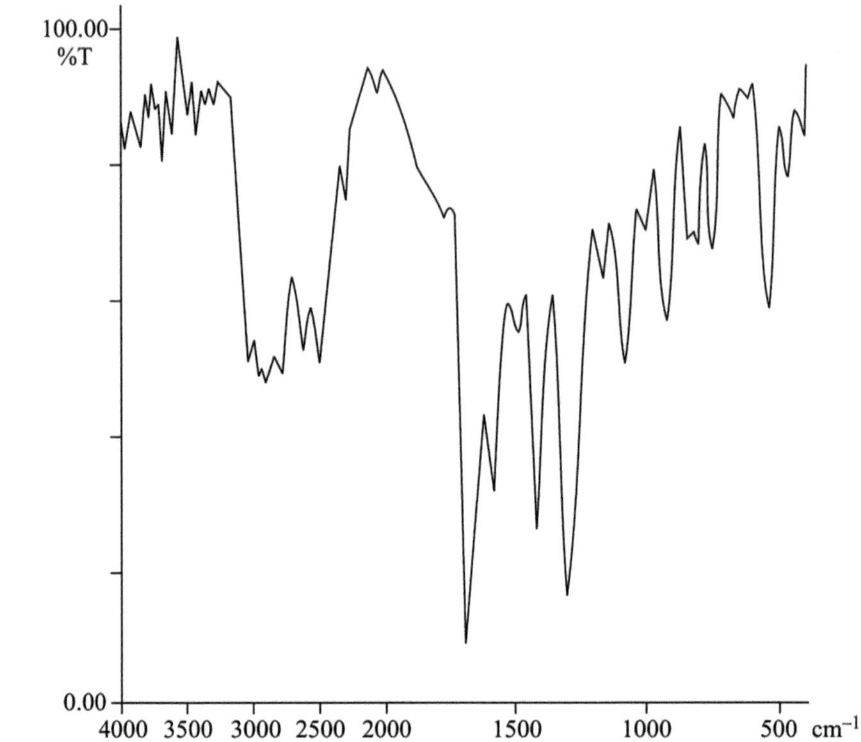

Fig. 11.9 IR spectrum of o-chlorobenzoic acid

Fig. 11.10 Alignment possibilities of proton in external magnetic field

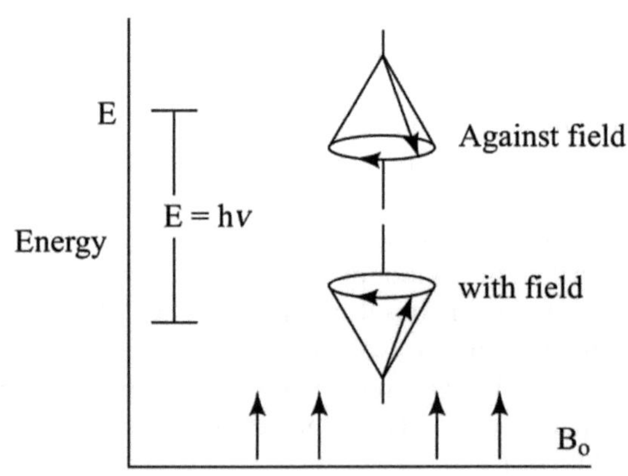

NMR phenomenon is, however, only possible for those nuclei which have spin quantum number greater than 0. Spin quantum number is associated with mass number and atomic number of nuclei as shown in Table 11.2.

Table 11.2 Association of mass number and atomic number on spin quantum number

Mass number	Atomic number	Spin quantum number	Example
odd	odd or even	$\frac{1}{2}, \frac{3}{2}, \frac{5}{2}$	^{1}H, ^{19}F, ^{31}P, ^{11}B, ^{79}Br, ^{13}C, ^{127}I
even	even	0	^{12}C, ^{16}O, ^{32}S
even	odd	1.2, 3	^{2}H, ^{14}N

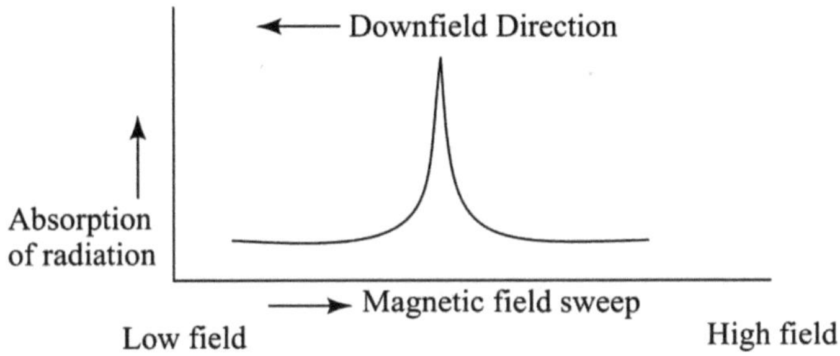

Fig. 11.11 Signal pattern of NMR

According to the magnetic field, a nuclei can take different orientation; the number of orientations is given by $(2I + 1)$. Thus, for nuclei having spin $\frac{1}{2}$, two orientations are possible.

In an applied magnetic field proton precess at a frequency υ (in Hz) given exactly as

$$\upsilon = \frac{\gamma B_0}{2\pi}$$

where B_0 = external magnetic field, γ = magnetogyric ratio, which is a ratio of nuclear magnetic moment, and the nuclear magnetic angular momentum, I.

In NMR technique, the radio frequency is kept constant but the magnetic field is varied. At an applied magnetic field the energy to flip the proton according to the energy of radiation depends on the nature of nuclei. As such, absorption takes place, and a signal is observed. Thus, signal or peaks are observed in an NMR as shown in Fig. 11.11.

11.9.2 Instrumentation

Earlier NMR instruments used permanent magnets or electromagnets corresponding to 60, 80, 90, or 100 MHz. Above 100 MHz instruments are based on helium cooled super conducting magnets (solenoids) and operate in pulse FT mode.

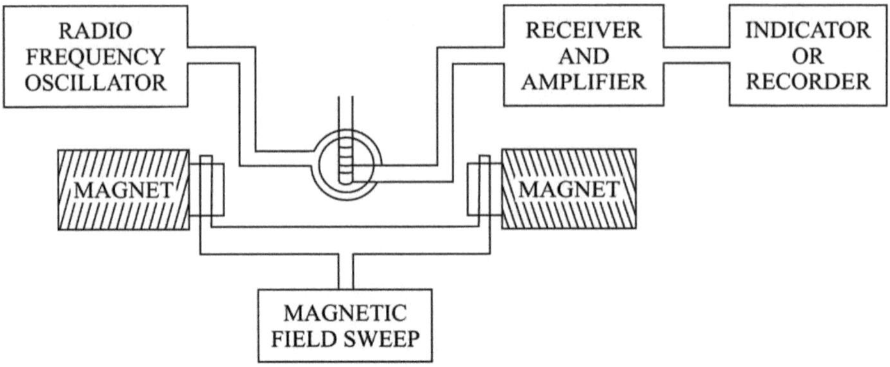

Fig. 11.12 A block diagram of NMR

The sample is placed in the probe having the transmitter, receiver, and a spinner to spin the tube. These elements are placed between the poles of an electromagnet or permanent magnet. The spectrum is obtained either by cw scan or pulse FT as a series of peaks whose areas are proportional to the number of protons. A block diagram of NMR instrument is presented in Fig. 11.12.

The NMR provides us the following information:

1. The number of signals obtained tells us how many different kinds of protons are present.
2. The position of the signals tells us about the type of protons.
3. The splitting of signals tells us about the nearby protons.
4. The intensity of signal tells us about number of protons.

Protons present in the same environment absorb at the same applied field, whereas, if environment is different for protons, they absorb at different field strengths. For example, consider compounds such as

$CH_3 - CH_2 - Cl$	$CH_3 - CHCl - CH_3$	$CH_3 - CH_2 - CH_2Cl$
Ethyl chloride	Isopropyl chloride	*n*-propyl chloride

In the first compound, the environment around protons of two carbons are different; hence, two NMR signals will be obtained. In isopropyl chloride, again two NMR signals will be obtained because the two CH_3 are identical, whereas in *n*-propyl chloride three NMR signals will be obtained as protons at three carbons have different environment.

11.10 Chemical Shift

In NMR spectrum, the number of signals tell us about the different kinds of protons present in the molecule. The position of the signal tells whether the proton is aliphatic or aromatic, primary, secondary, or tertiary, adjacent to or not to any other group. These all conditions give to different electronic environments which determine where a proton absorbs is NMR spectrum.

Electron spins under the influence of a magnetic field, and thus their own magnetic field (induced field) opposing the applied magnetic field is diminished. Thus, field actually experienced by the proton is diminished and in the proton is called as shielded. Instead, if the induced field reinforces the applied field, the proton is deshielded. Hence, shielding shifts the absorption upfield, while deshielding shifts it to downfield. Thus, shielding and deshielding cause shift in positions of NMR absorption, referred to as chemical shift. When change in absorption of a particular proton from that of a reference proton is considered, it is called as chemical shift of a proton. Protons in the same chemical environment have the same chemical shift, whereas protons in different chemical environment exhibit different chemical shift.

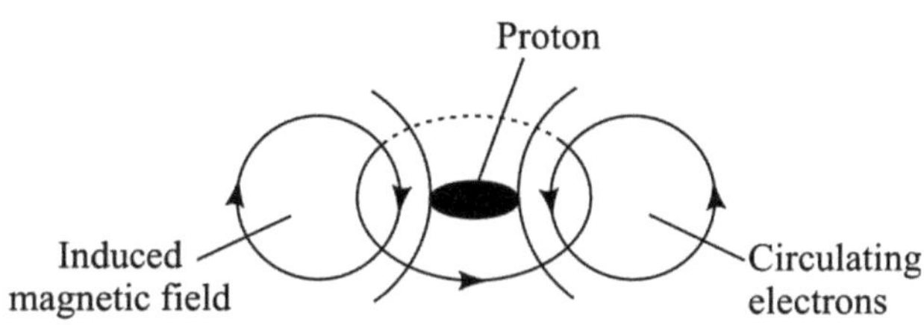

11.10.1 Measurement of Chemical Shift

Commonly, difference in frequency is measured with respect to some reference group of nuclei. Generally for proton and ^{13}C, tetramethyl silane (TMS) is used as a reference. TMS is chosen due to the following reasons.

1. It is chemically inert
2. It has low boiling point and hence is easily removed for recoverable sample.
3. It is soluble in most organic solvents and added to sample solution as an internal standard.

Since TMS is not soluble in water or in D_2O its sodium salt, 3-(trimethylsilyl) propane sulfonic acid is used. Essentially for measurement of peak position, there are three scales.

(a) The ppm or δ scale with TMS at 0
(b) The τ (tau) scale with TMS signal at 10
(c) The Hertz (Hz) scale

The units of chemical shift δ (delta) and τ (tau) are independent of magnetic field applied as the operating frequency of an instrument is directly proportional to the strength of applied magnetic field. Thus,

$$\delta = \frac{\text{observed shift (Hz)}}{\text{operating frequency (MHz)}}$$

Chemical shifts are generally expressed in units of δ (delta) which is proportional to difference in parts per million from a reference standard. The relationship between δ values (in ppm) and frequency (in Hz) can be expressed by the relations:

$$\delta x = \frac{\upsilon x - \upsilon_{\text{TMS}}}{\upsilon_0}$$

where

δx = chemical shift
υ_x and υ_{TMS} = frequencies of signals for sample x and TMS
υ_0 = operating frequency of the instrument

Relations between δ and τ (tau) is

$$\delta = 10 - \tau$$

11.10.2 Factors Affecting Chemical Shift

Electronegativity and Inductive Effect

Chemical shift position of CH_3 protons when methyl group is attached to highly electronegative groups has been studied. It was found that as the electronegativity of the group attached increases the value of δ increases. Thus, for the halo methanes, the value of δ was found as follows:

$$\underset{\delta\,4.26}{CH_3F} > \underset{\delta\,3.0}{CH_3Cl} > \underset{\delta\,2.82}{CH_3Br} > \underset{\delta\,2.16}{CH_3I}$$

This indicates that when an electronegative group or atom is attached, the shielding effect decreases because the atom or group attracts electron toward itself. Thus, electron density around the proton deceases, and hence deshielding takes place. Because of smaller deshielding, smaller values of applied field will be required to bring the proton to resonance. Thus, higher electronegativity results in greater deshielding effect.

11.10.3 Effect of Magnetic Anisotropy Space Effect

Circulation of π electrons about nearby nuclei generates an induced field which either opposes or reinforces with the applied field at the protons depending upon the position of proton in space. Shielding occurs in case of alkynes but in alkanes deshieding occurs according to the location of proton in space and hence causes space effect. The aldehydic protons and protons on benzene appear far downfield, in NMR (Table 11.3).

Downfield shift or paramagnetic shift of proton attached to alkenes, aldehydes, and aromatic protons is explained by considering the magnetic fields induced by the internal magnetic field B_0 on π electrons. Since the magnetic fields induced by π electrons are dependent on direction, it is referred to as anisotropic effect.

11.10.4 Hydrogen Bonding

It is found that hydrogen atom showing property of hydrogen bonding absorbs downfields in NMR, as the hydrogen bonded proton is attached to an electronegative group which attracts electrons. Thus, hydrogen bonding decreases the electron density and causes deshielding and thus absorption occurs downfield.

Table 11.3 Chemical shifts of protons for some groups

Gruop	*Approximate δ value*
aldehydic	9.5–10.0
benzene	6–9
alkene C = CH	4–8
alkyne C ≡ CH	1.5–3.5
R— OH	5
R—C(=O)—OH	11
CH_3—N(—)—Ar	2.6
CH_3—C(=O)—OR	2.0
CH_3Cl	3.0
CH_3 — O — R	3.3

Spin-Spin Splitting

When the spin of a proton interacts with that of another proton or protons attached to adjoining carbon, a secondary environment effect is superimposed on the chemical shift and is called as spin-spin splitting. Interaction involves the spins of all H—C, C—C, and C—H bonds. If the protons are in same environment, interaction is not manifested but when in different environments, splitting occurs. This splitting occurs as the applied field is either reinforced or opposed according to the local fields of the hydrogen nuclei attached to adjacent atom. For the purpose of reporting spin-spin splitting, a scale of hertz (Hz) is used.

Generally a chemical shift between two signals is reported. A range of chemical shifts observed for protons in different compounds is presented in Fig. 11.13.

When the spins of one set of nuclei exert upon the resonance behavior of another, multiple structures are obtained. This happens because there is small interaction of coupling between the two groups of protons.

Consider the case of an ethyl group CH_3CH_2. The three hydrogen atoms of methyl groups are equivalent. Thus, there are $(n + 1)$, i.e., $(3 + 1) = 4$ spin energy levels, two have $+1/2$ spin, and the remaining have $-1/2$. Now consider effect of methylene protons in ethanol (CH_3CH_2OH). The methylene protons can have four possible combination of spin states, which are

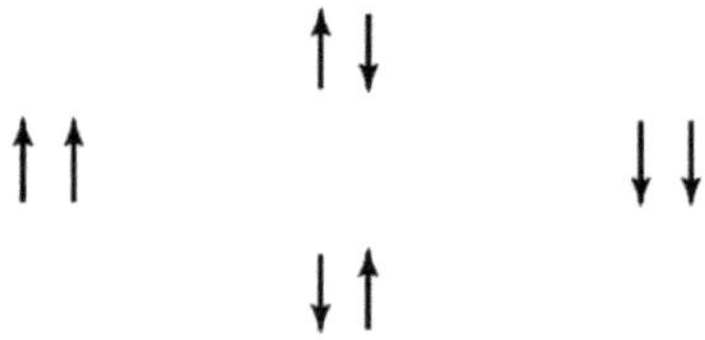

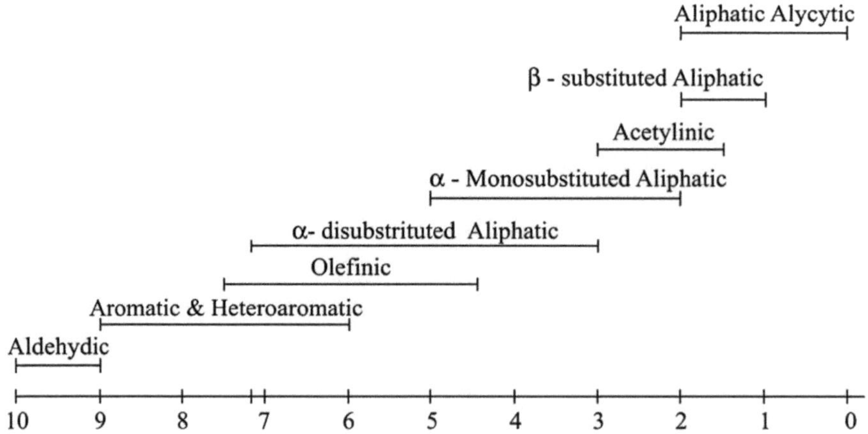

Fig. 11.13 Range of chemical shifts for some protons

Thus, if the spins are paired and opposed to external field, upfield shift results; otherwise, a downfield shift takes place. Opposed spins have no effect. Similarly eight possible combinations are possible if we consider the effect of three methyl protons on methylene peak.

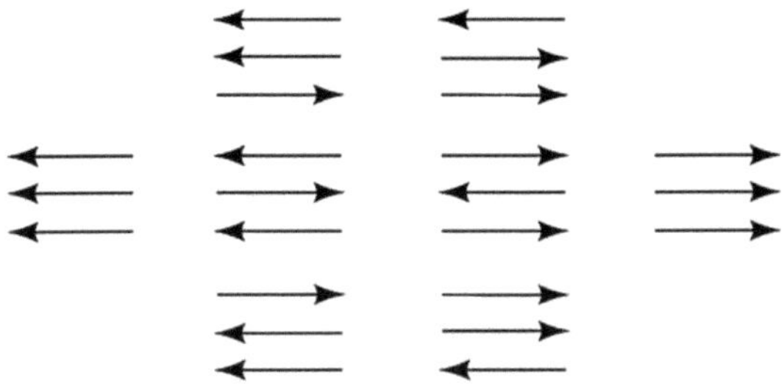

Among these, two groups have three combinations having equivalent magnetic effects. Hence, methylene is split into four peaks having areas in the ratio of 1 : 3 : 3 : 1. Shielding in NMR can be local or low range. In local shielding, field experienced by nucleus is modified by field due to induced circulation of π electrons localized on that nucleus. Instead in low-range secondary fields setup, the induced circulation of π electron (as in aromatic compounds), influence the fields of the nuclei and hence are not directly associated with π electrons.

A typical NMR spectrum of ethanol is presented in Fig. 11.14 and clearly indicates that methyl protons appear as triplet, methylene protons as quartet, and —OH proton as a singlet.

Coupling Constant (J)

The space between the adjacent peaks in a multiplet is called as spin-spin coupling constant (J) which is expressed in cycles per second. It must be realized that the J does not depend upon the magnetic field.

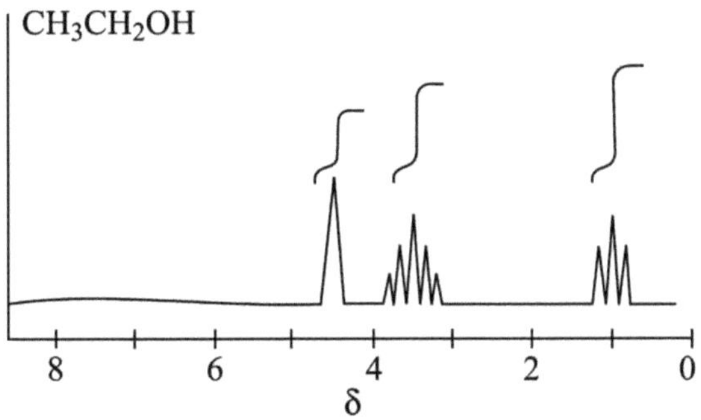

Fig. 11.14 NMR spectrum observed for ethanol

11.10.5 Applications

Study of NMR reveals:

1. The presence or absence of particular functional group.
2. Relative number of nuclei in the group.
3. Relative position of the group from its multiplicities.
4. Identification of compounds.
5. Hydrogen bonding present in compounds.
6. Keto-Enol Tautomerism.

11.11 Mass Spectrometry

Mass spectrometry is a technique using which accurate molecular weight of a compound can be determined. The technique is highly useful for the organic chemists for the determination of molecular weight of a compound under investigation. This determination helps in establishing the structure of the compound. The instrument used for this purpose is called as mass spectrometer and basically has three parts.

(i) Electron source/ionization source
(ii) Ion accelerating region
(iii) Ion collector/amplifier and detector

The vapors of a compound are passed through a high vacuum tube and is bombarded at right angles through an electron beam of high energy ($\sim$70 eV) as shown in Fig. 11.15. Such a bombardment causes ionization, and positive ions produced are forced to enter through accelerating slit using weak electrostatic field (6–100 V). The electrostatic field is then increased at successive slits so that ions can attain their final velocities and reach to the collector. The magnetic field is applied so that lighter ions reach faster and hit the detector before heavier ions arrive. Thus,

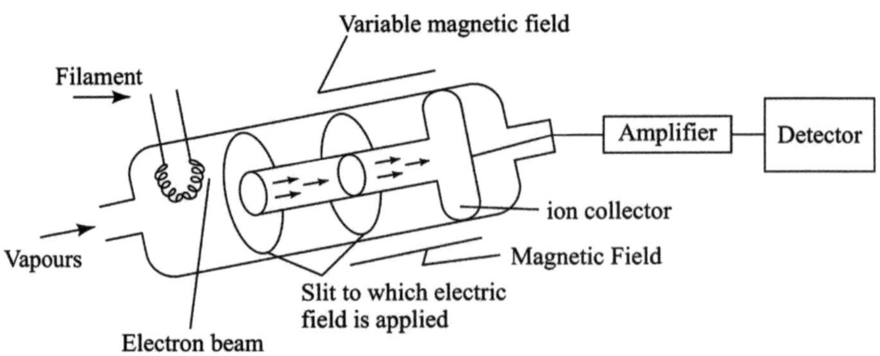

Fig. 11.15 Basic principle of mass spectrometry

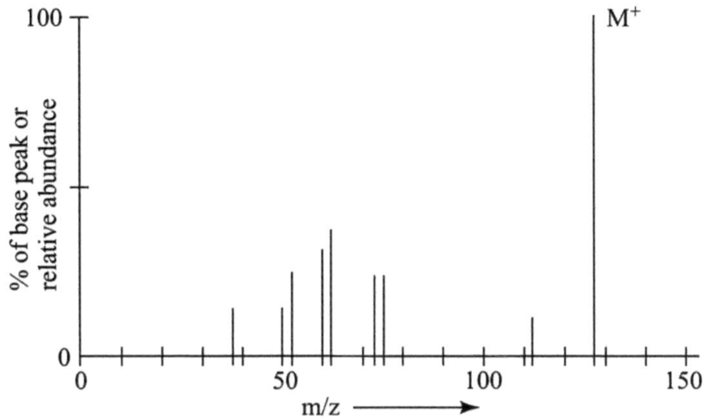

Fig. 11.16 Mass spectrum of naphthalene

electric field is used to accelerate the movement of ions, whereas magnetic field separates the lighter ions from heavier ions. Each type of ion has a particular mass to charge ratio represented as m/z, and as charge on most of the ions is unity, the m/z represents the mass of the ion. Nowadays, computer output provides bar graph in which relative peak abundance is plotted against the m/z as shown in Fig. 11.16 for naphthalene. The molecular ion peak ($m/z = 128$) is also the base peak in this case, and all other peaks are plotted against the base peak.

The peak heights in the bar graph are proportional to the number of ions of that mass. Thus, the instrument records the intensity of signal at each m/z and then plots it with respect to the base peak. The careful study of m/z values of different ions provides an idea about the fragmentation pattern of the molecule and thus helps in establishing its structure. Let us consider a simple case of methanol. When methanol is passed in vapor phase through the mass spectrometric tube, it produces cationic free radical on bombardment of electron beam energy of 70 eV.

$$CH_3OH + e \longrightarrow CH_3OH^{\bullet +} + 2e$$

The radical cation so formed disintegrate to give other ions in less than 10^{-3} s. For example,

$$CH_3OH^{\bullet +} \longrightarrow CH_2OH^+ + H^+$$
$$(m/z = 31)$$
$$CH_2OH^+ \longrightarrow CHO^+ + H_2$$
$$(m/z = 29)$$
$$CH_2OH^{\bullet +} \longrightarrow CH_3^+ + OH$$
$$(m/z = 15)$$

the radical ion of methanol is expected to give peaks at m/z 31, 29, 15, and so on. These peaks are actually noticed in the mass spectrum of methanol.

In many cases, the molecular ion peak is not clearly observed or may be very weak, and in such cases, chemical ionization spectrum is recorded which gives an intense peak at $M + 1$, i.e., one more than the molecular mass of the compound. In chemical ionization, methane (CH_4) is used as a reagent gas and ionization of the compound occurs from ion-molecule chemical interaction. The reactions now involve two steps:

1. In the first part, reagent gas (CH_4) is ionized due to impact of electrons giving ions such as $CH_4{}^+$or $CH_3{}^+$

$$CH_4 + e \longrightarrow CH_4^+ + 2e$$

These ions further reacts with other reagent gas molecules or other ions to give stable reagent gas ions:

$$CH_4^+ + CH_4 \longrightarrow CH_5^+ + CH_3^+$$

$$CH_3^+ + CH_4 \longrightarrow C_2H_5^+ + H_2$$

2. In the second type of reactions, the reagent gas ions react with sample molecule as shown below:

$$CH_3^+ + MH \longrightarrow CH_4 + M^+$$

$$CH_4^+ + MH \longrightarrow CH_4 + MH^+$$

$$CH_5^+ + MH \longrightarrow CH_4 + MH_2^+$$

This type of reactions are usually slow in nature and hence the molecular ion peak is clearly noticed in chemical ionization mass spectrum.

11.12 Atomic Absorption Spectrometry

Atomic absorption spectrometry (AAS) is a technique used for determining concentration of large number of elements in micromolar or nanomolar concentration levels. The basic theory of AAS can be explained as follows.

As we know every element has a fixed number of electrons associated with its nucleus. The number of electrons present depends on the atomic number of the element. The stable electronic configuration of an atom is called as “ground state” whereas by absorbing energy, the outermost electrons jump to higher energy levels. This less stable configuration of an atom is known as “excited state.” Now there are two possibilities.

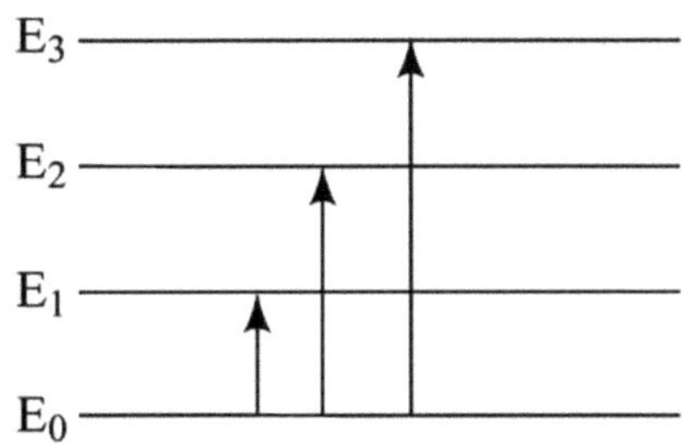

Fig. 11.17 Possible transitions of electron from ground level

11.12.1 Atomic Absorption

Let us assume an electron in the outermost orbit of atom. The ground state atom absorbs energy, and the electron enters into higher or excited energy levels as shown in Fig. 11.17. Out of various higher-energy levels, for example, E_1, E_2, E_3, the electron will enter into a level depending upon the energy absorbed by it. As per Bohr's equation, the amount of energy absorbed (ΔE) for the transitions between two energy levels (say from E_0 to E_1) is expressed by the relation

$$
\begin{aligned}
\Delta E &= E_1 - E_0 \\
&= hv \\
&= h \cdot \frac{c}{\lambda}
\end{aligned}
$$

where h is Planck's constant, c is the velocity of light, and λ is the wavelength of the radiation absorbed. Obviously the value of ΔE for transitions from E_0 to E_2 or E_0 to E_3 will be higher than for the transition E_0 to E_1.

In AAS, the atom absorbs light energy of a particular wavelength, and its electron enters into excited state. Thus, if a number of atoms in the path of light are more, absorption of light will be more. A quantitative determination of an element is thus possible by measuring the amount of light absorbed. A variety of elements can be determined by selecting proper wavelength and special light source.

11.12.2 Atomic Emission

In atomic emission, the excited state atoms are produced using a high-energy thermal environment. Normally a flame or more recently plasma is used to provide high-energy thermal environment. However, as mentioned earlier, the excited state is unstable; the electron jumps back to the ground state and emits light. A collection of emission wavelengths produce a spectrum called as emission spectrum. The intensity of emission line will increase as the number of excited atoms of the element increase. However, it must be realized that the emission spectrum of an element

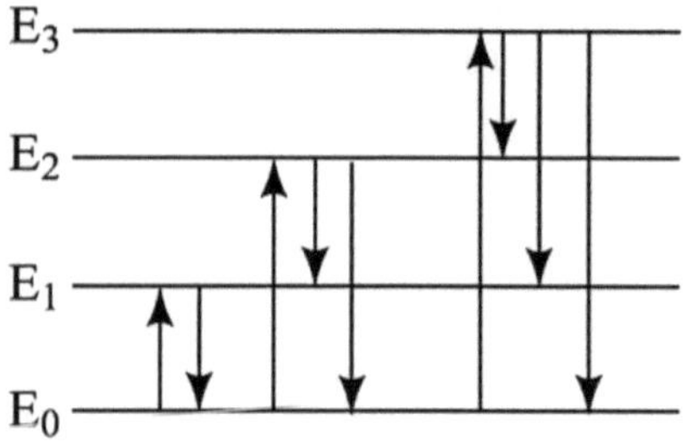

Fig. 11.18 Possible transitions from excited states to lower energy states

is complex because apart from the transitions from excited state to ground state (E_3, E_2, E_1 to E_0), transitions like E_3 to E_2, E_3 to E_1 or E_2 to E_1 are also possible as shown in Fig. 11.18.

Thus, the basic differences between atomic absorption and atomic emission are:

- In atomic absorption, the function of the flame is to convert the sample into atomic vapors, which then absorb light from the light source and electron jumps to give excited state of the element whereas
- In atomic emission, the flame serves two purposes

  (i) it converts the sample solution into atomic vapors and
  (ii) it provides thermal energy to excite electron from ground state to higher-energy levels.

When the excited atoms return to ground level, the energy is liberated which falls in the UV or visible region of the spectrum and is detected by the instrument. If the concentration of the element is higher in the solution, the intensity of the light emitted will also increase.

Most of the instruments designed for AAS are capable of measuring both atomic absorption as well as atomic emission. As we are more concerned with atomic absorption in this chapter, a detailed discussion on absorption aspects will be made.

It is observed that whenever energy is supplied, only a small fraction of the atoms are excited even under most favorable conditions. The expression for the relation between ground state and excited state population of atoms is described by the Boltzmann equation.

$$\frac{N_1}{N_0} = \left(\frac{g_1}{g_0}\right) e^{-\Delta E/kT}$$

where the various terms are

N_1 = number of atoms in excited state
N_0 = number of atoms in ground state
g_1/g_0 = ratio of statistical weights of the atoms in ground state and excited state
ΔE = energy of excitation
k = Boltzmann constant
T = Absolute temperature

Table 11.4 Effect of temperature on the ratio of N_1/N_0 for some elements

Element	ΔE/eV	N_1/N_0	
		2000 K	3000 K
Na	2.10	9.86×10^{-6}	5.88×10^{-4}
Ca	2.93	1.21×10^{-7}	3.69×10^{-5}
Fe	3.33	2.29×10^{-9}	1.31×10^{-6}
Cu	3.82	4.82×10^{-10}	6.65×10^{-7}
Zn	5.80	7.45×10^{-15}	5.50×10^{-10}

This equation clearly tells that the ratio of N_1/N_0 is dependent on the values of ΔE and T. The effect of increase in temperature noticed on the ratio of N_1/N_0 for various elements is presented in Table 11.4. The values of N_1/N_0 clearly suggest that a number of atoms excited are very small and hence the proportion of ground state to excited atoms is very high. The value of ΔE plays an important role in the ratio of N_1/N_0.

11.12.3 The Instrument

Atomic absorption spectrophotometer, like an ordinary spectrophotometer, uses the same principle of measuring absorbance (A), which is given by the logarithmic ratio of I_0/I_t, where I_0 is the intensity of incident light signal and I_t in the intensity of the transmitted light. The absorbance (A) is also

$$A = \log I_0/I_t = KLN_0$$

equal to KLN_0 where K = constant, L = path length through the flame in cm, and N_0 is the concentration of atoms in the flame, i.e., number of atoms per ml. Thus, the instrument measures light intensity and amplifies the signal. The atomic absorption spectrophotometer has four basic components:

(i) Light source
(ii) Absorption cell
(iii) Monochromator for high dispersion
(iv) Detector

Light Source

The block diagram of a simple single beam atomic absorption spectrophotometer is presented in Fig. 11.19. The light source is an important part of the spectrophotometer which emits a spectrum specific to the element of which it is made. Hollow cathode lamp or electrode less discharge lamps are common light source in AAS. A hollow cathode lamp has an anode wire along the outside of a cylindrical cathode

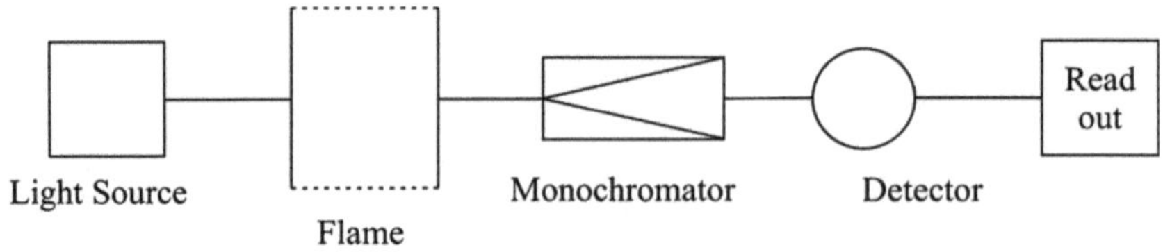

Fig. 11.19 Block diagram of AAS

Fig. 11.20 A hollow cathode lamp used in AAS

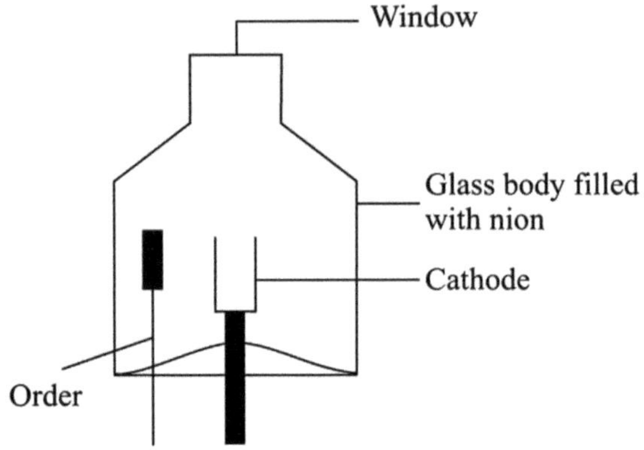

as shown in Fig. 11.20. The lamp is filled with neon or argon. The discharge occurs between the two electrodes on applying the voltage. Neon is the preferred gas for filling in hollow cathode lamps because it gives up to three times better intensity and has a tendency to suppress the ionic spectrum of some elements. For the determination of each element, a different type of hollow cathode lamp is needed which can provide narrow band emissions specifically required for that element.

Absorption Cell

The next part of the instrument is flame system where the sample solution is converted to mist or aerosol using a nebulizer system. Two common fuel combinations used these days are:

(i) An acetylene and
(ii) Nitrous oxide-acetylene mixture

The temperature of air-acetylene flame is about 2300°C, whereas that of nitrous oxide-acetylene flame is 2900°C. Air acetylene flame is used for the determination of nearly 35 elements, whereas, for the elements which form refractory oxides, nitrous oxide acetylene flame is used.

The sample solution is converted to aerosol in a vaporizing chamber, and the resulting mist is mixed with carrier gas and allowed to flow to the burner head. The flame path is usually 10 to 12 cm in length. The atoms absorb the light from the

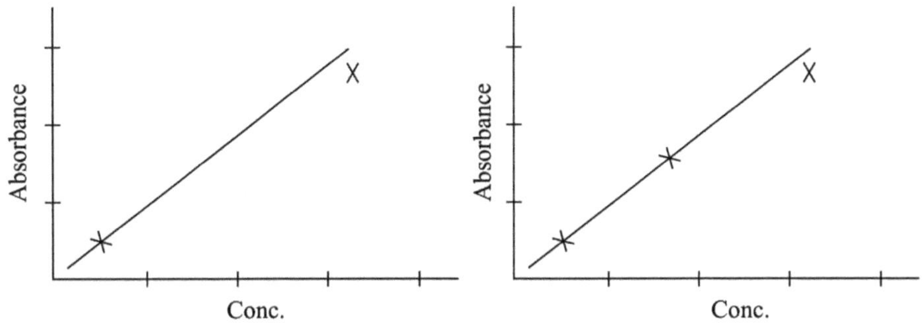

Fig. 11.21 Calibration plots used for determination by AAS

source and the monochromator, capable of high resolution, allows the selected light to reach to the detector (photomultiplier tube) which then provides output through electronic devices.

The most important aspect of the determination of an element by AAS is the preparation of calibration curve. Usually one or two standard samples of the element to be determined are taken and a calibration plot is prepared. However, it is noticed that more than two standard samples should always be analyzed for preparing calibration curves to get an accurate result. The effect of two and three standard samples a calibration curve of a sample is presented in Fig. 11.21.

Most of the elements can be easily determined using AAS technique in the range 1–600 μg/ml, and the technique has been successfully applied for the determination of Ca and Mg in tap water, lead in ferrous alloy, vanadium in lubricating oil, and many more.

11.13 Microwave (Rotational) Spectra of Diatomic Molecules

A diatomic molecule having a permanent dipole may be treated as a rigid rotator in which the two atoms are joined along the line of center by a bond. Let us assume that the length of bond is r_0 between the two nuclei.

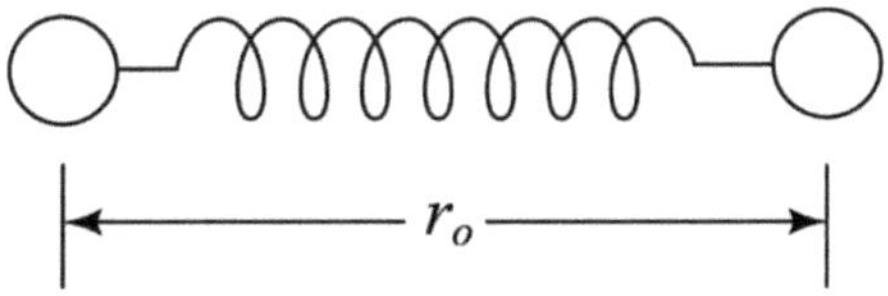

According to Schrodinger equation, the energy of such a system in any given rotational quantum level (J) is given by the expression

$$E = \frac{h^2}{8\pi^2 I} J(J+1) \tag{11.1}$$

where

h = Planck's constant
I = Moment of Inertia and
J = Rotational quantum number having any integer value such as 0, 1, 2, etc.

If the atomic masses of the atoms are m_1 and m_2, the moment of Inertia (I) is expressed as

$$\mathrm{I} = \frac{m_1 m_2}{m_1 + m_2} (r_0)^2 \tag{11.2}$$

in which the term $\frac{m_1 m_2}{m_1+m_2}$ is called as reduced mass (μ).

Let us assume that absorption of photon of appropriate energy, a transition occurs from rotational level of quantum number J' to J. The energy difference in such a case will be equal to

$$\Delta E = Ej - Ej'$$

$$\Delta E = \left[\frac{h^2}{8\pi^2 I}\right] J(J+1) - \left[\frac{h^2}{8\pi^2 I}\right] J'\left(J'+1\right)$$

$$\Delta E = \frac{h^2}{8\pi^2 I}\left[J(J+1) - J'\left(J'+1\right)\right] \tag{11.3}$$

The value of J can be higher or lower by one unit than J' and thus $\Delta J = \pm 1$. The positive sign indicates to absorption and negative sign to emission. The equation then becomes

$$\Delta E = \left[\frac{2h^2}{8\pi^2 I}\right] J \tag{11.4}$$

$$= \left[\frac{h^2}{4\pi^2 I}\right] J \tag{11.5}$$

As ΔE is also equal to hv, we can write

$$\Delta E = hv = h.c/\lambda = hc\bar{v}$$

Hence, the frequency of radiation absorbed or emitted will be equal to

$$\bar{v} = \left[\frac{h}{4\pi^2 Ic}\right] J \tag{11.6}$$

the term $\frac{h}{4\pi^2 Ic}$ is constant for a given molecule; hence, the rotational spectrum of a molecule will exhibit a number of equally spaced lines. The frequency of lines will be given by various values of J.

On rewriting Eq. (11.6) as

$$\bar{\upsilon} = \left[\frac{2h}{8\pi^2 Ic}\right] J$$

$$\bar{\upsilon} = 2B \text{ m}$$

where B is used for the term $\left[\frac{h}{8\pi^2 Ic}\right]$ and J is represented by 'm'. On putting the value of m as 1, 2, 3, 4, etc., the frequencies of successive spectral lines can be calculated. The term $2B$ represents spacing between successive lines and is called as frequency separation. Thus,

$$2B = \left[\frac{2h}{8\pi^2 Ic}\right]$$

and using this expression frequency separation can be calculated.

Exercises

1. Calculate E_{max} for a compound whose maximum absorbance (A) is 1.2. The cell length (I) is 1.0 cm and the concentration of solution is 1.9 mg per 25.0 cm^3. The molar mass of the compound is 100 gm mole^{-1}.
2. Calculate the approximate frequency of C—H stretching vibration from the following data:
   (i) Force constant $k = 5 \times 10^5$ gm sec^{-2}
   (ii) Mass of carbon atom $(m_1) = 20 \times 10^{-24}$ gm
   (iii) Mass of hydrogen atom $(m_2) = 1.6 \times 10^{-24}$ gm; $c = 9.3 \times 10^{13}$ cm sec^{-1}
3. The fundamental vibration frequency of CO is 2140 cm^{-1}. Calculate the force constant of the molecule. Given the atomic masses: $^{12}C = 19.9 \times 10^{-27}$ kg and $^{16}O = 26.6 \times 10^{-27}$ kg.
4. Prove that $k(\text{HCl}) < k(\text{HF})$;
   Given that observed vibrational frequency of HCI = 2886 cm^{-1} and HF = 3958 cm^{-1}. Reduced masses of HCl and HF are 1.63×10^{-27} kg and 1.57×10^{-27} kg, respectively.
5. The strongest absorption band in the IR spectrum of $^{12}C^{16}O$ (centered at 2143.26 cm^{-1}) is observed to split into two sets of closely spaced peaks, one on each side of the center. The separation between the peaks immediately to the right and left of the center is 7.655 cm^{-1}.

Calculate

(i) the force constant and
(ii) The bond length of the CO bond, from the above data.

6. A solution of thickness 3 cm transmits 30% incident light. Calculate the concentration of the solution, if extinction coefficient (ϵ) is 4,000 $dm^3 \ mol^{-1} \ cm^{-1}$.
7. The percentage transmittance of 4×10^{-5}M solution of an unknown compound at 300 nm is 20%. If the cell used is 2 cm and temperature 25°C. Calculate

(i) The absorbance (A)
(ii) The molar extinction or absorption coefficient, and
(iii) The percentage transmittancy of a 2×10^{-5}M solution in a 4 cm cell.

8. Predict the NMR spectrum of CH_2Cl—$CHCl_2$.
9. Predict the NMR spectrum you expect for CH_3CH_2OH.
10. How will you distinguish between benzene and anthracene by UV spectroscopy. Which one will absorb at longer wavelength?
11. How will you identify the following functional groups by IR spectroscopy:

(a) —OH
(b) —NH_2
(c) —CN
(d) $> C = O$

12. How would you distinguish between the following pairs of compounds by IR spectroscopy.

(a) $CH_3CH_2CH_2\,N\,(CH_3)_2$ and CH, $CH_2CH_2NH_2$
(b) $CH_3CH_2CH_2COOH$ and $CH_3CH_2COOCH_3$
(c) $CH_3CH_2COCH_3$ and $CH_3CH_2COOCH_3$

Hint:

(a) N—H absorption 3350 cm^{-1}
(b) O—H absorption 2500 – 3000 cm^{-1}
(c) C—O absorption 1050 cm^{-1}.

13. Two compounds (A) and (B), have the same molecular formula C_2H_6O. They have different IR spectra. Compound (A) shows a strong/broad absorption at 3400 cm^{-1}, while compound (B) does not. Suggest structure for (A) and (B) and explain which group accounts for the difference.
14. Draw the structure of a compound with each of the following molecular formulas that will show only one peak in its NMR spectrum:

(a) $C_3H_6Cl_2$
(b) C_5H_{12}
(c) C_2H_6O
(d) C_4H_6

15. How many NMR signals do you expect in each of the following compounds. Indicate also the splitting pattern of the various signals.

    (a) CH_3OCH_3
    (b) $CH_3OCH_2CH_3$
    (c) CH_3CH_2OH

16. How would you distinguish between the following pairs of compounds by NMR spectroscopy.

    a. 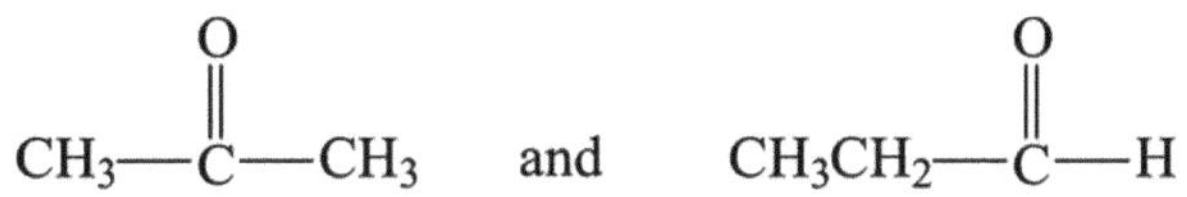

    b. 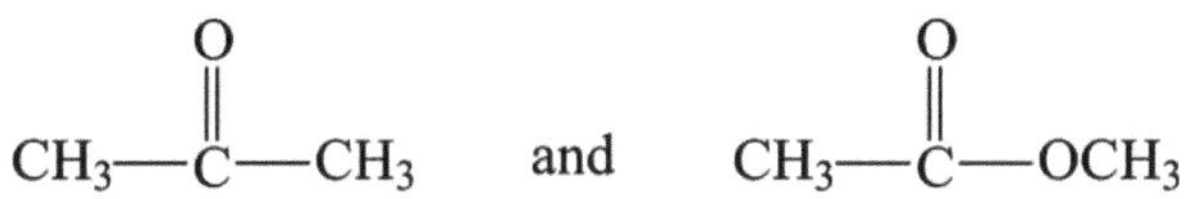

17. The NMR spectrum of a compound C_4H_9Br consists of a single sharp line. What is the structure of the compound.
    Hint: Try to write structure having identical position of protons
18. How will you distinguish between the three dibromobenzenes by their NMR spectra.
19. Give a structure consistent with the following NMR data: Molecular formula $C_3H_5Cl_3$

    (a) Singlet at $\delta 2.20$, 3H
    (b) Singlet at $\delta 4.02$, 2H

    Hint: CH_2Cl—CCl_2—CH_3.
20. Give a structure with the following NMR data:
    Molecular formula $C_{10}H_{14}$

    (a) Singlet at $\delta 1.30$, 9H
    (b) Singlet at $\delta 7.28$, 5H

    Hint: As only two singlets are obtained, the structure could be

    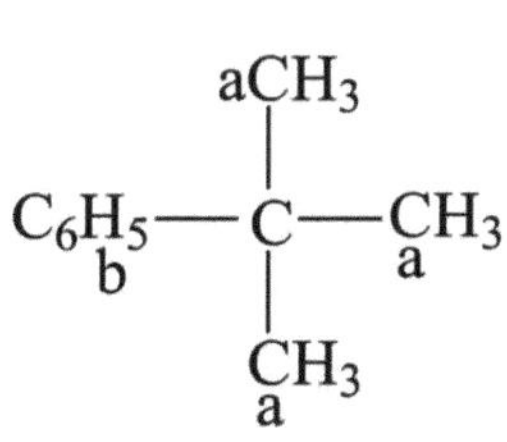

    t-Butylbenzene.
21. Give structure of a compound exhibiting following NMR data:
    Molecular formula C_9H_{18}

(a) Singlet at $\delta 6.789$, 3H
(b) Singlet at $\delta 2.25$, 9H

Hint: Consider a structure having only two types of protons $C_6H_9\,(CH_3)_3$ or

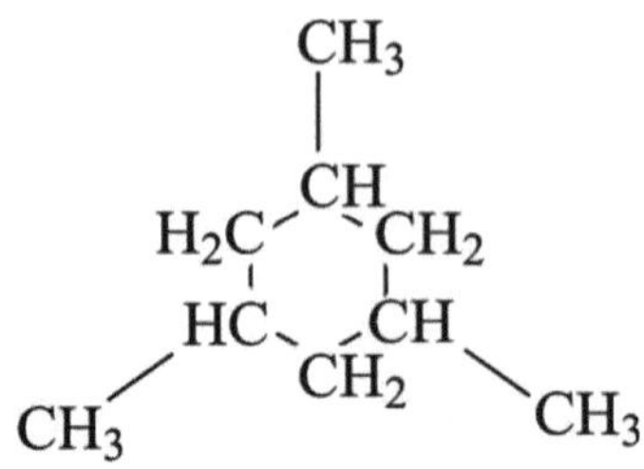

22. Distinguish between the following on the basis of their IR spectra:

    (a) Inter and Intra molecular H-bonding
    (b) 2-propanal and propanone
    (c) Ethyl alcohol and dimethylether.

23. Ultra violet spectrum of acetone shows two peaks at $\lambda_{max} = 189$ nm ($\varepsilon_{max} = 900$) and $\lambda_{max} = 273$ nm ($\varepsilon_{max} = 15$). Identify the electronic transitions for each peak.
    Hint:

    1. higher wavelength less energy to show $n \longrightarrow \pi^*$ transition, i.e., λ_{max} 189 nm.
    2. lower wavelength high energy to show $\pi \longrightarrow \pi^*$ transition, i.e., $\lambda_{max} =$ 273 nm.

    In energy $\pi \longrightarrow \pi^* > n \longrightarrow \pi^*$

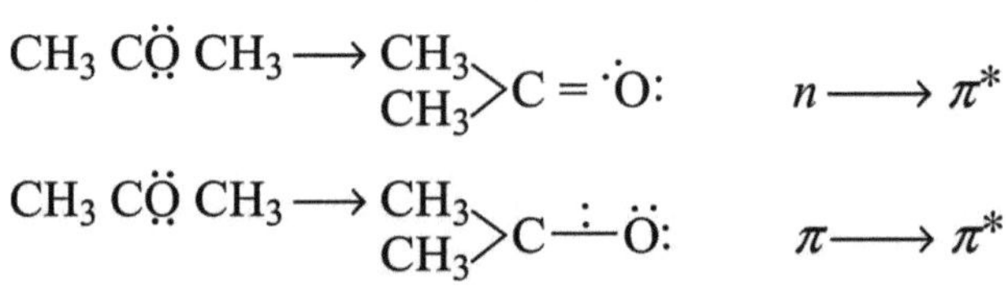

24. A 2×10^{-5}M solution of an organic compound placed in 1 cm tube showed an absorbance (A) 0.63 at $\lambda_{max} = 237$ nm. Calculate ε_{max}.
25. How will you compare the IR spectra of following compounds

    (a) CH_3CH_2CHO and $CH_2 = CH.CH_2OH$.
    (b) *o*-Hydroxy benzoic acid and *p*-hydroxy benzoic acid.

26. How will you distinguish between the following pairs of compounds using IR spectra:

    (a) Ethanol and propanone

(b) Methanol and ethanoic acid
(c) Cyclobutane and cyclohexanone
(d) Ethyl methyl ketone and methyl vinyl ketone.

27. A compound has molecular formula C_3H_9N and shows peaks at.

(a) 3012 cm^{-1}
(b) 3423 cm^{-1}
(c) 3236 cm^{-1}
(d) 1615 cm^{-1}

When compound is treated with nitrous acid a strong peak at 3430 cm^{-1} is observed. Identify the compound.
Hint:
N – H symm. stretching at 3236 cm^{-1}—shows $-NH_2$ exmp.
N – H asymm. stretching at 3423 cm^{-1}—shows $-NH_2$ exmp.
C – H stretching at 3012 cm^{-1}, N—H bending at 1615 cm^{-1}
Hence likely structure is $CH_3CH_2CH_2NH_2$.
On tracting with HNO_2, aliphatic $-NH_2$ converted to –OH to show peak at 3430 cm^{-1}

$$CH_3CH_2CH_2NH_2 + HNO_2 \longrightarrow CH_3CH_2CH_2OH + N_2 + H_2O$$

28. Predict the NMR signals of the following compounds:

(a) $CH_2 = CHCl$
(b) $CH_3 - CHCl - CH_2Cl$

Hint:

(i) Three signals

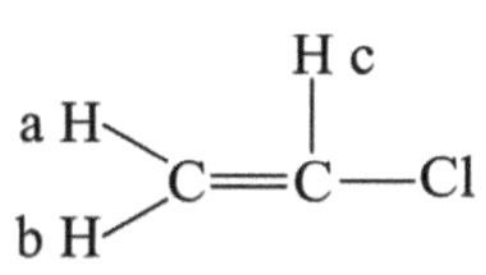

(ii) Four signals

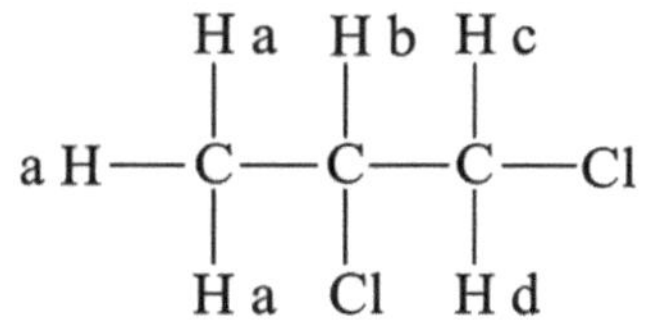

29. Two isomeric compounds *A* and *B* have the molecular formula C_6H_8. Both the compounds decolorize Br_2 solution in CCl_4 and alkaline $KMnO_4$. A gives absorption peak at 256 nm and *B* at 200 nm. Identify *A* and *B*.
Hint: *A* and *B* have two double bonds and both have added four hydrogen and both are cyclic. Due to decolorization of Br_2 in CCl_4 and in alkaline $KMnO_4$, the compounds will be cyclohexadiene. As *A* has λ_{max} = 256 nm higher than *B*, so *A* is conjugated diene, i.e., 1,3-cyclohexadiene and *B* is 1, 4-cyclohexadiene.

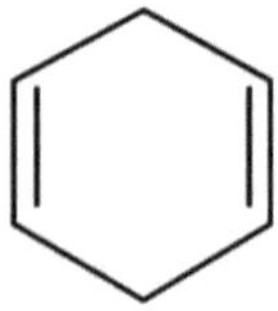

Exercises

1. What are absorption laws? What is the source used in ultraviolet region?
2. Explain π, σ and n-electrons. How do electronic energy transitions take place?
3. Describe the principle of UV spectroscopy in brief.
4. Are the terms electromagnetic radiation and light synonymous?
5. "Increase in polarity of the solvent shifts $\pi \longrightarrow \pi$ "band to longer wave length, but $n \longrightarrow \pi$ and $n \longrightarrow \pi^*$ bonds to shorter wavelength"? Comment on the statement.
6. Which is the region of greatest interest to organic chemists in UV and IR spectra? What is the cause of absorption in UV and IR regions?
7. Define infrared spectroscopy. Describe the various molecular vibrations in the technique.
8. Which of the following molecules will show IR spectrum and why?

$$N_2, O_2, H_2, HCl, CH_4, CO_2, H_2O_2,\ N_2O_4, C_2H_4$$

9. Indicate the various regions of the electromagnetic spectrum in which the photons will be:

   (i) $\lambda = 200$ nm
   (ii) $\bar{\upsilon} = 2000\ \text{cm}^{-1}$

10. Convert 15,000 cm^{-1} into frequency and wavelength.
11. Which of the following molecules will show infrared spectrum and why?

$$H_2, HCl, CH_4, CO_2, H_2O_2$$

   [***Hint:*** All except H_2]
12. Can ultraviolet spectral data be useful to distinguish the compounds of the following pairs? Give reasons.

   (i) Ethyl benzene and styrene.
   (ii) $CH_2 = CH - CH = CH_2$ and $CH_3 - CH = CH - CH = CH_2$

   Hint: (i) Extended conjugation in styrene results in bathochromic shift as compared to ethyl benzene.

13. Describe briefly the theory of NMR spectroscopy. What information can be obtained from the NMR absorption peaks?
14. Which of the following atoms will not exhibit nuclear magnetic resonance?

$$C^{12}, O^{16}, N^{14}, N^{15}, H^{2}, F^{17}, C^{13}, P^{31}$$

15. What is meant by the term chemical shift?
16. Select the compounds which will absorb UV radiations.

    (a) 1, 3-Butadiene
    (b) Cyclobutane
    (c) Nitrobenzene
    (d) Chlorobenzene
    (e) 1-Hexene
    (f) 1, 3-Cyclohexadiene

    [***Hint:*** (a), (c) and (f)]
17. What wavelengths are likely to be absorbed by the following groups?

    (i) –C = C–   (ii) 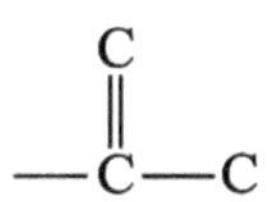   (iii) —C = C — C = C —

18. Mention one advantage of NMR spectroscopy over conventional chemical analysis.
    [***Hint:*** The sample may be recovered after the study.]
19. Calculate the (a) number of peaks and (b) area under each peak in the NMR spectrum of a sample of $CH_3OC_2H_5$.
    [***Hint:*** The ratio of areas of the three peaks will be 3 : 2 : 3]
20. What is TMS? What is its use in NMR spectroscopy?
    [***Hint:*** Trimethylsilane as reference to measure the chemical shift.]
21. State Beer's law. If the path length of a beam of light through the sample is doubled and the concentration is reduced to half, what will be the value of absorbance change?
22. The low resolution NMR spectrum of the compound $C_5H_{10}O$ exhibits two peaks with relative intensities of 2 and 3. Deduce its structure.
23. How many NMR signals will be observed in the spectrum of

    (a) CH_3COOCH_3
    (b)

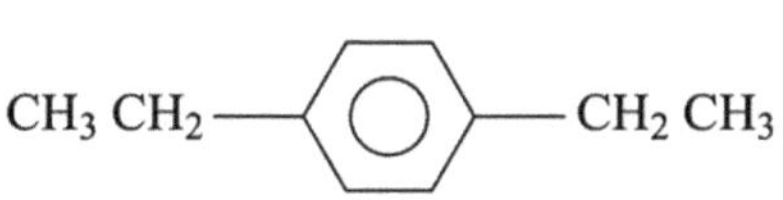

    (c) CH_3OCH_3
    (d) $CH_3CH_2CH_3$

(e) CH_3CH_3
(f) $CH_3COOCH\,(CH_3)_2$

24. The low resolution NMR spectrum of $C_4H_8O_2$ exhibits three peaks at 8.1, 5.2 and 1.3 ppm with relative intensity ratio 1 : 1 : 6. Deduce the structure.
Hint: Consider

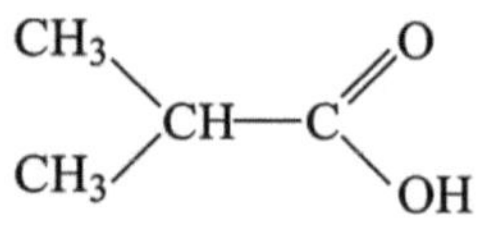

25. What is the ratio of peak areas due to $-CH_3$ group and—OH group in NMR spectrum?
26. Which of the following, CH_3COCH_3 and $CH_2 = CHCOCH_3$, will show higher value of λ_{max} in the visible -UV spectra? Give reasons.
27. An organic compound A with molecular formula C_7H_5OCl reacts with ammonia to give B. On treating with bromine and sodium hydroxide B gives C, which shows two absorption peaks at 3300 and 3400 cm^{-1} in the IR spectrum. Compound C reacts with one mole of CH_3I to give D, which reacts with nitrous acid to yield a yellow oil E. Give the structures of A to E and write the mechanism involved in the conversion of B to C.
Hint: (A) is C_6H_5COCl.
28. What do you understand by the following terms:

(i) $n \longrightarrow \pi^*$ transition
(ii) $\pi \longrightarrow \pi^*$ transition
(iii) $\sigma \longrightarrow \sigma^*$ transition
(iv) Finger print region of IR spectrum.

29. Place the following types of radiation in increasing order of energy per photon radar signals; radiation from microwave oven: red light from a neon sign; gamma rays from nuclear fission.
30. An AM radio station at 600 kH_Z broadcasts news at a frequency of $6.0 \times 10^{-5}\ s^{-1}\left(1\,kH_Z = 1000\ s^{-1}\right)$. Calculate the wavelength of the signal in meters.
31. The sodium lamps in streets emit light of wavelength 589.0 nm. What is the energy of 1 mole of photons of this light?

Chapter 12
Polymer Chemistry

A polymer may be defined as a high-molecular-weight compound, ranging into thousands and millions of grams. Approximately 75% of the organic chemical industry is devoted to the production of synthetic polymers. The use of polymers in daily life has become so common that one can see things made up of polymers in every sphere of life. Clothes made of polyester, Dacron or nylon are used by a common man. The Teflon coating on frying pans used in kitchen and the plastic bottles in laboratory are examples of polymers. Synthetic polymers are so widely used these days that one should not be surprised to know that almost 80 kg of such materials are made per person per year in the USA.

The development of polymer chemistry began in early 1920s when chemists, all over the world, were struggling in establishing the structure of many compounds. Also wood, rubber, cotton, gelatin, etc. were exhibiting unusual properties. For example, rubber (empirical formula C_5H_8) on dissolving in organic solvents was displaying unusual properties of high viscosity and low osmotic pressure. Such observations indicated the presence of higher-molecular-weight material, but the chemists were not ready to accept the idea. Thus, for a number of years, it was thought that rubber consisted of small molecules of C_5H_8 attached with intermolecular forces. This misconception was removed by a German Chemist Hermann Staudinger,[1] who clearly showed that the aggregates of small molecules are, in fact, giant molecules of high molecular mass in which small molecules are attached to one another through covalent bonds. Thus, a better definition of polymer is that—*A polymer is a high-molecular-weight compound formed by the combination of a large number of one or more types of small molecules (low molecular weight).* The small unit of which a polymer is made is known as monomer. Thus, for example, the monomer of polyethylene is ethylene:

[1] Staudinger was awarded Nobel Prize in Chemistry in 1953.

© The Author(s), under exclusive license to Springer Nature Switzerland AG 2026

R. N. Goyal, H. Goel, *Textbook of Engineering Chemistry*,
https://doi.org/10.1007/978-3-032-22068-4_12

$$\underset{\substack{\text{monomer} \\ \text{ethylene}}}{n\,CH_2 = CH_2} \longrightarrow \underset{\substack{\text{polymer} \\ \text{polyethylene}}}{\left[CH_2 - CH_2 \right]_n}$$

Polymers may be divided into two broad classes:

*Natural and *Synthetic

The common examples of natural polymers are cellulose, protein, nucleic acids, etc., whereas a wide variety of synthetic polymers are available in market these days. The natural polymers that exist in human system are also called as biopolymers.

12.1 Natural Polymers

Proteins are well-known example of polymers and consist of amino acids as monomer. The living system consists often amino acids and all of them are α-amino acids which signifies that the amino group is attached to α-carbon atom beginning from −COOH group. The names of these amino acids are:

- Alanine
- Aspartic acid
- Cysteine
- Glutamic acid
- Histidine
- Glycine
- Lysine
- Phenylalanine
- Serine
- Valine

and the general formula is

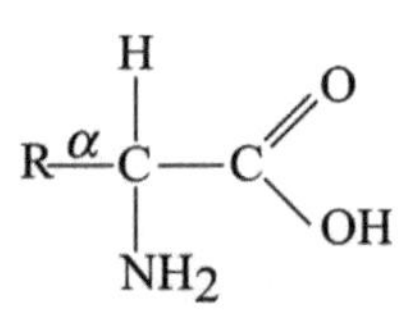

As the α-carbon of these amino acids possesses four different groups, they have a chiral center and thus exist in L- and D-forms. Interestingly, only one of the enantiomeric forms of a given amino acid (L form) exists in nature and thus living organisms utilize only "L" form. Figure 12.1 represents an easy way to understand the "L" form of an amino acids.

If one has to climb the stairs to touch the NH_2 end in the molecule, and group R comes to your left, the acid is in "L" form. If group R comes to your right side, the acid is in "D" form.

To form proteins from amino acids, several steps are involved. The first step is the condensation reaction of two amino acids in which loss of water molecule takes place. For example

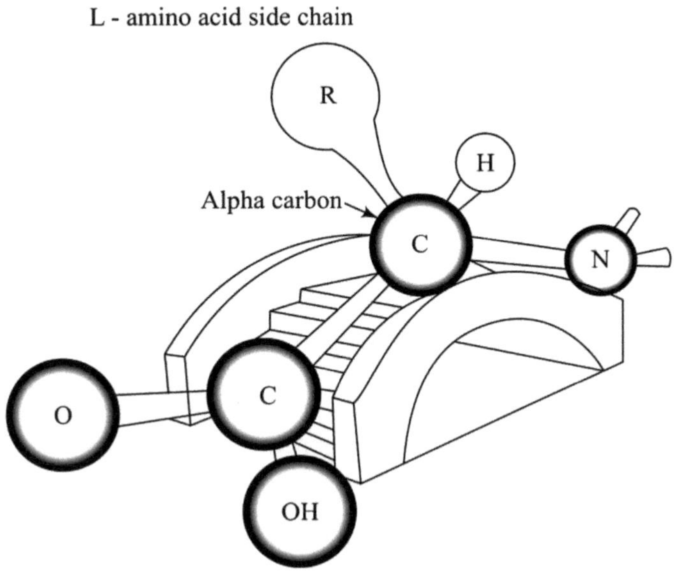

Fig. 12.1 A simple representation to understand L- and D-forms of an amino acid

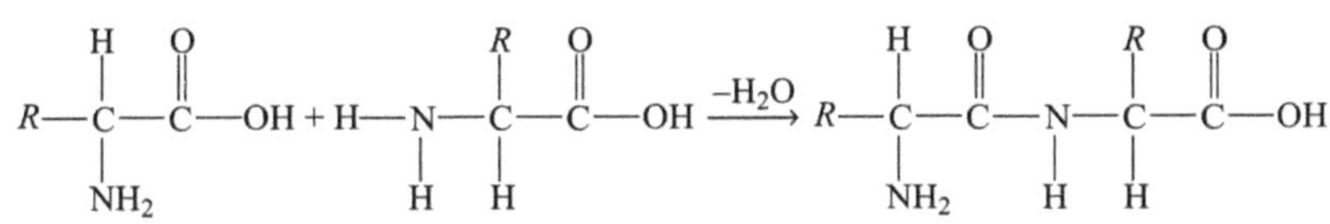

When two molecules of amino acid are condensed, a water molecule is eliminated between the carboxylic group of one amino acid and amino group of another amino acid. The result is the formation of a dipeptide molecule having –CO – NH– linkage, which is called as peptide linkage. The dipeptide has an amino group at one end and carboxylic group at another; hence, it can easily react with other molecules of amino acids, and tripeptide, tetra peptides, etc. will be obtained. Thus, many molecules of amino acids can combine to give a polypeptide or protein. In other words protein is a polymer of amino acids in which amino acids are linked together in a particular sequence and with particular geometry.

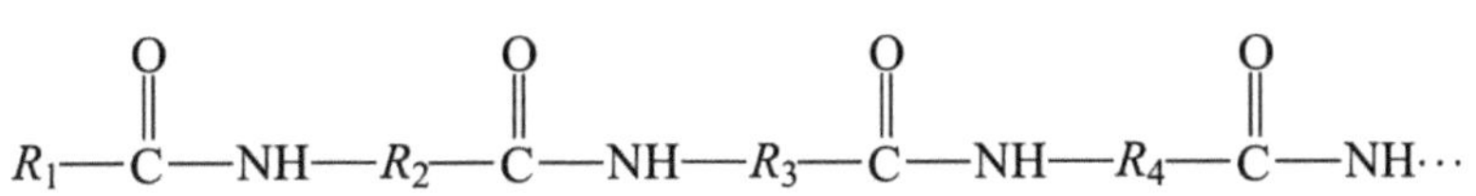

Proteins occur in a variety of sizes in human system. A common protein hemoglobin contains four protein chains, two of them are identical having 141 amino acids linked to each other. The other two chains are again identical with 146 amino acids. Thus, amino acids can be combined in different ways to form proteins. The human system is thought to contain about 1,00,000 different proteins; however, the structure of only few of them is completely known. An important question that comes to the mind at this stage is whether the protein chains are straight or folded up in a specific fashion. The resonance structure of peptide bond indicates that C—N bond has partial double bond character and hence it has a flat configuration. The positive charge on N and negative charge on oxygen atom also opens the possibility of H-bonding between different molecules of amino acids.

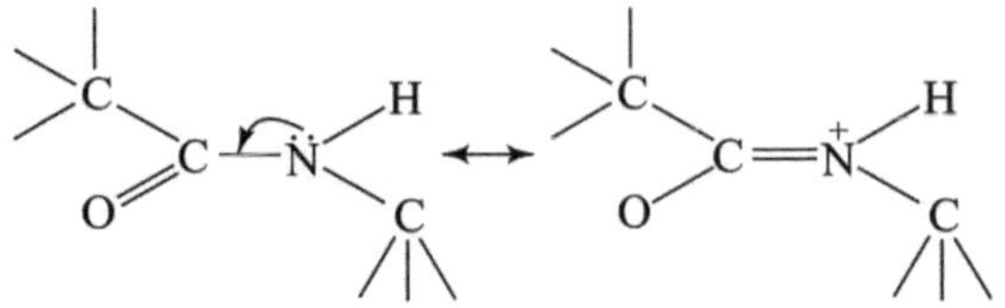

Considering the above two characteristics, Linus Pauling suggested that proteins can have two common structure, viz., (i) α-helix and (ii) β-pleated sheet. If the H bonding is intramolecular, the proteins acquire α-helix structure in which the helix always has a right-handed arrangement. To further understand α-helical structure, hold a wooden or iron rod in your right hand so that your thumb points along the axis of the rod; the finger curl in the same direction as right-handed α-helical structure as shown in Fig. 12.2. Most of the fibrous proteins possess α-helical structure, and the common examples are hair, skins, beaks, claws, nails, and so on. In a β pleated sheet structure, the polypeptide chains are fully extended and H-bonding between two polypeptide chains occur to form a sheet-like structure. The structure is not elastic and hence cannot be stretched. If the sheets are pulled, the covalent bonds and many H-bonds break, holding the individual protein strands in the sheets.

12.1.1 *Carbohydrates*

The literal meaning of the word carbohydrate is "hydrate of carbon," and their general formula is $C_x\,(H_2O)_y$ where x and y are integers. Carbohydrates are widely distributed in nature and glucose ($C_6H_{12}O_6$), fructose ($C_6H_{12}O_6$), sucrose

Fig. 12.2 Illustration of α-helical and β-pleated sheet structures

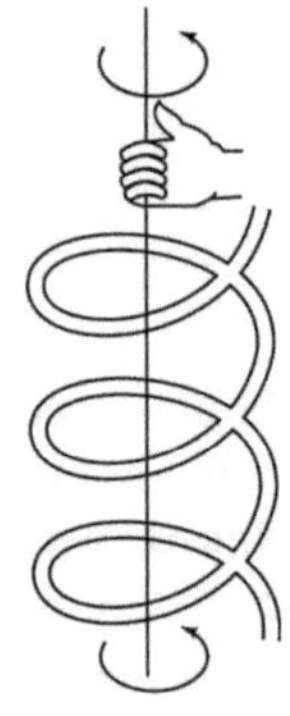

($C_{12}H_{22}O_{11}$), starch, and cellulose are some common carbohydrates. Starch and cellulose when eaten by animals break into glucose unit. Discovery of many other sugars having molecular formula $C_6H_{12}O_5$ (rhamnose) indicated that the ratio of hydrogen and oxygen atom is not 2:1 as found in water and hence cannot be called hydrate of water. Thus, carbohydrates are defined by another more accurate definition according to which "*carbohydrates are polyhydroxyaldehydes or polyhydroxyketones*," and some of their chemical properties are similar to alcohols, aldehydes, or ketones.

In nature, carbohydrates are produced in plants by the process of photosynthesis in which carbon dioxide, water, and energy from the sun is utilized:

$$xCO_2 + yH_2O \xrightarrow{h\nu} C_x\,(H_2O)_y + xO_2$$

or

$$6CO_2 + 6H_2O \xrightarrow{h\nu} C_6H_{12}O_6 + 6O_2$$

The carbohydrate so produced by the plants serves as the ultimate source of food for animals. Cellulose, a linear polymer of β-glucose units, is the main structural component of plants and is responsible for their texture, rigidity, shape, and durability. Wood, paper, cotton, and many other important materials, derived from natural sources, also contain cellulose.

12.1.2 *Monosaccharides: Simple Sugars*

Monosaccharides are colorless, sweet compounds which are soluble in water. In the general formula $(CH_2O)_n$, the value of n in these compounds can vary from 3 to 6. Two important monosaccharides are ribose having $n = 5$ and glucose ($n = 6$). In the open chain structure, these compounds are written as

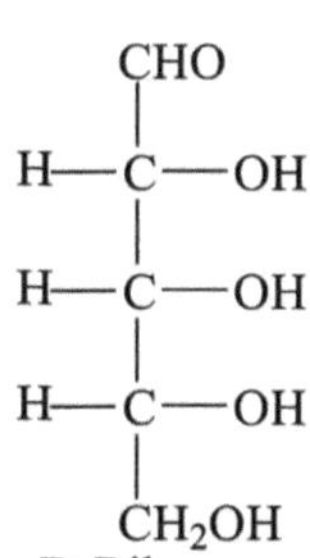
D-Ribose

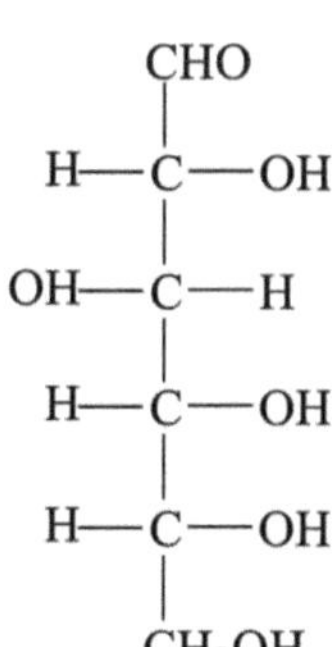
D-Glucose

The cyclic structure established for these compounds is as follows:

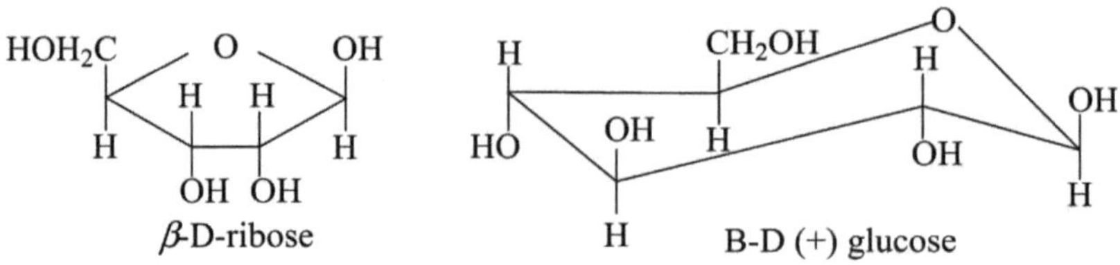

12.1.3 *Disaccharides and Polysaccharides*

Two monosaccharides units can easily condense with an elimination of water to give disaccharides. Sucrose (cane sugar) is a common disaccharide obtained by the condensation of one molecule of β-D-glucose and one molecule of β-D-fructose:

$$\text{Monosaccharide} + \text{Monosaccharide} \longrightarrow \text{Disaccharide} + H_2O$$

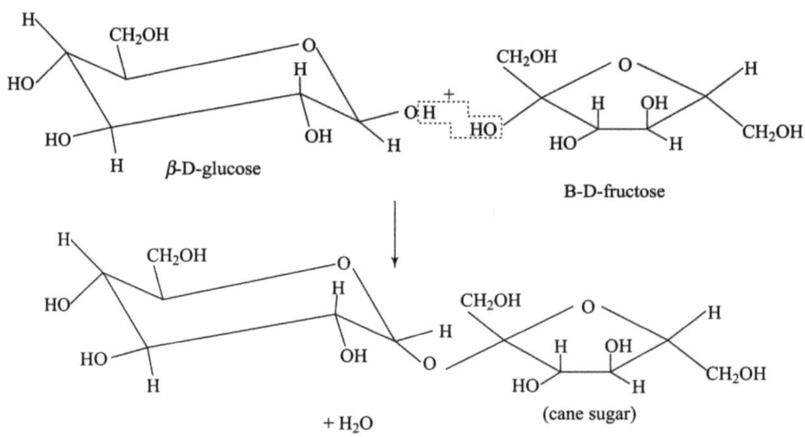

In such a condensation glucoside linkage is formed. Sucrose obtained from sugar cane or sugar beets is a disaccharide. Honey is a mixture of two monosaccharides glucose and fructose and is widely used as a natural sweetener.

12.1.4 *Polysaccharides or Biopolymers*

Cellulose, starch, and glycogen are three important polysaccharides, and all possess glucose as a monomer. Cotton is a pure form of cellulose and is the most abundant organic compound on earth. Cellulose is also found as woody part of the trees; however, human system cannot digest cellulose.

Human system depends on another polysaccharide starch as a source of glucose. Starch occurs in two forms, amylose and amylopectin, and both are polymers of glucose in which glucose units are attached through α-1,4-linkage. Out of the two starch forms, amylopectin is a common one in which about 1000 units of glucose are present and is having a branched chain structure. Branching occurs at an interval of nearly 20–25 units and involves a OH group of position 6 of the main chain with C-l of the branched chain unit. Human system has "amylase" in saliva and pancreatic juices which can easily convert starch into glucose.

Starch

Starch is a polysaccharide obtained by many α-glucose units joined together by removing one water molecule. It is a mixture of two polysaccharides: the one is a linear macromolecule called amylose, and the other is branched chain amylopectin. Its molecular formula is $(C_6H_{10}O_5)_n$.

The structure of starch is as given below

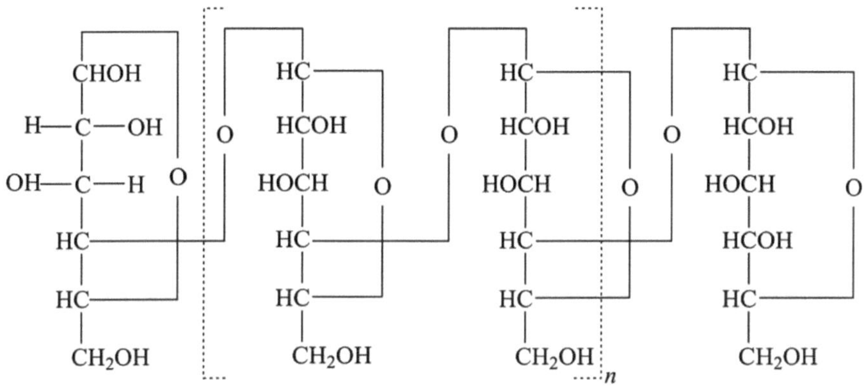

or we can also write starch as

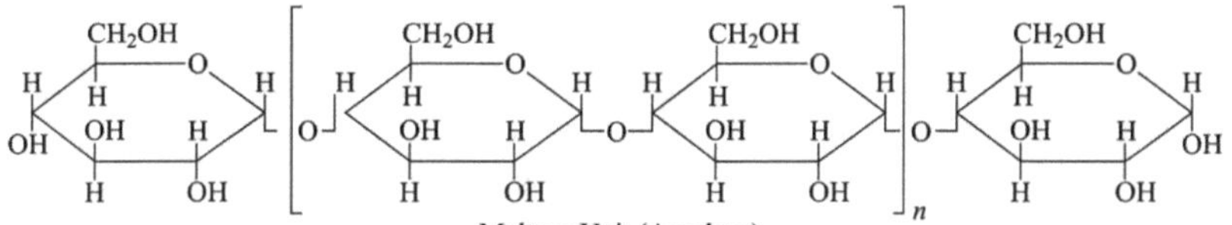

Maltose Uuit (Amylose)

Starch is amorphous, white powder, slightly soluble in water, and gives blue color with iodine. It has two polysaccharides:

(i) Amylose - present in starch cells.
(ii) Amylopectin - present in cell wall.

Structure of amylose (soluble starch)

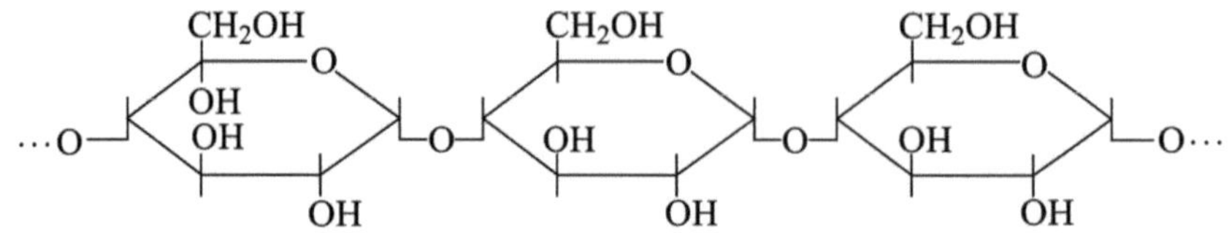

Its uses:

1. as food products
2. in paper and textile
3. in the formation of alcohol by fermentation.

Structure of Amylopectin (Insoluble starch)

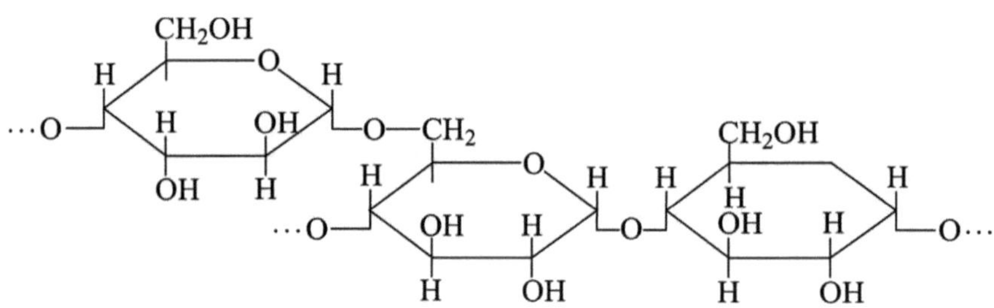

Nucleic Acids and Nucleotides

RNA (ribonucleic acid) and DNA (deoxyribonucleic acid) are two common nucleic acids found in biological systems. RNA controls the protein synthesis and transmits the genetic information, whereas DNA acts as the storage center of genetic information and controls the synthesis of RNA in the nucleus of the cell.

Nucleic acids possess three basic units in their structure:

1. Purine or pyrimidine base
2. Sugar
3. Phosphate molecules-one or more in number.

Two purines (adenine and guanine) and three pyrimidine bases (cytosine, uracil and thymine) are soul of nucleic acids. Their structures are as follows:

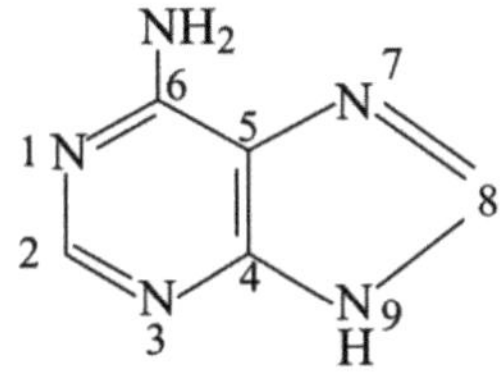

(Adenine)

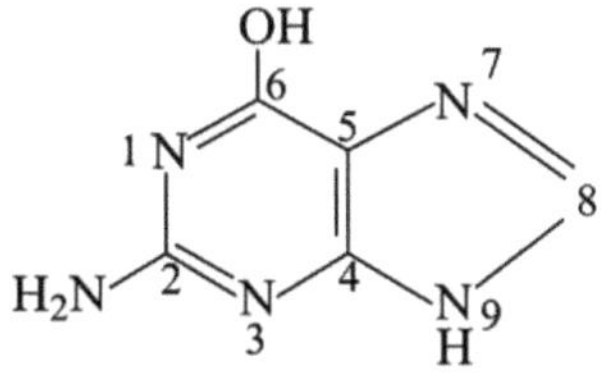

(Guanine)

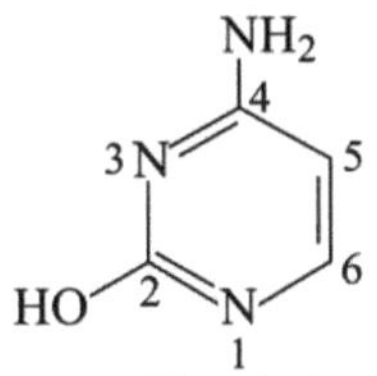

(Cytosine)

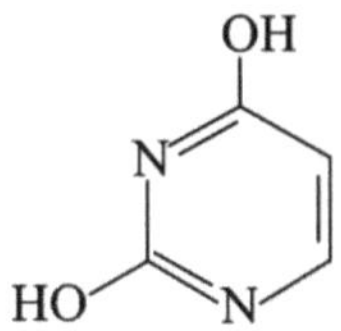

(Uracil only in RNA)

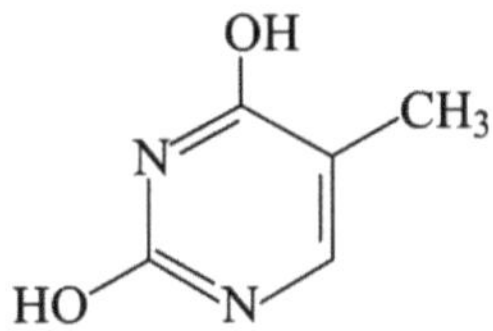

(Thymine only in DNA)

The basic difference between a purine and pyrimidine system, which can be easily judged from these structures, is that purines (adenine and guanine) possess a pyrimidine ring fused with imidazole ring. In RNA, the sugar present is ribose, whereas DNA possesses deoxyribose.

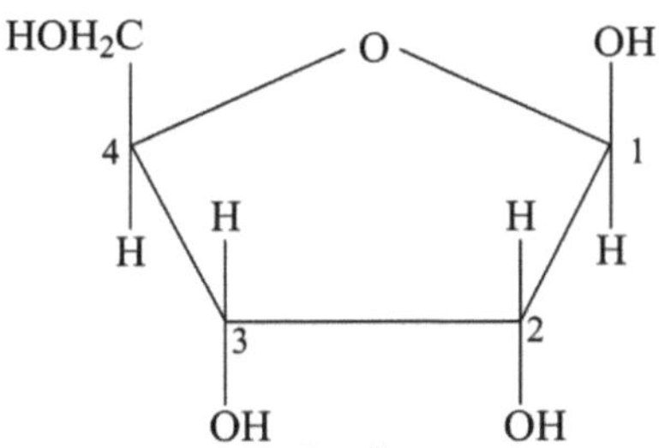

D - ribose

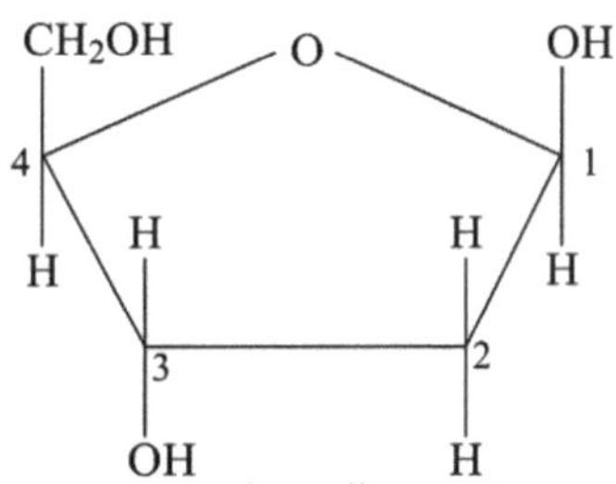

D - deoxyribose

When purine or pyrimidine bases are bound to C-l of the sugar unit, a nucleoside is formed. Such a binding in purine molecules occurs through position 9, whereas in pyrimidine, it occurs through position 1. When –OH group of the sugars forms ester with phosphate ion, a nucleotide is obtained.

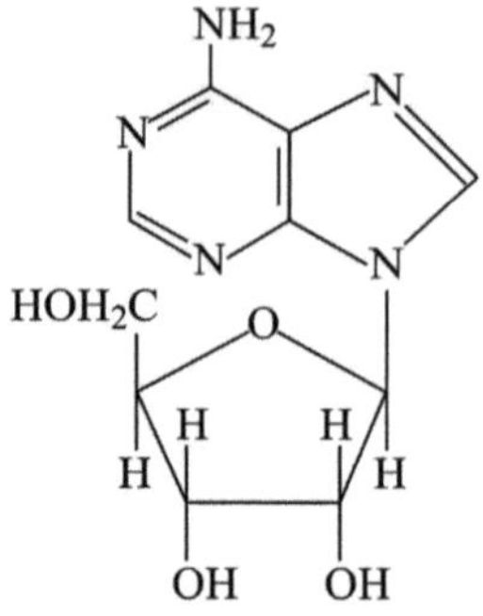

(Adenosine – a nucleoside)

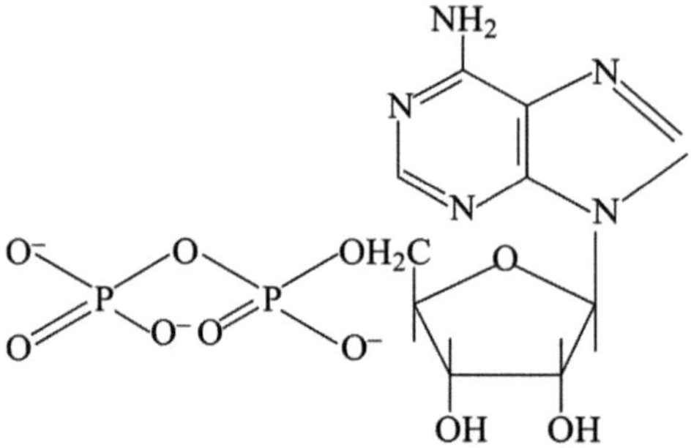

(Adenosine diphosphate ADP)

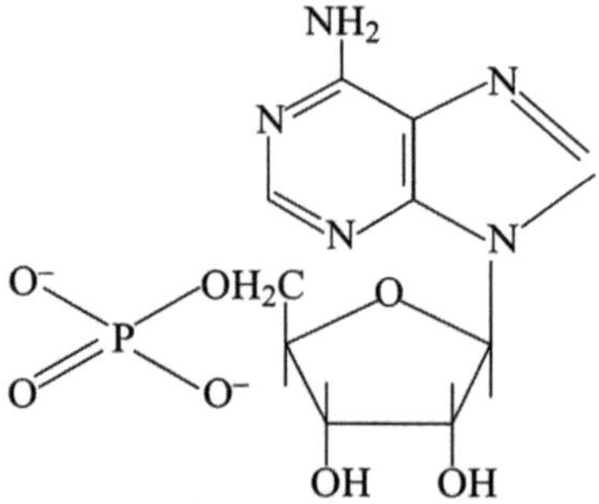

(Adenosine monophosphate AMP)

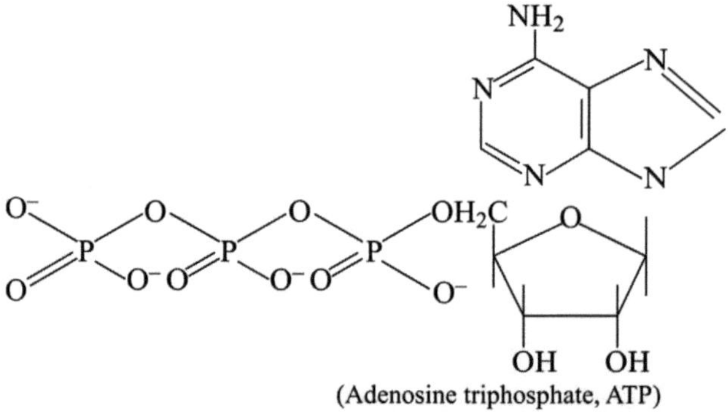

(Adenosine triphosphate, ATP)

ATP, a nucleotide of adenine, serves as an energy storage system in humans and is known as energy-rich compound. The formation of polymer takes place when the phosphate present at C-5 of the sugar unit binds to another molecule of ribose through phosphate. The backbone of DNA consists of such bindings through 5′–3′ units. The genetic information is coded in these bases. DNA composed of two such polymeric chains is coiled into a double helix in such a way so that the base units are pointed into the interior of the helix. The base of the two chains are attached with each other through H-bonding and keep the two chains together. A purine base of one chain is always found to attach with pyrimidine base of other chain. It is noticed that due to the size and shape of the bases, adenine is always paired with thymine, and guanine is paired with cytosine. The two sugar phosphate chains run in opposite direction as shown below:

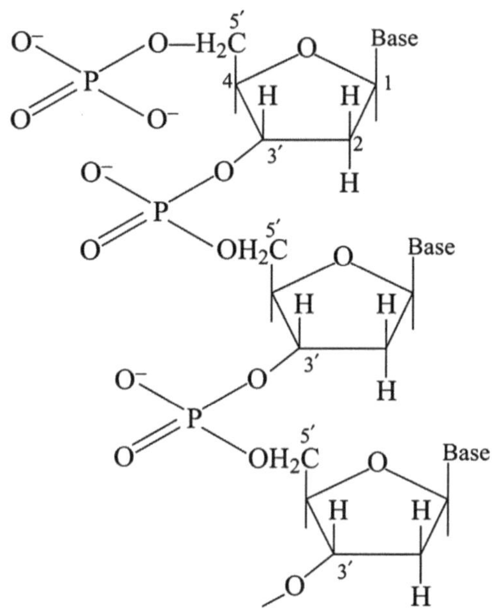

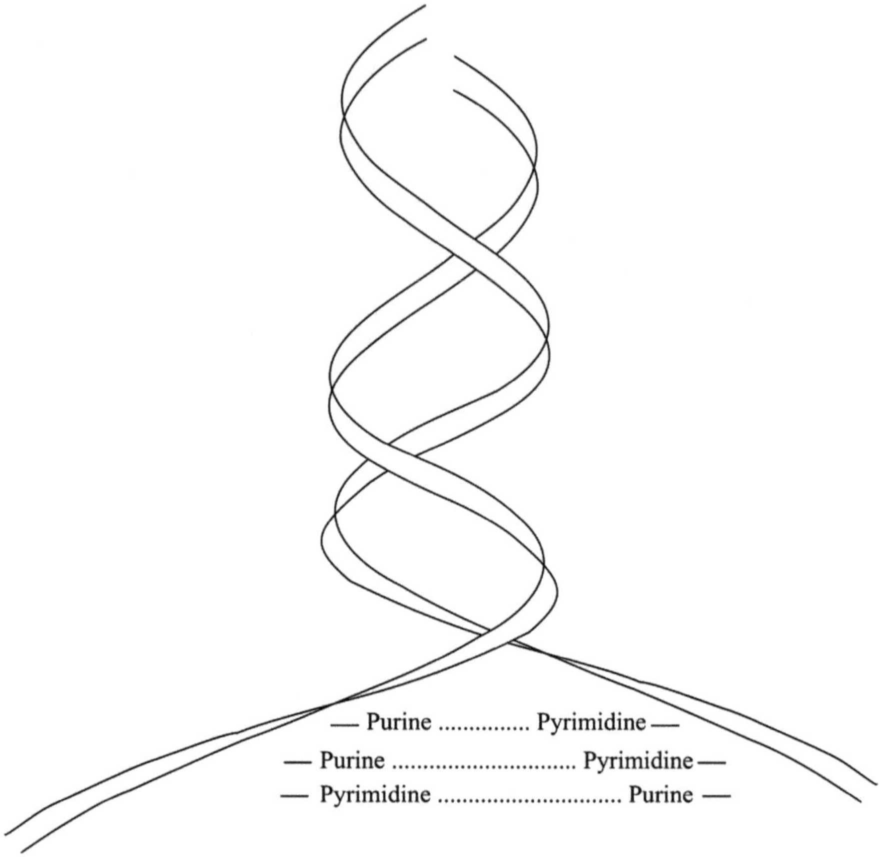

12.2 Synthetic Polymers

In polymer industry, polymers are classified as elastomers, plastics, and fibers, and this classification is based on their elasticity. The elasticity of elastomers is high, and they can be stretched anywhere between 100 and 1000%, whereas fibers are least elastic and can be stretched less than 10%. The elasticity of the plastics is in between fibers and elastomers. Polymers can also be classified on the basis of their response to heating. Polymers which become soft on heating and are chemically unchanged are named as *Thermoplastics*, whereas polymers which degrade or decompose on heating are named as *Thermosettings*. It is observed that polymers having linear chains are thermoplastics, whereas polymers having one chain linked to another are thermosettings. Before we discuss individual polymers, let us see how polymers are formed.

Some common monomers used for addition polymerization.

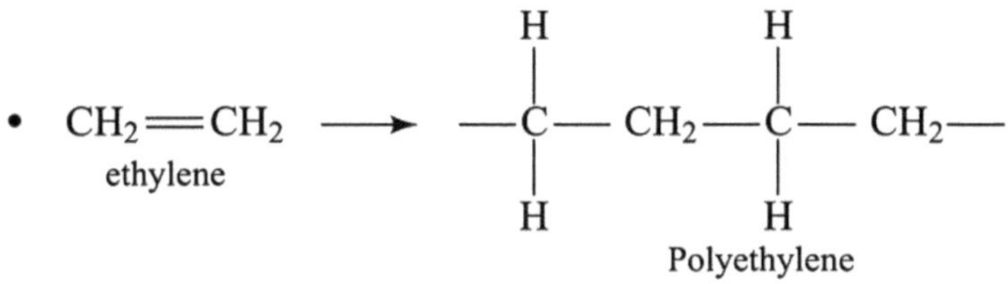

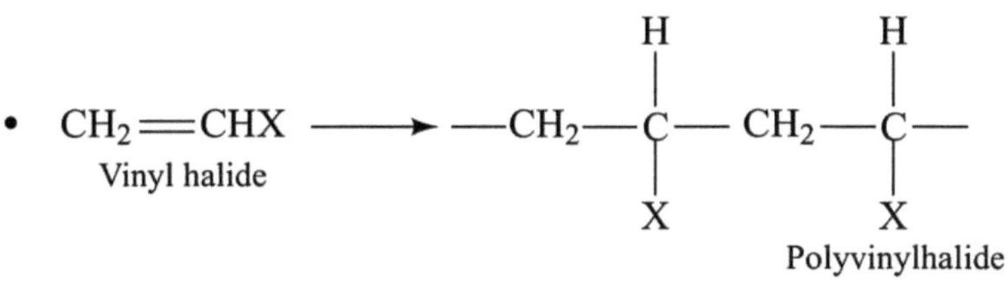

X = Cl; vinyl chloride

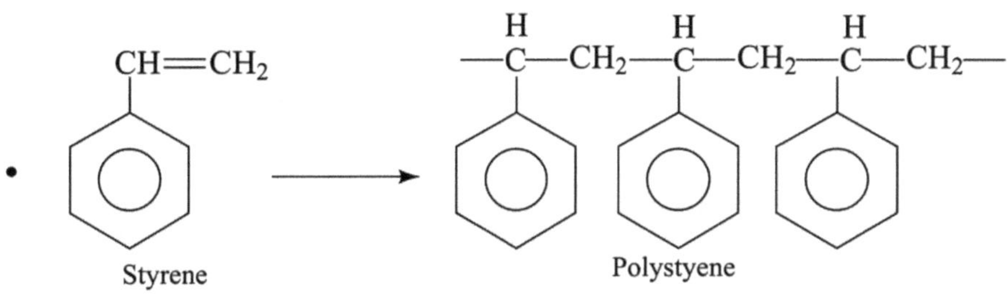

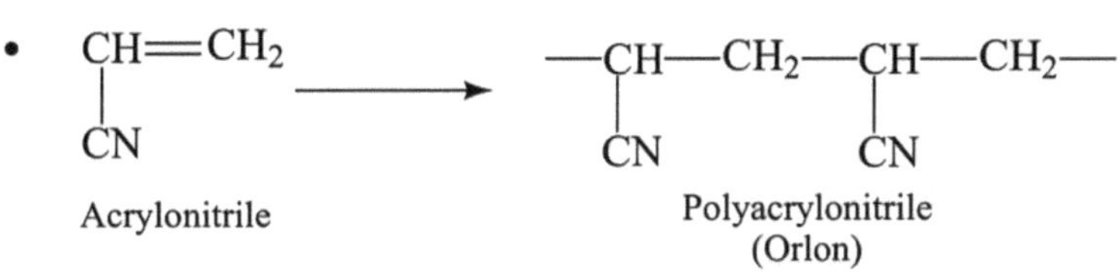

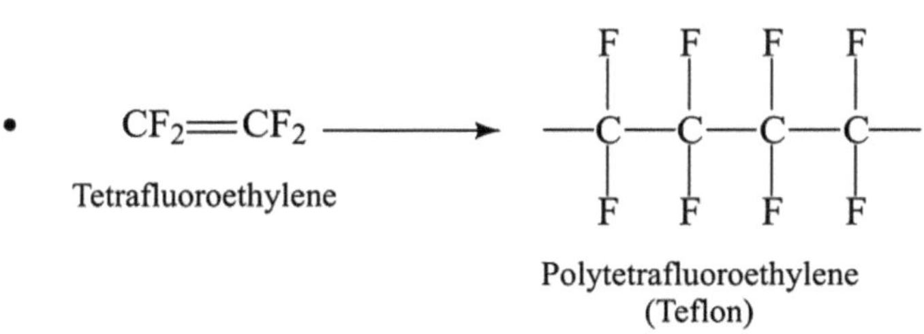

Monomers, the low-molecular-weight compounds of which polymer is made, can combine in two different ways to give polymer.

(i) **Addition:** in which self-addition of monomers takes place without loss of simple molecules like water.

(ii) **Condensation:** in which elimination of small molecules like H_2O, CH_3OH etc., takes place during condensation of monomers.

12.2.1 *Addition Polymerization*

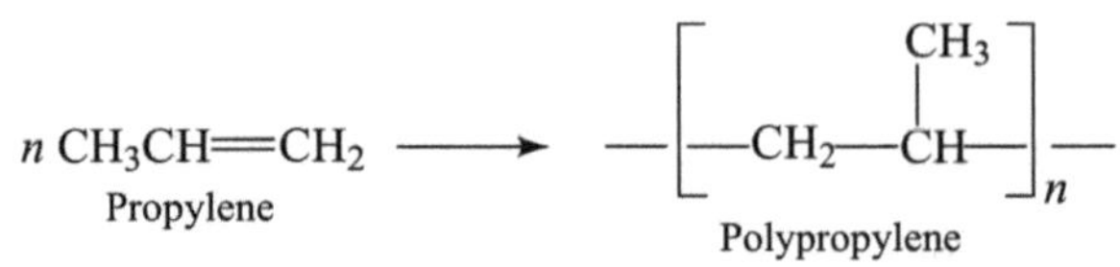

Addition polymerization usually occurs in the monomers containing carbon-carbon double bond and a large number of polymers such as polyethylene, polypropylene, polystyrene, polyvinyl chloride, etc., are produced by this method.

Addition polymerization proceeds by a chain reaction consisting of three distinct steps:

(a) Initiation
(b) Propagation
(c) Termination

Three active centers are possible which can cause initiation of polymerization. They are free radicals, cationic, and anionic species. Let us discuss them one by one.

Free Radical Addition Polymerization

As initiation in this type of polymerization occurs through free radicals. Reagents which can easily give free radicals are used as initiators.

Initiation A common reagent frequently used for initiation is benzoyl peroxide which can readily give free radicals

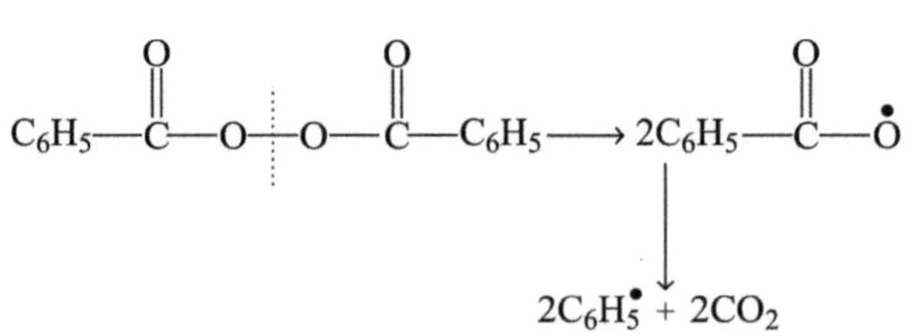

Similarly persulfates also decompose readily to give free radicals

$$K_2S_2O_8 \longrightarrow 2KSO_4^{\bullet}$$

The free radical formed rapidly reacts with the monomer to produce another free radical. As mentioned earlier

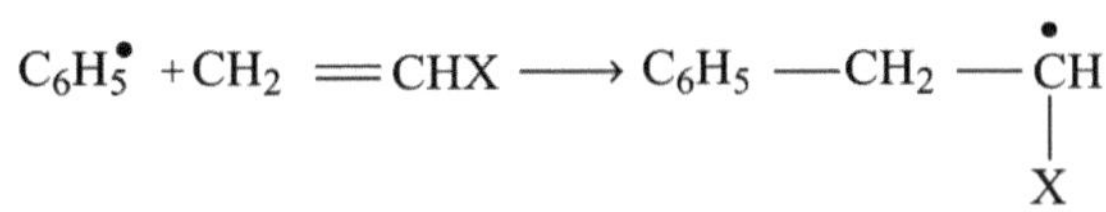

The monomers in this type of polymerization are unsaturated and possess $C = C$ linkage.

Propagation The free radical generated in the initiation step readily combines with another molecule of a monomer forming a new free radical

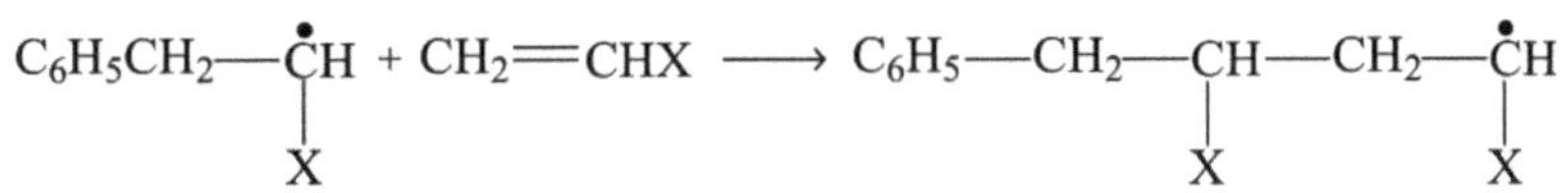

An important aspect during propagation is the way in which free radical attacks on the monomer. For example, the free radical can attack in two different ways in the above reaction.

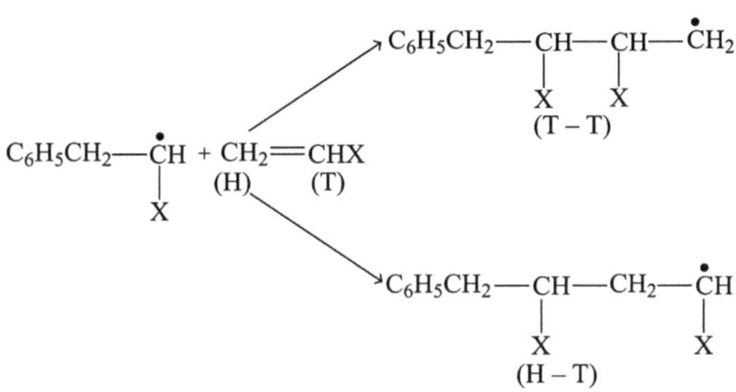

Thus, if combination is tail to tail (T – T), the new free radical generated is primary in nature, whereas if combination is head to tail or tail to head (H – T or T – H), the new radical generated is secondary in nature. As secondary free radicals are more stable than primary, the chain grows with H – T or T – H arrangement. In special conditions, the chain can also grow in H – H or T – T arrangement. Thus, propagation proceeds until all the molecules of monomer are used

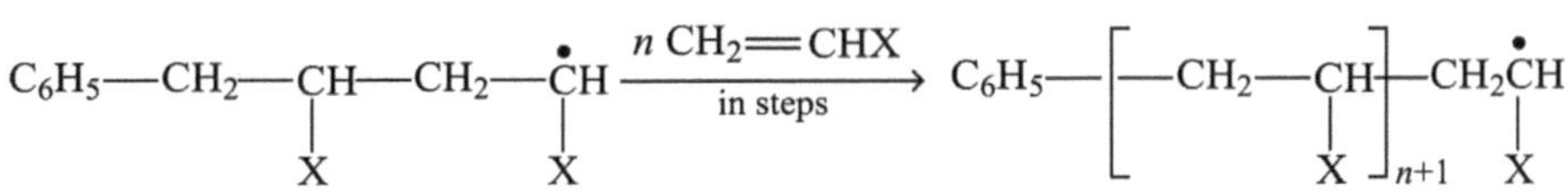

Termination Free radical chains are usually terminated in the end by radical-radical coupling or disproportionation reaction.
For example:

– ***Coupling***

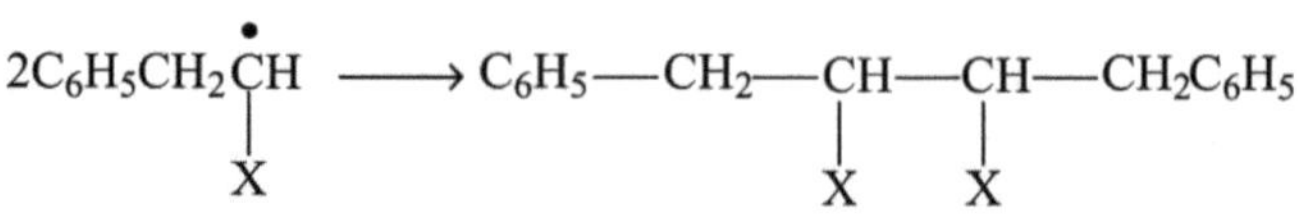

In this case two free radicals combine and active center is lost.

– ***Disproportionation***

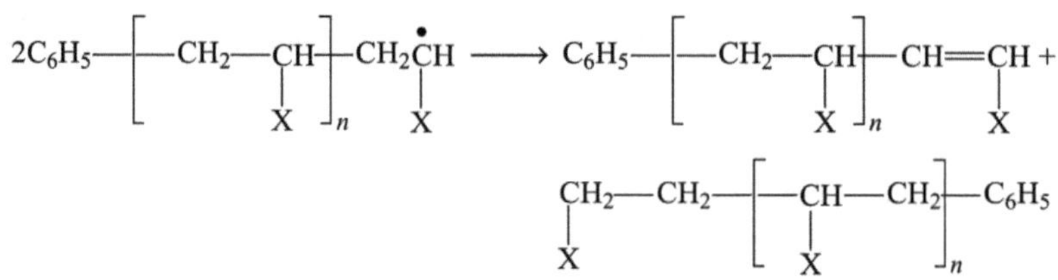

Sometimes the chain is also terminated by the impurity present in the solvent or by the monomer itself. However, in the latter case, the active center is deactivated by the production of a new free radical.

– ***Termination by Impurity***

$$—CH_2—\dot{C}HX + CCl_4 \longrightarrow -CH_2—CHXCl + \dot{C}Cl_3$$

$$\dot{C}Cl_3 + \text{Impurity} \longrightarrow \text{deactivaion}$$

– ***Termination by monomer***

$$—CH_2—\dot{C}HX + CH_2=CHX \longrightarrow —CH_2—CH_2X + CH_2=\dot{C}X$$

Many substances reduce the rate of polymerization by terminating the radical chains and are known as *Inhibitors*. They are used for controlling the rate of polymerization and molecular weight of the desired product. One such substance commonly used is phenol, which can easily transfer a proton to the free radical chain, and the resulting phenoxide ion stabilizes due to resonance and hence does not add to the monomer molecule.

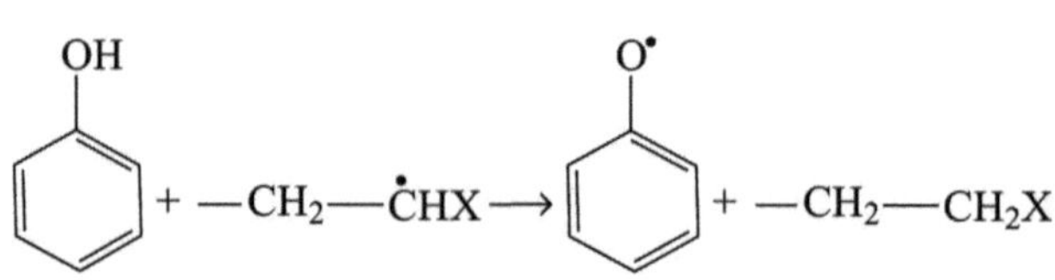

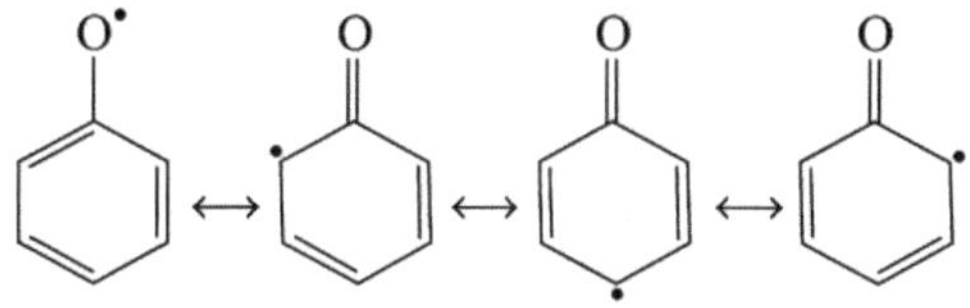

If a polymer is obtained from only one type of monomer molecules, it is called as homopolymer, and the process is known as homopolymerization, whereas polymers desired these days are obtained from two different type of monomers. Such polymers are called as copolymers, and the process is known as copolymerization, which is of great industrial importance. The properties of a copolymer not only depend on the relative concentrations of the two monomers but are also dependent on the reactivity of the free radicals of two monomers. A more reactive monomer will certainly incorporate easily into a copolymer. Saran Wrap, an important copolymer used for wrapping the food items, is prepared from vinylidine and vinylchloride monomers.

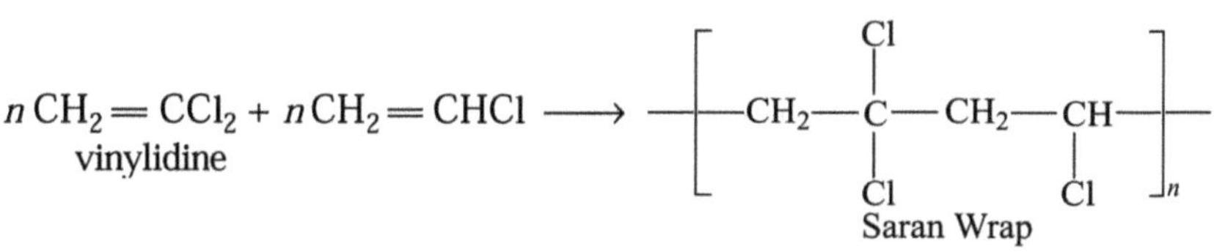

Polystyrene methyl methacrylate is a copolymer obtained from styrene and methyl methacrylate as monomers.

Example 12.1 1,3-Butadiene on free radical polymerization gives a polymer having two types of units

$$\ldots\ldots CH_2CH = CH - CH_2 \ldots\ldots \text{ and } \ldots CH - \underset{\vdots}{C}H - CH = CH_2$$

the ratio of which depends on the temperature used. Explain the reason.

Solution The free radical polymenization proceeds by the attack of free radical on monomer. The radical produced can resonate to give two structures.

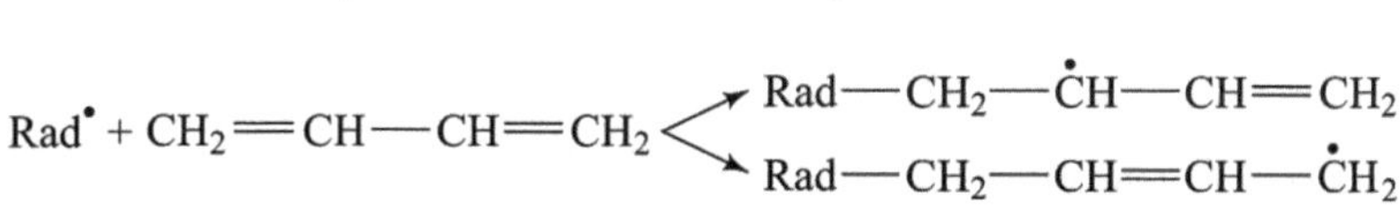

Now chain can form *via* 1,2-addition or 1,4-addition.

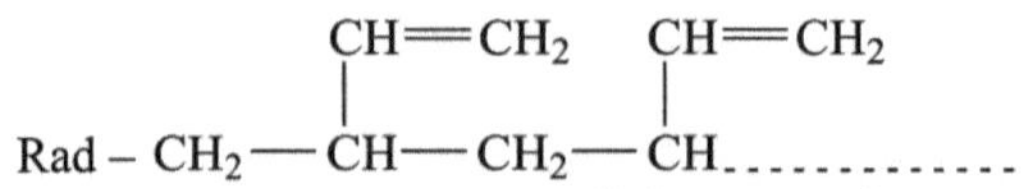

(1,2 – product)

$$\text{Rad} - CH_2\text{-}CH = CH - CH_2 - CH = CH - CH_2 \ldots\ldots$$

(1, 4 – product)

Ionic Addition Polymerization

The chain carriers in ionic addition polymerization may be (i) anionic or (ii) cationic. However, this type of polymerization has limited industrial importance due to complex steps involved in initiation, propagation, and termination.

Anionic Polymerization As the name indicates the chain carriers are negatively charged species, and to stabilize the various species formed during propagation step, it is necessary that the monomer must possess an electron withdrawing group such as nitro, phenyl, carboxylic, vinyl, etc. Typical initiators used in anionic polymerization are alkali metal amides and Grignard reagents. Consider polymerization of styrene under these conditions.

Initiation

$$KNH_2 \longrightarrow K^+ + NH_2^-$$

$$-NH_2 + CH_2 = CHC_6H_5 \longrightarrow H_2NCH_2\overline{C}H - C_6H_5\, K^{\oplus}$$

Thus, potassium amide or lithium amide produces a negatively charged species NH_2^-, which attacks on the monomer of styrene to give an anionic species. The phenyl group present in the monomer helps in its stabilization.

Propagation In the next step, the anionic active center attacks on another molecule of styrene to give a new anion, and the chain increases due to combination of several monomers.

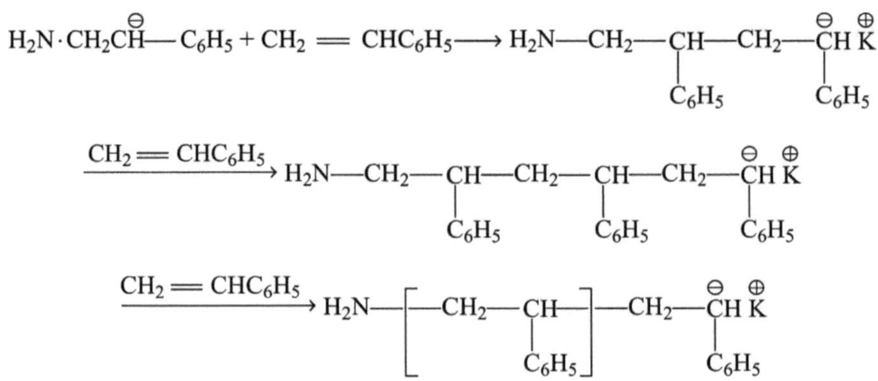

Termination The active center terminates by a hydride transfer or by the reaction with Lewis acid or solvent.

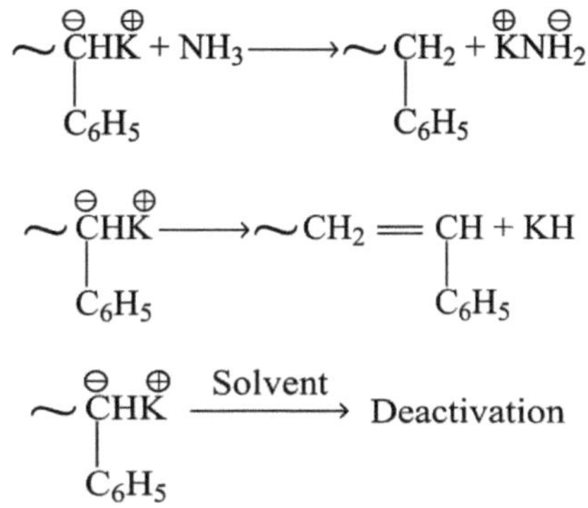

Cationic Addition Polymerization The chain carriers in this polymerization are positively charged species, and this type of polymerization is initiated by acids, Lewis acids, etc. Some common Lewis acids used as initiators are $AlCl_3$, $TiCl_4$, BF_3, and $SnCl_4$. As carbocations (carbonium ions) are the active centers, the presence of electron releasing substituents in the monomer helps in stabilization of the positive charge. Let us consider polymerization of Isobutylene.

Initiation Boron triflouride and water are used for initiation and water acts as co-catalyst.

$$BF_3 + H_2O \rightleftharpoons BF_3 \cdot H_2O$$

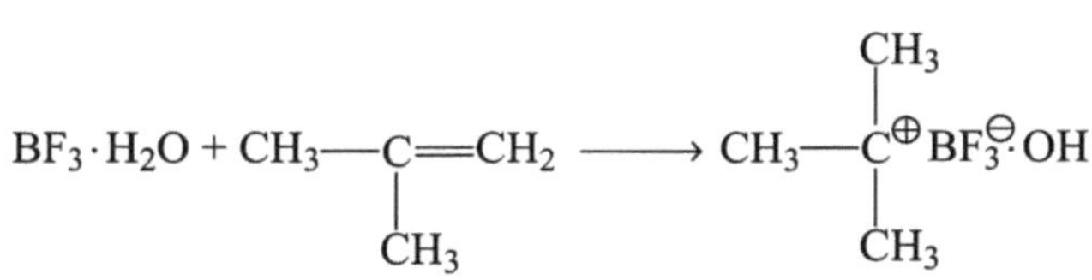

Propagation The active center formed rapidly reacts with other molecules of monomer to produce a chain

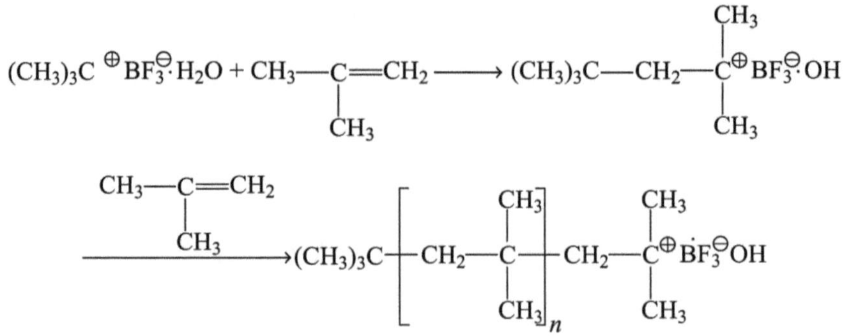

Termination The chain is rapidly terminated by a proton transfer forming double bond as shown.

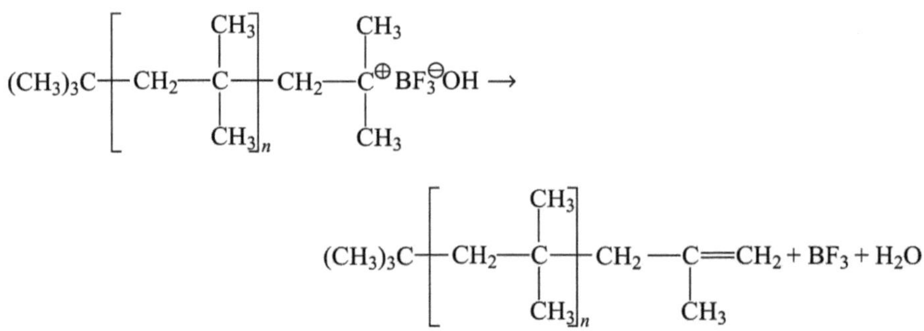

Cationic addition polymerization usually takes place at a very high rate even at low temperatures. For example, polymerization of isobutylene can be achieved in few seconds at −70 °C temperature.

The polymers obtained by more than one methods of polymerization were found to exhibit different properties such as strength, reactivity with solvents, softening point, etc. Till the late 1950s, the most common method used for polymerization was free radical polymerization. However, in the early 1960s, scientists Zeigler and Natta proposed a new method for carrying out polymerization to achieve polymers with unique properties.

Zeigler-Natta Polymerization

The usefulness of Zeigler-Natta polymerization can be realized by the fact that ethylene or propylene which requires high temperature and pressure for polymerization by free radical polymerization can be polymerized at a moderate temperature and pressure in presence of an inert solvent (hexane or heptane) by this method. The catalyst and cocatalyst used in Zeigler-Natta polymerization are titanium tetrachloride ($TiCl_4$) and triethyl aluminum $[(C_2H_5)_3Al]$. Occasionally $LiAlH_4$ is also used as co-catalyst. An important point noticed in this polymerization is that the product is stereospecific to a large extent.

Polythene obtained by Zeigler-Natta polymerization has a molecular weight between 10,000 and 100,000, and a linear polymer is formed. The product is more crystalline and inert to acids, alkalies, and organic solvents and has a very high melting point.

$$n\,CH_2{=}CH_2 \xrightarrow[\text{heptane}]{TiCl_4 + (C_2H_5)_3Al} \underset{\text{polythene}}{-\!\!\!\!+\, CH_2 - CH_2 \,+\!\!\!\!-_n}$$

The mechanism of Zeigler-Natta polymerization is still not fully known because of complex structure of solid catalyst and co-catalyst used in this type of polymerization. The most acceptable mechanism is that the growing chain has a C - Ti bond. The monomers are then added by coordination with titanium through π-bond, followed by an intramolecular rearrangement.

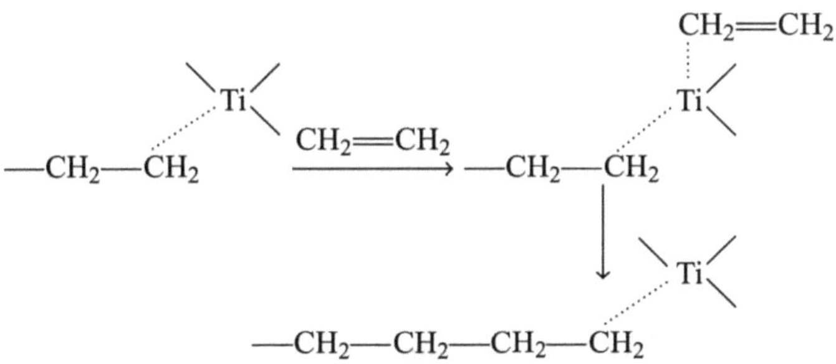

Stereospecific Polymers

During propagation step, the monomers can attach in three different ways to give different polymers. If the monomer is vinyl halide ($CH_2 = CHX$), the polymer can have three possibilities of arrangement of monomers

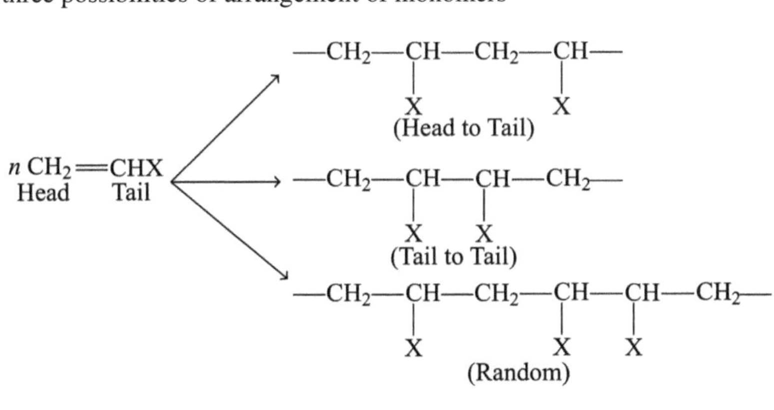

Thus, monomers can attach in H − T, T − T, or H − H and in a random order. This attachment of monomers has been described stereochemically as:

(i) ***Isotactic:*** In isotactic polymers, all X atoms have identical configuration. For a better representation, tetrahedral carbons are considered.

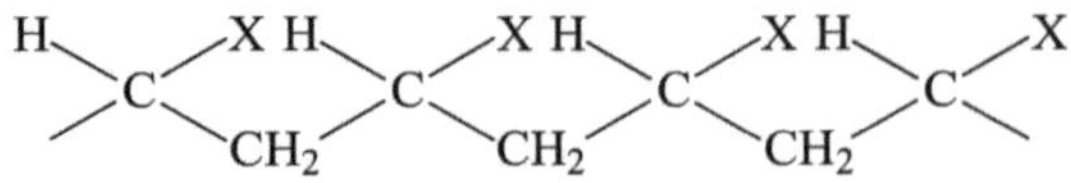

(ii) ***Syndiotactic:*** The arrangement of X atoms change alternately. Thus,

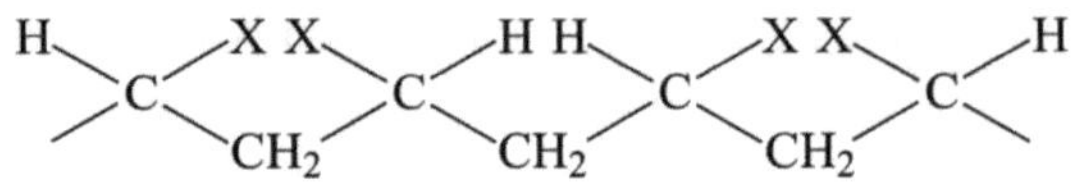

(iii) ***Atactic:*** The arrangement of X atoms in these polymers is random.

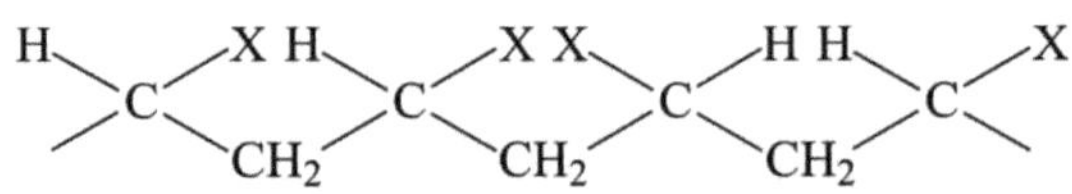

Using Zeigler-Natta polymerization, it is now possible to polymerize isoprene to a material virtually identical to natural rubber.

12.2.2 Condensation Polymerization

As mentioned earlier, small molecules like H_2O, CH_3OH, etc. are evolved when two monomers condense in this type of polymerization. The essential requirement of the monomers for this process is that they should possess bi-or polyfunctional groups. If there is only one functional group, the reaction will stop at dimer stage. For example, if monomers are $R - \mathrm{OH}$ and $R' - \mathrm{COOH}$

$$R - \mathrm{OH} + \mathrm{HOOC} - R' \longrightarrow \underset{\text{(dimer)}}{R\ \ \mathrm{O}\ \ - \mathrm{CO} - R'}$$

As dimer has no free functional group, reaction will not proceed further. However, if monomers are bifunctional

$$\mathrm{HO} - \mathrm{CH_2} - \mathrm{OH} + \mathrm{HOOC} - \mathrm{CH_2} - \mathrm{COOH}$$

$$\xrightarrow{-\mathrm{H_2O}} \underset{\text{(dimer)}}{\mathrm{HO} - \mathrm{CH_2} - \mathrm{O} - \mathrm{CO} - \mathrm{CH_2} - \mathrm{COOH}}$$

the dimer has free functional groups and hence can easily react with other monomers to give polymer. A common polymer obtained by this method is Nylon 66, which can be prepared from the monomers hexamethylene diamine $H_2N(CH_2)_6NH_2$ and adipic acid $HOOC(CH_2)_4COOH$. As both the

$$H_2N(CH_2)_6NH_2 + HOOC\,(CH_2)_4\,COOH$$
$$\longrightarrow H_2N{-}(CH_2)_6 - CO{-}(CH_2)_4 - COOH$$

monomers have six carbon atoms in their structure, the polymer is named as Nylon 66 and has structure

$$\left[HN{-}(CH_2)_6{-}NH{-}\overset{\overset{\displaystyle O}{\|}}{C}{-}(CH_2)_4{-}\overset{\overset{\displaystyle O}{\|}}{C} \right]_n$$

The fibers of Nylon 66 are tough, resistant to friction and possess great tensile strength and elasticity. The fibers are chemically inert and used in bristles, textile fibers, sheets, etc.

(i) 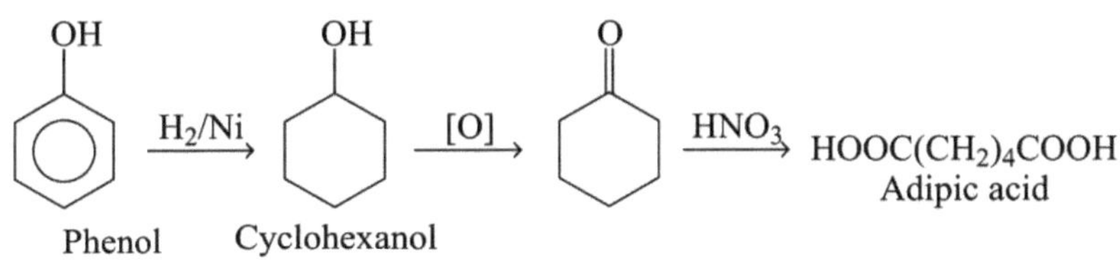

(ii) 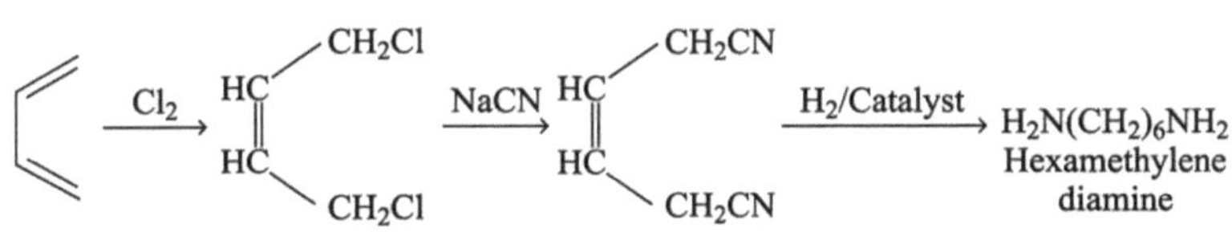

Hexamethylene diamine and adipic acid required for the preparation are commercially obtained from phenol and 1, 3-butadiene in a series of reactions as shown.

12.3 Some Important Polymers

12.3.1 Polyethylene

Polyethylene is obtained from the polymerization of ethylene. The nature of final product obtained depends on the method of polymerization used. For example, at high temperature (250 °C) and high pressure (100–3000 atm. pressure) in presence of free radical initiator, ethylene gives polyethylene which is chemically inert and tough and is used for making insulation wire, pipe, molded toys, and other products.

$$n\mathrm{CH_2} = \mathrm{CH_2} \longrightarrow \left[\mathrm{CH_2} - \mathrm{CH_2}\right]_n$$

In presence of Ziegler-Natta catalyst, ethylene can be polymerized at low temperature and low pressure (6, 7 atm. pressure). The product obtained possesses ethylene in linear chain and the ratio of main chain to side chain is less than $1/200$. The polyethylene obtained by this method is known as high-density polyethylene and is more crystalline in comparison to the one prepared by free radical polymerization. It is used in the preparation of containers and house ware materials.

12.3.2 Teflon

Teflon is prepared by the polymerization of tetrafluoroethylene at high pressure and controlled temperature. The polymerization is carried out using free radical polymerization, and the product obtained is highly crystalline having melting point 327 °C. It is resistant to heat, acid, alkali, and almost all the organic solvents and is widely used in the chemical laboratories.

$$n\mathrm{CF_2} = \mathrm{CF_2} \longrightarrow \left[\mathrm{CF_2} - \mathrm{CF_2}\right]_n$$

Teflon is called as polytetrafluoroethylene, PTFE.

12.3.3 Polystyrene

Styrene can be polymerized under different conditions to give polystyrenes having different properties. Polystyrene obtained through free radical polymerization is usually an atactic polymer which becomes soft at 85 °C. However, if polymerization of styrene is carried out using Zeigler-Natta catalysts, a polymer having isotactic arrangement is obtained. This polymer has a much higher melting point (~ 250 °C). Polystyrene is widely used for making television coils, caps of bottles, and novelty items.

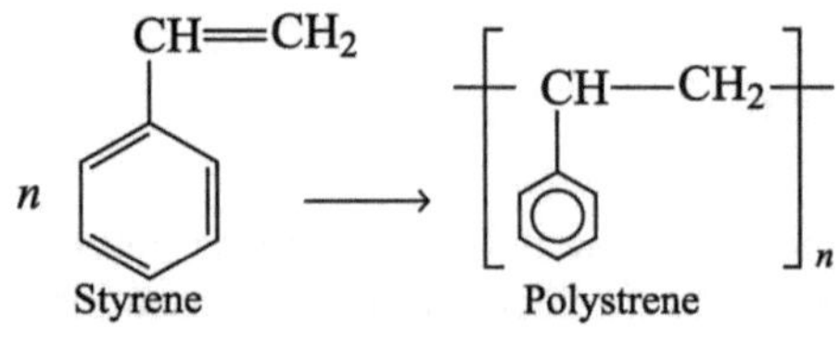

12.3.4 PVC

Poly vinyl chloride is obtained by the polymerization of vinyl chloride in presence of free radical initiators. The product obtained is tough plastic material having good flame resistance. The polymer is widely used for making molded pipes bottles, etc.

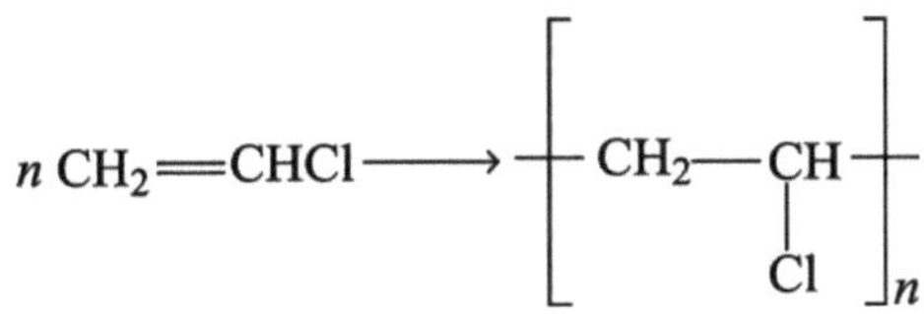

The tough PVC material can be made softer by mixing it with low-molecular-weight aliphatic polyesters, commonly known as *Plasticizers*. The material so obtained is used for making vinyl leather and plastic raincoats.

12.3.5 Orlon

Orlon is a polymer of acrylonitrile, and the polymerization is carried out in water. It is a fibrous material which is insoluble in common solvents but is soluble in nitrophenol, dimethylformamide, etc. It has very high stability in dry as well as in wet states and has excellent resistance to sunlight. It can be blended with wool after dissolving in dimethyl formamide.

$$n\,(CH_2 = CHCN) \longrightarrow {+}CH_2 - CHCN{+}_n$$

12.3.6 Phenol-Formaldehyde Resins

Phenol and formaldehyde undergo condensation polymerization to give important polymers known as phenol-formaldehyde resins. The nature of polymer obtained depends on the ratio of formaldehyde and phenol used for polymerization. If the ratio is less than one, a thermoplastic polymer called as novolak is obtained. It is widely used in the manufacture of varnish, laminated material, etc.

$$n\,CH_2{=}CHCN \longrightarrow \left[CH_2-\underset{\displaystyle CN}{\underset{|}{CH}} \right]_n$$

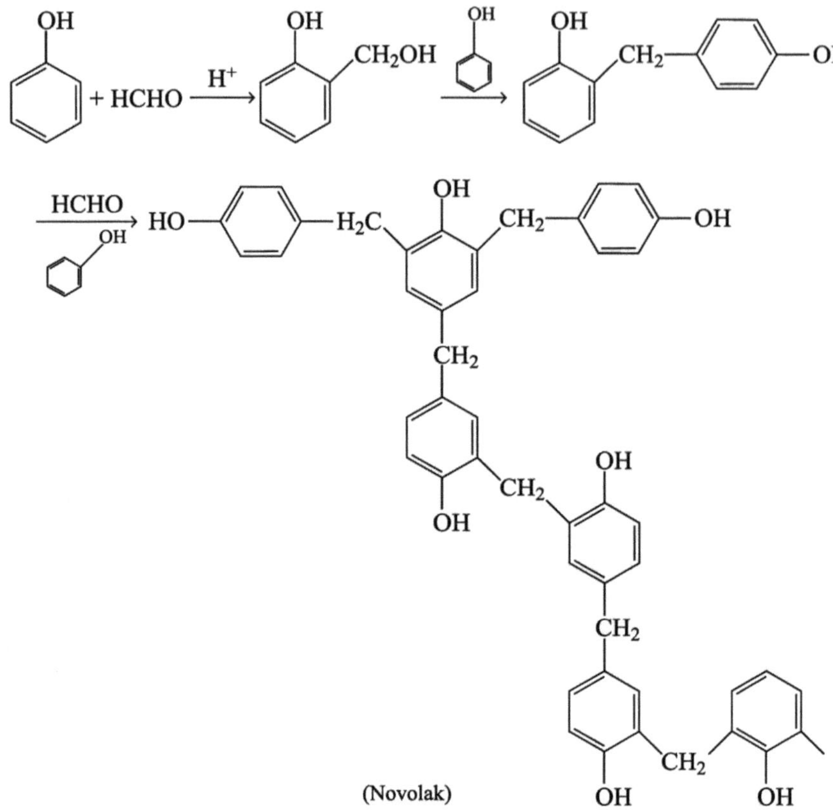

(Novolak)

On using molar ratio of formaldehyde to phenol as one or more a polymer having thermosetting property is obtained and is known as Bakelite. It is a three-dimensional polymer in which all the three positions (2-ortho and 1-para) of phenol are attached with formaldehyde. It is bad conductor of electricity and is widely used for the manufacture of electrical switches, plugs, and other appliances.

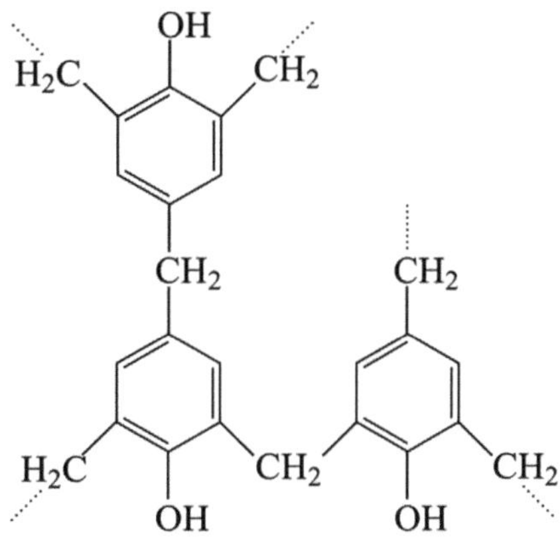

(Bakelite)

12.3.7 *Polyurethanes*

Polyurethanes are of considerable use these days because of their use as urethane foam, cushions, and padding. An isocyanate on reacting with substance containing active hydrogen yields urethanes.

$$\underset{\text{Isocynate}}{R{-}N{=}C{=}O} + R'OH \longrightarrow \underset{\text{Urethane}}{RNH\overset{\displaystyle O}{\overset{\|}{C}}{-}OR'}$$

In the preparation of polyurethane foam, a prepolymer having OH group at the end is produced. The molecular weight of this prepolymer varies between 1000 and 2000. Toluene diisothiocyanate is then added to form polyurethane.

(i) Preparation of prepolymer is achieved using ethylene glycol and adipic acid as monomers.

$$\underset{\text{Ethylene glycol}}{HOCH_2CH_2OH} + \underset{\text{Adipic acid}}{HO\overset{\displaystyle O}{\overset{\|}{C}}{-}(CH_2)_4{-}\overset{\displaystyle O}{\overset{\|}{C}}{-}OH} \longrightarrow$$

$$HO{-}CH_2CH_2O{-}\left[\overset{\displaystyle O}{\overset{\|}{C}}{-}(CH_2)_4{-}\overset{\displaystyle O}{\overset{\|}{C}}{-}OCH_2CH_2O\right]_n{-}H$$

Prepolymer

(ii) The prepolymer so prepared is condensed with toluene diisothiocyanate to give polyurethanes. Polyurethanes are linear molecules without interconnecting linkages between them.

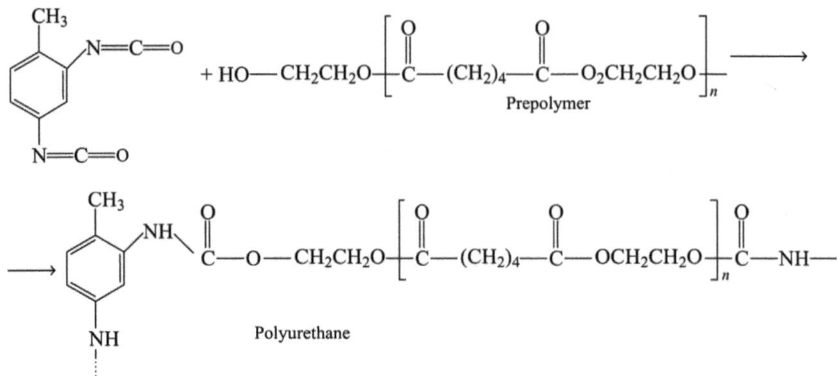

12.3.8 *Dacron*

Dacron is a polyester fiber which is obtained by the condensation of either terephthalic acid or dimethyl terephthalate with ethylene glycol. The condensation is usually carried out at 200 °C. The fibers are generally used in a blend with cotton fibers for preparing garments and are extensively used in the textile industry.

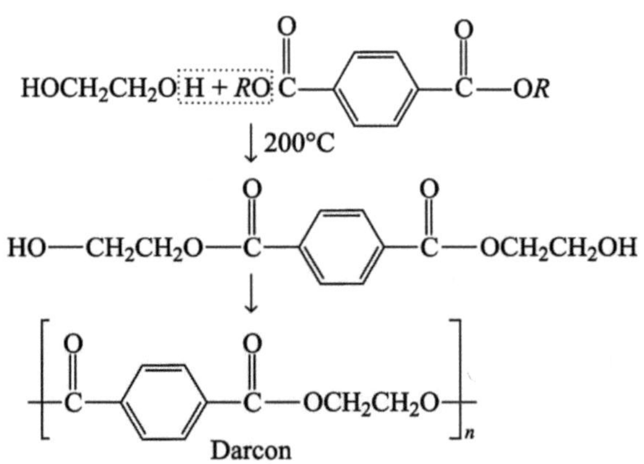

Darcon

12.3.9 Nylon 6

Nylon 6 - is a polymer of major importance these days in synthetic fiber industry. It is prepared from caprolactum which is synthesized from cyclohexane in a series of reactions

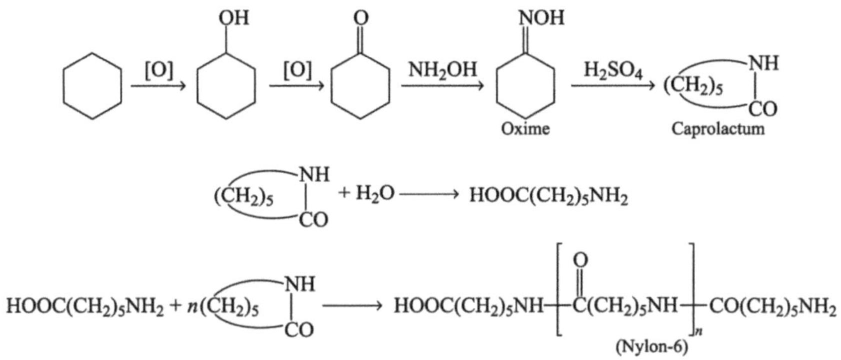

The polymerization of caprolactum is initiated by its hydrolysis to give amino caproic acid, which rapidly reacts with other molecules of caprolactum to give Nylon 6. The linear chains of Nylon 6 are held together by intermolecular H-bonding.

12.3.10 Poly (Methyl Methacrylate)

The monomer of this polymer is methyl methacrylate, and the polymer is prepared by free radical polymerization. It is also called as Plexi Glass and Lucite.

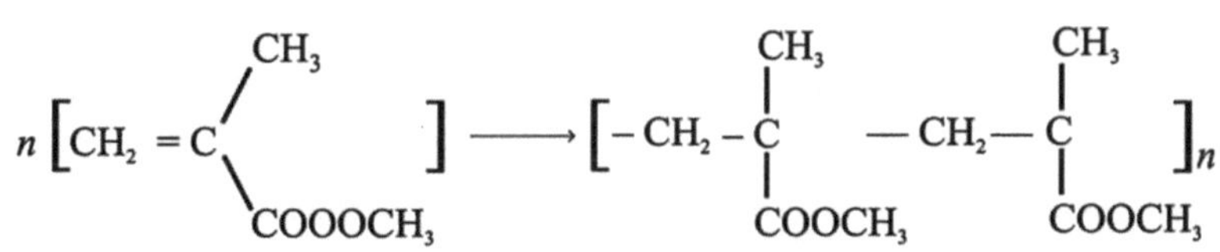

It has fiber properties and is used in the formation of fabrics.

Molecular Mass

Polymers are heterogeneous in molecular mass because molecules of different masses remain present at different places of polymers. Hence, three types of molecular mass of polymer have been recognized.

(i) Number average,
(ii) Weight average and
(iii) Viscosity average

Out of these first two are common.

(i) *Number average molecular mass* is commonly represented by $\overline{M}_n$ and is determined by using deviation in boiling point or depression in freezing point. Other method such as lowering of vapor pressure is a common method used for determining $\overline{M}_n$. The number average molecular mass ($\overline{M}_n$) of a polymer is defined as the total mass of all the molecules divided by total number of monomers used.

$$\overline{M}_n = \frac{\omega}{\sum N}$$

where ω is the total mass of all the molecules present and ΣN represents the sum of total number of monomers present in the polymer. The value of $\overline{M}_n$ provides an idea about the impact and tensile strength of the polymer.

(ii) *Weight average molecular mass* of a polymer is determined by using light scattering technique or ultra-centrifugation technique. Basically these techniques provide molecular size of the polymer. The weight average molecular mass is defined as

$$\overline{M}_\omega = \frac{\sum \omega_i M_i}{\sum \omega_i}$$

where ω_i represents weight fraction of the molecules having mass M_i. The expression has further been simplified as

$$\overline{M}_\omega = \frac{\sum N_i M_i^2}{\sum N_i M_i}$$

where N_i is number of molecules of mass M_i. For a better understanding, let us consider an example.

Example 12.2 A polymer has been found to possess the population of various molecules as follows:

(i) 5 molecules of molecular mass each 5,000.
(ii) 10 molecules of molecular mass each 6,000.
(iii) 20 molecules of molecular mass each 10,000.
(iv) 20 molecules of molecular mass each 15,000.
(v) 10 molecules of molecular mass each 25,000.

Calculate its number average and weight average molecular mass.

Solution The number-average molecular mass will be

$$\overline{M}_n$$

$$= \frac{\sum N_i M_i}{\sum N_i}$$

$$= \frac{(5 \times 5000)+(10 \times 6000)+(20 \times 10,000)+(20 \times 15,000)+(10 \times 25,000)}{5 + 10 + 20 + 20 + 10}$$

$$= \frac{25,000 + 60,000 + 200,000 + 300,000 + 250,000}{65}$$

$$= \frac{835 \times 10^3}{65}$$

$$= 12.84 \times 10^3$$

The weight-average molecular mass will be

$$\overline{M}_\omega = \frac{\tau \sum N_\iota M_i^2}{\sum N_i M_i}$$

$$\overline{M}_\omega = \frac{\left[5 \times (5000)^2\right] + \left[10 \times (6000)^2\right] + \left[20 \times (10,000)^2\right] + \left[20 \times (15,000)^2\right] + \left[10 \times (25,000)^2\right]}{835 \times 10^3}$$

The denominator ΣN_i, M_i, will be same as calculated for $\overline{M}_n$ above.

$$\overline{M}_\omega = \frac{13235 \times 10^6}{835 \times 10^3}$$

$$= 15.85 \times 10^3$$

Rubber

Rubber is a natural substance obtained from nearly five hundred different plants, but the main source is the tree Hevea brasiliensis. It is obtained from the plants by making cuts in the bark, and the latex is collected, which possess 25–40% of rubber hydrochloride along with other impurities such as fatty acids, proteins, resins, etc. On treatment with acetic acid, latex gives cold rubber which is coagulated in the solution and is composed of 90–95% rubber hydrocarbon (C_5H_8); 2–4% protein and 1–4% resins. The destructive distillation of rubber gives isoprene (C_5H_8) as the major product and thus suggests that rubber is a polymer of isoprene C_5H_8, which is 2-methyl-l, 3-butadiene. The natural rubber obtained from trees is hard, brittle, and non-elastic and is therefore of little use. To achieve the outstanding elastic property of the rubber, it is necessary to make some minor changes in its structure.

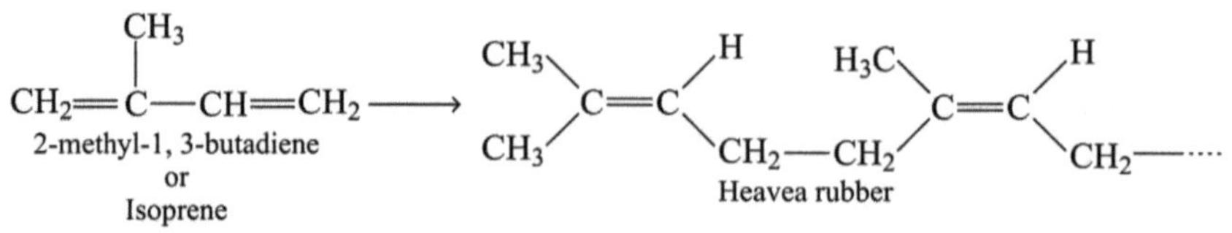

The credit of achieving major changes in the properties of rubber by minor changes in its structure goes to Charles Goodyear in which rubber is heated with sulfur or sulfur compounds at about 150 °C. Such a heating causes formation of —C—S—S—C— bonds, and the process is known as vulcanization. Two polymer chains can be easily linked by vulcanization, forming —C—S—S— bonds.

Thus,

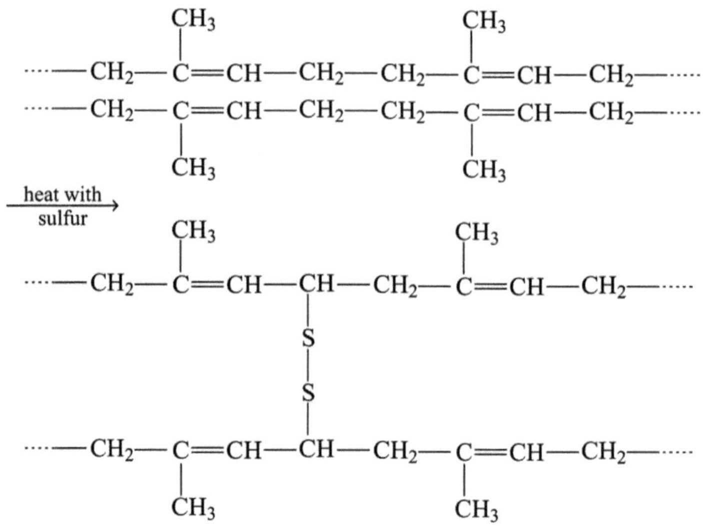

Such a cross-linking makes the rubber harder, stronger, and more useful. With increase in demand of rubber worldwide, it became necessary to use other methods to obtain isoprene for the preparation of rubber.

Butadiene and its derivatives were developed. Chloroprene (2-chloro-1, 3-butadiene) was found to rapidly polymerize to give neoprene rubber.

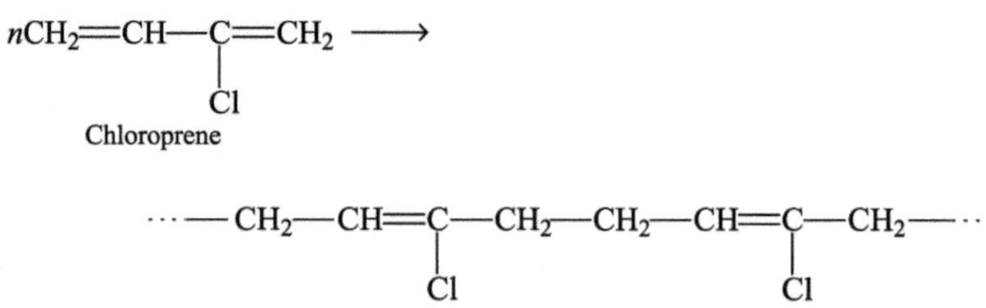

The chlorine content of neoprene imparts some useful properties to this polymer like resistance to chemicals, oils, heat, and flames.

An important synthetic rubber widely used in manufacturing of rubber materials, tires, etc. is a copolymer of 75% butadiene ($CH_2 = CH - CH = CH_2$) and 25% styrene ($C_6H_5 - CH = CH_2$). The polymer obtained is commonly known as BUNA-S, and the polymerization is carried out in an emulsion system at 50 °C in presence of peroxide catalyst.

As there are ethylenic double bonds present in the chain, it can also be vulcanized using sulfur to give cross-linked polymer. BUNA-S or SBR (styrene-butadiene rubber) has a greater resistance to mechanical abrasions.

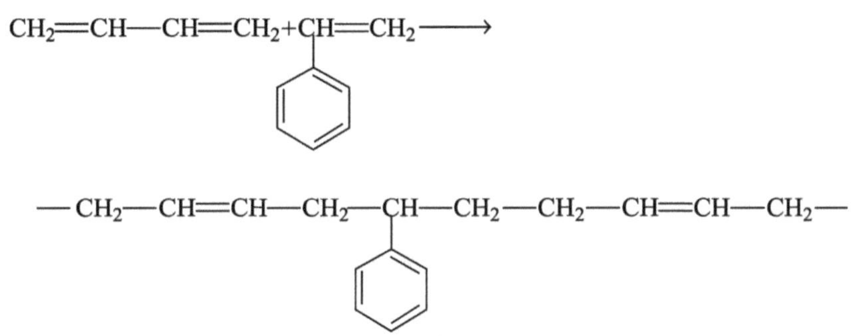

(BUNA-S)

12.4 Conducting Polymers

Most of the polymers are poor conductors of electricity due to non- availability or less availability of free electrons. The polymers, which can conduct electricity, are called as conducting polymers. However, one must understand that polymers are not as conductive as metals. Broadly, conducting polymers are classified as follows:

1. **Intrinsic Conducting Polymers (ICP):** In these polymers, the associated groups consist of delocalized electron pair which are responsible for electrical charge. When electric field is applied, the conjugated electrons of the polymer are excited and results into the formation of valence bonds as well as conduction bands. This increases the conductivity to a large extent. Some of the polymers which show this type of property are polymers of polyacetylene, polyaniline, polythiophene, polyphenantherelene, etc.
2. **Doped Conducting Polymers:** These polymers are obtained by exposing an ICP polymer to a charged transfer agent either in gas or liquid phase. These polymers have low ionization potential and high electron affinity and hence can be easily oxidized or reduced. This property increases the conductivity by creating positive or negative charges on polymer backbone. Similar to semi-conductors, the doping in polymers is of two types:

(i) ***P-diping:*** This type of doping is obtained by treating ICP with Lewis acids, which create positive charge on polymer backbone. Some Lewis acids commonly used are $FeCl_3$, I_2 etc.

$$(C_2H_2)_n + 2FeCl_3 \longrightarrow (C_2H_2)_n^+ FeCl_4^- + FeCl_2$$

$$2\,(C_2H_2)_n + I_2 \longrightarrow 2\left[(C_2H_2)_n^+ I_3^-\right]$$

(ii) ***N-doping:*** In this case ICP polymers are treated with Lewis base to create negative charge on polymer backbone, for example, lithium, sodium, calcium, etc.

$$\left[CH=CH-CH=CH\right]_n + \text{Base} \longrightarrow \left[CH=CH-\underset{B^+}{\underset{|}{C}}=CH\right]_n$$

3. **Extrinsically Conducting Polymers:** These polymers exhibit conductivity due to externally added materials. These can be divided into two types.

   (a) ***Conducting element-filled polymers:*** The polymers filled with conducting elements like carbon, metallic fibers, metal oxides, etc., come under this class. In these polymers, the polymer acts like a binder to hold together conducting elements. They have good conductivity, low cost, and high weight.

   (b) ***Blended conducting polymers:*** This type of polymers is combination of a conventional polymer and conducting polymer either by physical or chemical change. They possess better physical, chemical, and mechanical properties.

4. **Coordination Conducting Polymers or Inorganic Polymers:** This type of polymers is obtained by combining a metal with a poly dentate ligand. Generally this type of polymer has low degree of polymerization.

12.4.1 Applications of Conducting Polymers

Conducting polymers are widely used due to their light weight, easy preparation, and good mechanical properties. Some common applications are:

1. In rechargeable light weight batteries (perchlorate doped poly acetylene lithium batteries which are nearly ten times lighter than Pb batteries)
2. In aircrafts and aerospace
3. In telecommunication.
4. In antistatic coatings or clothing
5. In LCD display
6. In electronic devices such as diodes etc.
7. In solar cells

8. In drug delivery systems of human body
9. In optical display devices like LED.

12.5 Biodegradable Polymers

The last two or three decades have seen an enormous growth is the use of polymers, most of which are nonbiodegradable. This leads to the problem that they may last for many years in rubbish dumps. The main reason for clogging of drain pipes is accumulation of materials such as poly bags, etc. Such polymers are non-reactive and hence remain in the environment for decades. When such polymers are used in landfill sites, they are preserved for hundreds of years. If they are burnt, they produce very harmful gases. The best example is, if we leave a plastic bag on the floor, it would not decay for centuries. Most of the common polymers widely used these days, including poly(ethene) and poly(propene), are not biodegradable, and their improper disposal leads to variety of problems such as clogging of drain line. In country like India, many cities have tried to put ban on the use of polyethylene carry bags; still more efforts are needed in this direction. Nowadays, biodegradable polymers are synthesized, which means that bacteria can break them down on disposal. The addition of substances such as cornstarch, cause the polymer to break down more quickly. In many developed countries, carrier bags and refuse bags made from such biodegradable polymers are in use.

A general definition of biodegradable polymers is that they are specific type of polymers, which after their intended use breaks down in the environment. A complete biodegradation means that no oligomer or monomer is left. The breakdown results to natural by-products such as common gases (nitrogen, CO_2, etc.), water, biomass, etc. A large number of natural polymers and synthetic polymers are biodegradable, and they chiefly consist of amide, ester, or ether functional groups in their structure. The properties and breakdown mechanism of biodegradable polymers depend on their exact structure. However, a basic requirement of such polymers is that should be stable and durable for use in their particular applications, and on disposal, they should easily breakdown. The two mechanism responsible for degradation of polymers are hydrolysis and photodegradation. The second process can take place in aerobic and anaerobic conditions.

Biodegradable polymers possess strong backbone of carbon chain, which is difficult to break. Hence, the degradation in such polymers starts from the end groups. Minimum cross-linking or branching is preferred in such polymers. As the degradation of these polymers begins at the end, it results in large surface area, which has easy access for chemical, light, or organism for its degradation. These polymers are hydrophilic in nature because hydrophobic polymers prevent an enzyme from easy interacting. Nowadays, biodegradable polymers are of significant interest to the scientific community and, hence, are widely used in a variety of fields including medicine, agriculture, and packaging. However, the primary requirements for their medicinal use are as follows: (i) they should be nontoxic and (ii) should have controlled rate of degradation. One of the most active areas of research, which attracted considerable attention in the last two decades, is in controlled drug delivery and release. They are able to slowly release the drug into the body, and degradation depends on location in body, pH, amount of water, and so on.

12.5.1 Some Biodegradable Polymers

Polylactic acid (PLA) is a common biodegradable polymer used for packaging purposes. The production of PLA has several advantages, the most important of which is the ability to tailor the physical properties of the polymer through processing methods. PLA is used for the preparation of a variety of films, wrappings, and containers (including bottles and cups). In the year 2002, the Food and Drug Administration department (FDA) of the USA ruled that it is safe to use PLA in food packaging industries. It can be prepared by direct polymerization of L-lactic acid or through lattice as shown below.

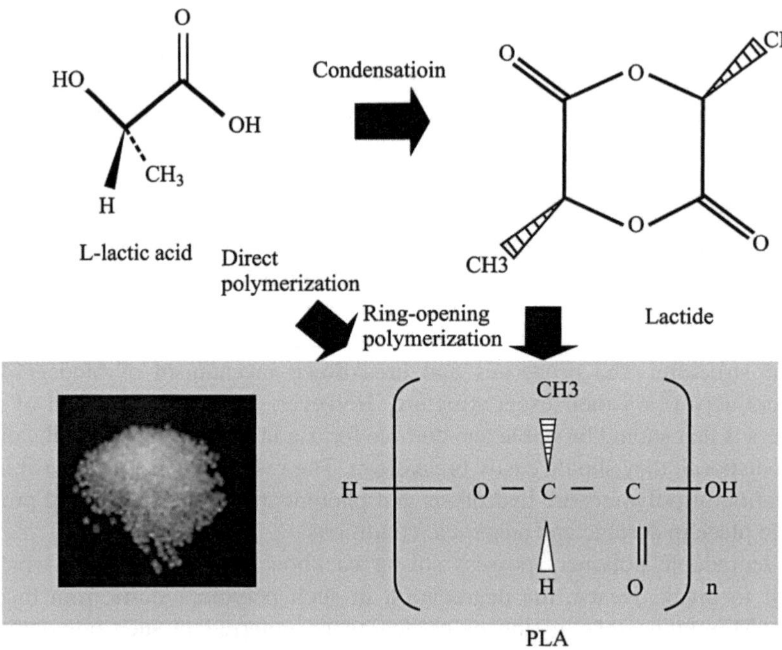

12.5.2 Examples of Biodegradable Polymers

(i) ***Polyhydroxy butyrate (PHB):*** Polyhydroxy butyrate (PHB) is obtained from the condensation of 3-hydroxy butanoic acid.

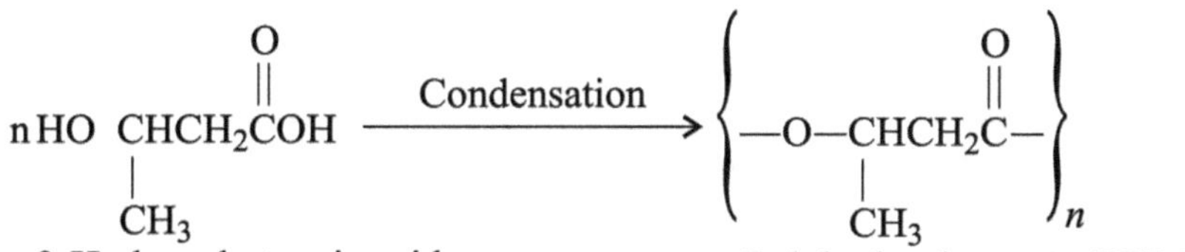

3-Hydroxybutanoic acid   Polyhydroxbutyrate (PHM)

(ii) ***Poly-hydroxybutyrate-co-b-hydroxy valerate (PHBV):*** It is copolymer of 3-hydroxy butanoic acid and 3-hydroxy pentanoic acid, in which the monomer units are joined by ester linkages.

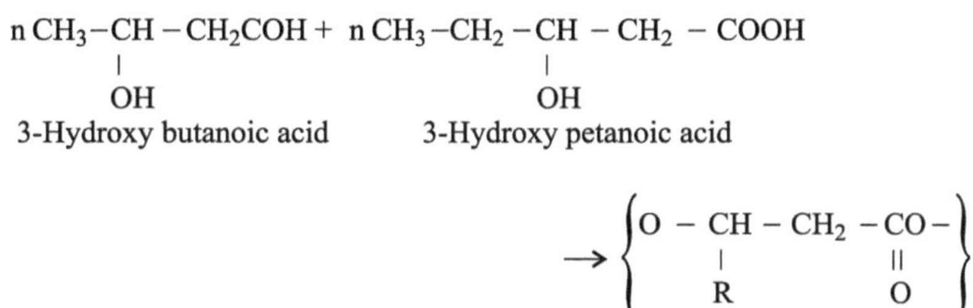

3-Hydroxy butanoic acid   3-Hydroxy petanoic acid

PHBV

$R = CH_3, C_2H_5$

The properties of PHBV can be controlled by varying the ratio of acids. On one hand, 3-hydroxy butanoic acid provides stiffness, whereas 3-hydroxy pentanoic acid gives flexibility to the copolymer.

(iii) ***Polyglycolic acid (PGA):*** Polyglycolic acid is obtained by the chain polymerization of cyclic dimer of glycolic acid.

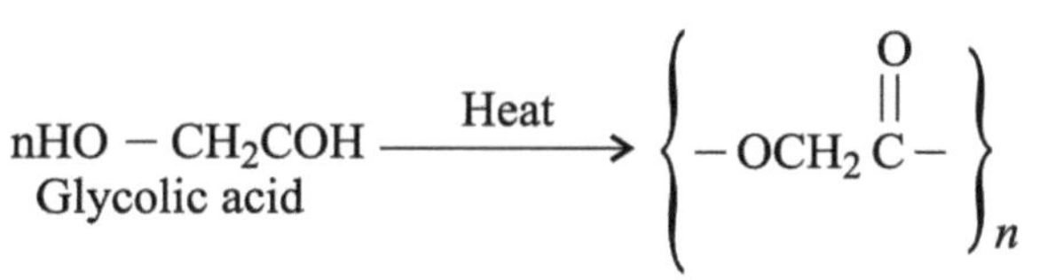

Glycolic acid

Polyglycolic acid (PGA)

Exercises

1. Write the names and structures of the monomers used for the preparation of Bakelite. In what ratio the monomers should be used. Write why Bakelite is used in the preparation of electrical switches, plugs, handles of appliances, etc.

2. In the preparation of Nylon-6. Caprolactam is used as a monomer. How you will prepare caprolactam using Beckmann rearrangement? Which ketone you will take as the starting material.
3. Write the two important advantages of Ziegler-Natta polymerization. Write the chemicals used in the preparation of Ziegler-Natta catalysts.
4. Write the name and structure of monomer used for the preparation of Teflon. Why it is used widely in chemical laboratories.
5. Write the monomers likely to be used in the preparation of the following polymers:

(i) $-CF_2-CF_2-CF_2-CF_2-CF_2-CF_2-\cdots\cdots$

(ii) $-CH_2-\underset{\displaystyle Cl}{\underset{|}{CH}}-CH_2-\underset{\displaystyle Cl}{\underset{|}{CH}}-CH_2-\underset{\displaystyle Cl}{\underset{|}{CH}}\cdots\cdots$

(iii) $-CH_2C(Cl_2)-CH_2C\,(Cl_2)-CH_2C\,(Cl_2)\cdots\cdots$

(iv) $-CH_2-CH(CN)-CH_2\,CH(CN)\cdots\cdots$

(v) $-CH_2-CH\,(C_6H_5)-CH_2-CH\,(C_6H_5)\cdots\cdots$

(vi) $-\underset{\displaystyle O}{\underset{\|}{C}}\,CH_2\,(CH_2)_4NH-\underset{\displaystyle O}{\underset{\|}{C}}\,CH_2\,(CH_2)_4NH\cdots\cdots$

6. A polymer of Teflon had 2000 unit of its monomer. What is the expected molecular mass of this polymer? Determine the number-average and weight average molecular mass of the polymer.
7. Write the structure of a tripeptide obtained from glycine molecules.
8. Aspartic acid

$$HOOC-CH_2-CH\,(NH_2)-COOH$$

forms a dipeptide with glycine. Write the possible structure.

9. Write the structures of four dipeptides that can be obtained if alanine and serine are mixed

$$\text{alanine} = CH_3-CH\,(NH_2)\,COOH$$
$$\text{serine} = HOH_2C-CH\,(NH_2)\,COOH$$

10. What do you understand by the term "copolymer"? Give an example.

Chapter 13
Stereoisomerism and Mechanism of Organic Reactions

13.1 Stereochemistry of Organic Compounds

The type of isomerism in which the position of atoms or groups are different in space is called as stereoisomerism. These compounds have a chiral character, i.e., such compounds cannot be superposed on their mirror images. The word chirality in Greek means "hand." If a left hand is placed against a mirror, then the mirror image will be a right hand which is not superimposable on its image. The two non-superimposable molecules are not identical molecules and are called as stereoisomers or enantiomers.

The stereo isomerism is of two types

- Optical isomerism and
- Geometrical isomerism

13.2 Optical Isomerism

The two enantiomers behave differently with plane polarized light. Ordinary light rays have vibrations in all directions, perpendicular to the direction of propagation, while plane polarized light has vibration in only one plane or to the direction of propagation. The plane polarized light is obtained by passing ordinary light through Nicol prism (tourmaline crystal).

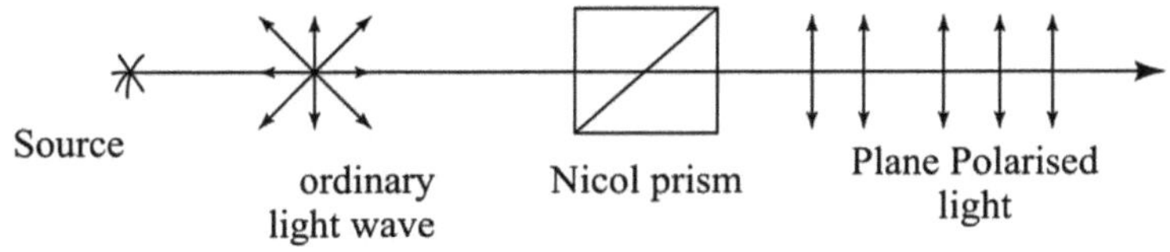

© The Author(s), under exclusive license to Springer Nature Switzerland AG 2026

R. N. Goyal, H. Goel, *Textbook of Engineering Chemistry*,
https://doi.org/10.1007/978-3-032-22068-4_13

When a beam of plane polarized light is passed through a solution of an optically active compound, the plane polarized light is rotated or diverted to a fix angle either to left or right. This is known as optical activity, and the phenomenon is known as optical isomerism. If the plane of polarization is twisted to the right, the isomer is referred to as dextrorotatory (d), whereas if the plane is twisted to left, it is called as levorotatory (l).

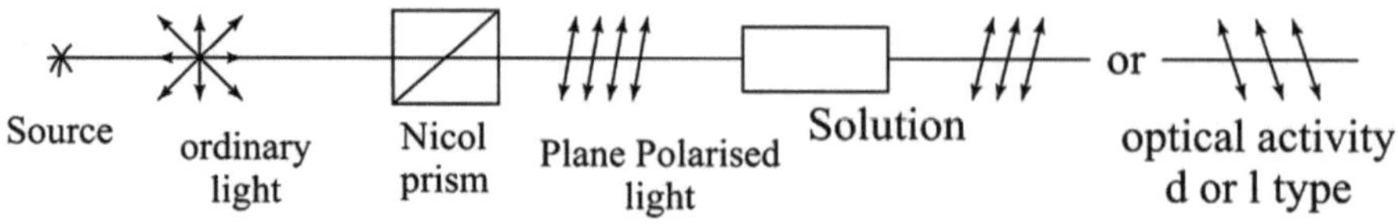

The rotatory power of a given compound depends upon

1. The length of solution in the tube
2. Temperature
3. Concentration
4. Nature of solution
5. Wave length of light.

The optical rotation of the solution at a constant temperature and wavelength can be measured by the formula

$$[\alpha]_{\mathrm{D}}^{t^{\circ}C} = \frac{\theta}{l \times d}$$

where

$$\theta = \text{angle of rotation}$$
$$l = \text{length of solution}$$
$$d = \text{density of solution}$$

13.2.1 Reason of Optical Isomerism

According to Van't Hoff, a compound will exhibit optical isomerism if it has

a. A tetrahedral structure
b. It should have at least one asymmetric carbon atom.

The optical isomerism is due to asymmetric carbon to form enantiomorphs image (mirror image) which is not super imposable, but the explanation is not fool proof as some compounds do not have any asymmetric carbon, yet they are optically active, and some compounds which have asymmetric carbon are not optically active, e.g., diphenyl amine is optically active but meso tartaric acid is not.

Hence, presently two new conditions are considered for presence of optical activity in a molecule. These are the following.

Molecular Dissymmetry or Asymmetry

Molecular dissymmetry means that the molecule does not possess any elements of symmetry, like plane of symmetry, center of alternating axis of symmetry, etc. Out of these even if one is present, the molecule is symmetric.

A plane of symmetry divides a molecule in such a way that the two halves of compound will produce a mirror images of each other, e.g., tartaric acid has two similar chiral centers, and hence one structure can be written with plane of symmetry and is called as meso form.

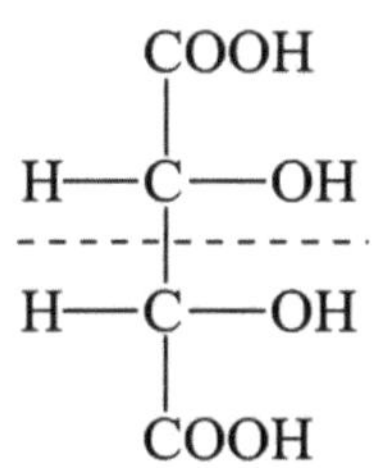

meso - tartaric acid

Plane of symmetry

The center of symmetry is a point from which if any line is drawn to some element/group of the structure, on extending in the opposite direction encounter an identical element/group at equal distance. It is found in ring compounds.

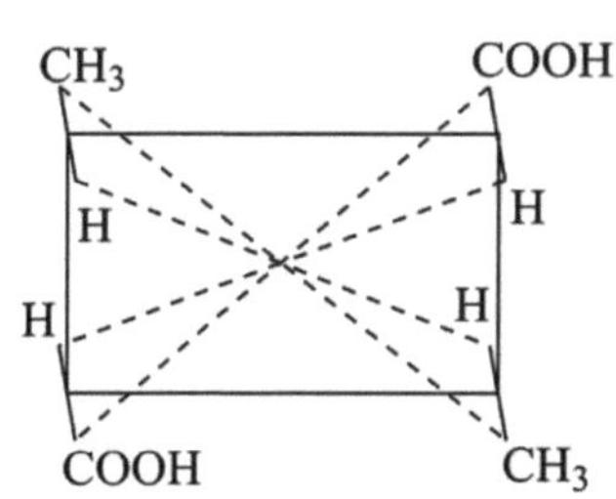

Dimethyl cyclobutane dicarboxylic acid

Center of Symmetry

The molecules should be enantiomorphs of each other.

13.2.2 *Examples of Optical Isomerism*

1. **With one asymmetric carbon atom:** If compound has only one asymmetric carbon atom, then number of stereoisomer will be $2^n = 2^1 = 2$, e.g., lactic acid has two isomers dextro and levo, and they are mirror image of each other.

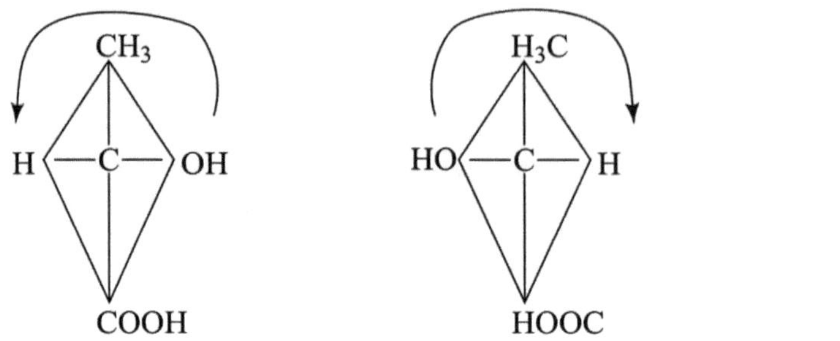

d - lactic acid (Anti clock) l - lactic acid (clockwise)

When *d* and *l* forms are mixed in equimolecular amount, then the mixture will be optically inactive and is known as racemic mixture or (±).

2. **Two similar asymmetric carbon containing compounds:** The number of stereoisomers in this case will be $2^n = 2^2 = 4$. Which are d, l, dl and meso forms.
For example: Tartaric acid exists in three stereoisomeric forms

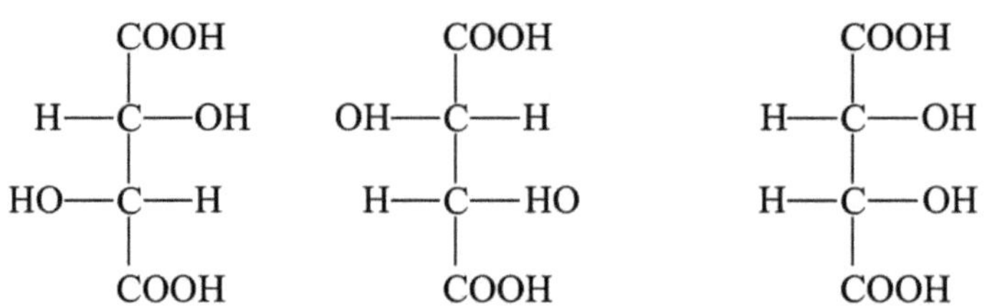

leavo form dextro from meso (optically inactive)

The fourth is optically inactive racemic form, containing equal amount of dextro and levo forms.
The difference between racemic mixture and meso isomer is that racemic mixture can be separated in dextro and levo forms, whereas meso form is optically inactive due to internal compensation.

3. **Two different asymmetric carbon atoms containing compounds:** The number of stereoisomeric forms will be $2^n = 2^2 = 4$. For example, bromochlorosuccinic acid

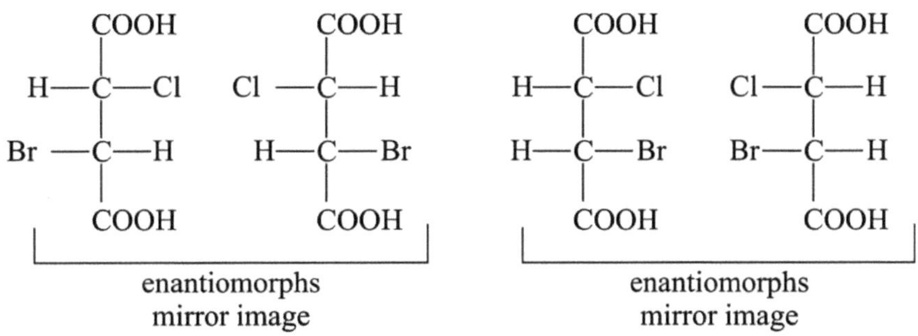

A simple way to represent tetrahedral chiral carbon is Fischer projection in which the chiral carbon is represented by a cross and the four groups are then written on them. The point of intersection of the two lines represents carbon. Thus, lactic acid is written as

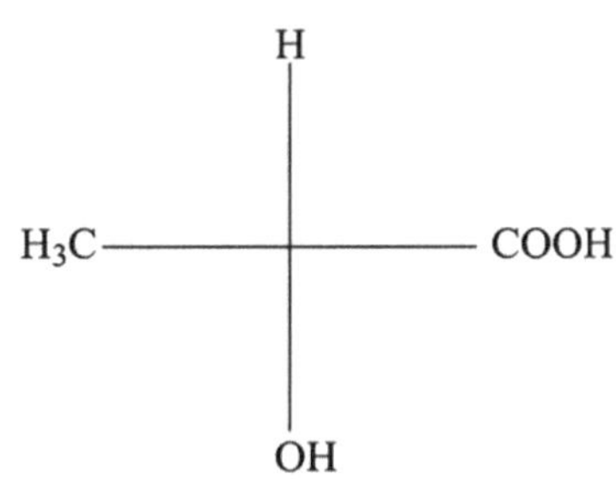

The horizontal line of the cross represents bonds that project forward, whereas vertical line represents bonds that project in the rear. For representing enantiomers by Fischer projection any two groups should be interchanged. If the remaining two groups are also changed, the original molecule is obtained, although the groups may occupy different positions than the original. Let us consider 2-chlorobutane. On interchanging two groups, structure (2) is obtained which is enantiomer of (1). Further interchanging of the two groups give (3),

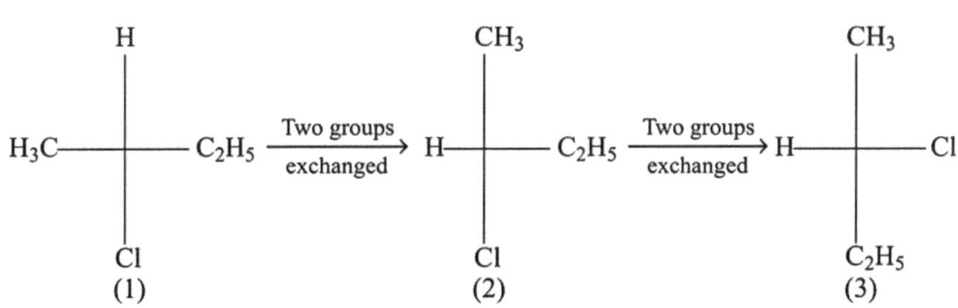

which is identical to 1, though groups are at different positions. A basic principle to remember in Fischer projections is that rotation of the molecule by 90° gives an enantiomer whereas rotation by 180° gives the original isomer. We will make

Fischer projections more clear by considering various structures of (1). Thus, a student can write (1) in many ways. Structures 4 to 7 are identical and represent only one enantiomer of 2-chlorobutane. The other isomer for the structures will be obtained only on rotating the molecule by 90°. Thus, if we rotate the molecule (4) by 90°, structure (9) is obtained and is the enantiomer of (4). Structure (8) is obtained by rotating (4) by 180° and

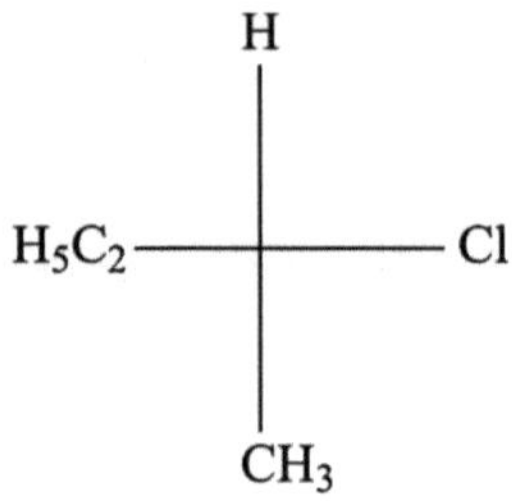

(4)

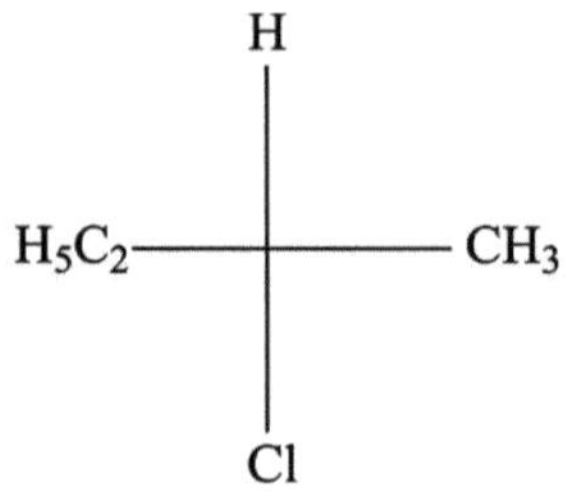

(5)

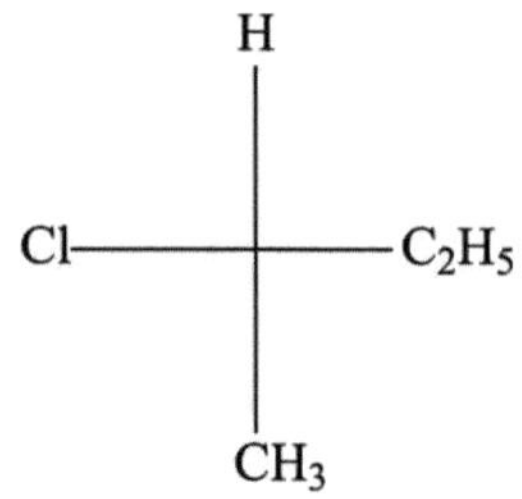

(6)

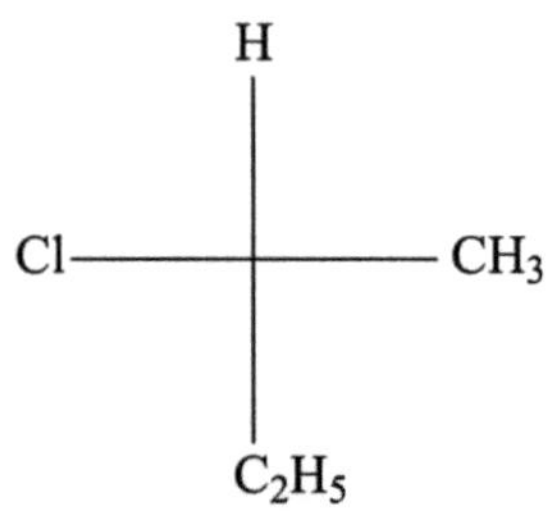

(7)

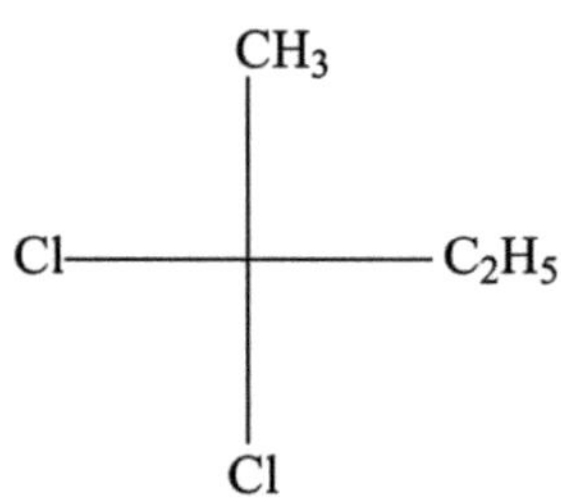

(8)

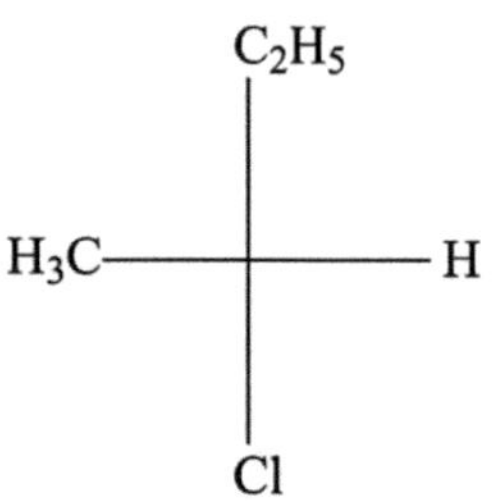

(9)

hence is similar to (4).

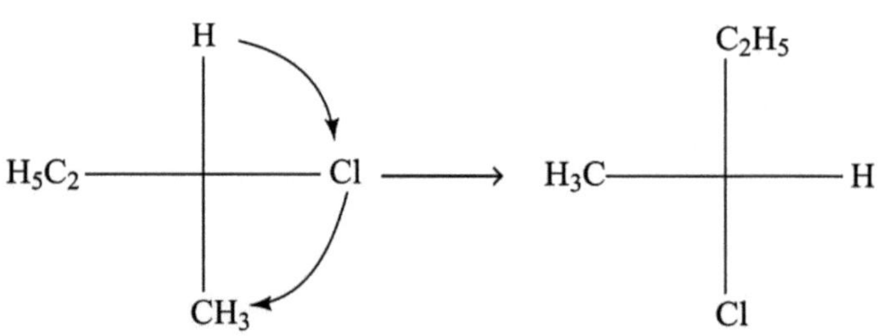

(4) (9)
enantiomer of (4)

13.2.3 *R and S System of Nomenclature*

To describe absolute configuration of a chiral center, R and S nomenclature was suggested by R.S. Cahn, C. Ingold, and V. Prelog. The basic rules for R and S nomenclature are as follows:

1. The atoms present at the chiral center are assigned priority numbers depending on their atomic numbers so that $1 > 2 > 3 > 4$. If groups are present, only the atom attached to chiral center is considered. In the case of first similar atoms, next atoms are considered to decide the priority. Thus, if a molecule has H, CH_3, OH, and C_2H_5 groups, the priority will be 1 for OH, 2 for C_2H_5, 3 for CH_3, and 4 for H. In the case of C_2H_5, priority is higher than CH_3 because carbon in C_2H_5 is bonded to two hydrogens and one carbon, whereas in CH_3, carbon is bonded to three hydrogens.
2. A three-dimensional structure of the molecule is then considered in a way so that the group of lowest priority is far away. Thus, the molecule shown above will be viewed as presented below.

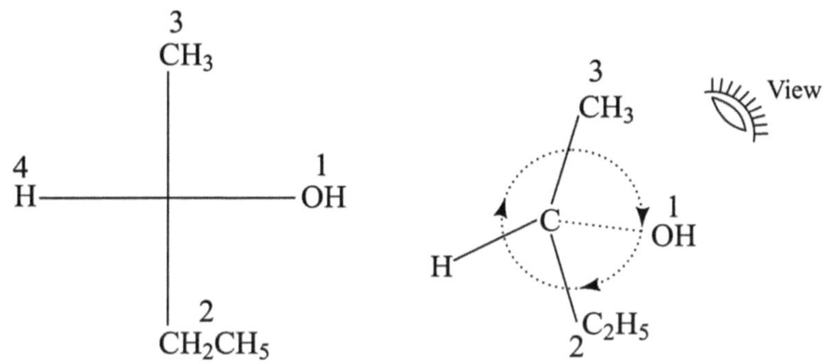

3. Now in order to decide the sequence $1 \longrightarrow 2 \longrightarrow 3$, if the movement is clockwise, the configuration is called as *R*, whereas if the movement is anticlockwise, the configuration is called as *S*.

The symbols *R* and *S* are taken from the Latin and mean rector and sinister meaning right and left, respectively.

Example 13.1 For the given structure of bromo-chloro flouromethane, assign *R* or *S* configuration.

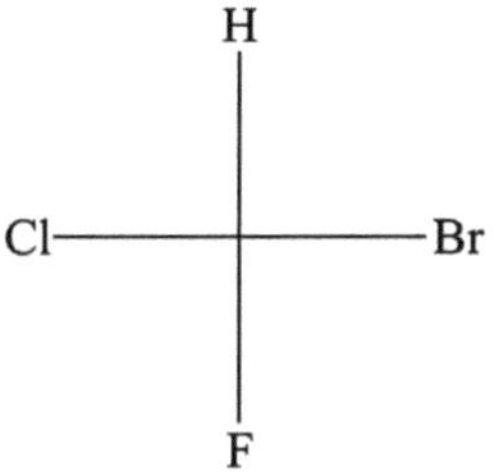

The priority of different atoms is assigned, and depending on their atomic numbers, it is Br(1), Cl(2), F(3), and H(4). Now look at the structure so that H is at the side opposite to the view.

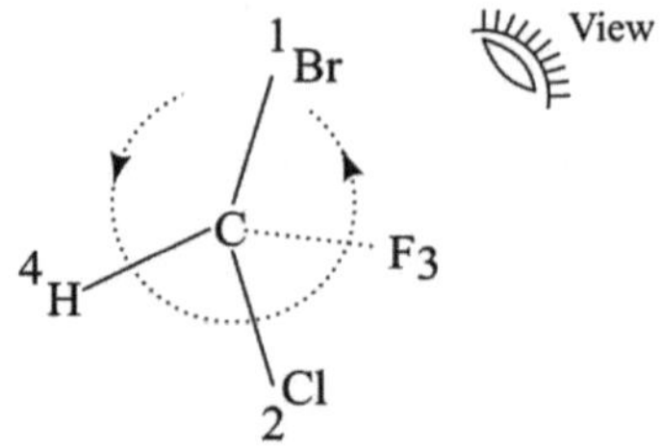

Anti Clockwise (S)

As movement is anticlockwise, the configuration is *S*.

An easy way to assign *R* and *S* configuration to the molecules is that if the group of lowest priority (4) is on the vertical line (no matter at the top or bottom), the sequence to going 1 → 2 → 3 gives the correct configuration. For example, consider the case of glyceraldehyde, and on assigning priorities, we can write 1 to OH, 2 to CHO, 3 to CH_2OH and 4 to H. As H, group of lowest priority is on vertical line, let us consider the direction of going 1 → 2 → 3. As the direction of movement is clockwise, the configuration is (*R*).

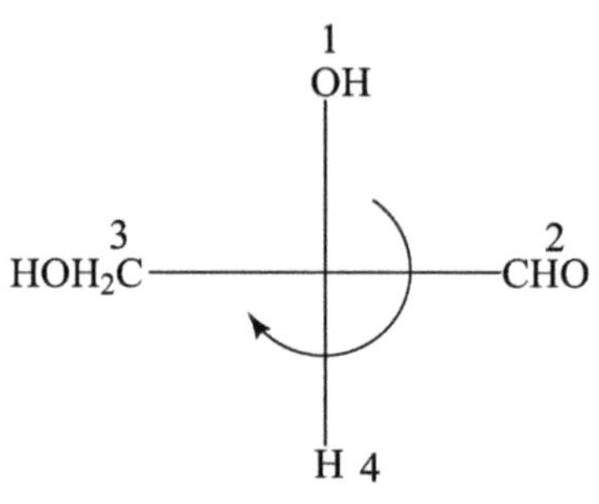

Clockwise (R)

If the group of lowest priority is on horizontal line as shown below for glyceraldehyde, the configuration obtained for the sequence 1 → 2 → 3 is reversed than that of obtained. Thus, the sequence for the structure is clockwise and should be *R*, but it is to be reversed and hence must be stated as *S*.

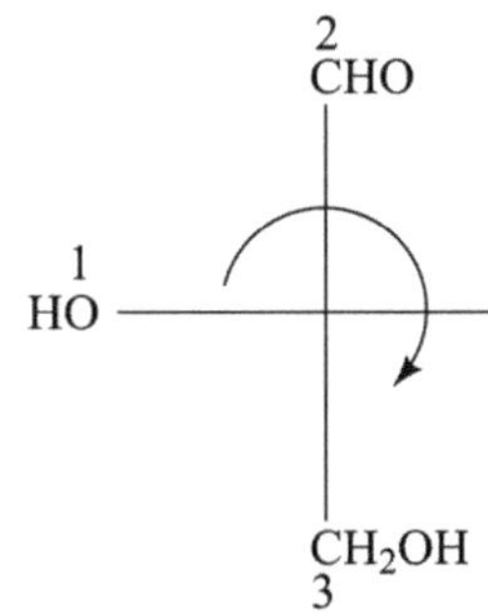

The order for assigning priority to some groups is as shown below

$$I > Br > Cl > SO_3H > F > OCH_3 > OH > NO_2 > NHCOCH_3 > NHCH_3$$
$$> NH_2 > COCl > COOH > CONH_2 > CHO > C_2H_5 > CH_3 > H$$

13.3 Geometrical Isomerism

To consider geometrical isomerism, let us look at C—C and C=C bonds to see how intramolecular movement is restricted which affects the molecular geometry. A simple molecule having C—C linkage is ethane in which both the carbon atoms are sp^3 hybridized and a single sigma bond joins the two carbon atoms. The rotation of the molecule occurs around this sigma bond.

In the case of ethylene, two carbon atoms are joined by a sigma bond. However, in addition to the sigma bond, there is a pi bond between the two carbon atoms formed by the sidewise overlapping of p_z-orbitals. Now the rotation will move two p_z-orbitals out of alignment and hence will totally or partially destroy the pi bond. Such a rotation requires energy of the order of 270 kJ/mol, and hence the rotation around C=C bond is considered as restricted, and therefore more than one stable geometry is possible. Such molecules are called as *geometrical isomers* and possess similar molecular formulas but different structural formulas. A simple example is of butene-2, which can be written in two forms as

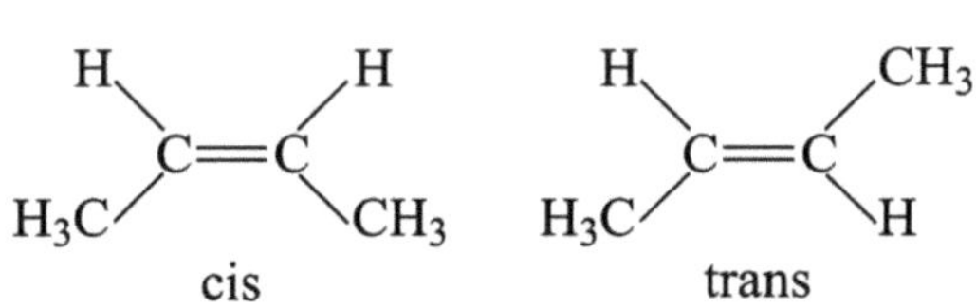

In the *cis* isomer, the CH_3 groups are close to each other, whereas, in the *trans* isomer, they are on the opposite side. The physical properties such as melting point, boiling point, etc. of the two isomers are different. The dipole moments of the two isomers are also different and can be used as a method to distinguish *cis* and *trans*

isomers. The *trans* isomer has a dipole moment zero because of cancellation of effect whereas *cis* isomer always shows some value.

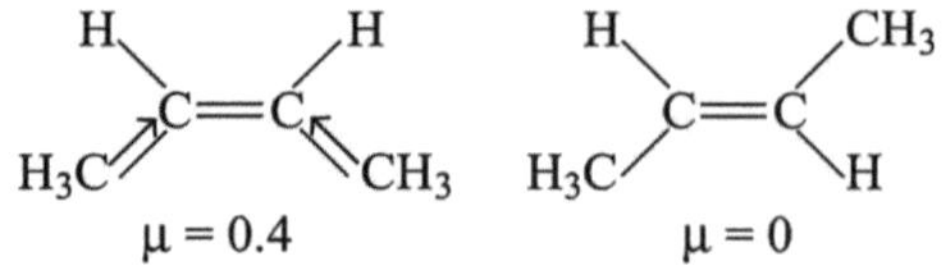

The chemical reactivity is also different for both the isomers. However, the conversion of one isomer to other is possible by providing heat or irradiation with UV light.

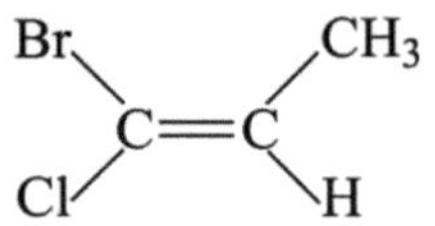

The naming *cis-* or *trans-* becomes difficult when the substituent on the carbon atoms are not identical. For example, if we consider an alkene having four different groups, it is not possible to assign *cis-trans* configuration as there are no two identical groups.

In such cases, a different nomenclature known as $E - Z$ nomenclature is used for assigning a structure to isomer and will be discussed later in this chapter.

Geometrical isomerism can also be observed for substituted derivatives of cyclic compounds like cyclopropane, cyclobutane, cyclopentane, etc. In these compounds, the rotation is hindered as atoms joined in ring are not free to rotate in presence of substituents. For example, if we consider 1, 2-dichlorocyclopropane, the two chlorine atoms can be attached in a plane above or below the plane of the ring, assuming the ring plane is almost horizontal. Thus, in the *cis*-form, the two chlorine atoms will be on same side and in *trans* form on the opposite side.

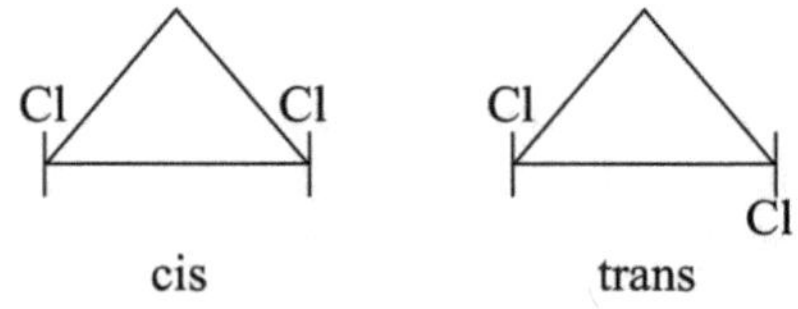

Similarly we can consider 1, 2-dimethylcyclopentane which also exists in *cis* and *trans* forms.

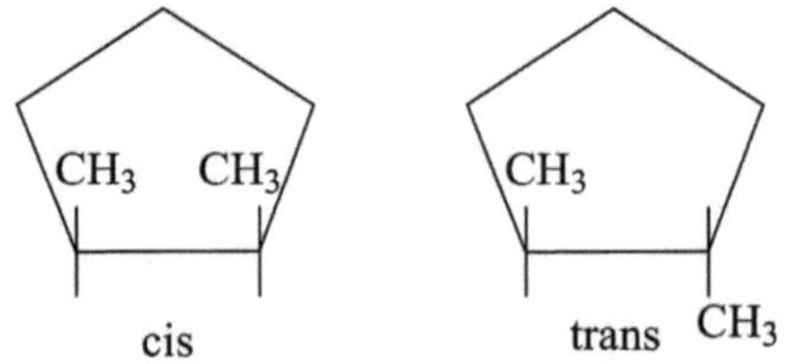

If these compounds possess chiral centers, the stereoisomerism becomes important because optical isomers are also possible in such cases. In both the examples considered above, two asymmetric carbon atoms are present. However, if we consider their mirror images, the *cis* isomer will form a mirror image which will be superimposable on the original compound due to plane of symmetry, whereas the *trans* isomer will form non-superimposable mirror image, and hence it will exhibit *d* and *l-isomers.*

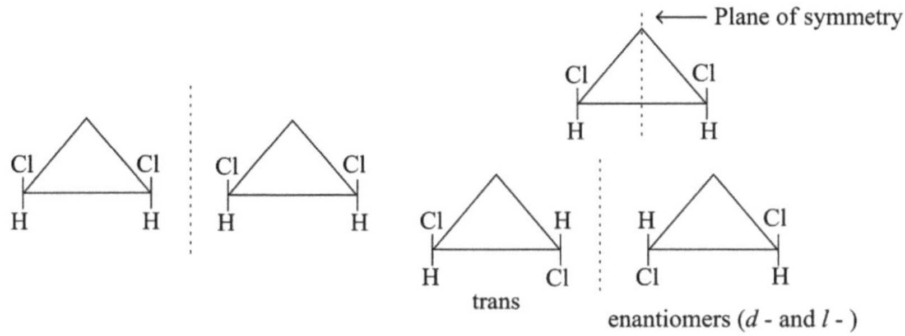

13.3.1 Reason for Geometrical Isomerism

When two carbon atoms are attached through a single bond, the rotation around a single bond is free; hence, the two forms are interconvertable, but as soon as the two atoms have a double bond between them, the rotation will be restricted so the formation of isomers of same or opposite side will depend upon the condition of propagation.

The double bond is a type of banana bond which does not permit the rotation around C=C bond.

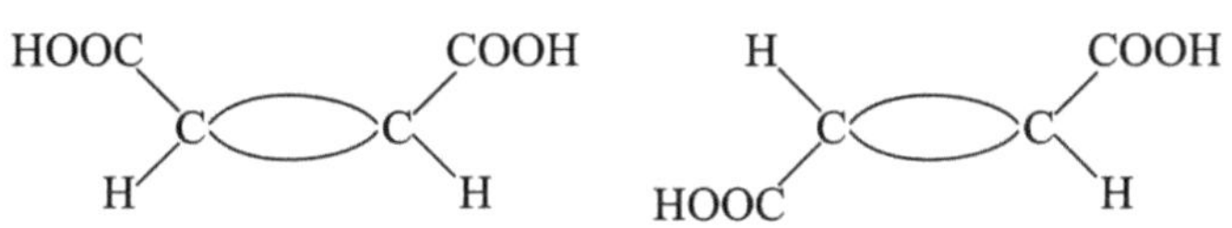

The double bond is formed by maximum overlapping of σ and π orbitals and thus gives the shape like banana bond.

13.4 *E* − *Z* Nomenclature

According to this nomenclature, the two groups attached to each carbon atom are assigned priority as is done in the case of *R* and *S* system of nomenclature. Thus, the atom with higher atomic number is assigned priority 1 and then the second atom or group as 2. The atoms or groups of both the carbon atoms are assigned priority 1 and 2. Now when the atoms or groups of priority 1 are on the same side of double bond, the isomer is named as Z-isomer (Z is derived from the German word Zusammen meaning together). When the atoms of priority 1 are on the opposite side of the double bond, the isomer is called as E-isomer (E is derived from the German word entgegan meaning opposite). The following example will make $E-Z$ nomenclature more clear.

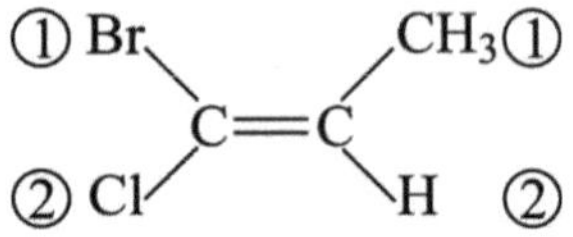

Consider, at carbon atom 1, the two groups present are Br and Cl. As Br has higher atomic number than Cl, Br is assigned priority 1 and Cl as 2. Similarly at carbon atom 2, the groups are $-CH_3$ and H. As carbon has higher atomic number than H, the group $-CH_3$ is assigned priority 1 and H as 2.

Now look at the molecule, and as groups or atoms of higher priority are on the same side of double bond, the configuration is Z.

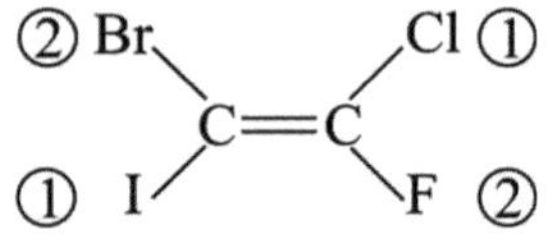

Similarly if we consider a molecule, as shown above, the priorities of the groups are I (1) and Br (2) on one carbon atom and Cl (1) and F (2) on the other carbon atom. Since the higher priority groups are on the opposite side of double bond, the molecule has E configuration.

13.5 Isomerism in Coordination Compounds

Coordination compounds can also exhibit *cis-trans* isomerism. The well-known case of a platinum complex $Pt(NH_3)_2Cl_2$ used in chemotherapy is interesting. Out of the two isomers, only *cis* isomer has been found physiologically active. The complex is formed from two ammonia molecules, two chloride ions, and Pt^{2+} to give a square planar structure in which metal is situated at the center and groups are attached at four corners making on angle of 90° to each other. Thus, two structures are possible.

(1) in which two identical groups make an angle of 90° and
(2) in which two identical groups make an angle of 180°.

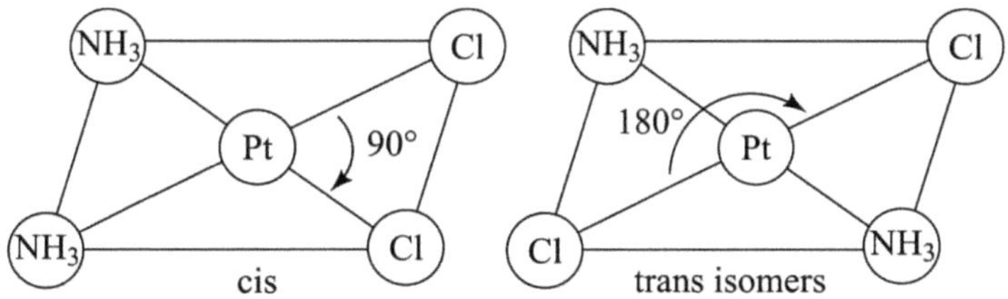

The first case when the two identical groups are adjacent to each other is called *cis* isomer and when they are at opposite side is called as *trans* isomer. Remember, if in the above complex, three similar ligands are present, say $[Pt(NH_3)Cl_3]^-$, *cis–trans* isomers are not possible because no matter how you draw the structures, two Cl will be adjacent to each other and two Cl at opposite sides making an angle of 180°. Hence, it is necessary that in such complexes, two groups or atoms should be the same and other two may or may not be similar.

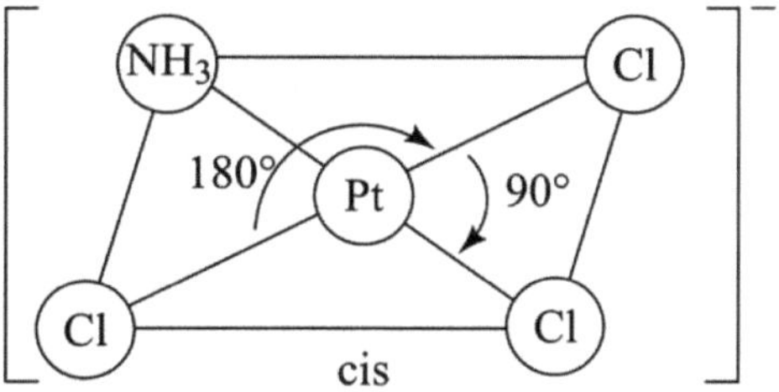

Let us consider another complex $[Pt(NH_3)_2(Cl)NO_2]$ in which two similar groups attached are NH_3. The *cis-trans* isomers are possible for this complex because in the *cis* form, two NH_3 molecules are present on the same side making an angle of 90° to each other, whereas in the *trans* isomer, the two NH_3 are present on opposite side at an angle of 180°.

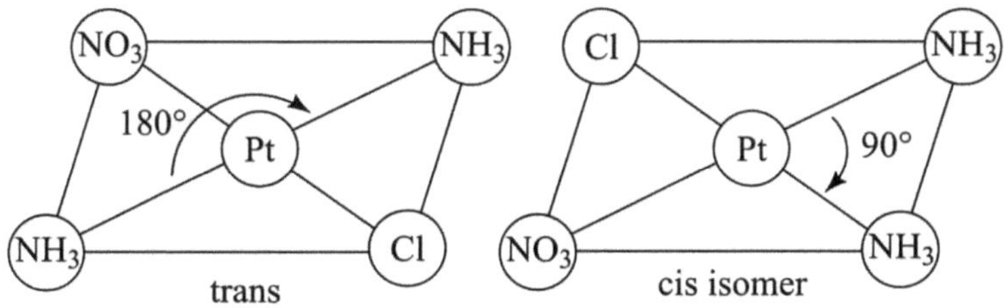

Similarly, octahedral complexes exhibit geometrical isomerism. Consider a complex $[Pt(NH_3)_4Cl_2]^{2+}$ in which platinum is attached to four NH_3 molecules and two Cl^-. This Pt^{4+} complex has an octahedral structure, and two isomers can be written in which two Cl^- ions can attach making an angle of 90° or 180° with each other.

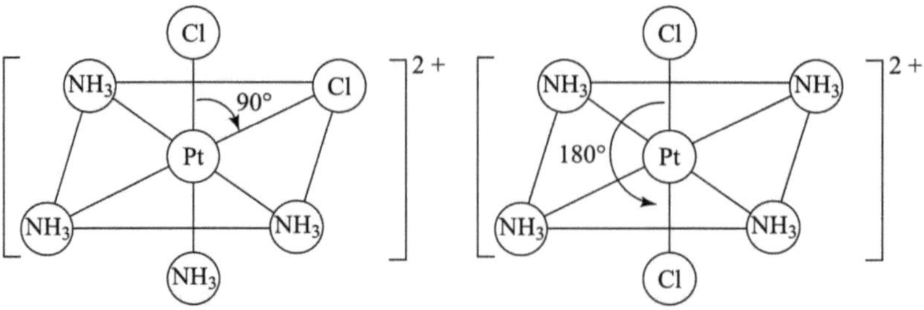

cis trans isomer

However, even if three out of six groups or atoms become identical in octahedral complexes, geometrical isomerism is still possible because three ligands may be present in the same plane or they can be at the corners of an equilateral triangle as shown below in the case of $[Cr\,(NH_3)_3\,Cl_3]$.

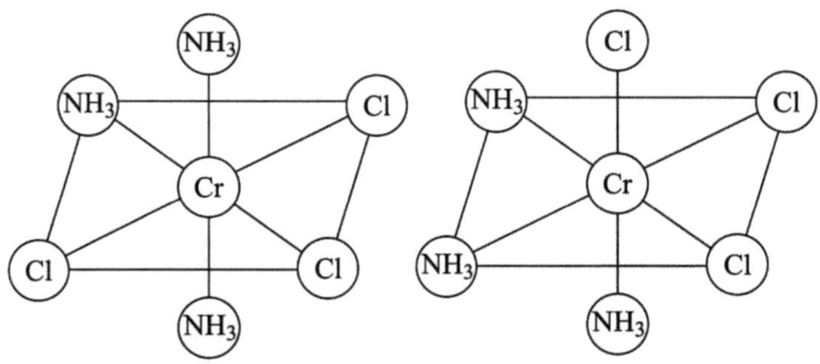

mer isomer fac isomer

The isomer in which three similar groups or atoms are present on the meridians is often called as mer isomer, whereas when three $Cl^-$are present on the face of octahedron is called as fac isomer.

Four coordinate complexes do not exhibit *cis-trans* isomerism because they have tetrahedral geometry and the angle between them is 109°, so all ligands are effectively adjacent. Thus, isomerism is an important aspect of organic and coordination compounds because it has been found that in physiological systems, most of the times, only one isomer shows activity.

Organic chemistry basically deals with the study of carbon compounds. You may be surprised that why a single element requires a separate branch of chemistry. The answer is that more than 90% of the compounds ever synthesized contain carbon as one of the elements. The next question then comes to the mind is—"Why there are so many compounds of this element carbon." The reason is that carbon has a unique ability to bond together to form long chains or rings. Further, carbon is also capable of forming bonds with other elements.

The major area of organic chemistry is the development of new synthetic methods for the production of organic compounds. In addition to this, organic chemists also study the mechanism of organic reactions and suggest pathways for the formation of products.

13.5.1 Bonding

The electronegativity of carbon (2.5; Pauling) is roughly intermediate between alkali metals and halogens. This value of electronegativity is similar to that of elements to which carbon often form bonds, such as N, O, S, P, etc. In general the bonds between carbon and other elements are polar, and the bonds formed between carbon-carbon are highly covalent. The stronger bond between two elements is formed if the overlap from each of the participating atom is greater.

The relative overlapping powers (overlap integral) of various orbitals have been calculated and taking overlapping power of s-orbital as 1.00, the values are

$$s = 1.00$$
$$p = 1.72$$
$$sp^1 = 1.93$$
$$sp^2 = 1.99$$
$$sp^3 = 2.00$$

Thus, the hybrid orbital results in the formation of stronger bonds. A simple example is of methane in which one carbon atom combines with four hydrogen atoms to form stronger bonds. Interestingly, if two atoms come sufficiently close to each other, the two atomic orbitals are replaced by two molecular orbitals which spread over both the atoms. The energy of one molecular orbital is lower than the energy of two original atomic orbital, whereas the other one has higher energy as shown below in Fig. 13.1. The molecular orbital with lower energy than atomic orbitals is called as bonding molecular orbital and the other as non-bonding orbital.

13.5.2 Breaking of Bonds

A covalent bond between two atoms can be broken in three different ways:

$$A{-}B \rightarrow A\bullet + B\bullet$$
$$A{-}B \rightarrow A^+ + B^-$$
$$A{-}B \rightarrow A^- + B^+$$

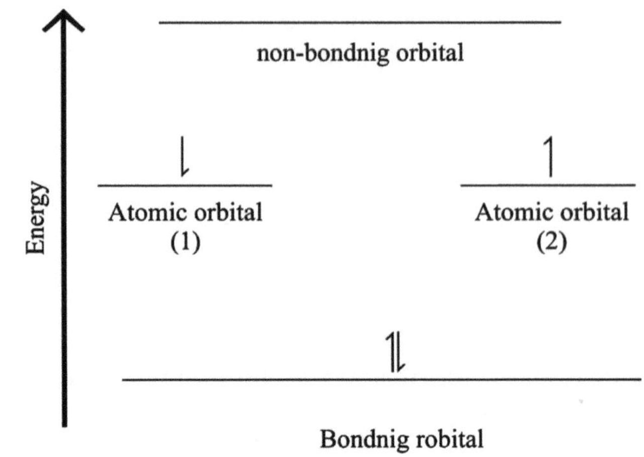

Fig. 13.1 Energy levels of bonding and non-bonding orbitals

- In the first case, the covalent bond between A and B breaks in such a way so that both the atoms take one electron leading to the formation of reactive free radicals ($A^\bullet$ and $B^\bullet$). This type of bond breaking is called as *homolytic fission.*
- In second and third cases, the pair of electrons between A and B shifts to B or A producing corresponding B^- or A^-. The other atom becomes positively charged. Either of these processes is called as *heterolytic fission.*

Such radicals or ions are formed transiently in an organic reaction as an intermediate. Free radicals are formed in reactions which occur in gas phase and in solution in non-polar solvents. Such reactions are catalyzed by light and by the addition of other free radicals. In polar solvents ionic intermediates are formed due to ease of separation of charge. The free radicals as well as ionic intermediates are often stabilized by delocalization of radical or charge over other carbon atoms or atoms of different elements. A simple example of propene −2−ol-1 (allyl alcohol) will explain it more clearly.

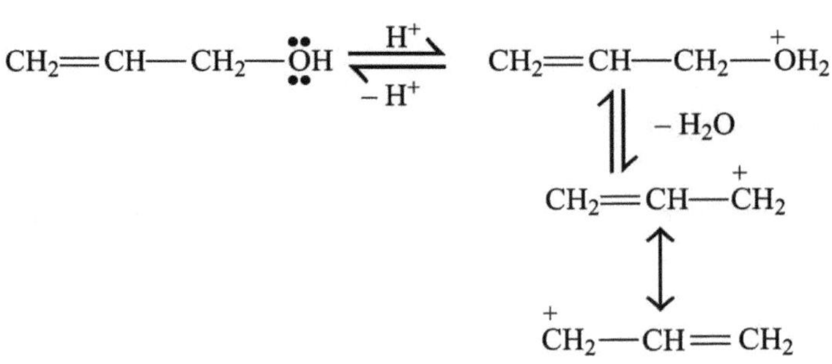

If carbon atoms acquires positive charge, it is called as *carbocation*, and if negative charge, the entity is known as carboanion, commonly called as *carbanion.* The mechanism of an organic reaction basically depends on these reactive intermediates in spite of the fact that these species are formed only transiently.

Example The bond length of N—O bonds in nitrobenzene is found as 1.21 Å. The nitro group is commonly represented as

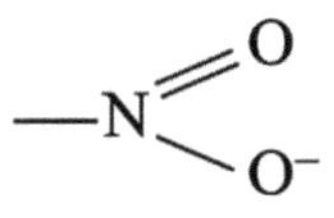

The usual lengths of N = O and N—O bonds are 1.18 and 1.36 Å. On the basis of this observation, explain which should be a better representation of nitro group.

The observation indicates that bonds between N—O are not truly single or double bond. Thus, nitro group should be represented as

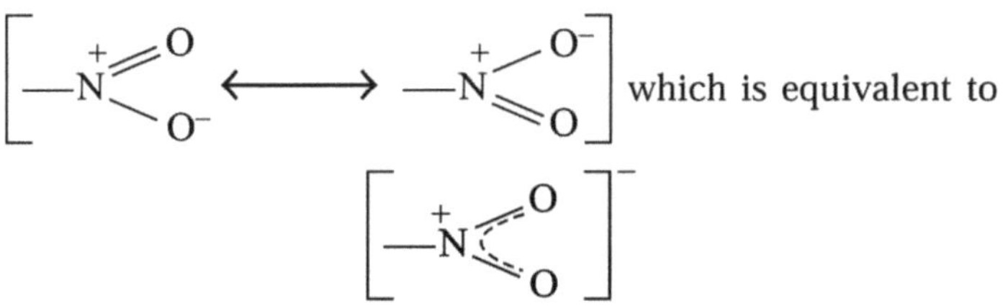

The availability of electrons between two atoms or at particular atom is influenced by several factors such as nature of groups present in the vicinity of atoms, reagent, etc. The position of high electron availability will be attacked by positively charged reagent easily, whereas position of low electron density will be readily attacked by negatively charged reagent. Some common effects recognized due to nature of atoms or groups present in a molecule can be divided into two types and cause the displacement of electrons.

(i) Time variable effects or temporary effects
(ii) Permanent effects

Let us discuss these effects in little detail.

(i) **Time variable effects:** These types of effects are temporary and are produced at the demand of attacking reagent. The distribution of electrons reverts back to the original ground state, if the reagent is removed without the reaction being allowed. These effects are *inductomeric* and *electrometric* effects.

- ***Inductomeric Effect:*** It is a polarizability effect. The permanent effect such as inductive effect may be temporarily increased due to the approach of reagent. It has been found that it is almost impossible to distinguish between permanent and time variable effect.
- ***Electrometric Effect:*** It involves transfer of pair of electrons between two atoms to one of them on the requirement of attacking reagent. For example consider a reaction between ethylene and bromine.

$$CH_2 = CH_2 + Br_2 \longrightarrow CH_2Br - CH_2Br$$

Ethylene which is a non-polar compound acquires polar nature on the approach of Br_2 molecule which is represented as $Br^+ - Br^-$. Thus,

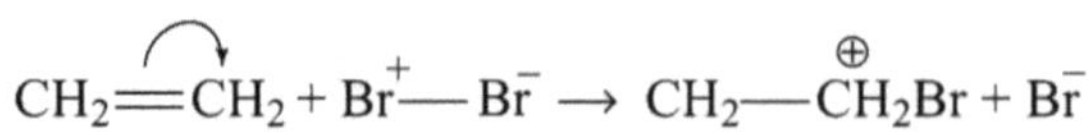

with approach of Br_2, the electron pair between two carbon atoms shifts to the right side as shown above. Such a shift produces negative charge on that carbon atom and Br^+combines with it to give

$$\overset{+}{C}H_2 - CH_2Br$$

In the next step Br^-attacks to give $CH_2Br - CH_2Br$.
Such an effect is frequently observed in compounds having double or triple bonds or in conjugated systems

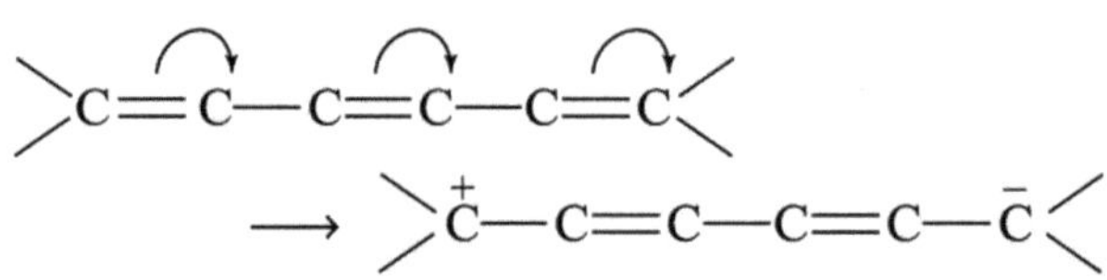

(ii) **Permanent Effects:** Such effects cause displacement of electrons between two atoms and are called as polarization effects. Two common effects are

- ***Inductive Effect:*** A covalent bond between two similar atoms is shared equally between the two atoms and thus lies midway between the two nuclei. Due to such a sharing of electrons, the molecule is non-polar. However, if the atoms are different the shared electrons shift toward move electronegative atom. Say alkyl chlorides in which C—Cl bond is present. As Cl is more electronegative than carbon, the shared electrons shift toward Cl, and thus Cl acquires slight negative charge, and C acquires slight positive charge.

$$>CH-Cl \longrightarrow >\overset{+\delta}{CH}-\overset{-\delta}{Cl}$$

The displacement of shared electrons in one direction is called as inductive effect. Most atoms or groups exhibit inductive effect in the same direction as that of chlorine and the exception is alkyl groups, which are electron donating and causes electron shift in the opposite direction. Thus, inductive effect is of two types: + I effect and −I effect.
When an atom or group attracts electron more strongly than hydrogen it is said to poses −I effect whereas, if it attracts less strongly than hydrogen than it is called as + I effect.

$\overset{-\delta}{C}—\overset{+\delta}{CH_3}$	C—H	$\overset{+\delta}{C}—\overset{-\delta}{Cl}$
(+ I effect)	(no I effect)	(– I effect)

The relative Inductive effect of various atoms/groups in decreasing order is given as

$$\xrightarrow[\text{decreasing} - \text{I effect}]{-NO_2,\ -F,\ -Cl,\ -Br,\ -I,\ -OCH_3,\ -C_6H_5,\ H}$$

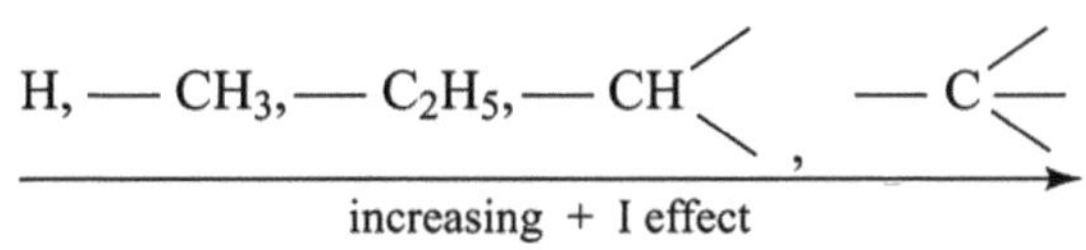

Thus, the relative magnitude of the inductive effect for alkyl group increases with increase in number of alkyl groups.

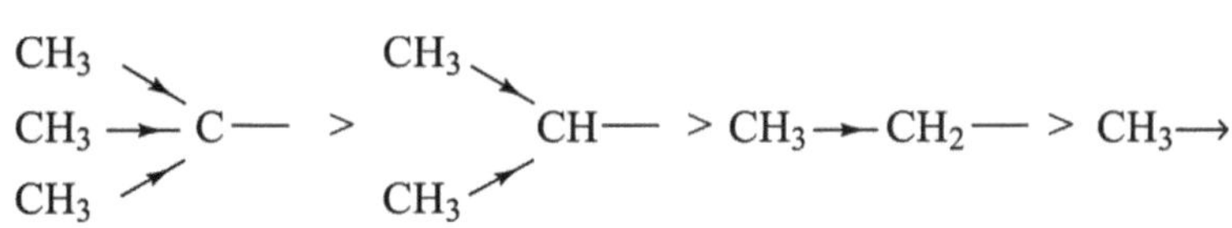

However, if these groups are attached to an unsaturated system, for example, a double bond or benzene nucleus, the electron donating tendency of these groups is disturbed and methyl groups ($-CH_3$) release electron much more than ethyl group. This behavior is explained on the basis of an effect called as hyperconjugation.

- ***Hyperconjugation:*** The exactly opposite behavior of alkyl groups when connected to unsaturated systems is explained on the basis of extension of conjugative effect. If a $-CH_3$ group is attached to propene-1, the delocalization may occur as follows:

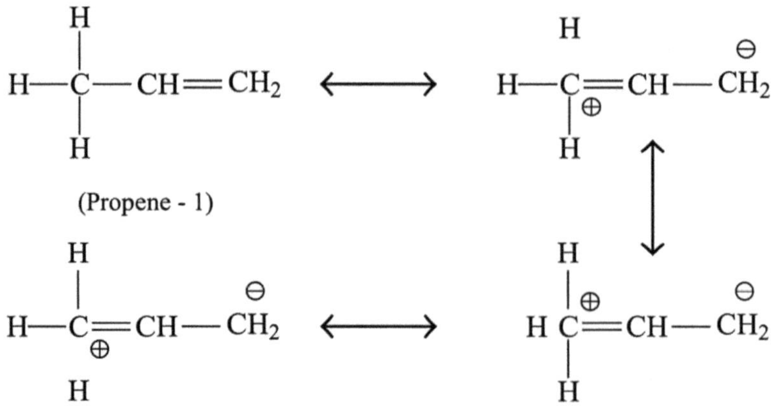

It appears from these structures that a proton becomes free; however, it is not so. The delocalization is so fast that proton remains always attached with carbon. Similarly in toluene, the delocalization of electrons increases electron density at ortho and para positions.

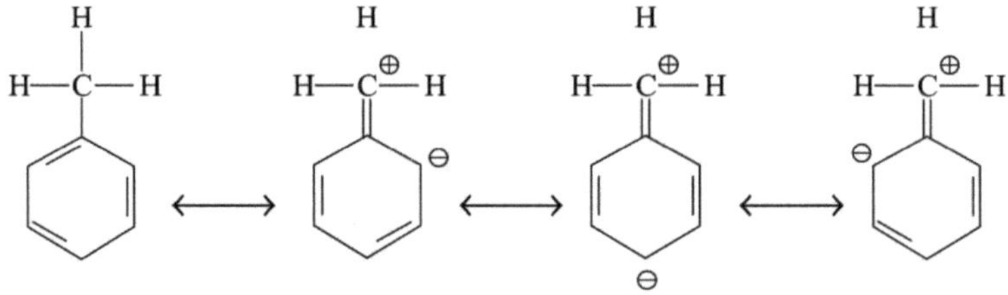

Similar structures can be written by removing other C—H bonds of $-CH_3$ group. Hence, the increased electron donating capability of $-CH_3$ group can be explained on the basis of hyperconjugation. Such an effect is negligible for $(CH_3)_3C-$.

- ***Mesomeric Effect:*** The electron distribution through unsaturated or conjugated systems is called as mesomeric effect. The simplest example is of carbonyl group.

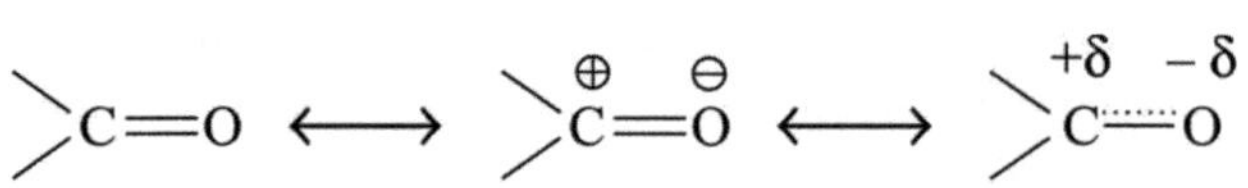

(a) (b) (c)

The complete polarization of structure (a) will give (b), whereas the intermediate stage between (a) and (b) is (c). In structure (c), there will also be inductive effect due to difference in electronegativity of carbon and oxygen; however, σ bonds are less polarizable than π electrons. In the case when such group is attached to unsaturated or conjugated system, this effect is transmitted from one end of the molecule to other as shown below.

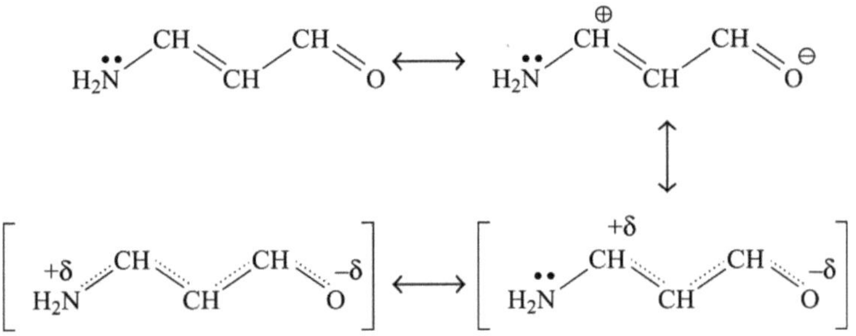

The actual structure of these molecules lie intermediate between the two extremes. Mesomeric effect is represented by M and is a permanent polarization effect. +M and −M effects are due to donating of electrons and withdrawing of electrons. Some of the groups showing the effect are

+ M effect: − OH, $-NH_2$, $-OCH_3$, −SH, −Cl

− M effect: − NO_2, −COOH, −CHO, $-COCH_3$

Example 13.2 Explain why phenol is acidic in nature

Solution The phenoxide anion formed is stabilized by the delocalization of charge and hence proton is released causing acidity of phenol.

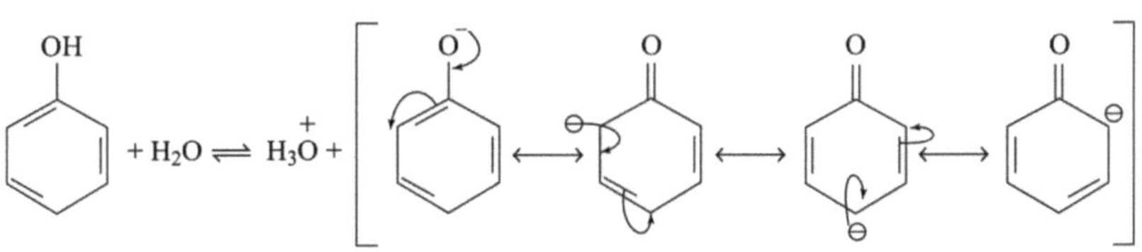

Similarly, we can write various structures for nitrobenzene

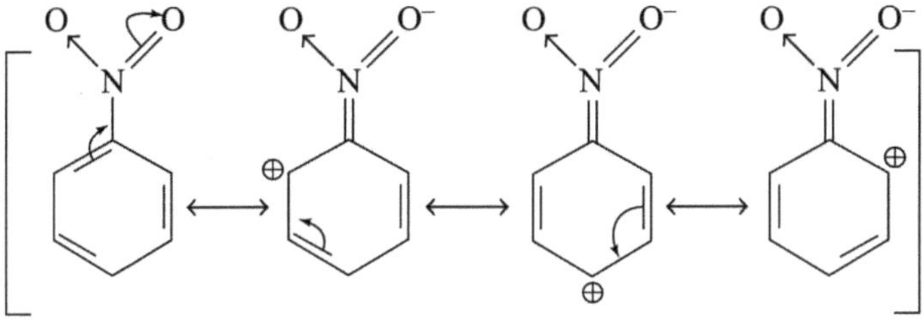

Thus, −M effect causes (+) charge in the phenyl ring whereas +M effect produces (−) charge in the phenyl ring. In both the cases, however, only ortho and para positrons are affected and meta position always remain unaffected.

The essential difference between inductive effect and mesomeric effect is that inductive effect can operate in saturated as well as unsaturated compounds, whereas mesomeric effect is possible only in unsaturated compounds. Further, inductive effect is operative only up to 2 to 3 carbon atoms and vanishes after that, whereas mesomeric effect can be transmitted from one end of the molecule to the other if conjugation is available.

An important example of a group which exhibits mesomeric effect as well as inductive effect is chloro group. The lone pair of chlorine shifts toward phenyl group in chlorobenzene and increases electron density at ortho, para positions. At the same time, the inductive effect of chlorine pulls electrons toward itself. The combined effect of +M and −I is its ortho, para directing with deactivation.

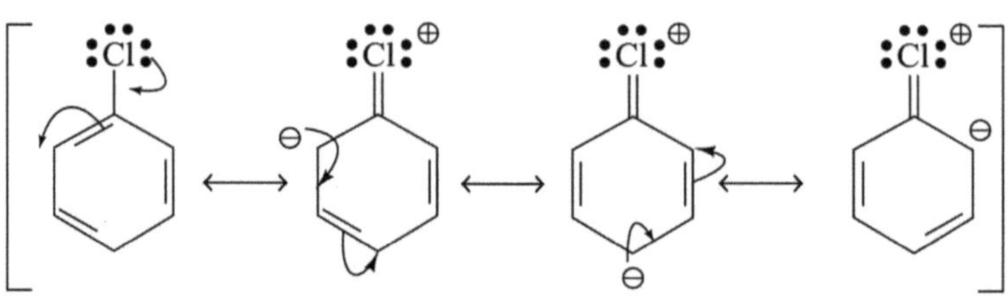

Two main types of attacking reagents involved in ionic reactions are

(i) Electrophilic reagents.
(ii) Nucleophilic reagents.

(i) **Electrophilic reagents:** The reagents which are electron deficient or positively charged species are called as electrophilic reagents. Some common reagents of this type are

$$H_3O^+, SO_3H^+, NO_2^+, AlCl_3, BF_3, FeCl_3, Br^+$$

As these are electron deficient, these reagents attack at the position of high electron density. The reactions in which electrophiles substitute the atom are called as, *electrophilic substitution reactions.*

(ii) **Nucleophilic reagents:** This type of reagents are electron rich or negatively charged species. They readily attack at the position of low electron density. Some common reagents are

$$OH^-, Br^-, CN^-, -\ddot{N}H_2, >S:$$

The reactions in which nucleophiles substitute the atom of other molecule are called as *nucleophilic substitution reactions.*
To understand these reactions in little detail, let us consider the mechanistic aspects of these reactions *nucleophilic substitution* (S_N)
These reactions are designated as S_N reactions, where S stands for substitution and N represents nucleophilic. Consider the attack of base on an alkyl halide. The product formed is an alcohol and

$$R - X + OH^- \longrightarrow R - OH + X^-$$

thus X atom is replaced by OH^-. Such a substitution of X by OH^-can occur by two different mechanisms

- Unimolecular (S_N^1)
- Bimolecular (S_N^2)

In unimolecular (S_N^1) reactions, the mechanism proceeds in two distinct steps, i.e.,

$$R - X \longrightarrow R^+ + X^-$$
$$R^+ + OH^- \longrightarrow R - OH$$

The first step is a slow step, whereas the second step is fast. The rate of reaction thus depends only on the concentration of [R − X] and is independent of the concentration of base.

$$\text{Rate } \alpha \ [RX]$$

As alkyl halide ionizes in the first step, an optically active alkyl halide yields racemic mixture because alkyl halide on ionization will give a carbonium ion which is planar and now the attack of OH^-will take place from both the sides, giving equal amount of *d*- and *l*-forms

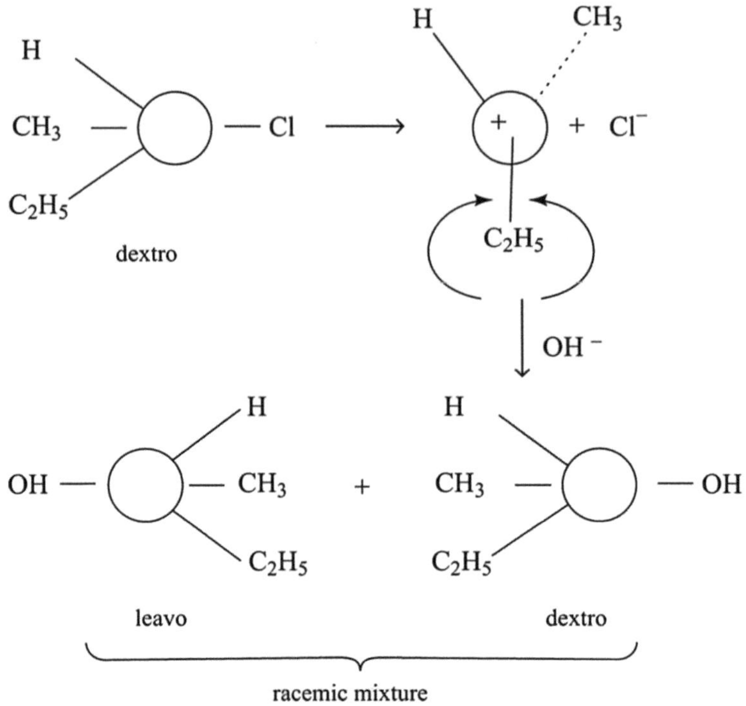

In S_N^2 mechanism, the rate of reaction depends on the concentration of alkyl halide as well as base. Thus, in the rate-determining step, both are involved. The complete ionization of alkyl halide does not

$$\text{Rate}\ \alpha\ [\text{alkyl halide}]\ \left[OH^-\right]$$

take place in this case; rather from one end, halide atom starts leaving the molecule, and from the other end, base approaches the molecule.

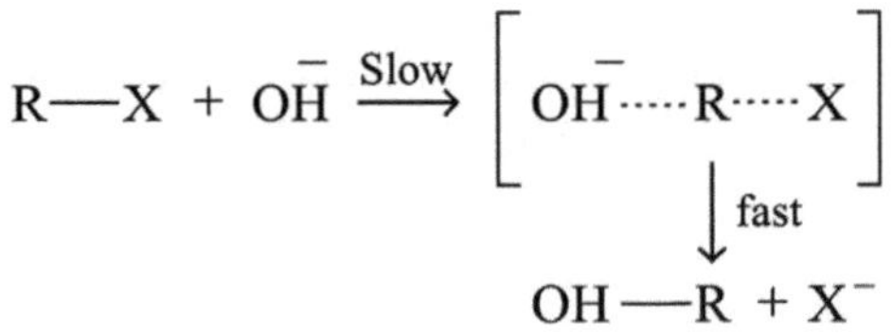

The base completely attaches with alkyl group and halogen atom leaves the molecule as halide. As OH^- or base attacks from the side opposite to the leaving group the inversion of configuration will take place. Consider an optically active alkyl halide which reacts with a base.

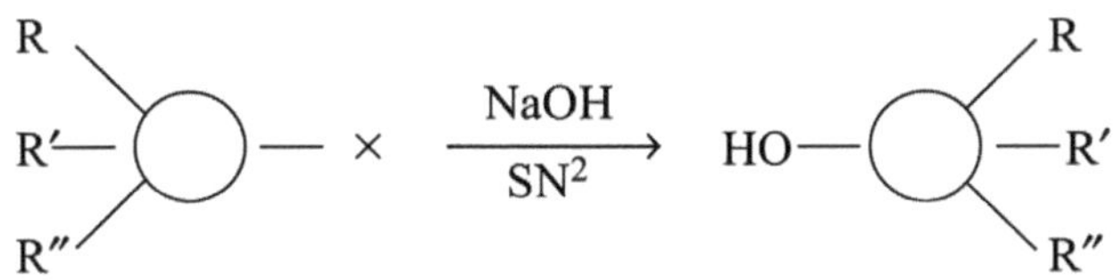

dextro leavo

The transition state can be written in which carbon is partially bonded to both X and OH, i.e., the bond between C – X is not completely broken and bond between C – OH is not completely formed. The OH and X are located as far apart as possible, and the most convenient situation is that both are at the opposite side to each other. Remember, this is the reason for the complete stereochemical inversion in an S_N^2 reaction.

Example 13.3 (-) 2-Bromo octane of 100% optical purity is allowed to react with NaOH. Assuming second-order kinetics $\left(S_N^2\right)$, write the transition stage and the product of the reaction. Comment on the optical activity of the final product formed.

Solution (-) 2-Bromo octane can be written as

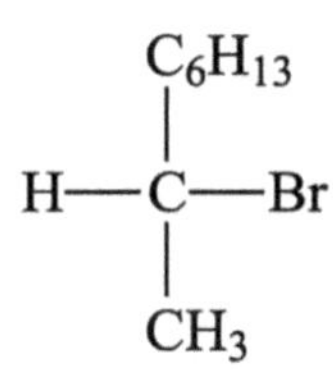

on undergoing S_N^2 mechanism hydrolysis with base; the transition stage will be

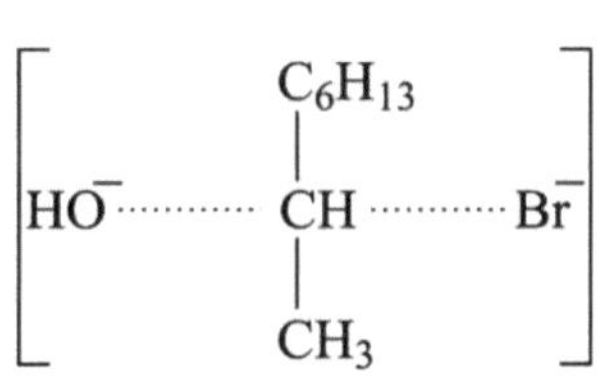

As inversion of configuration will take place in S_N^2, the product will be (+) 2-butanol.

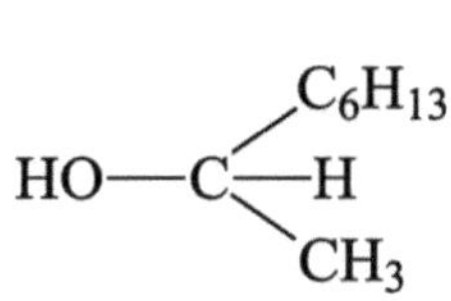

It has been observed that reactivity of substrates in S_N^2 reaction follows the order

$$CH_3- > CH_3CH_2- > (CH_3)_2\,CH- > (CH_3)_3\,C-$$

Thus, methyl halide or primary alkyl halide will have a tendency to undergo hydrolysis by S_N^2 mechanism, whereas tertiary alkyl halide would early go by S_N^1 mechanism. Secondary alkyl halides can undergo hydrolysis by both the mechanisms.

13.5.3 Electrophilic Substitution

The well-known examples of electrophilic substitutions are in benzene or aromatic compounds where H atom of the ring is substituted by other atom or group. The attacking reagent is an electrophile which attacks at the position of high electron density. Let us consider a reaction in which an electrophile (E^+) attacks the benzene ring to give substituted product

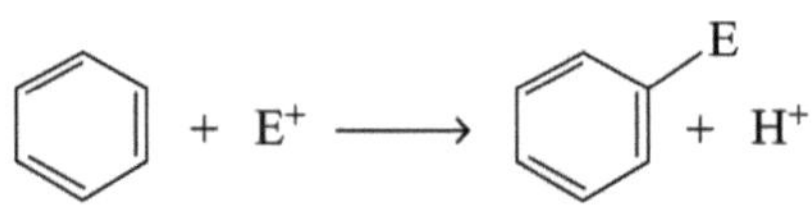

The reaction essentially proceeds in two steps.

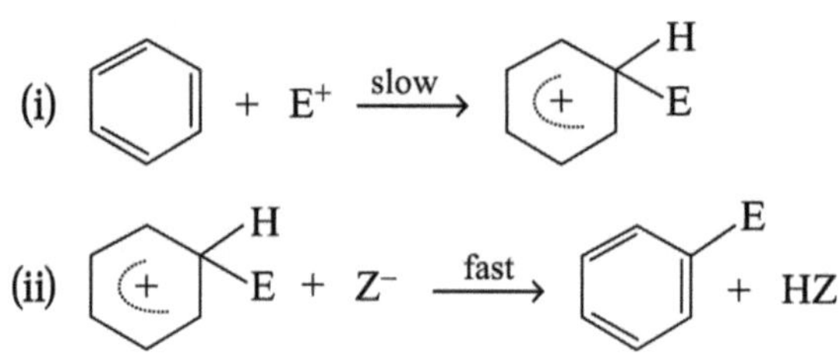

It has been established that electrophilic substitution actually proceeds in two stops. If hydrogens of the aromatic ring are replaced by deuterium or tritium and then such compounds are subjected to electrophilic substitution, the rate of substitution of deuterium or tritium is similar to hydrogen. Thus, breaking of C—H bond is not involved in electrophilic aromatic substitution. Some common examples of electrophilic aromatic substitution are:

- **Nitration:** Nitration of aromatic compounds is carried out by reacting them with a mixture of concentrated HNO_3 and H_2SO_4. Such a mixture is called as nitrating mixture and rapidly produces a nitronium ion, which attacks at the ring

$$HNO_3 + H^+ \longrightarrow \left[H_2NO_3\right]^+ \longrightarrow H_2O + NO_2^+$$

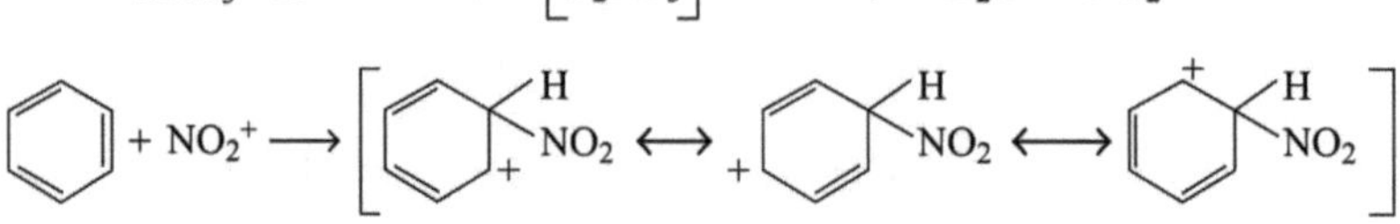

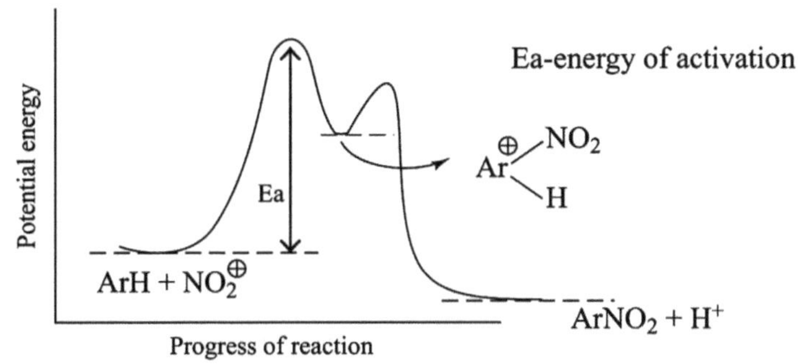

Fig. 13.2 Energy profile diagram for nitration of benzene

and now a proton is abstracted by HSO_4^-

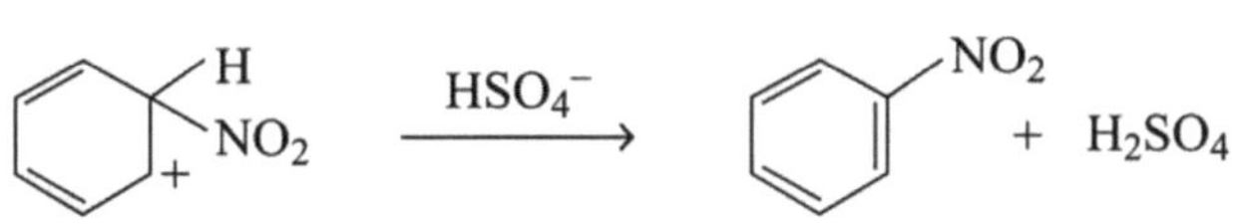

If further nitration of nitrobenzene is carried out, m-dinitrobenzene is obtained as the product; however, the conditions are more drastic. Third nitro group cannot be introduced in the ring because two nitro groups deactivate the ring (nitro group is electron withdrawing), and hence further attack of electrophile is not possible. The energy profile diagram of nitration can be represented as shown in Fig. 13.2.

- **Halogenation:** Halogens, such as chlorine, bromine, etc., are not capable of substituting benzene or other aromatic compounds themselves. However, in presence of Lewis acids such as $AlCl_3$ or $FeBr_3$, halogens are polarized, and positively changed Cl^+ or Br^+attack the molecule.

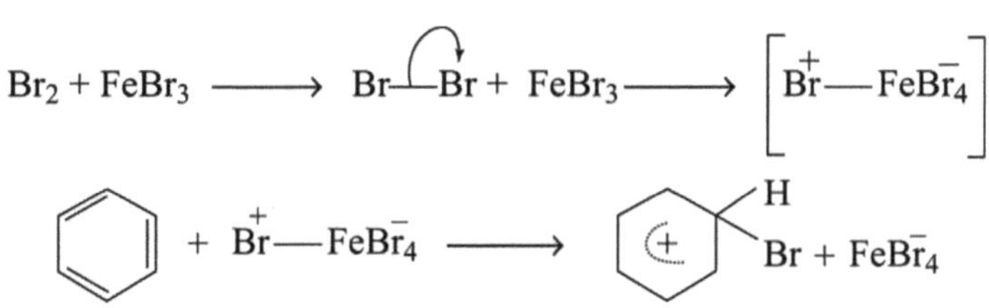

Similarly in the case of chlorination

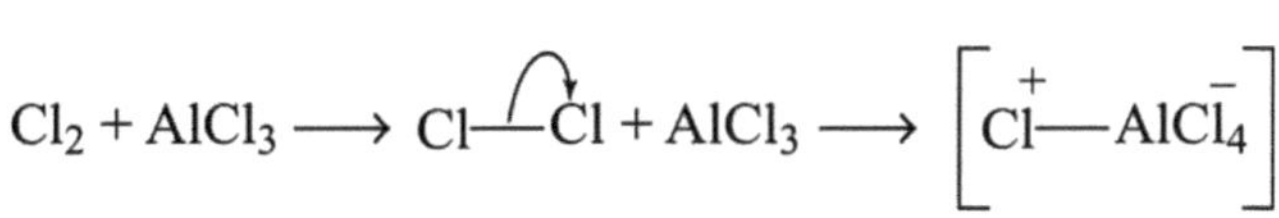

Hypochlorus acid is also effective chlorinating agent because it is capable of producing Cl^+ ions

$$HOCl \longrightarrow [\overset{-\delta}{HO} - \overset{+\delta}{Cl}] \longrightarrow HO^- + Cl^+$$

Iodination is often assisted by the presence of base or some oxidizing agent which tend to produce iodonium ion $I^{\oplus}$. Interhalogen compounds such as I – Cl or Br – Cl are also effective for carrying out electrophilic substitution.

$$I{-}Cl \longrightarrow \overset{\oplus}{I} + \overset{\ominus}{Cl}$$

$$Br{-}Cl \longrightarrow \overset{\oplus}{Br} + \overset{\ominus}{Cl}$$

- **Sulphonation:** Sulfonation of benzene is carried out using hot concentrated H_2SO_4 still the reaction is slow; however, use of oleum significantly enhances the rate. Earlier it was believed that the electrophile responsible for sulfonation is SO_3H^+, which is generated in the reaction as follows.

$$H_2SO_4 + H^+ \rightleftharpoons [H_3SO_4]^+ \longrightarrow SO_3H^+ + H_2O$$

However, sulfonation is now related to SO_3. Though SO_3 is a neutral molecule, but three oxygens make sulfur atom a powerful electrophile.

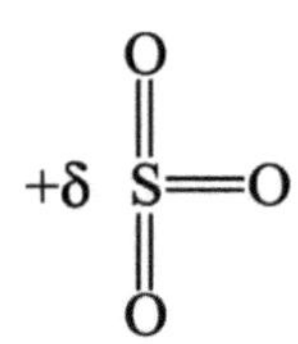

and SO_3 is produced as follows:

$$2H_2SO_4 \rightleftharpoons H_3O^+ + HSO_4^- + SO_3$$

which then carries out sulfonation

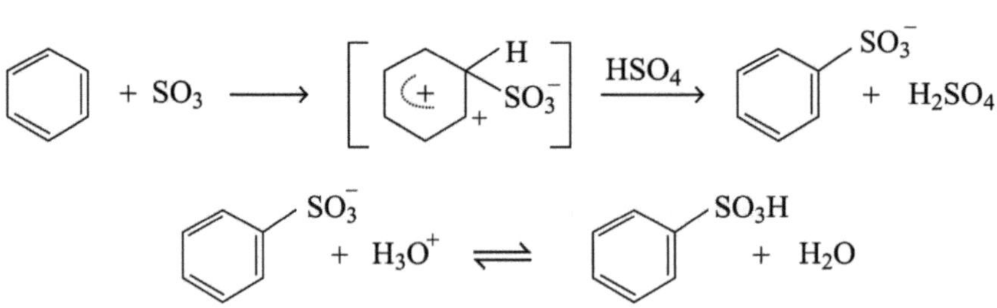

Example 13.4 Nitration of aromatic compounds can also be carried out by using only conc. HNO_3, however, rate of nitration is very slow. Explain the reason. Write a mechanism for the formation of NO_2^+.

Solution The mechanism is

$$HONO_2 + HONO_2 \rightleftharpoons \left[H_2\overset{+}{O}NO_2\right] + ONO_2^-$$

one molecule of nitric acid accepts a proton from the other molecule and $[H_2NO_3^+]$ is generated

$$[H_2NO_3^+] \rightleftharpoons H_2O + NO_2^+$$

However, the rate of nitration is very slow in this case. Use of very strong acid such as sulfuric acid cause nitric acid to ionize like OH^-—NO_2^+rather than H^+—ONO_2^-. As strongly basic OH− group is a poor leaving group, the nitration is very slow in absence of H_2SO_4.

The mechanism of some important name reactions and their industrial importance are discussed below.

13.6 Beckmann Rearrangement

Conversion of ketoximes to acid amide in presence of acids like H_2SO_4, H_3PO_4, *etc.* is known as Beckmann rearrangement. The reaction is of industrial importance as it is involved in the preparation of Nylon 6 in which oxime is converted to caprolactam.

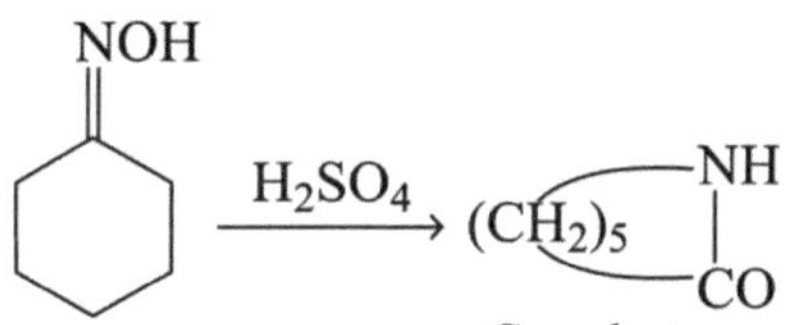

Caprolactum

The Beckmann rearrangement is an example of intramolecular rearrangement involving 1, 2-shift. For sake of simplicity consider a simple oxime, viz., acetophenone oxime. In the first step, protonation of OH group occurs in acidic medium and water is removed causing migration of group from carbon to nitrogen atom.

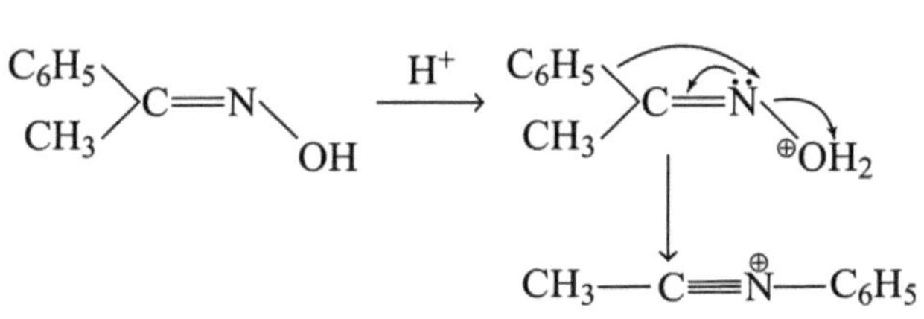

It has been successfully demonstrated that it is always the group at antiposition to $-OH$, which migrates in Beckmann Rearrangement. Thus, if the oxime is as shown below having $-CH_3$ and $-C_6H_5$ groups at the carbon, the group that migrates is CH_3 which is anti to $-OH$ and not C_6H_5.

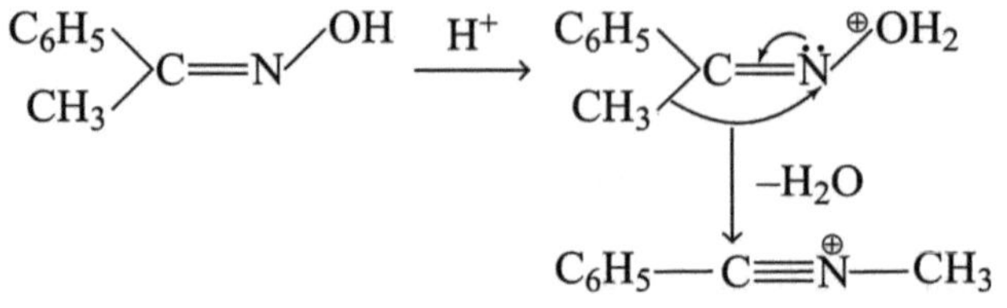

The product obtained on hydrolysis gives hydroxylated molecule which rapidly rearranges to give acid amide.

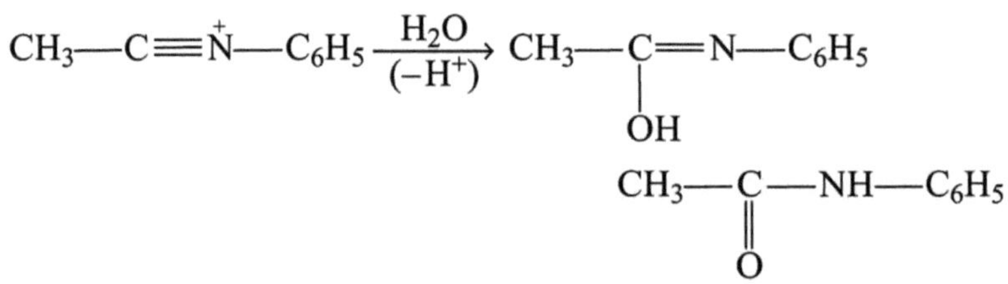

Acetanilide

The Beckmann rearrangement can be used to identify the configuration of oximes by characterizing the acid amide obtained as the final product. During synthesis of caprolactam (monomer of Nylon 6), Beckmann rearrangement is used in the step when cyclohexanone oxime is converted to caprolactam. The conversion mechanism involved in this step is as follows:

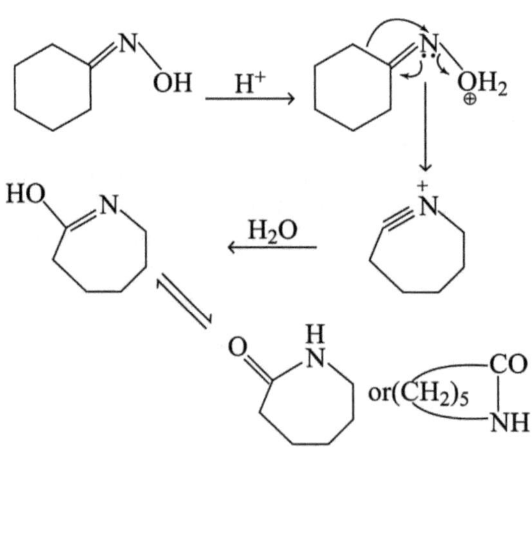

13.7 Cannizzaro Reaction

Two molecules of aliphatic or aromatic aldehydes (having no α hydrogen atoms) undergo disproportionation in presence of aqueous base to give one molecule of primary alcohol and one molecule of corresponding carboxylic acid. Thus, one molecule of aldehyde is oxidized on expense of the other, which undergoes reduction. For example,

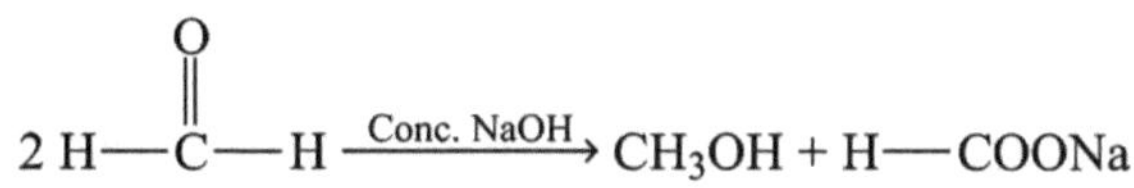

The mechanism of Cannizzaro reaction involves three distinct steps. In the first step, the nucleophilic attack of OH^- takes place on carbonyl carbon atom.

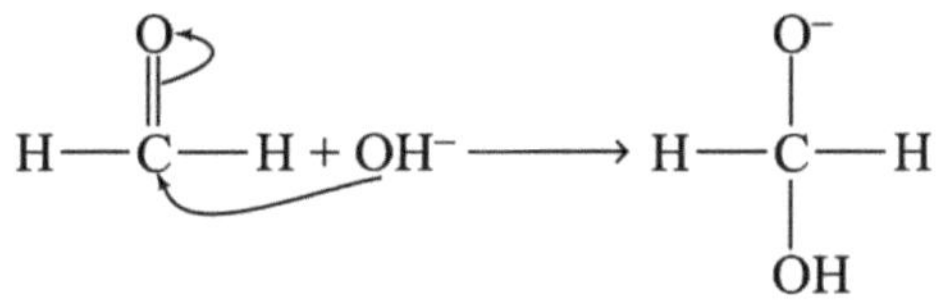

In the second step, hydride ion is shifted from the intermediate formed in the first step to the second molecule of aldehyde.

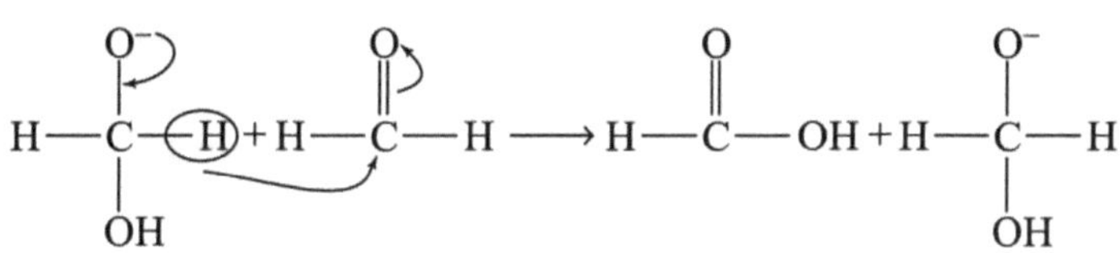

Finally in the third step, acid base reactions occur to give one molecule of alcohol and another molecule of acid as shown below:

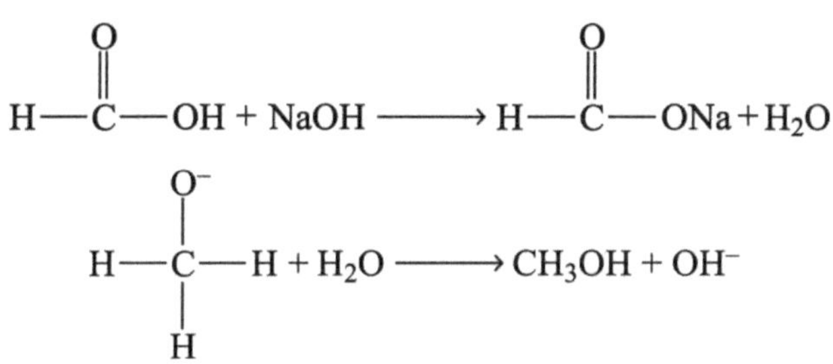

Similarly benzaldehyde also undergoes reaction with NaOH to give one molecule of benzyl alcohol and another molecule of benzoic acid.

$$\underset{\text{Benzaldehyde}}{C_6H_5CHO} + C_6H_5CHO \xrightarrow{\text{NaOH}} \underset{\text{Benzyl alcohol}}{C_6H_5CH_2OH} + \underset{\text{Sod. Benzoate}}{C_6H_5COONa}$$

Two different aldehydes when undergoing Cannizzaro reaction all possible products are formed and in such cases the reaction is known as Crossed Cannizzaro reaction.

Cannizzaro reaction is widely used in the preparation of variety of alcohols and acids in industries. Some common acids and alcohols prepared using this reaction are glycolic acid, mandelic acid ($C_6H_5CHOH \cdot COOH$), furfuryl alcohol, etc.

13.8 Diel's-Alder Reaction

It is a useful reaction for the synthesis of cyclic compounds. In Diel's-Alder reaction, a conjugated diene is heated with an alkene to give a cyclic product. The reaction is a one-step process in which transition state involves a cycle of six orbitals similar to benzene. The alkene used for reaction is called as dienophile and generally possesses an electron withdrawing substituent. For example when 1, 3 butadiene reacts with ethylene, the product obtained is cyclohexene.

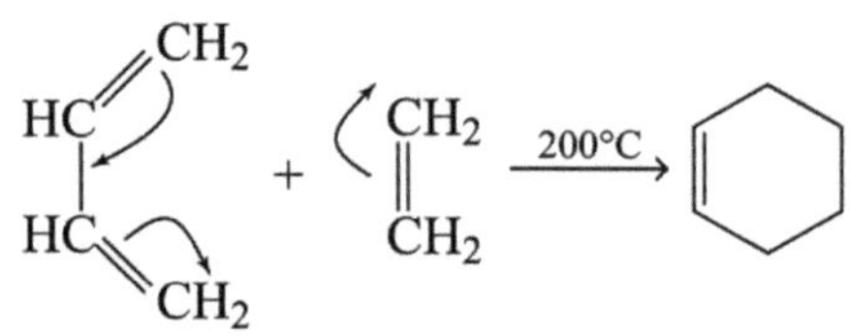

In the presence of substituents such as $-CHO$, $-COOH$, $-COOR$, $-NO_2$, etc., on dienophile, the reaction becomes fast and can be carried out at lower temperatures. As the reaction does not involve free radical or ionic intermediates, it is unaffected by the presence of polar or non-polar solvents.

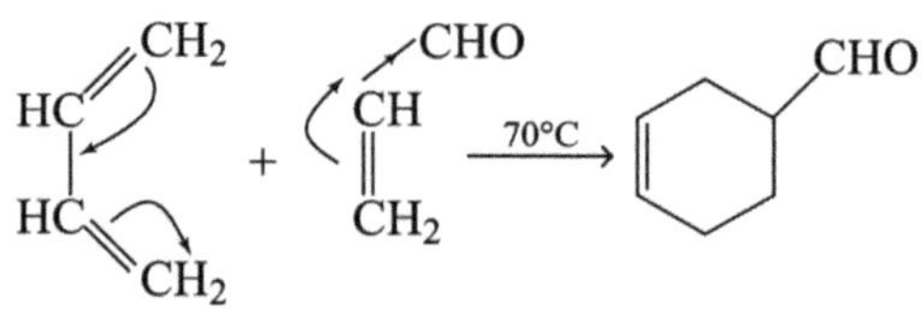

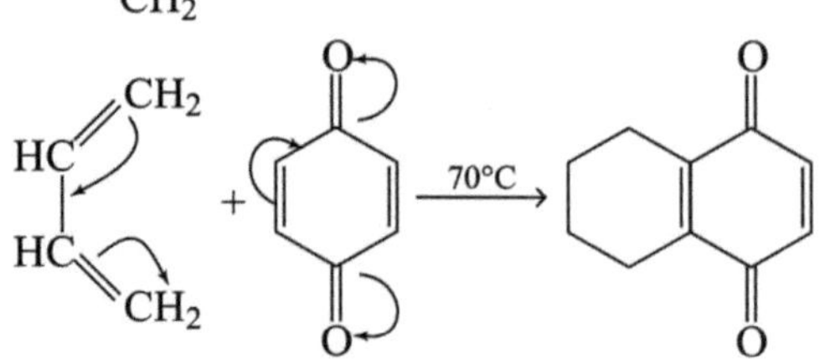

Benzoquinone

Dienophiles having a triple bond also shows the reaction leading to a cyclic product. For example,

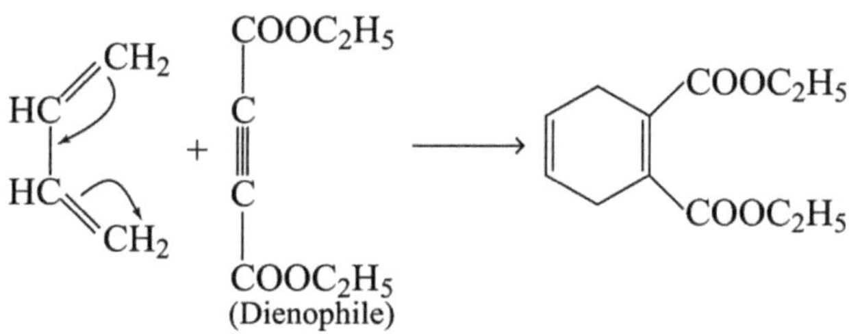

In the Diel's-Alder reaction, no ionic or free radical intermediates have been detected. It is suggested that the reaction proceeds in a concerted manner involving a cyclic transition intermediate with simultaneous formation and breaking of bonds. The reaction is used in the synthesis of variety of compounds.

13.9 Hofmann Reaction

The reaction of converting an amide into amine with the loss of carbonyl carbon atom is known as Hofmann reaction or Hofmann degradation. In many books, it is also mentioned as Hofmann rearrangement. The conversion is carried out with solution of bromine in alkali; however, chlorine in alkaline medium is also effective in carrying out the reaction. The general reaction can be written as

$$RCONH_2 + X_2 + 4NaOH \longrightarrow RNH_2 + 2NaX + 2H_2O + Na_2CO_3$$

where

$$X_2 = Cl_2 \quad \text{or} \quad Br_2$$

The reaction has an industrial significance because it can be used for the synthesis of a lower homologue from the higher one and works with aliphatic or aromatic amides. Thus, reaction of benzamide under similar conditions gives aniline

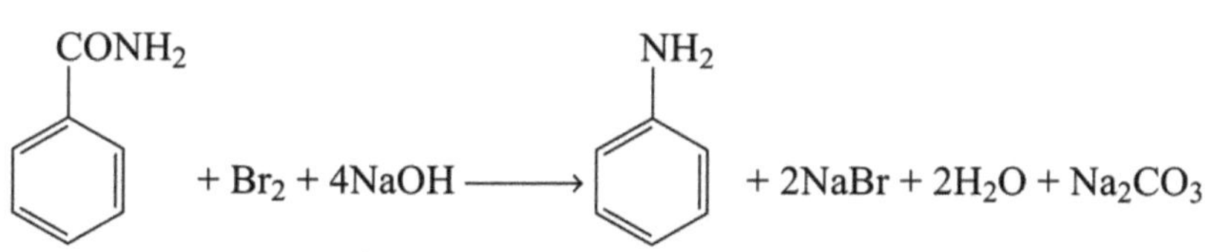

The mechanism of the Hofmann reaction has been established by isolating many intermediates of the reaction. The first step of the mechanism is the formation of N-bromoamide, which occurs by the attack of base on amide.

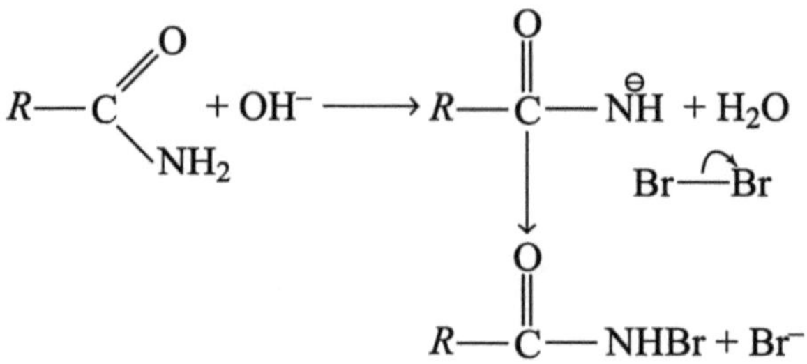

N-Bromoamide

The loss of proton from the NH_2 group gives an intermediate having negatively charged nitrogen. The attack of bromine then rapidly gives N-bromoamide.

In the second step, attack of the base used on *N*-bromoamide gives anion as shown below:

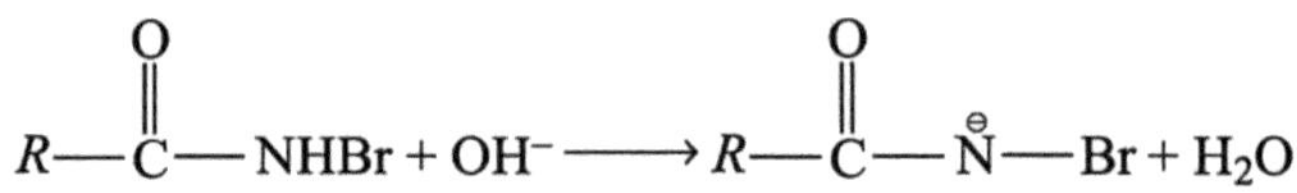

The anion so formed rapidly loses bromide ion, and an active species nitrene is formed. The important point is that nitrogen has only six electrons in nitrene and hence is electron-deficient species.

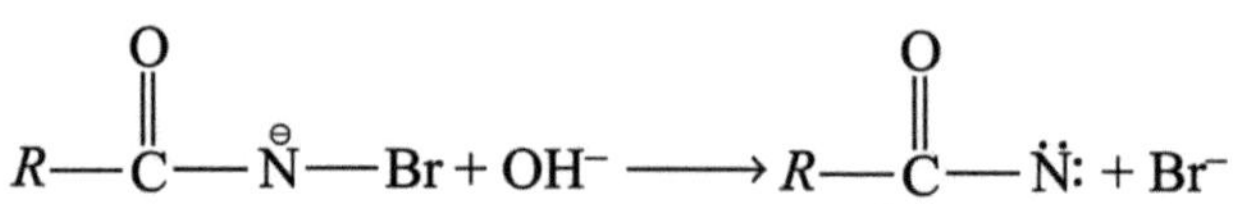

nitrene

As soon as nitrene is formed, the group *R* migrates from carbon to nitrogen with its electrons and the result is the formation of isocyanate. The migration of group from carbon to nitrogen in this case is similar to 1, 2-shift usually observed in the rearrangement of carbonium ions. It is believed that the

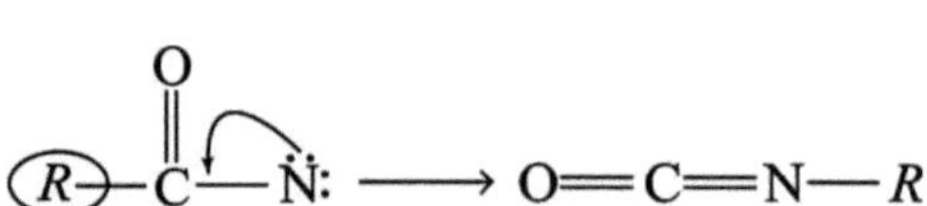

Nitrene Isocynate

formation of nitrene and shift of group *R* from carbon to nitrogen occurs simultaneously. The hydrolysis of isocyanate gives the final product of the reaction, *i.e.*, primary amine:

$$O = C = N - R \xrightarrow{H_2O} R - NH_2 + CO_2$$

Hofmann reaction has been widely used for the preparation of number of aliphatic as well as aromatic amines. It has been found that electron donating group in the aromatic ring increases the rate of reaction. Thus, the order of reactivity for conversion increases tremendously in the case of $-OCH_3$ and decreases with NO_2 group present in the ring.

13.10 Reimer-Tiemann Reaction

Phenols on treatment with chloroform in basic solution give corresponding aldehydes, and the reaction is known as Reimer-Tiemann reaction. It involves attack of electrophile (dichlorocarbene: CCl_2) on highly reactive phenoxide ring. The electrophile of the reaction is generated by the reaction of chloroform with base.

$$CHCl_3 + OH^- \longrightarrow H_2O + :CCl_3^{\ominus}$$
$$:CCl_3^{\ominus} \longrightarrow \underset{\text{Dichlorocarbene}}{:CCl_2} + Cl^-$$

Dichlorocarbene is electrically neutral; however, it acts as an electrophile because it has only six electrons and is thus electron deficient. The attack of dichlorocarbene on phenoxide then occurs, and a dichloromethyl derivative is formed. The rapid hydrolysis of the dichloromethyl derivative finally

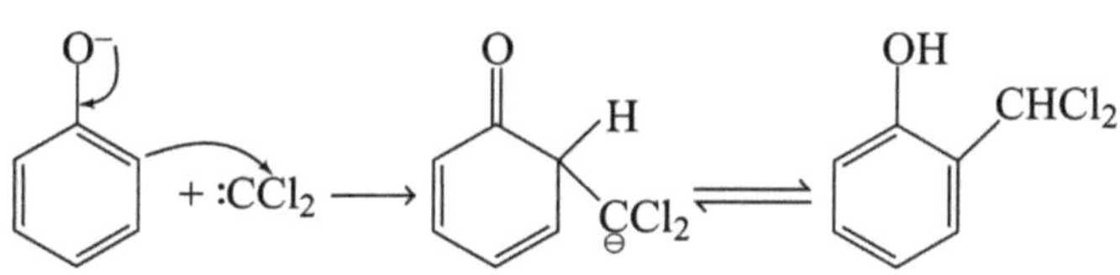

gives salicylaldehyde

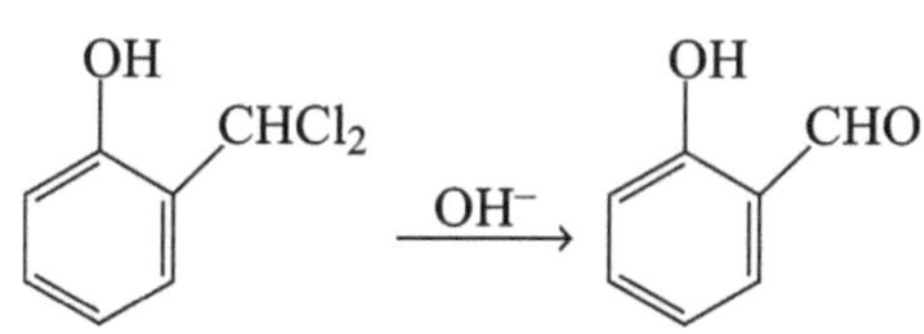

The mechanism of reaction has been well established by the fact that the formation of dichlorocarbene (CCl_2) has been observed in a large number of other reactions also. An important point noticed in Reimer-Tiemann reaction is that it is always the ortho isomer which predominates in the reaction. The para isomer formed as a minor product can be separated using steam distillation, where only ortho isomer is volatile. One would certainly think at this stage as to why the ortho isomer is a major product. The greater stability of the ortho isomer seems to be the reason which is achieved due to intramolecular hydrogen bonding.

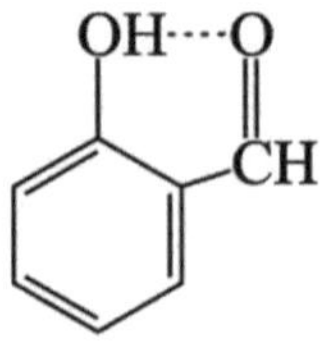

13.11 Skraup Synthesis

By using this reaction, a quinoline skeleton can be easily synthesized in which a benzene ring is fused with pyridine ring. A mixture of aniline, glycerol, nitrobenzene, concentrated H_2SO_4, and ferrous sulfate are heated to give quinoline. A primary aromatic amine having an ortho position vacant can be used to obtain substituted quinolines. In this reaction, concentrated H_2SO_4 acts as condensing agent, nitrobenzene, as oxidizing agent, whereas ferrous sulfate is used to make reaction less violent. Thus, the basic reaction is

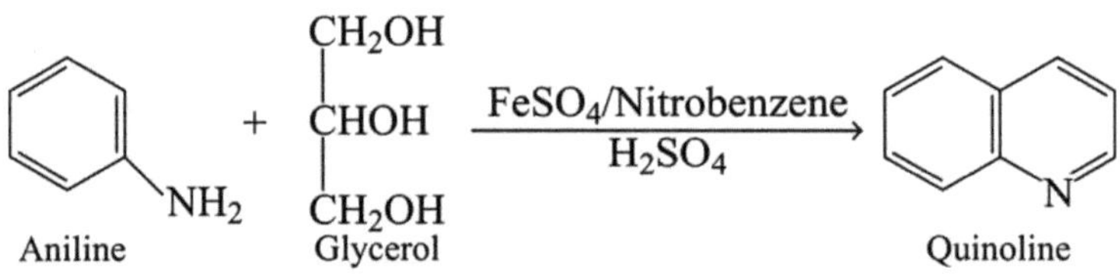

The mechanism of Skraup's synthesis has been studied in detail and found to involve the following steps:

(i) The first step is dehydration of glycerol to give acrolein, which rapidly undergoes

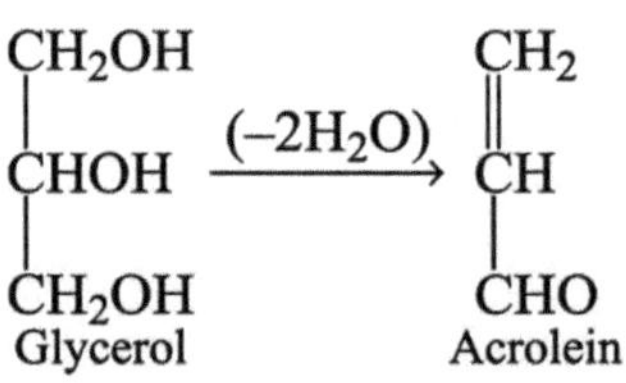

1, 4-addition by aniline.

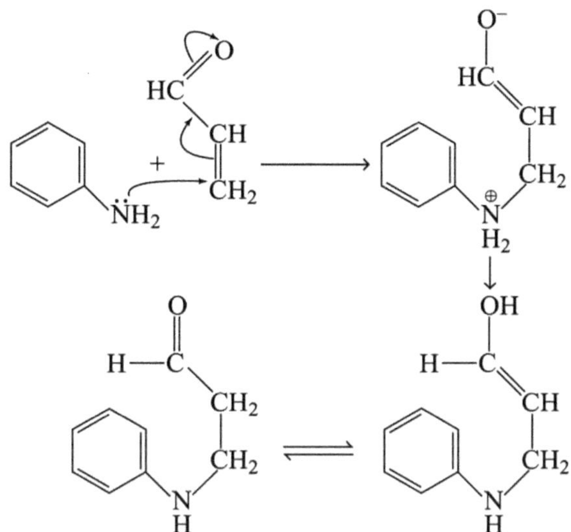

β-anilinopropionaldehyde

(ii) β-anilinopropinaldehyde becomes protonated in acidic medium and then cyclization occurs by loss of proton, which on dehydration gives dihydroquinoline.

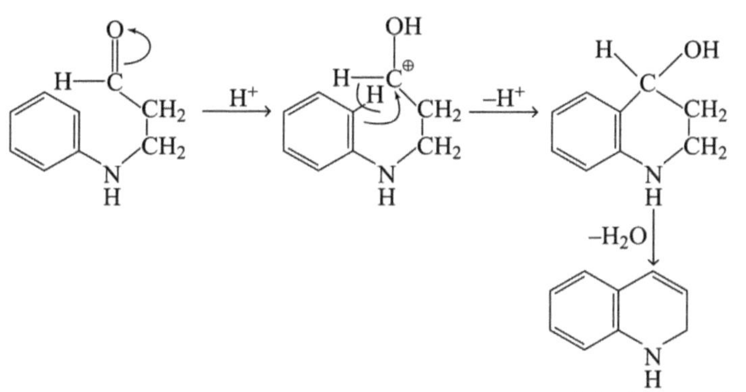

Dihydroquinoline

(iii) The dihydroquinoline so obtained is oxidized by nitrobenzene to give quinoline. The nitrobenzene used for the oxidation is reduced to aniline.

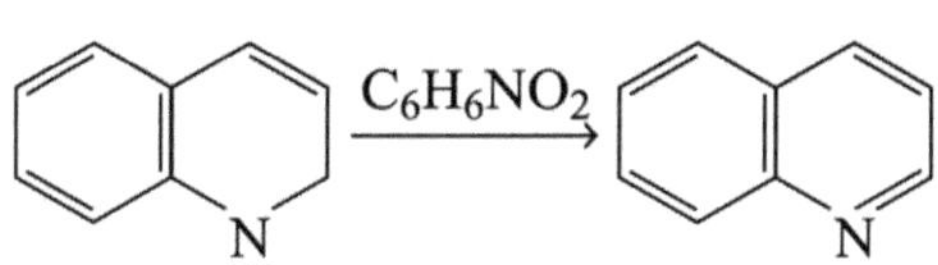

Quoniline

Quinoline and its derivatives are of special importance because quinoline skeleton is the basic unit of large number of antimalarial and analgesic drugs.

13.12 Friedel-Crafts Reaction

It consists in introducing alkyl or acyl group in the aromatic hydrocarbons in presence of anhydrous $AlCl_3$. Various other catalysts like $FeCl_3$, $ZnCl_2$, BF_3 etc., have also been used, but they are not as efficient as $AlCl_3$. For example,

1.

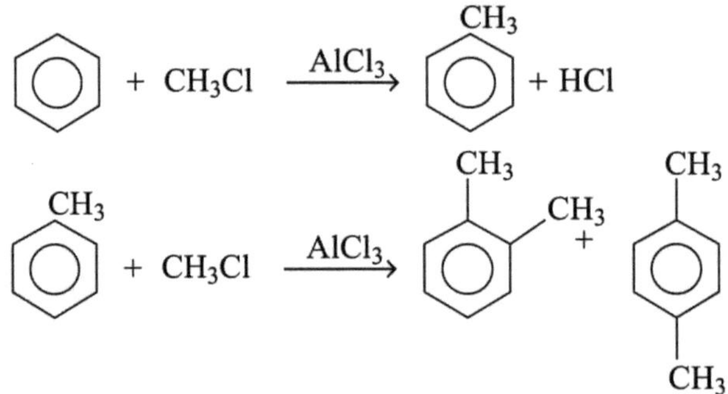

2.

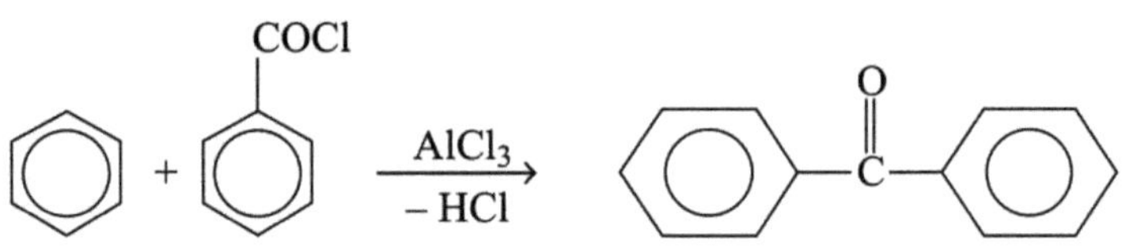

3.

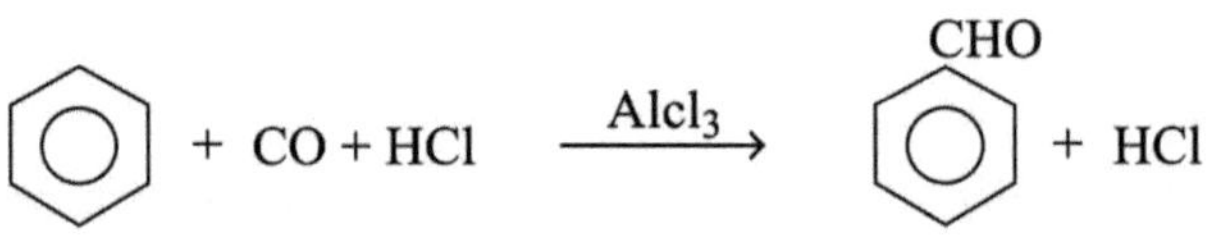

The reaction can be divided into two types —alkylation and acylation.

13.12.1 Alkylation

In alkylation reactions, alkyl group is introduced in the aromatic compounds. The first step of the reaction is formation of 1 : 1 complex of alkyl halide and anhydrous $AlCl_3$.

$$CH_3Cl + AlCl_3 \longrightarrow CH_3^{\oplus}\, AlCl_4^{\ominus}$$

Such type of complexes have actually been isolated in solid state at −78 °C. The alkyl part of the complex acts as an electrophile and attacks the aromatic ring to give intermediate moiety.

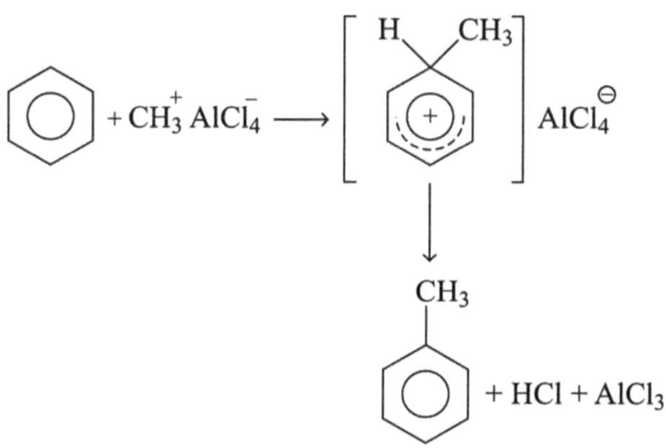

The formation of intermediate commonly known as Wheland intermediate has been actually established by isolation at low temperature. The intermediate then gives toluene (alkylbenzene), HCl, and $AlCl_3$. The rate of the reaction is dependent on all the three components.

$$\text{Rate} = k[\text{ArH}][\text{RX}][\text{AlX}_3]$$

Thus, presence of Lewis acids is an essential requirement to carry out this reaction. $AlCl_3$ is an effective Lewis acid, and the order of reactivity of some halides is as shown below

$$AlCl_3 > FeCl_3 > BF_3 > TiCl_3 > SnCl_4$$

The presence of Lewis acid also leads to several complications such as:

- The presence of alkyl group activates ortho and para positrons in the benzene ring, and hence dior tri alkylated products are also formed.

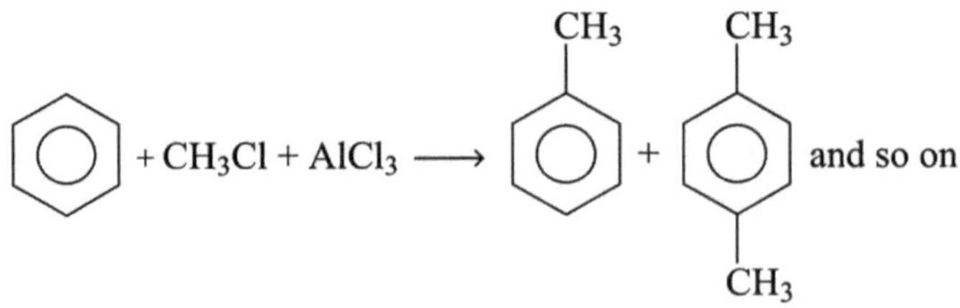

- Many times dealkylation also occurs

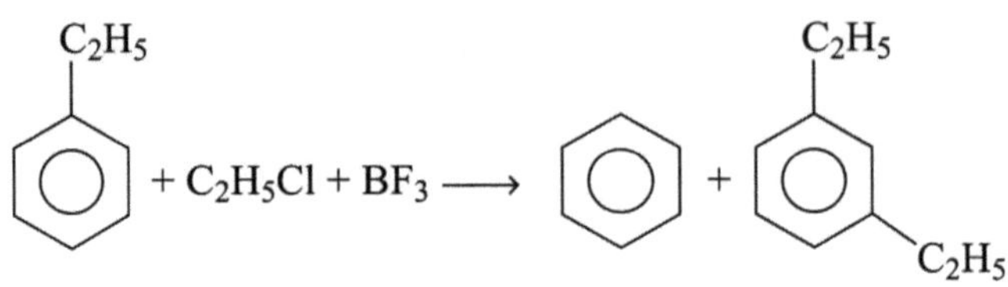

- Rearrangement of the less stable carbocation occurs to give more stable carbocation. This leads to rearranged products.

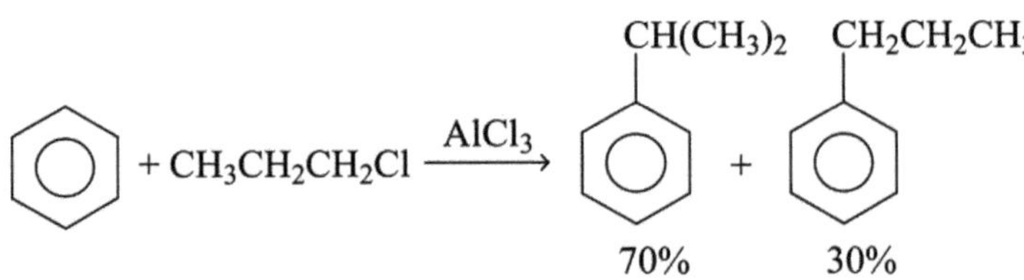

13.12.2 Acylation

In acylation reactions, benzene is allowed to react with acyl halides in presence of anhydrous $AlCl_3$.

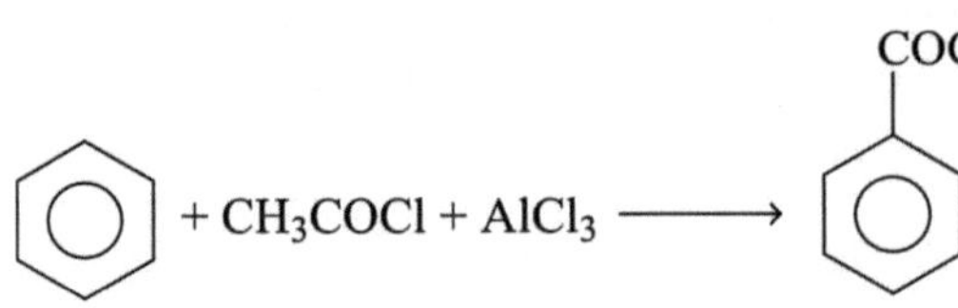

Like alkylation, the species responsible for acylation is believed to be

$$R\overset{+}{C}{\equiv}O\ \overset{\ominus}{AlCl_4}\ \text{or}\ R{-}\overset{+\delta}{\underset{\underset{Cl}{|}}{C}}{=\!=}O{-\!-}\overset{-\delta}{AlCl_3}$$

and the mechanism is similar to alkylation. However, the essential difference between alkylation and acylation is that in the case of acylation more than one mole of Lewis acid is required. The Lewis acid forms complex with the ketone and removes it from the sphere of the reaction.

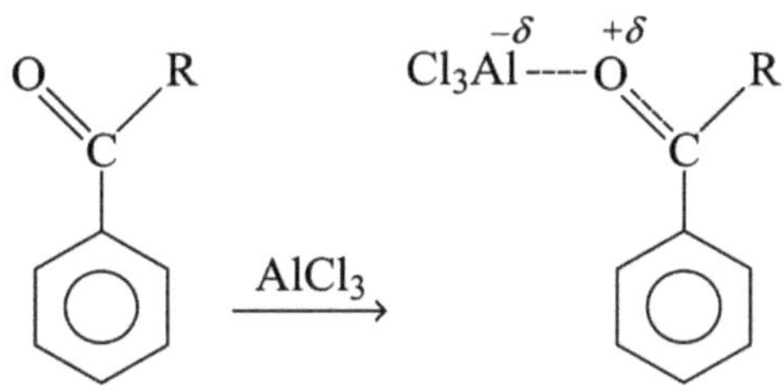

As ketones are less reactive, hence no polyacylation occurs in the reaction.

13.13 Aldol Condensation

When two molecules of a carbonyl compound having α-hydrogen reacts in presence of base, a β hydroxy carbonyl compound is obtained as the product. For example, two molecules of acetaldehyde on condensation in presence of dilute alkali gives 3-hydroxybutanal commonly known as aldol.

$$CH_3CHO + CH_3CHO \xrightarrow{\text{dil NaOH}} \underset{\text{aldol}}{CH_3CH(OH)CH_2CHO}$$

The mechanism of the reaction can be explained in three distinct steps:

Step 1: Formation of carboanion or carbanion due to abstraction of proton by dil alkali.

$$\overset{H}{C}H_2CHO + \overset{-}{O}H \longrightarrow \overset{\ominus}{C}H_2CHO + H_2O$$

Step 2: The carbanion so formed attacks at the position of low electron density in other molecule of acetaldehyde.

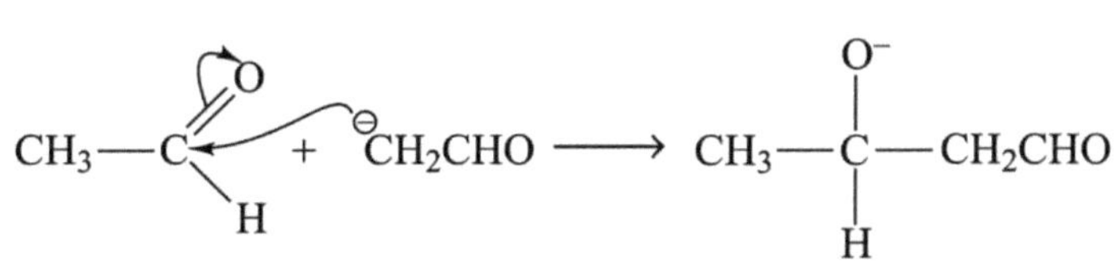

Step 3: Is the abstraction of proton.

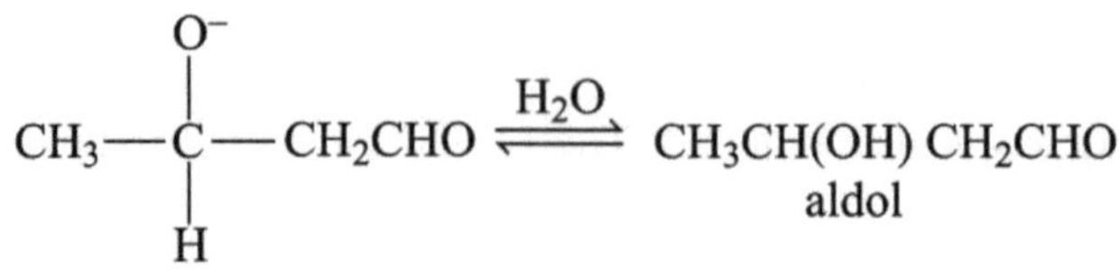

A large number of carbonyl compounds exhibit formation of aldol in presence of dilute alkali.

13.14 Pinacol-Pinacolone Rearrangement

The conversion of 1,2-diols (Vic. Diols) to ketones or aldehydes by slight heating with dilute mineral acids, acid chloride, or $ZnCl_2$ or any other electrophilic reagent is known as pinacol-pinacolone rearrangement. The reaction involves the protonation of OH group followed by elimination of water. The name was given from the classical example of the conversion of pinacol (2,3-dihydroxy2,3-dimethylbutane) to pinacolone (3,3-dimethyl-2-butanone, a ketone).

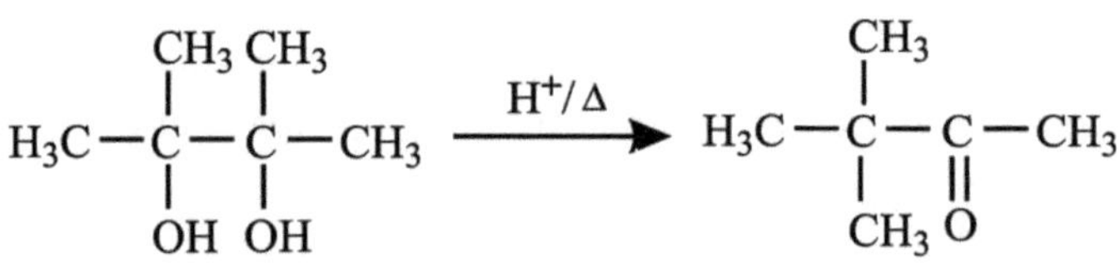

13.14.1 Mechanism

The reaction involves four steps as shown below:

Step 1: Protonation of one hydroxyl group of diol. The acid used causes protonation of OH group. If 1,2-diol is symmetrical, then any of OH can be protonated. However, if diol is unsymmetrical, then protonation of OH, which has high electron density, will be protonated.

Step 2: Loss of water molecule will take place to form a carbonium ion.

Step 3: The shift of alkyl or aryl group from adjacent carbon will take place to form a rearranged carbonium ion. Remember, the stable carbonium ion will be predominant

Step 4: Elimination of proton to form carbonyl compounds takes place.

Generally steps 2 and 3 are considered as rapid and simultaneous. It is suggested that the migrating group does not become completely free; rather it remains partially bonded from leaving carbon and partially attached with adjacent carbon as shown below. Hence, intramolecular type of rearrangement takes place.

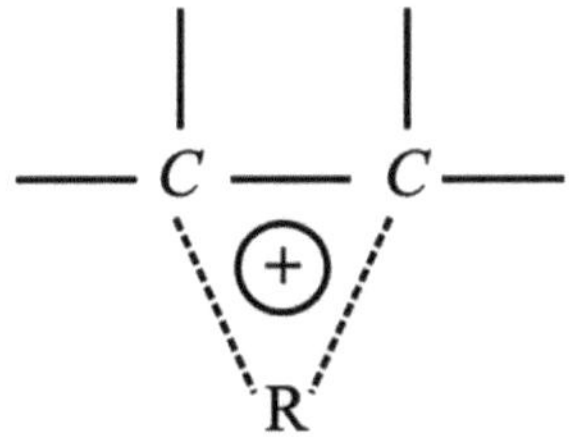

To understand the mechanism clearly, let us consider an unsymmetrical 1,2-diol. Now treatment with acid can cause protonation of any of the two OH group. Thus, we have two course of action (a) and (b). In course (a), protonation of OH present at carbon having two CH_3 groups takes place, whereas, in course (b), protonation of OH at primary carbon takes place. You can see that after elimination of water molecule, a tertiary carbocation is formed in course (a), whereas a primary carbocation is formed in course (b). As tertiary carbocation is more stable than primary, the reaction proceeds by course (a) to give 2-methylpropanal.

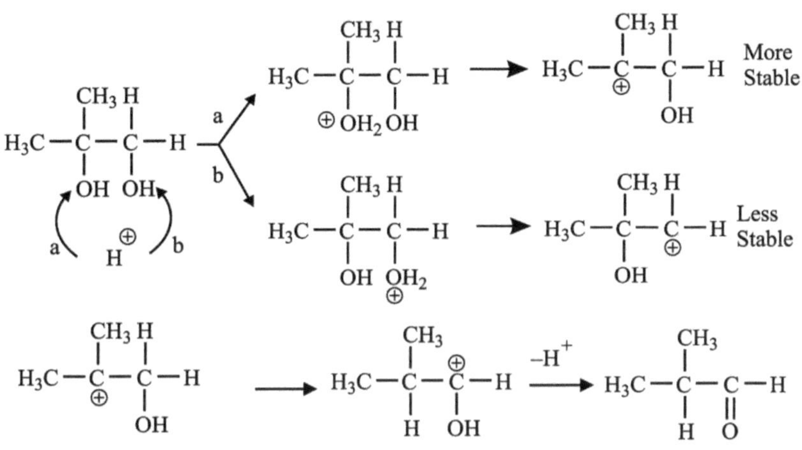

13.15 Effect of Substituents

A number of 1,2-diols have been studied for pinacol-pinacolone rearrangement. It is observed that initially protonation of OH takes place. The migration of the group then takes place with its electron pair. Generally the migration order is H > Aryl > Alkyl. If there is a competition between the groups C_6H_5, p-$OCH_3C_6H_4^-$, p-$CH_3C_6H_4^-$, p-$ClC_6H_4^-$, the more nucleophilic groups are expected to migrate easily. Hence the order of migration will be

$$p-OCH_3C_6H_4- > p-CH_3C_6H_4- > C_6H_5- > p-ClC_6H_4-$$

If electron withdrawing groups are present, the migration of group slows down. The ortho substituted groups migrate slowly due to steric hindrance. Pinacol-pinacolone has variety of applications in synthetic organic chemistry such as synthesis of carbonyl compounds from alkenes, preparation of ketones from cyclic diols, and so on.

Exercises

1. State differences between enantiomers and diastereoisomers with suitable examples.
2. Define the terms: plane of symmetry, center of symmetry, and alternating axis of symmetry.
3. The specific rotation of a solution of coniine containing 0.35 g/ml measured in 10 cm tube is +16°. What will be the observed rotation of its enantiomer?
4. Explain optical isomerism of chiral organic compounds.
5. What are the essential conditions for a compound to show geometrical isomerism?
6. Define the following terms:

   (a) Racemization
   (b) Plane of symmetry
   (c) Center of symmetry
   (d) Resolution
   (e) Diastereoisomerism

7. Write three-dimensional formulas for the following compounds.

   (a) CHClBrF (R configuration)
   (b)

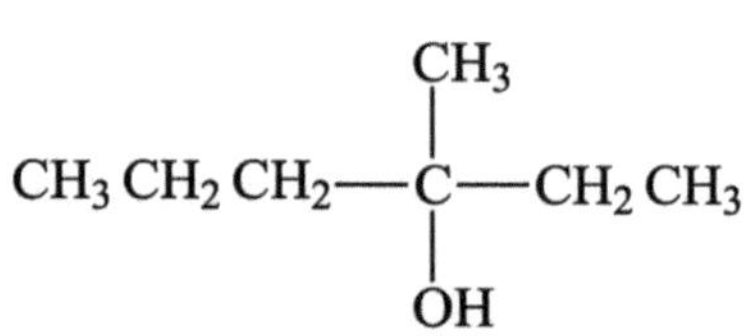

(S configuration)

   (c) 3 R, 4 S-Dimethylhexane.

8. The specific rotation of a pure enantiomer is +12°. What will be its observed rotation if it is isolated from a reaction with:

   (a) 20% racemization and 80% retention and
   (b) 80% racemization and 20% inversion?

9. (a) Calculate the specific rotation of coniine, a toxic component of poison hemlock. If the solution containing 0.75 g/10ml is placed in a 1 – dm tube and its observed rotation at 25 °C in polarimeter is +1.2°.
   (b) What is the specific rotation of enantiomer of coniine?
10. (a) Find out the observed rotation of solution of coniine containing 0.35 g/ml as measured by 5.0 cm tube in polarimeter. The specific rotation is +16°.
    (b) What is the observed rotation if:
       (i) Concentration is doubled and
       (ii) The length of the tube is doubled
       (iii) What is the value of specific rotation if concentration or length of polarimeter tube is doubled?
11. How many stereoisomers are possible for 2-bromo-4-hydroxycyclo-hexane carboxylic acid?
12. Draw all possible stereoisomers for the following and state which are optically active and inactive and are meso form:

    (a) MeCHOHCHOHMe
    (b) $MeCHOHCH_2OH$
    (c) MeCHOHCOCHOHMe
    (d) MeCHClCOOH
    (e) $HOOC{\cdot}CHBrCH_2$ COOEt
    (f) $CH_2 = CHMe$
    (g) MeCH = CHEt
    (h) $MeCHBrCH = CH_2$
    (i) $(Me)_2C = C = CH_2$
    (j) CMeCl = C = CHMe

13. Assign *R* or *S* configuration to each of the following compounds:

(a) 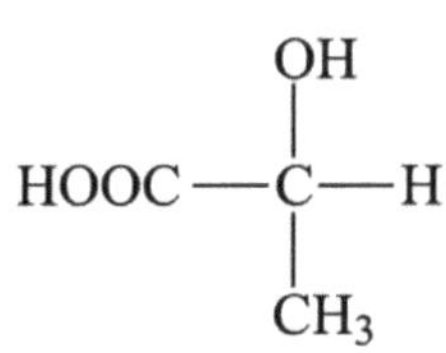

(b) 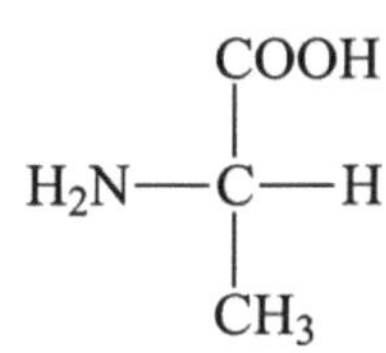

14. Allocate the symbol *E* or *Z* to each of the following compounds:

(a) 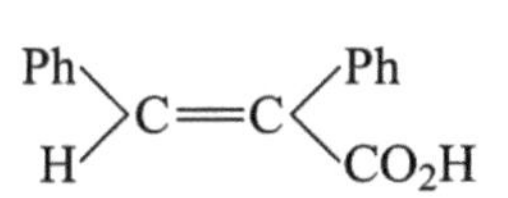
($Ph{=}Phenyl{=}C_6H_5$)

(b) 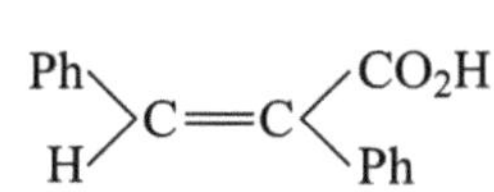

(c) 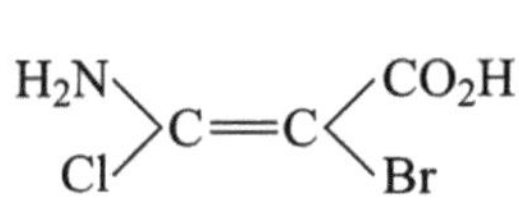

(d) 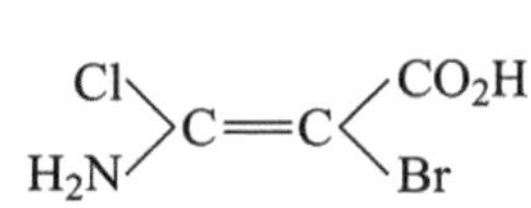

15. The structure of five isomeric compounds are shown:

(a) 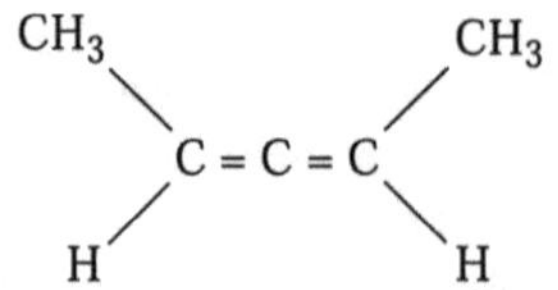 (b) 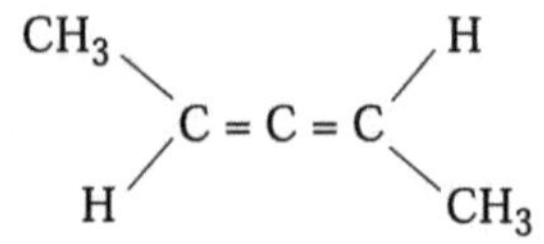

(c) 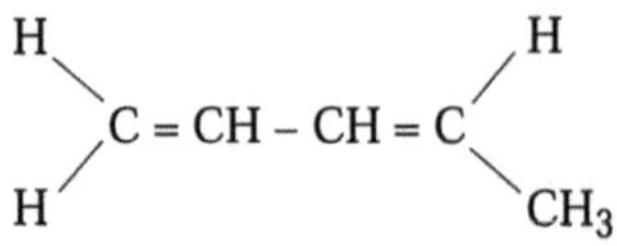 (d) 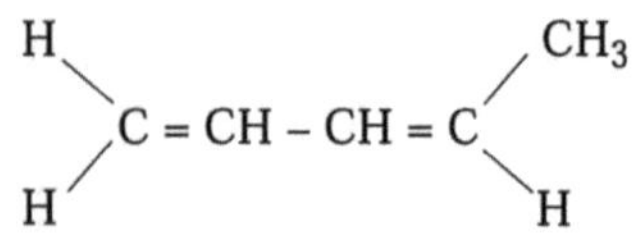

(e) $CH_2 = CH - CH_2 - CH = CH_2$

(i) Do you expect (a) and (b) to have same boiling point.
(ii) Do you expect (a) and (b) to have same effect on plane polarized light.
(iii) Do you expect (c) and (d) to have same effect on plane polarized light.
(iv) Do you expect (c) and (d) to have same boiling point.
(v) How many stereoisomers exist for (e)?

16. Assign R or S configuration to each of the following compounds:

(a) 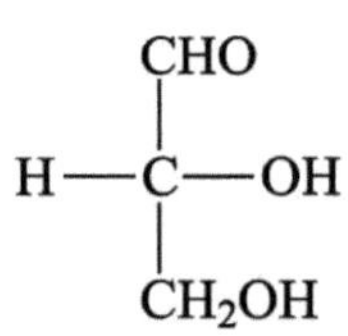 (b) 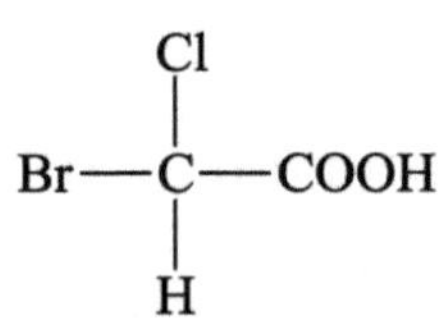

17. Assign E or Z configuration to each of the following compounds:

(a) 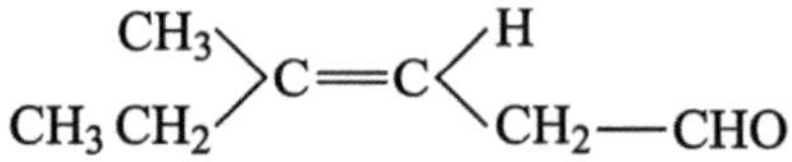 (b) 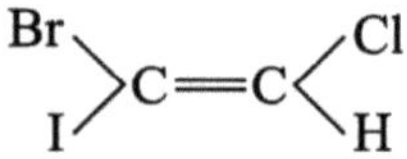

Chapter 14
Coordination Chemistry

In inorganic chemistry, there are three types of salts:

1. **Simple salts:** These can be completely ionized in solution state, for example, $NaCl$, KCl, $BaCl_2$, $MgCl_2$, etc.
2. **Double salts:** These are obtained by the combination of two or more simple salts, and they ionize completely in solution. For example,
   Alum: $K_2SO_4Al_2SO_424H_2O$
   Mohr salt: $FeSO_4\ (NH_4)_2\ SO_412H_2O$
3. **Complex salts:** These salts are not completely ionized in solution and give complex ion. For example,

$$K_4Fe(CN)_6 \rightleftharpoons 4\ K^+ + \underset{\text{complex ion}}{\{Fe(CN)_6\}^{4-}}$$

The salts which do not lose their identity completely are called complex compounds or coordination compounds.

14.1 Werner's Theory of Coordination Compounds

The bonding in complex compounds is explained by Werner's theory. The principles of theory are

1. All metals exhibit two types of linkages

   (i) Primary linkage
   (ii) Secondary linkage

© The Author(s), under exclusive license to Springer Nature Switzerland AG 2026
R. N. Goyal, H. Goel, *Textbook of Engineering Chemistry*,
https://doi.org/10.1007/978-3-032-22068-4_14

Primary linkage is ionizable valency to satisfy the ionic valency of metal, whereas secondary linkage is non-ionizable valency to satisfy coordination valency of metal.

2. Every metal must satisfy both types of valencies.
3. Every metal has a fixed secondary valency known as coordination valency of metal.
4. Secondary valency is directional while primary valency is ionic.
5. Secondary valency is filled by anions or neutral molecules which are called ligands. The number of ligands attached is called coordination number.
6. The total charge on complex ion is the combination of algebraic sum of charges on metal and ligand. For example,

$$Cr^{3+}\,(NH_3)_6 \longrightarrow \left[Cr\,(NH_3)_6\right]^{3+}$$

$$Fe^{2+}\left(CN^-\right)_6 \longrightarrow [Fe(CN)_6]^{4-}$$

Application of Theory On the basis of Werner's theory, different cobalt amines are designated as $Co - NH_3$ complexes. Such as

(a) $CoCl_36NH_3$
(b) $CoCl_35NH_3$
(c) $CoCl_34NH_3$

having different complex structures.

With $AgNO_3$ solution compound

(a) gives three moles AgCl as precipitate
(b) gives two moles of AgCl as precipitate
(c) gives one mole of AgCl as precipitate

With K_2SO_4

(a) gives $Co_2\,(SO_4)_3\,12NH_3$
(b) gives $Co_2\,(SO_4)_3\,10NH_3$
(c) gives no sulfate.

On the basis of above facts, Werner suggested the structure of these complexes as

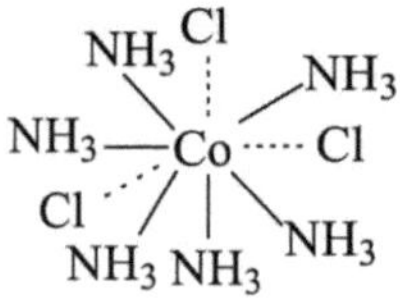

(a)

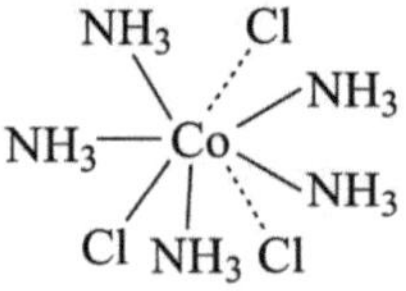

(b)

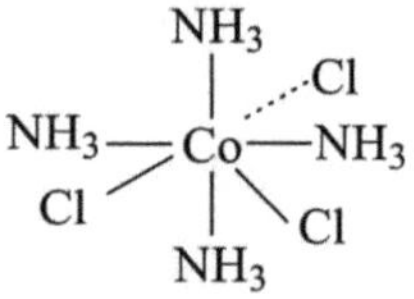

(c)

14.2 Sidgwick's Theory for Electronic Interpretation of Coordination

Sidgwick accepted the Lewis concept of the two electron covalent bond between two atoms in a molecule and introduced a new concept of coordinate bond which, in fact, is an extension of Lewis concept of covalent bond.

According to this concept, the ligands attached with the central metal ion have atoms which have at least one unshared electron pair. When the ligands are attached with the central metal ion, they donate the electron pair to the central metal ion. The atoms which donate the electron pair to the central metal ion are called donors, and the central metal ion which accepts the electron pair is called an acceptor.

The bond between the ligands and the central metal ion is a coordinate or semipolar bond, which is not very different from a covalent bond. The only difference is in the mode of its formation, which indicates that the donor group (ligand) has donated an electron pair to the metal ion (acceptor).

Thus, on the basis of Sidgwick's concept, the structure of $[Co\,(NH_3)_6]^{3+}$ can be shown as

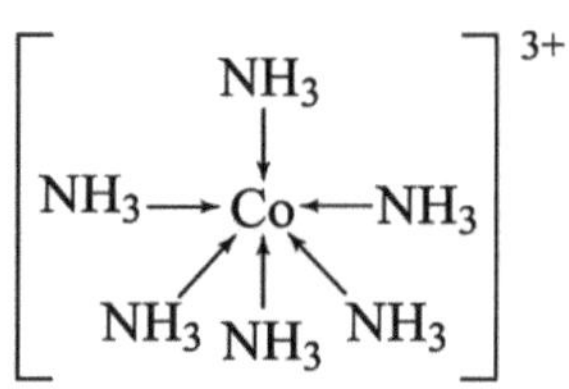

14.3 EAN Concept of Sidgwick Theory

Sidgwick gave another concept to explain the stability of complex which is based on effective atomic number (EAN), which is equal to the atomic number of next inert gas.

$$\text{EAN} = \begin{pmatrix}\text{Atomic Number}\\ \text{of metal}\end{pmatrix} - \begin{pmatrix}\text{Oxidation state of}\\ \text{metal in compound}\end{pmatrix} + \begin{pmatrix}\text{Total no. of electrons}\\ \text{gained by metal}\end{pmatrix}$$

Example 14.1 EAN of $[Co\,(NH_3)_6]\,Cl_3$ can be calculated as

$$\text{Atomic number of Co} = 27$$

$$\text{Oxidation state of Co} = +3$$

$$\text{electron donated by ligand } = 6 \times 2 = 12$$

EAN $= 27 - 3 + 12 = 36$ which is equal to atomic number of inert gas Kr. However, this rule is not applicable to all the complexes and has several exception such as $\left[Fe^{III}Cl_4\right]^-$ or $\left[Fe^{III}(CN)_6\right]^{3+}$ etc.

Chelating Complex When a ligand occupies two or more coordination positions with a metal atom in a complex, then the ligand is called chelating ligand and complex is chelating complex. Ligands containing 2, 3, 4, 5 and six donor sites are termed as bidentate, tridentate, tetradentate, pentadentate, and hexadentate ligands. For example, EDTA (ethylene diamine tetraacetic acid) is represented as

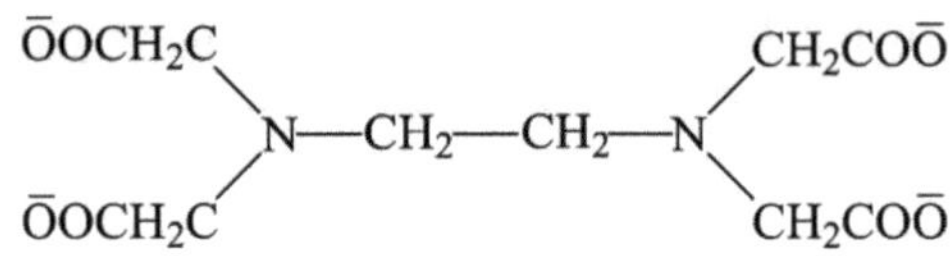

Similarly ethylenediamine $\left[\ddot{N}H_2 - (CH_2)_2 - \ddot{N}H_2\right]$ and glycine $[H_2N - CH_2 - COOH]$ are represented as "en" and "gly," respectively.

14.4 Nomenclature of Coordination Compounds

The first nomenclature for coordination compounds was suggested by A. Werner and later modified by IUPAC. However, fundamental rules are same. The following rules are applied for naming the coordination compound.

1. The cation is named first than anion. Neutral complexes are given a single name.
2. The contents of coordination sphere ligand are called first then metal.
   (a) The ligands are named as (i) −ve ligand (ii) neutral (iii) +ve ligend
   (b) The nature of ligand is defined as
      (i) for +ve ligand suffix used is ium –
      (ii) for −ve ligand suffix used is O –
      (iii) no suffix is used for neutral ligand

   This will be more clear with the name of some ligands such as $NH_2NH_3^+$ is called hydrazinium.

   – NO^+ Nitrosylium.
   – CH_3COO^- is acetato
   – NH_3 is amine
   – H_2O is aquo
   – Cl^- is chloro.

(c) Several ligands of same type are denoted by di, tri, tetra, etc.
(d) For several ligands of different types, they are listed alphabetically. For example, $\left[Pt\,(NH_3)_4\,NO_2Cl\right]SO_4$. Chloro nitro tetramino Pt (IV) sulfate.
(e) Finally name of the metal is given with its oxidation state in Roman letters.
(f) If coordination sphere is −vely charged, then suffix -ate is used, for example, $Na^+\,\{Co(CO)_4\}^-$sodium tetracarbonylcobaltate.
(g) No suffix for positive or neutral spheres is used.

3. If the complex contains two or more metals, the ligand which is attached with metal atom is called bridge group and separated by suffix (−) and suffix (μ)

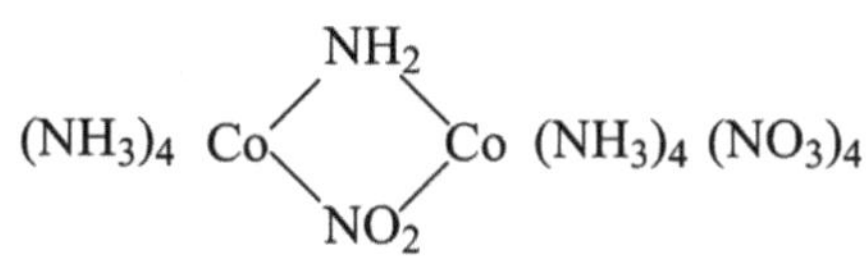

Octa ammine − μ – amido − μ – nitro dicobalt (III) nitrate.

14.5 Valence Bond Theory in Coordination Compound

According to Pauling, the features of the theory are

1. Central metal in complex has a number of empty orbitals to form coordinate bonds, which are called coordination number.
2. The empty orbital available in metal hybridize to give quantized orbitals of equal number; these orbitals describe the geometry of complex as

   sp hybridized orbital imparts linear geometry
   sp^2 hybridized orbital imparts trigonal geometry and
   sp^3 hybridized orbital imparts tetrahedral geometry, etc.

3. Each ligand has at least one lone pair of electron.
4. The empty hybrid orbital overlaps with lone pair sigma orbital to form ligand-metal coordinate bond.
5. The central metal's d-orbitals should be inner $(n-1)d$ or outer nd to give low spin and high spin complexes, respectively.
6. If complex possesses unpaired electrons, then it exhibits paramagnetic property; otherwise, it shows diamagnetic property.

14.6 Applications of Valence Bond Theory for Complexes

1. ***Square planer complex:*** For square planer complexes, the coordination number of central metal is 4 which gives dsp^2 hybridization. The geometry of such complex will be planar. For example, $(Ni(CN)_4)^{2-}$.

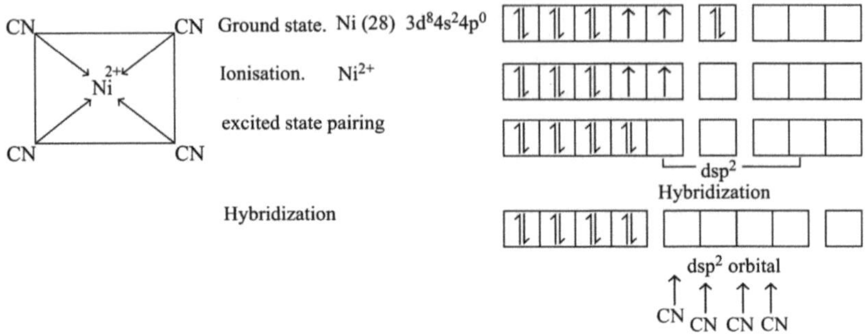

Due to CN^- the pairing of electrons in 3*d*, shell is forced and the complex is diamagnetic as there is no unpaired electron

2. ***Tetrahedral complex:*** They also have coordination number 4, but the hybridization is sp^3. This type of hybridization is shown in $[Ni(CO)_4]$, $(ZnCl_4)^{2-}$, $[MnCl_4]^{2-}$ etc.

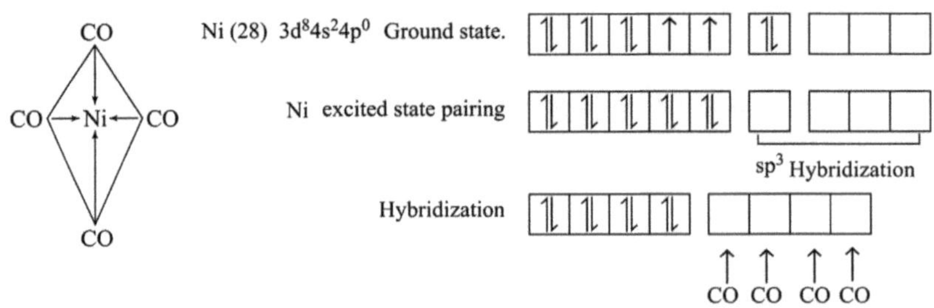

3. ***Octahedral complex:*** The coordination number of central metal in these complexes is 6 having d^2sp^3 hybridization. The complexes formed, if they have inner *d* orbitals, are called low spin complexes and if they have outer *d* orbitals are called high spin complexes. The inner *d* orbitals are diamagnetic or less paramagnetic in nature; hence, they are called low spin complexes with less magnetism. While outer *d* orbitals will be paramagnetic to have high magnetism; hence, they are called high spin complexes. For example,

   Low spin complexes – $[Fe(CN)_6]^{3-}$, $[Cr(NH_3)]^{3+}$, $[Cr(H_2O)_6]^{3+}$ etc.

   High spin complexes – $(CoF_3)^3$, $[Ni(NH_3)]^{2+}$, $[Cu(NH_3)_6]^{3+}$, $[Fe(H_2O)_6]^{3+}$

1. $[Fe(CN)_6]^{3-}$ (Low spin complex)

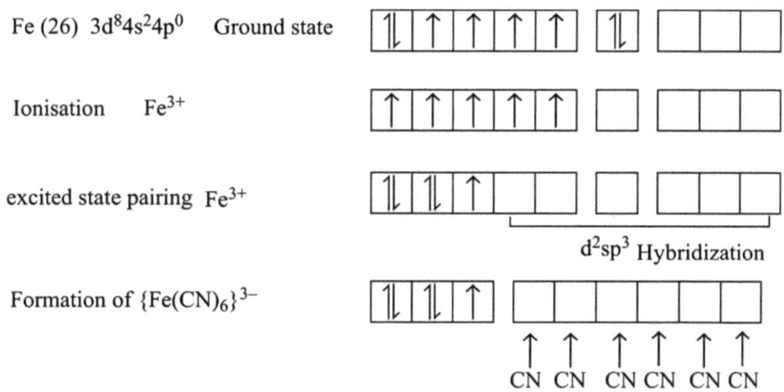

The complex is having only one unpaired electron; hence, it will be low spin complexes.

2. $(CoF_3)^{3-}$ (high spin complex)

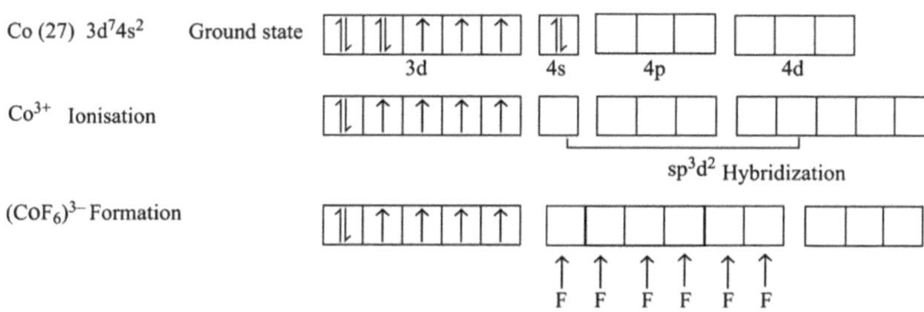

Due to the involvement of outer *d* orbitals, the pairing of inner *d* orbital electrons will not take place, due to which there are large number of unpaired electrons to give high spin complexes.

14.7 Limitations of Valence Bond Theory

1. The theory is unable to explain the color of complexes.
2. The theory is unable to clearly explain low and high spin complexes.
3. The theory is unable to explain as to why some complexes are more labile (ligand can be replaced by other strong ligand) than others and some complexes are inert (ligand cannot be replaced by other ligands).

14.8 Crystal Field Theory

The theory suggests the bond between metal and ligand as purely electrostatic, i.e., 100% ionic. This theory was given by H. Bethe and Vleck in 1929 to explain the limitations of valence bond theory like colors and magnetic properties of crystalline solids. That is why the theory is known as "crystal field theory"? The main postulates of the theory are as follows:

1. The central metal is surrounded by ligands which contain one or more lone pairs of electrons.
2. The ionic ligands are point charge $\left(CN^-, Cl^-, F^-\right)$ and neutral ligands are dipole (H_2O as $H^+{-}OH^-$). For neutral ligand, the −ve end will be toward metal cation.
3. The metal-ligand bond will be purely ionic.

14.9 Explanation of Color and High Spin/Low Spin Properties of Complexes

In octahedral complexes, the metal ions have five degenerated orbitals, and the six ligands are placed around the octahedron.

The two ligands are approached by the axis, and the electrons of d orbital metal cations are repelled by −ve point charge or −ve end of dipole of ligands. The energy of all d-orbitals will be increased by this repulsion. If all the ligands approaching the central metals are at an equal distance from each of the d orbitals, the increase in energy of each d orbital will be same. But hypothetically the lobs of dz^2 and $dx^2 - y^2$ orbitals lie directly in the path of ligands to increase their energy more than dxy, dyz and dzx orbitals.

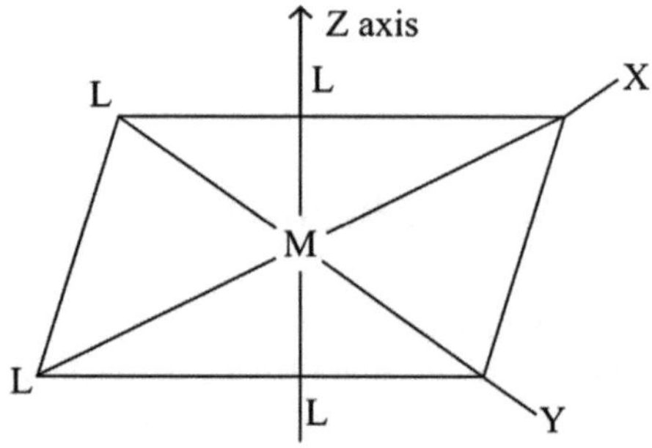

Hence, the five d orbitals split as e_g $\left(dz^2 \text{ and } dx^2 - y^2\right)$ and t_{2g} $(dxy, dyz$ and $dzx)$ where e_g has high energy then t_{2g}. This splitting is due to the ligand interaction with metal called crystal field splitting.

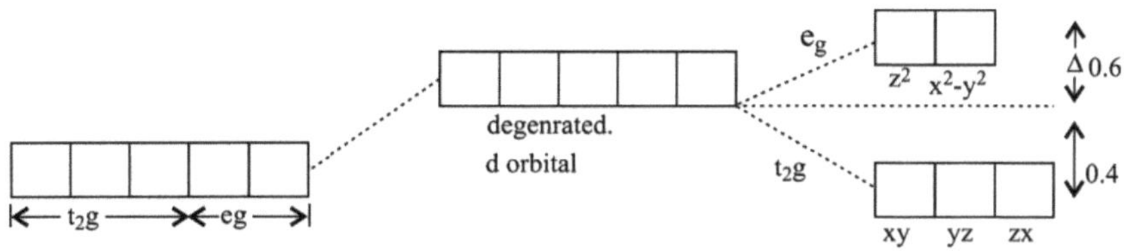

In octahedral complexes with weak ligands, the energy difference Δ_0 between t_{2g} and e_g is relatively small, and hence the energy of five d orbital is approximately the same, and Hund's rule is applicable to all five d orbitals, due to which pairing of electrons is less probable. This gives greater number of unpaired electrons. So weak field complexes are high spin or spin-free complexes (paramagnetic).

In the opposite way, complexes with strong ligands have splitting of d orbitals to t_{2g} and e_g causing high possibility of pairing of electrons, due to which the number of unpaired electrons will be less. So high field complexes are low spin or spin paired (diamagnetic) complexes.

14.10 Explanation of Coloring in Complexes

The complexes having one or more unpaired electrons in d orbital are colored. It is due to splitting of fire d orbitals to t_{2g} and e_g due to ligand interaction. When white light falls on such complexes, the unpaired electrons of t_{2g} absorb the energy equal to Δ_0 and excited to e_g to $d - d$ transition. When this absorbed energy is released back with different wavelength, it imparts color to complexes.

Exercises

1. What are the reasons for the formation of complexes by transition metal cations.
2. Explain Werner's coordination theory.
3. Give the differences between double salts and coordination compounds.
4. Which complex has larger crystal field splitting in the following pairs?

   (a) $[Co(CN)_6]^{3-}$ or $\left[Co\,(NH_3)_6\right]^{3-}$
   (b) $\left[Co\,(H_2O)_6\right]^{2+}$ or $\left[Co\,(H_2O)_6\right]^{3+}$
   (c) $\left[Co\,(NH_3)_6\right]^{3+}$ or $\left[Rh\,(NH_3)_6\right]^{3+}$

5. What are chelating complexes? Explain.
6. Write the factors that determine the stability of a complex?
7. Write the IUPAC names of the following:

   (a) $K_4\,[Fe(CN)_6]$

(b) $[Ag(NH_3)_2]Cl$
(c) $[Cr(en)_3]Cl_3$
(d) $[CoCl(NH_3)_5]Cl_2$
(e)

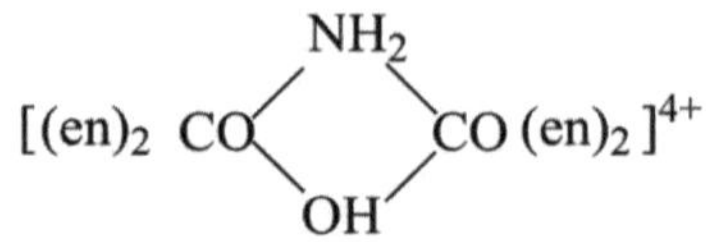

(f) $K(PtCl_3NH_3)$
(g) $[Fe(H_2O)_6]SO_4$
(h) $Na_2[CrF_4]$
(i) $K_3[Fe(CN_5)NO]$
(j) $Na_3[Ag(S_2O_3)_2]$
(k) $[Au(CN)_4]^-$
(l) $[Cr(NH_3)_6)]Cl_3$
(m) $[Cu(NH_3)_4]^{2+}$
(n) $[Pt(NH_3)_4Cl_2]^{+2}$
(o) $Na_2[CO(H_2O)(OH)_4]$
(p) $Pt[(Cl_4)(NH_3)_2]$
(q) $[Zn(SCN)_4]^{2+}$
(r)

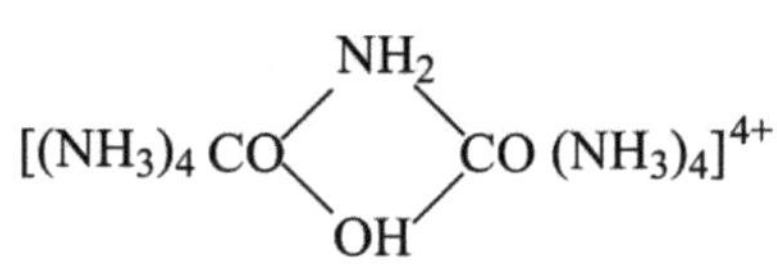

(s) $(NH_4)_3[Co(C_2O_4)_2]$
(t) $[Co(en)(NH_3)_2ClBr]NO_3$

8. (a) How does crystal field theory explain the magnetic character of coordination complexes. Give the number of unpaired electrons in the following octahedral complexes:

   (i) Cr (III) low spin
   (ii) Ni (III) low spin

   (b) Discuss the factors which affect the magnitude of Δ_0 or 10 Dq.
9. Write the names of following complexes according to IUPAC system:

   (a) $[Fe(NH_3)_3Cl_3)]$
   (b) $K_3[Al(C_2O_4)_3]$
   (c) $K_4[Ni(CN)_4]$
   (d) $NH_4[Cr(NH_3)_2(NCS)_4]$

(e) $[Pt(NH_3)_4][PtCl_4]$
(f) $[Co(en)_3Cl_3]$

10. Explain crystal splitting in octahedral complexes.
11. We have three complex compounds, namely, $[Cr(H_2O)_6]Cl_3$, $[Cr(H_2O)_5Cl]Cl_2.H_2O$ and $[Cr(H_2O)_4Cl_2]Cl.2H_2O$. Answer the following questions (i) What is the oxidation number of Cr– atom in each complex compound (ii) Write down the coordination number of the central Cr-atom in each case (iii) Write the name of each complex.
12. Explain the geometry and magnetic behavior of the following complexes on the basis of valence bond theory $[FeF_6]^{-3}$ and $K_4[Fe(CN)_6]$.
13. Explain why complexes of Co(III) like $[Co(NH_3)_6]^{3+}$, $[Co(en)_3]^{3+}$ and $[Co(NO_2)_6]^{3-}$ are orange yellow in color while $[CoF_6]^{3-}$ and $[Co(H_2O)_6]^{3+}$ are blue in color.
14. On the basis of crystal field theory, write the number of unpaired electrons for the following complex ions:

(a) $[Fe(CN)_6]^{6-}$
(b) $[Fe(CN)_6]^{3-}$

Note: [The correct names of some common ligands are given below, and students are expected to learn them]

Cl^-	Chloro	NH_2^-	Amido	$S_2O_3^{2-}$	Thiosulphatc
Br^-	Bromo	N_3^-	Azide	NO_3^-	Nitrato
I^-	Iodo	ONO^-	Nitrito	O_2^-	Oxo
OH^-	Hydroxo	CN^-	Cyano	O_2^{2-}	Peroxo
CN^-	Cyano⁻	NCS^-	Isothiocyanato	NC^-	Isocyano
SO_2^{4-}	Sulphato	SCN^-	Thiocyanato	SO_3^{2-}	Sulphito
CO_3^{2-}	Carbonato	NO_2^-	Nitro		
H_2O	Aquo	NO	Nitrosyl	CH_2NH_2	
				| Ethylenediamine (en)	
NH_3	Ammine	CO	Carbonyl	CH_2NH_2	
NO^+	Nitrosonium	$C_5H_5N^+$	Pyridinium		
H_3O^+	Hydronium	$NH_2NH_3^+$	Hydrazinium		

Chapter 15
Quantitative Analysis

Quantitative analysis involves the measurement of substances in terms of volume or weight. Accordingly, it is known as volumetric analysis or gravimetric analysis.

15.1 Volumetric Analysis

It is a method to determine the concentration of solutions, involving accurate measurements of volumes of liquid.

15.2 Gravimetric Analysis

It involves accurate measurements of weights by separating the constituents as precipitate.

15.2.1 Volumetric Analysis Involves the Following Terms

- **Titrate:** The reagent of known concentration which is taken in burette and its concentration is known.
- **Titrant:** The unknown solution whose concentration is to be determined by titration. It is taken in conical flask.
- **Titration:** It is the process of reaction of the titrant with titrate to get an idea of completion of reaction.
- **Indicator:** It is a reagent which helps to detect the completion of end point by change of color at the completion of reaction.

© The Author(s), under exclusive license to Springer Nature Switzerland AG 2026

R. N. Goyal, H. Goel, *Textbook of Engineering Chemistry*,
https://doi.org/10.1007/978-3-032-22068-4_15

- **End point:** The point at which addition of last drop of titrate causes change in color is called end point.
- **Equivalence point:** The point at which the amount of reagent and unknown substance being determined is chemically equivalent.
- **Titration error:** For an ideal titration, the end point and equivalence point should be equal, but for change of color of indicator, there is a difference between end point and equivalence point, i.e., an indicator changes its color a little before or after the equivalence point. It is called as titration error. To minimize this error, the selection of indicator is very important, which can give a sharp change in color at the end point.
- **Standard solution:** A solution containing a known weight of solute in a constant volume of solvent/solution is called standard solution. The concentration or strength of solution is defined in different ways :

(1) **In physical units**

(i) **g/litre:** Strength or concentration is defined as grams of solute per liter of solution.

(ii) **ppm:** The grams of solutes in 10^6 milliliters of solution or mg of solute per liter of solution is defined.

(iii) **% volume:** If solute is liquid then ml of solute per 100 ml of solutions is reported.

(iv) **% weight:** If solute is solid, then grams of solute in 100 g of solution are reported

(2) **In chemical units**

(i) **Normality:** It is the number of gram equivalents of solute in one liter of solution; mathematically, it can be represented as

$$\text{Normality} \times \text{Equivalent weight} = \text{g/litre}.$$

(ii) **Molarity:** It is the number of gram molecules of solute in 1 liter of solution, i.e.,

$$\text{Molarity} \times \text{Equivalent weight} = \text{g/litre}$$

(iii) **Molality:** It is the number of gram molecules of solute in 1000 grams of solvent, i.e.,

$$\text{Molality} \times \text{Molecular weight} = \text{g/1000 g of solvent}$$

15.3 Classification of Volumetric Analysis

The volumetric analysis can be classified into the following types:

(i) **Neutralization titrations:** When acid and base are titrated to give feebly ionized water molecules, then the titration is called neutralization titration, for example, strong acid-strong base titrations.
(ii) **Oxidation-reduction titrations:** When the chemical reactions proceed with transfer of electrons, then titration is called oxidation reduction titration. For example, titration of ferrous sulfate to ferric sulfate by reduction of $KMnO_4$ in acid medium.

$$MnO_4^- + 8H^+ + 5e^- \rightleftharpoons Mn^{2+} + 4H_2O \qquad \text{Reduction}$$

$$\left\{Fe^{2+} \rightleftharpoons Fe^{3+} + e^-\right\} \times 5 \qquad \text{Oxidation}$$

$$MnO_4^- + 8H^+ + 5Fe^{2+} \rightleftharpoons Mn^{2+} + 5Fe^{3+} + 4H_2O$$

These titrations are also called redox titrations.
(iii) **Precipitation titrations:** When titrate and titrant react in such a way that the precipitates are formed, then the titration is called precipitation titration, for example, titration of $AgNO_3$ with Cl^- ions.
(iv) **Complex formation titration:** When titrate and titrant react to give complex compound, for example,

$$Ca^{2+} + \text{EDTA} \longrightarrow \text{Ca(EDTA) complex}$$

$$Mg^{2+} + \text{EDTA} \longrightarrow \text{Mg(EDTA) complex}$$

15.4 Calculation of Equivalent Weight for Different Compounds

Equivalent weight of a substance is the number of parts by weight of substance, which combines with or displaces eight parts of oxygen by weight or 1.0078 part of hydrogen by weight. When equivalent weight is expressed in grams, then it is called gram equivalent weight.

Equivalent weight is not a constant quantity. A compound may have different equivalent weights in different experimental conditions. It may be explained in following cases.

(i) Gram equivalent weight of an acid

$$\frac{\text{Mol. wt. of Acid}}{\text{Basicity (Number of replaceable H)}}$$

(ii) Gram equivalent weight of base

$$\frac{\text{Mol. wt. of Base}}{\text{Acidity (Number of replaceable (OH) group)}}$$

(iii) Equivalent weight of a salt

$$\frac{\text{Mol. wt. of Salt}}{\text{Number of electrons exchange in salt}}$$

For example, equivalent weight of

$$Na_2CO_3 = \frac{\text{Mol. wt. of } Na_2CO_3}{2} = \frac{106}{2} = 53.0$$

(iv) Equivalent weight in oxidation reduction reactions

(A) Oxygen concept

(a) **Eq. wt. of** $K_2Cr_2O_7$**:** In acidic medium $K_2Cr_2O_7$ supplies oxygen according to the equation

$$K_2Cr_2O_7 + 4H_2SO_4 = K_2SO_4 + Cr_2\,(SO_4)_3 + 4H_2O + 3O$$

Thus, one mole of $K_2Cr_2O_7$ gives three atoms of oxygen. Hence, the eq. weight of $K_2Cr_2O_7$ corresponding to 8.0 gm of oxygen will be $K_2Cr_2O_7 \equiv 3O$

$$\frac{294.2}{48} \times 8 = \frac{294.2}{6} = 49.03 \quad 294.4 \equiv 3(16)$$

(b) **Eq. wt. of** $KMnO_4$**:** In acidic, neutral, and basic media, $KMnO_4$ supplies oxygen hence, so it has different equivalent weight in each medium.

(i) in acidic medium

$$2KMnO_4 + 3H_2SO_4 \equiv K_2SO_4 + 2MnSO_4 + 3H_2O + 5O$$

$$2KMnO_4 \equiv 5\{O\}$$

$$2 \times 158 \equiv 5 \times 16$$

The eq. wt. of $KMnO_4$ corresponding to 8.0 grams of oxygen will be

$$\frac{2 \times 158}{5 \times 16} \times 8 = \frac{158}{5} = 31.6$$

(ii) In neutral medium

$$2KMnO_4 + H_2O = 2MnO_2 + 2KOH + 3[O]$$
$$2KMnO_4 \equiv 3O$$
$$2 \times 158 \equiv 3 \times 16$$

The eq. wt. of $KMnO_4$ to 8.0 grams of oxygen will be

$$\frac{2 \times 158}{3 \times 16} \times 8 = 52.66$$

(iii) In alkaline medium

$$2KMnO_4 + 2KOH = 2\ K_2MnO_4 + H_2O + O$$
$$2KMnO_4 \equiv [O]$$
$$2 \times 158 = 16$$

The eq. wt. of $KMnO_4$ in alkaline medium

$$\frac{2 \times 158}{16} \times 8 = 158$$

(B) Electron concept

(a) eq. wt. of $K_2Cr_2O_7$: In acid medium, the ionic equation for dichromate ion is

$$Cr_2O_7^{2-} + 14H^+ + 6e^- \rightleftharpoons 2Cr^{3+} + 7H_2O$$

therefore

$$\text{eq. wt. of } K_2Cr_2O_7 = \frac{\text{Mol. wt. of } K_2Cr_2O_7}{6}$$
$$= \frac{294.2}{6} = 49.03$$

(b) eq. wt. of $KMnO_4$: In acidic medium, the ionic equation of permanganate ion is

$$MnO_4^- + 8H^+ + 5e^- \rightleftharpoons Mn^{2+} + 4H_2O$$

therefore

$$\text{eq. wt. of } KMnO_4 = \frac{\text{Mol. wt. of } KMnO_4}{5} = \frac{158}{5} = 31.6$$

(C) Oxidation concept

$$\text{eq. wt.} = \frac{\text{Mol. wt.}}{\text{Change in oxidation number/molecule}}$$

(a) eq. wt. of $KMnO_4$: In acidic medium, $KMnO_4$ converts to $MnSO_4$, in which the oxidation state of Mn in $KMnO_4$ is "+7" and in $MnSO_4$ is "+2". Thus, change in oxidation state is "+5"

$$\text{eq. wt. of } KMnO_4 = \frac{158}{5} = 31.6$$

(b) eq. wt of $K_2Cr_2O_7$: In acidic medium, $K_2Cr_2O_7$, reduce to $Cr_2\ (SO_4)_3$, in which OS of Cr in $K_2Cr_2O_7$ is "+12" and in $Cr_2\ (SO_4)_3$ is +6. Thus, change in oxidation state is "+6".

$$\text{eg. wt. of } K_2Cr_2O_7 = \frac{294.2}{6} = 49.03$$

15.5 Equivalent Weight of Reducing Agent

(A) Available Hydrogen Concept: The gram equivalent weight of a reducing agent is that weight of it which contains 1.0078 gm of available hydrogen or which reacts with 8.0 gm of available oxygen.

(i) eq. wt. of ferrous ammonium.
Sulfate ($FeSO_4\ (NH_4)_2\ SO_4 \cdot 6H_2O$)

$$2FeSO_4 + H_2SO_4 + O \longrightarrow Fe_2\ (SO_4)_3 + H_2O$$

$$2\ (FeSO_4) \equiv O$$

$$2 \times 392 = 16$$

$$\text{eq. wt. of ferrous ammonium sulfate} = \tfrac{2\times392}{16} \times 8 = 392.0$$

(ii) eq. wt. of oxalic acid

$$C_2O_4H_2 + O \longrightarrow 2CO_2 + H_2O$$
$$(C_2O_4H_2) \equiv O$$
$$126 \equiv 16$$
$$\text{eq. wt. of oxalic acid} = \frac{126}{16} \times 8 = 63.0$$

(B) Electron Concept

(i) Eq. wt. of ferrous ammonium sulfate: The ferrous ion oxidizes to ferric ion as

$$Fe^{2+} \longrightarrow Fe^{3+} + e^-$$

Hence, eq. wt. of $FeSO_4 = \frac{\text{Mol. wt.}}{1} = \frac{392}{1} = 392$

(ii) Eq. wt. of oxalic acid: The oxalate ion oxidized to CO_2 as

$$C_2O_4^{-2} \longrightarrow 2CO_2 + 2e^-$$

Hence, eq. wt. of oxalic acid $= \frac{\text{Mol. wt.}}{2} = \frac{126}{2} = 63.$

(C) Oxidation number concept

$$\text{eq. wt.} = \frac{\text{Mol. wt.}}{\text{Change in oxidation number per molecule}}$$

(i) Eq. wt. of ferrous ammonium sulfate: The oxidation state of iron in ferrous ion is +2 but in ferric ion is +3. Thus, oxidation of ferrous to ferric will cause change in oxidation number by +1, hence,

$$\text{eq. wt. of ferrous. Amm sulfate} = \frac{392}{1} = 392$$

(ii) Eq. wt. of Oxalic Acid: The oxidation state of carbon atom in oxalic acid is +6 and in $2CO_2$ is +8

$$H_2C_2^{+6}O_4 \longrightarrow 2C^{+4}O_2$$

Hence, the change is +2

$$\text{and eq. wt.} = \frac{\text{Mol. wt}}{2} = \frac{126}{2} = 63$$

15.6 Acid-Base Titrations

When an acid is treated with base, then salt and water are formed. The end point of titration is being obtained by change of color of indicator. These indicators are called pH indicators because they are pH sensitive and change its color with pH.

15.7 Theory of pH Indicators

15.7.1 Ostwald Theory

pH indicators are weak acid or base and their undissociated form has a different color than dissociated form.

For example, phenolphthalein (HP) is a weak acid and dissociates as

$$\underset{\text{Colourless}}{HP} \rightleftharpoons H^+ + \underset{\text{Red}}{P^-}$$

The undissociated form is colorless and dissociated form is red in color. According to law of mass action

$$K_{in} = \frac{[H^+][P^-]}{[HP]}$$

where K_{in} is ionization constant of indicator.

$$[P^-] = \frac{K_{in}[HP]}{[H^+]}$$

This equation shows that $[P^-]$ is inversely proportional to H^+ ion concentration. Hence, phenolphthalein in an acid medium will not dissociate; hence, no color change will occur. However, in basic medium $[P^-]$, ions predominate to give red color.

Similarly, methyl orange is a weak base dissociated as

$$\underset{\text{Yellow}}{MeOH} \rightleftharpoons Me^+ + \underset{\text{Red}}{OH^-}$$

MeOH represents methyl orange. The undissociated form is yellow in color, whereas on losing OH^-, the color becomes red.

Applying Law of Mass Action

$$K_{in} = \frac{[Me^+][OH^-]}{[MeOH]}$$

$$[H^+] \times [OH^-] = [H_2O] = K_w \text{ Ionic product of water}$$

therefore

$$[Me^+] = \frac{K_{in}}{K_w} \times [MeOH][H^+]$$

$[Me^+]$ is proportional to H^+ ion concentration. When methyl orange is added to alkali solution, then MeOH will not dissociate properly and remains as yellow colored. On adding to acid dissociated Me^+, ion predominates to impart red color.

15.7.2 Quinonoid or Structural Theory

Ostwald's theory gives explanation for appearance of color of indicator in acidic or alkaline medium, due to loss of H^+or OH^-, but it is unable to explain the color change of indicator. According to Quinonoid theory, the color change of an indicator is due to structural changes with pH. The colorless or light color is due to benzenoid structure, while dark color is due to quinonoid structure. It appears that the color arises from the production of quinonoid form and the resonance between two or more structures.

For example, in phenolphthalein on adding alkali, following structural changes occur.

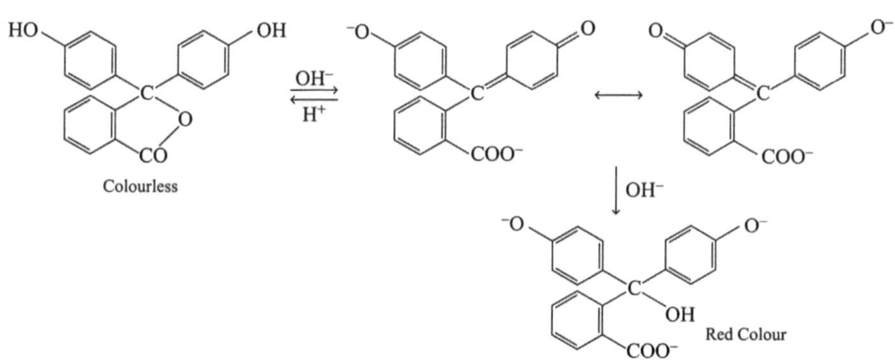

In case of methyl orange, in acidic medium, there is Zwitter ion. The forms are

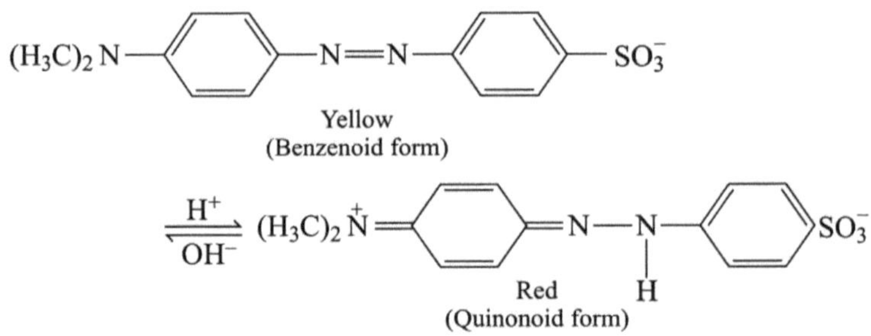

Range of pH Indicators The pH range in which the indicator changes from one color to another color is called range of indicator.

For example, pH range for methyl orange is 3.1–4.4-it means below 3.1 the methyl orange will be red and above 4.4 it will be yellow.

For some common indicators, the pH range is given in table below.

S. No.	Name of indicator	Range of pH	Colour change	
			Acid form	Alkaline form
1	Methyl orange	3.1–4.4	Red	Orange
2	Bromo-Phenol blue	3.0–4.6	Yellow	Blue
3	Methyl red	4.2–6.3	Red	Yellow
4	Litmus (azolitmin)	5.0–8.0	Red	Blue
5	Phenolphthalein	8.3–10.0	Colorless	Red

The indicator for acid-base titration is most suitable when its pH range is included in the pH change of the solution near the end point. The change of pH depends upon the kind of acid or alkali titrated. There are different cases for acid-base titrations.

(a) **Strong acid with strong base:** In such titrations, the pH at equivalence point will be ~ 7 and pH before and after end point is 4 and 10, respectively. Hence, the indicator having color change between pH 4 and 10 can be used. Phenolphthalein (8.3–10), methyl orange (3.1–4.4) and methyl red (4.2–6.3) are suitable for this type of titrations.

(b) **Weak acid and strong base titration:** The pH of equivalence point is about 8.5. It is due to the hydrolysis of the salt formed by reaction of weak acid and strong base. The pH before equivalence point (0.1%) is 7.5 and after end point is 10. Therefore, phenolphthalein (8.3–1.0) is most suitable in these cases.

(c) **Weak base with strong acid:** The pH of equivalence point is about 5.5 which is due to the hydrolysis of the salt formed during titration. The pH before equivalence point (0.1%) is 6.5 and after equivalence point is 4.5. Hence, methyl orange (3.1–4.4) or methyl red (4.2–6.3) is suitable for this type of titration.

(d) **Weak acid with weak base:** The pH of equivalence point is about 7 and pH before and after equivalence point is about 6.5–7.5, respectively. As there in no sharp change of pH, hence no proper indicator can be used for detection of end point.

15.8 Some Solution Preparations for Acid-Base Titrations

(1) **Primary:** The standard solution of solute should have the following properties:

   (i) It should be pure
   (ii) It should not be hygroscopic, oxidizable, etc. and should be stable at ordinary temperature
   (iii) It should be readily soluble
   (iv) It should be solid.

   Some common standards are Na_2CO_3, oxalic acid, borax, potassium hydrogen phthalate, etc.

(2) **Standardization of a solution:** The concentration of solution can be determined by titrating it with a primary standard solution called standardization. For example, standard solution of HCl, H_2SO_4, NaOH, etc. are obtained by suitably titrating them with primary standards.
   for acids: Primary standards are Na_2CO_3, borax, etc.
   for base: Primary standards are oxalic acid, potassium bi-phthalate, etc.

15.9 Preparation of Solutions

(a) **Hydrochloric acid:** Conc. HCl available in laboratory is about 11 N, and as it is liquid, we cannot prepare the solution by weighing. Hence, the solution can be prepared by volumetric dilution of conc. HCl. The conc. of HCl so prepared is not accurately known. Therefore, first an approximate conc. of HCl is prepared, and exact strength is obtained by titrating it against standard Na_2CO_3 solution using methyl orange indicator.

(b) **Sulfuric acid:** The normality of conc. H_2SO_4 is about 36 N. Because the strength of conc. H_2SO_4 is not accurately known, therefore, an approximate solution is prepared and exact strength is obtained by titrating against standard Na_2CO_3 solution using methyl orange indicator.

(c) **Na_2CO_3** (eq. wt. 53): It is a primary standard but may have some moisture; hence, before using, it should be dried at 260–270 °C for an hour and cooled in moisture-free desiccator.

(d) **Oxalic acid:** $C_2O_4H_2\ 2H_2O$ (eq. wt. 63) It is also primary standard.

(e) **Sodium Hydroxide:** (eq. wt. 40).

NaOH is deliquescent have moisture and impurity of Na_2CO_3. Hence, standardization of NaOH is required which is carried out by titrating it with solution of primary standard oxalic acid.

The NaOH is first prepared of slightly high concentration than the desired concentration. For example, weigh 40 g of NaOH pellets in a beaker; add some water and put it for 4–5 hours. Filter the solution to get a clear solution. Now dilute it to 1 liter and now titrate it with standard oxalic acid solution to get its exact strength. The desired strength can be prepared by dilution method.

15.10 Different Types of Titrations

15.10.1 Oxidation-Reduction Titrations

Oxidation is a loss of one or more electrons by a substance, and reduction is the gain of one or more electrons.

In a redox reaction, simultaneous loss and gain of electrons take place among the reacting substances. Such titrations are called redox titrations.

Indicators for Redox Titrations

There are three kinds of indicators commonly used in redox titrations.

1. **Self-indicators:** When one of the reactants in the titration acts as an indicator, then it is called as self-indicator. For example, $KMnO_4$ acts as self-indicator in oxalic acid titration. In the presence of sulfuric acid, $KMnO_4$ remains colorless with oxalic acid. At the end point, in the absence of oxalic acid, $KMnO_4$ gives purple color.

$$5C_2O_4H_2 + 2KMnO_4 + 2H_2SO_4 \longrightarrow K_2SO_4 + 2MnSO_4 + 10CO_2 + 8H_2O$$

   When oxalic acid is completely utilized, the permanganate pink color indicates end point.
2. **External indicators:** When in a redox titration the indicator has not been added in the titration vessel but used from outside, then indicator is called external indicator. For example, titration of ferrous salt with $K_2Cr_2O_7$ in H_2SO_4 medium. In this titration, potassium ferricyanide is used as an external indicator. The $K_3\,[Fe(CN)_6]$ is not added to titration cell due to the reason that Fe^{2+} ions react with it to produce deep blue permanent complex.

$$Fe^{2+} + [Fe(CN)_6]^{3-} \longrightarrow \left[Fe^{2+}Fe^{3+}(CN)_6\right]^{-}$$

Ferro ferricyanide complex

Prussian Blue

To use such indicators, drops of the indicator are placed on glazed tile. The drops of the titrating mixture are then touched with the indicator using a glass rod. The end point is detected when the indicator does not change its color.

3. **Internal indicators or redox indicators**: These indicators can undergo oxidation and reduction reversibly, and both forms have different colors. For example,

$$\left[Fe\,(C_{12}H_8\,N_2)_3\right]^{+++} + e^- \longrightarrow \left[Fe\,(C_{12}H_8\,N_2)_3\right]^{++}$$

oxidized ferroin Reduced ferroin

(Pale blue) (Red)

Preparation of Solutions for Redox Titration

1. **$KMnO_4$ Solution:** For $N/10$ solution, weigh approximately 3.2 g and dissolve it in 150 ml of distilled water in a beaker. Heat the solution to boil and heat for 2 hours on steam bath. Cool the solution and filter and dilute it to 1 liter. Allow to stand and again filter through glass wool. Now determine its exact strength by titration against standard ferrous ammonium sulfate solution.
2. $K_2Cr_2O_7$ **(eq. wt. 49):** It is a primary standard and hence can be prepared by weighing requisite amount and dissolving it in water.
3. **Ferrous ammonium sulfate,** $FeSO_4\,(NH_4)_2\,SO_4 \cdot 6H_2O$ **(eq. wt. 392):** It is a primary standard compound, and its solution can be prepared by weighing exact amount.
4. **Oxalic acid** ($C_2O_4H_2 \cdot 2H_2O$ **eq. wt. 63):** It is a mild reducing agent in acid medium. It is a primary standard and can be prepared by weighing the required amount.

15.10.2 Iodimetry and Iodometry Titration

Iodimetry

The determination of reducing agents by direct titration with standard iodine solution is called iodimetry. Reducing agents like sulfite, thiosulfate, arsenate, etc. can oxidized easily by iodine solution. For example,

$$As_2O_3 + 2I_2 + 2H_2O \longrightarrow As_2O_5 + 4HI$$

Iodometry

The indirect determination of oxidizing agent by titration of evolved iodine with standard sodium thiosulfate solution is called iodometry. Oxidizing agents like $K_2Cr_2O_7$, $KMnO_4$, $CuSO_4$, etc. are reduced with KI, and equivalent amount of iodine is liberated.

$$CuSO_4 + \underset{\text{excess}}{4KI} \longrightarrow 2CuI + 2\,K_2SO_4 + I_2$$

$$I_2 + \underset{\text{Hypo}}{2Na_2\,S_2O_3} \longrightarrow \underset{\text{Na-tetrathionate.}}{Na_2\,S_4O_6} + 2NaI$$

Preparation of Solution for Iodometry Titration

Sodium Thiosulfate $Na_2\,S_2O_3 \cdot 5H_2O$ (eq. wt. 248) is a mild reducing agent. It is efflorescent in nature due to which there is no confirmation regarding the water content in the salt. It is not very stable and decomposes on standing.

For preparing $\frac{N}{10}Na_2\,S_2O_3$, weigh about 6.2 g of hypo and transfer to 250 ml volumetric flask. Add 50 ml of water, and dissolve the hypo, and then make up the flask up to the mark with water. The solution is not exactly 0.1 N; hence, titrate it with standard $K_2Cr_2O_7$ or iodine solution.

If the solution is to be stored for long time, add 1 ml of $CHCl_3$/liter.

Soluble Starch Indicator In iodometry, starch is a good indicator. Starch reacts with I_2 in the presence of KI to form blue colored starch $-I_2$ adsorption complex. In the presence of KI, I_2 molecules are adsorbed by colloidal macromolecules of starch to produce blue color. Starch consists of two fractions α-amylose and β-amylose. Out of these α-amylose is water soluble and gives blue color with iodine.

Preparation of 1% Starch Solution Take 10 g of soluble starch and 5 g of KI. Make a paste with water. Add the paste drop by drop to 1 liter of boiling water to get a clear solution.

Argentometric Titration It is a type of precipitation titration in which $AgNO_3$ is used as titrant. This titration is used for anions which are sensitive with silver ions like halides, cyanide, thiocyanate etc.

$$AgNO_3 + NaCl \longrightarrow AgCl + NaNO_3$$

15.11 Reagents of Argentometry

1. Silver nitrate ($AgNO_3$ eq. wt. 169.9) Silver nitrate is a primary standard and can be prepared directly by dissolving a weighed amount in chlorine-free distilled water.

2. Chloride solutions:

   (a) NaCl (eq. wt. 58.45)—it is a primary standard but should be used after drying.
   (b) KCl (eq. wt. 74.55)—it is primary standard but should be used after drying.

3. Indicators for argentometry titration

   (a) K_2CrO_4 is used as an indicator in chloride and silver nitrate titration as silver nitrate is sensitive largely toward chloride ion precipitation. As all the chloride ions are converted to AgCl, then at the end point, silver nitrate reacts with K_2CrO_4 to give red precipitate of Ag_2CrO_4.

$$NaCl + AgNO_3 \longrightarrow AgCl + NaNO_3$$

$$2AgNO_3 + K_2CrO_4 \longrightarrow Ag_2CrO_4 + 2KNO_3$$

   (b) Ferric alum is used as an internal indicator in $AgNO_3$ and KSCN titration. On adding potassium thiocyanate (KSCN) in acidic medium, silver ion is removed as silver thiocyanate (precipitate), and after removal of all silver ions, excess of KSCN gives red color due to the formation of ferric thiocyanate complex.

$$AgNO_3 + KSCN = AgSCN + KNO_3$$

$$Fe^{3+} + SCN^- \rightleftharpoons \underset{\text{Red colour}}{[Fe(SCN)]^{2+}}$$

Chapter 16
Phase Rule

To predict quantitatively the effect of change of pressure, temperature, and concentration on a heterogeneous system in equilibrium. Willard Gibbs, an American physicist, in 1874 put forward a rule according to which:

> *"In a heterogeneous system in equilibrium the number of degrees of freedom plus the number of phases is equal to the number of component plus 2." This relationship is called phase rule.*

Mathematically, the rule can be expressed as

$$F + P = C + 2$$

where F denotes the number of degree of freedom; C, the number of components; and P, the number of phases present in the system.

16.1 Phase

"*The homogeneous but physically distinct and mechanically separable parts of the heterogenous system are called* ***phases.***"
For example

(i) Air which is a mixture of nitrogen, oxygen, carbon dioxide, water vapor, etc. constitutes a single phase only. It is not possible to mechanically separate them.
(ii) Solution of a substance in a solvent constitutes a homogeneous system. It consists of one phase only, for example, solution of sugar in water or solution of common salt in water.
(iii) If two liquids are completely miscible such as alcohol and water, it forms only one phase. If the liquids are immiscible such as benzene and water, it forms two distinct liquid phases.

© The Author(s), under exclusive license to Springer Nature Switzerland AG 2026

R. N. Goyal, H. Goel, *Textbook of Engineering Chemistry*,
https://doi.org/10.1007/978-3-032-22068-4_16

(iv) A solid mixture consists of as many phases as are solids present.
(v) For a reaction at equilibrium

$$\underset{\text{solid}}{\mathrm{CaCO_3}} \rightleftharpoons \underset{\text{solid}}{\mathrm{CaO}} + \underset{\text{gas}}{\mathrm{CO_2}}$$

has three phases (i.e., two solids and one gas)
(vi) Freezing water consists of three phases.

$$\underset{\text{solid}}{\text{Ice}} \rightleftharpoons \underset{\text{liquid}}{\text{Water}} \rightleftharpoons \underset{\text{gas}}{\text{Vapour}}$$

16.2 Component

The number of component 'C' is defined as "*The smallest number of independently variable constituents by means of which the composition of each phase present in the system can be expressed directly or in the form of an equation.*" To make it more clear, let us consider some examples.

Example 16.1

(i) If a water system consists of three phases

$$\underset{\text{solid}}{\text{Ice}} \rightleftharpoons \underset{\text{liquid}}{\text{Water}} \rightleftharpoons \underset{\text{gas}}{\text{Vapour}}$$

the number of component is one as each phase is essentially same, i.e., a different physical form of the same chemical compound H_2O.
(ii) The sulfur system consisting of four phases—rhombic, monoclinic, liquid, and vapor—is also a *one* component system. The reason is that each phase is simply different physical forms of sulfur. You must remember that if a system consists of only one chemical individual, it is a one component system.
(iii) If dissociation of ammonium chloride takes place according to the reaction in a closed vessel.

$$\underset{\text{solid}}{\mathrm{NH_4Cl}} \rightleftharpoons \underset{\text{gas}}{\mathrm{NH_4Cl}} \rightleftharpoons \underset{\text{gas}}{\mathrm{NH_3}} + \underset{\text{gas}}{\mathrm{HCl}}$$

The number of phases in the reaction at equilibrium is solid and gas (two phases). As both the phases can be built up from NH_4Cl alone, the number of components is one. In case the amount of NH_3 and HCl is not chemically equivalent, then the number of components would be different.
(iv) For dissociation of $CaCO_3$

$$\underset{\text{solid}}{\mathrm{CaCO_3}} \overset{\Delta}{\rightleftharpoons} \underset{\text{solid}}{\mathrm{CaO}} \rightleftharpoons \underset{\text{gas}}{\mathrm{CO_2}}$$

None of the three constituents is able to represent the composition of all the phases. However, two of the constituents are able to represent and hence it is a two component system. For example, if we consider CaO and CO_2 as the components, all these constituents can be represented.

$$\text{Phase} \quad CaO + CO_2 \longrightarrow CaCO_3$$
$$\text{Phase} \quad CaO + oCO_2 \longrightarrow CaO$$
$$\text{Phaseo} \quad oCO + CO_2 \longrightarrow CO_2$$

16.3 Degree of Freedom

The minimum number of independent variable factors pressure, temperature, and composition of the phase which must be arbitrarily fixed in order that the conditions of the system may be completely defined is known as degree of freedom. For example,

(i) For a water system

$$\underset{\text{solid}}{\text{Ice}} \rightleftharpoons \underset{\text{liquid}}{\text{Water}} \rightleftharpoons \underset{\text{gas}}{\text{Vapour}}$$

the three phases solid, liquid, and gas may exist in equilibrium only at one definite temperature and at one definite pressure. The system, thus, has *no* degree of freedom (invariant system) because if temperature or pressure is changed, the three phases of water will not remain in equilibrium and one of the phases will disappear.

(ii) Consider a system in which water is in contact with its vapors

$$\underset{\text{liquid}}{\text{Water}} \rightleftharpoons \underset{\text{gas}}{\text{Water vapor}}$$

For such a system, there is one degree of freedom (univariant system). If temperature is kept constant, the vapor pressure is automatically fixed, because pressure depends upon temperature.

(iii) For a gas if pressure and temperature are fixed, it will always occupy a definite volume. Thus, for completely defining a gaseous system, two factors have to be fixed up and hence have *two* degrees of freedom (bivariant system).

16.4 Heating and Cooling Curves

If we take 1.0 mole of water in the form of ice at $-10\,^\circ$C and warm it until it becomes steam at $110\,^\circ$C, the amount of heat necessary can be determined. Further,

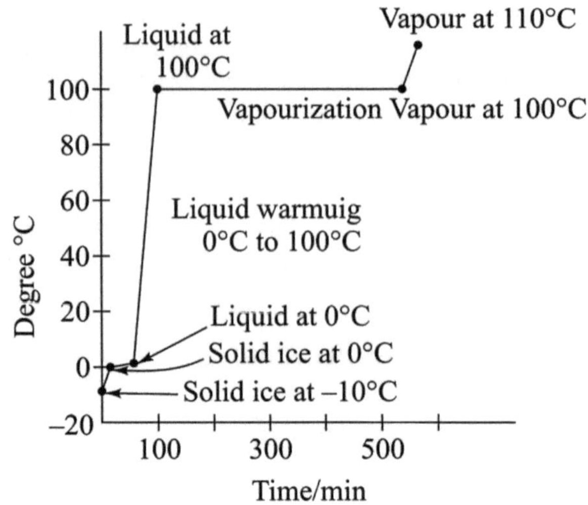

Fig. 16.1 Heating/cooling curve for water. One mole of water is heated from $-10\,°C$ (where it is in the form of ice) to $+110\,°C$. Heating rate is assumed at 100 J/min

if the heat is added at the constant rate of 100 J/min, the time required can also be calculated. Figure 16.1 presents the changes expected on heating 1.0 mole of water from $-10\,°C$ to $110\,°C$ at a constant heating rate of 100 J/min.

The specific heat of ice is 2.09 J/g.K which means raising the temperature of 1.0 mole of ice by 1 degree will require

$$= 1 \text{ mol of ice } \times 2.09$$

$$= 18 \times 2.09 = 37.6\text{ J}$$

37.6 J of heat. Since heat is added at a constant rate of 100 J/min, it will require 0.376 min. For 10 degree rise in temperature, the heat required will be 376 J and the time needed will be 3.76 min. The lowest segment of the graph in Fig. 16.1 represents this fact. When the ice has reached 0 °C, it can be converted to water at 0 °C by further giving heat equal to heat of fusion (6.01 kJ/mol). The temperature of the system will remain zero until all the ice is melted to give water at 0 °C.

The time required for this conversion will be

$$= \frac{6010 \times 1}{100} = 60.1\text{ min}$$

Now, the water at 0 °C can be heated to water at 100 °C.

Using the specific heat of water as 4.18 J g.K, one mole of water will require heat

$$= 1 \text{ mole of water } \times 4.18 \times (100 - 0)$$

$$= 18 \times 4.18 \times 100$$

$$= 7524\text{ J}$$

at the same heating rate, it will require 75.2 min. In the next step, further supply of heat will completely convert water vapors at 100 °C to vapors at 110 °C and will require further 40.7 kJ heat (the molar heat of vaporization) for 1 mole of water. Finally the heat required for converting water vapors at 100 °C to 110 °C will require heat.

$$= 1 \text{ mole of water } \times \text{ specific heat of steam } \times (110 - 100)$$

$$= 18 \times 2.0 \times 10$$

$$= 360\ \text{J}$$

It is clear from Fig. 16.1 that when a solid or liquid is heated, there is rapid rise in temperature up to the point of phase change, and the rise in temperature practically stops at phase change. When the change in phase is completed, the rise in temperature starts again. The rate at which the temperature rises is inversely proportional to the specific heat of the phase, which is smaller for solid and gases in comparison with liquids.

In Fig. 16.1, the energy given in the plateau or phase change region is essentially used to separate molecules, and very little is used to raise the temperature.

The oceans and lakes exhibit enormous effect on weather of that area due to high specific heat of water. In winter, the temperature of the atmosphere falls and hence the temperature of ocean water or lake water also falls. Such a fall in ocean or lake water gives up heat to the atmosphere for every degree drop in temperature. Since drop in water temperature is gradual, the temperature of the ocean or lake water is generally higher than the average temperature of the atmosphere.

16.5 Phase Rule and Water System

The water system is *one* component system as H_2O is the only chemical compound involved in all three phases. The three phases are

$$\underset{\text{solid}}{\text{Ice}} \rightleftharpoons \underset{\text{liquid}}{\text{Water}} \rightleftharpoons \underset{\text{gas}}{\text{Vapour}}$$

The number of phases that can exist in equilibrium at any time depends upon the conditions of temperature and pressure and can be known by phase rule equation. A phase diagram depicts how the phase of a system is affected by the changes in pressure and temperature. Each line in the diagram represents all the conditions of temperature and pressure at which equilibrium exists between the two phases on either side of the line. These conditions have been determined experimentally as well and the phase diagram obtained by plotting pressure against temperature as shown in Fig. 16.2.

The diagram consist of three curves OA, OB, and OC all meeting at the point O which is known as triple point. These curves divide the diagram into three areas:

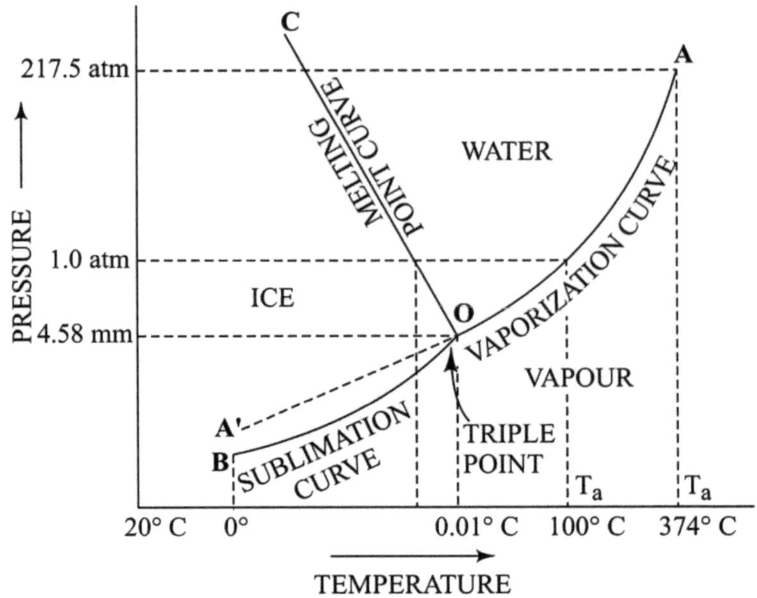

Fig. 16.2 Phase diagram obtained by plotting pressure against temperature for water

AOB, AOC, and BOC. The curves AO, OA′, BO, and OC can be explained as follows:

OA: Is the *vaporization curve* obtained by joining the points showing the value of pressure of saturated vapor at various values of temperature. The curve starts from O, the freezing point of water, and ends at A, the critical temperature T_c (374 °C) beyond which the two phases liquid water and water vapor merge into each other. At any point on this curve for a fixed value of pressure, the temperature has also a fixed value. If by keeping the pressure constant, an attempt is made to raise the temperature, the liquid phase will disappear, and if the temperature is lowered, the vapor phase will disappear. Thus, any point on this curve has one degree of freedom or is *univariant*. This can also be shown from phase rule equation.

$$F = C - P + 2$$
$$= 1 - 2 + 2$$

or

$$F = 1$$

OA': The curve AO can be further extended to A′ by carefully supercooling the liquid and represents a *metastable state*. On a slight disturbance, the supercooled water at once changes into solid ice and the curve merges into OB.

OB: Represents *sublimation curve* of water and shows the variation of vapor pressure of ice at various temperatures. The curve starts at O and ends at B at 0 °C and zero pressure at which no vapor can be present and only ice and vapor are in equilibrium. This also has one degree of freedom.

OC: Is the *melting point curve* of water and shows the effect of pressure on melting point of ice. The slope of the curve OC indicates that melting point of ice is lowered by increase of pressure. At any point of the curve, two phases, i.e., ice and liquid water, are in equilibrium and the system has again only one degree of freedom

$$F = C - P + 2 = 1 - 2 + 2 = 1$$

16.6 Triple Point O

It is the point where the three curves OA, OB, and OC meet. This point is unique one as all the three phases—solid ice, liquid water, and water vapor—are in equilibrium. This point denotes a definite temperature (0.0098 °C or 0.01 °C) and definite pressure (4.58 mm). If either of the parameters, pressure or temperature, is changed, then one of the phases will disappear and hence it is a *zero variant* system, i.e., has no degree of freedom. This can also be confirmed from phase rule equation

$$F = C - P + 2 = 1 - 3 + 2$$
$$F = 0$$

The areas (AOB, BOC, and COA) of the Fig. 16.2.

Any point lying outside the curves discussed above represents the areas, AOB, BOC, and COA. In these areas only one phase—water vapor, ice, or liquid water—respectively, is present. These areas are bivariant because to locate any point in the area the temperature as well as the pressure must be fixed. This also follows from phase rule equation:

$$F = C - P + 2 = 1 - 1 + 2$$
$$F = 2$$

Hence the phase diagram of water can be summed up as

System	Phase present	Degree of freedom
Curve (OA, OA′, OB, OC)	2	1
Areas (AOB, BOC, COA)	1	2
Triple point O	3	0

The number of component is one at all the points.

One component system with more than one solid phase: (Sulphur System)
A simple example of one component system comprising more than one solid phase is *sulfur system.*
It consists of *four* phases

Rhombic sulfur S_R	Solid phase
Monoclinic sulfur S_M	Solid phase
Liquid sulfur S_L	Liquid phase
Vapor sulfur S_V	Vapor phase

The number of phases that can exist in equilibrium at any time depends upon the conditions of temperature and pressure and can be predicted by using phase rule equation:

$$F = C - P + 2$$
$$= 1 - 4 + 2 = -1$$

This negative degrees of freedom being meaningless, it may be said that such an equilibrium, i.e., existences of four phases together in one component system, is not possible. It can, however, be possible for three or lesser number of phases to exist together in each system depending upon the degrees of freedom.

These conditions have been determined experimentally in case of sulfur system, and the phase diagram obtained by plotting pressure against temperature is presented in Fig. 16.3.

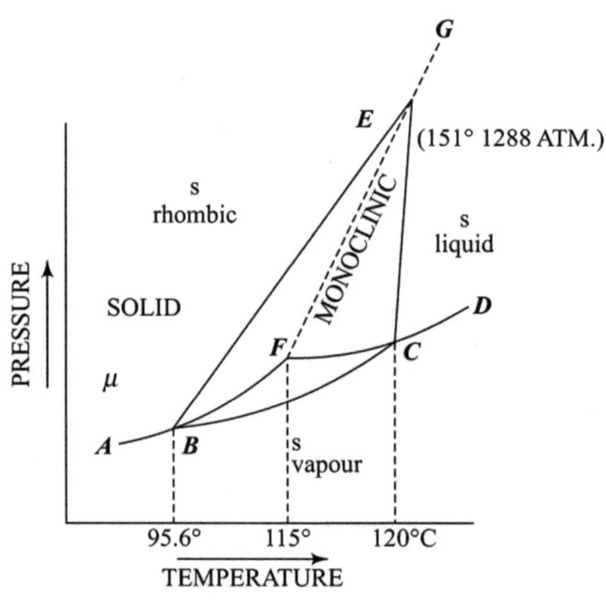

Fig. 16.3 Phase diagram of the sulfur system

The phase diagram consists of six stable curves, AB, BC, CD, BE, CE, and FE, and four metastable curves. These curves divide the diagram into four areas: areas ABEG, ABCD, GECD, and BEC. Let us discuss them one by one.

Curves (AB, BC, CD, BE, CE, EF)

AB: Is the *vapor pressure curve of rhombic sulfur* S_R. It shows the variation of vapor pressure of S_R, at different temperatures. The curve starts at B (95.6 °C), the transition temperature of S_R to S_M, and ends at (50 °C) below which the vapor pressure of S_R is not measurable. At any point on the curve, the two phases S_R and S_V are in equilibrium. The system is monovariant, i.e., it has one degree of freedom, as for fixed value of pressure the temperature has also a fixed value. The degree of freedom can also be calculated from the phase rule equation:

$$\begin{aligned} F &= C - P + 2 \\ &= 1 - 2 + 2 \\ F &= 1 \end{aligned}$$

The curve AB can be further extended to F (115 °C) which is the melting point of S_R by rapid heating. In this case transition of S_R to S_M does not take place at B. The curve BF presented is *metastable* curve of S_R, and an important point is that the slight disturbance in heating will cause the curve merge into BC, the vapor pressure curve of S_M.

BC: is the *vapor pressure curve of* S_M, i.e., shows the variation of vapor pressure of S_M at various temperatures. The curve starts at C, the melting point of S_M (120 °C), and ends at B which is the transition temperature of S_R. At the point B, three phases S_R, S_M and S_V are in equilibrium, while at any other point on the curve, two phases S_V and S_M are in equilibrium. The system as above has only one degree of freedom.

The curve BC can be extended further to H, the metastable vapor pressure curve S_M, and the transition of S_R to S_M is avoided.

CD: This line represents *vapor pressure curve of liquid* sulfur S_L; the curve starts at C, the melting point of S_M, and ends at D, the critical point beyond which out of the two phases S_L and S_V which are present at any point on this curve, only S_V can exist. Since at any point on the curve for a fixed value of pressure, temperature has a fixed value, the system like other curves has one degree of freedom.

The curve CD can be extended further to F by carefully supercooling the liquid, and curve CF obtained is metastable with phases S_R, S_L and S_V in equilibrium.

BE: presents the *transition curve* of S_R to S_M, i.e., it shows the effect of pressure on the transition point which is raised with increase of pressure. The curve ends at E beyond which S_M disappears.

CE: is the *melting point curve* of S_M and shows the effect of pressure on melting point of S_M. The melting point is raised by increase of pressure. At any point on the curve, the two phases S_M and S_L are in equilibrium and also have one degree of freedom.

EF: is the *melting point curve* of metastable S_R which means the two phases are metastable S_R and S_L are in equilibrium. The curve passes through the point E (151 °C) and becomes a stable melting point curve EG of S_R.

Triple Point There are *three* stable triple points B, C, and E in the phase diagram presented in Fig. 16.3. Since at these points there are three phases in equilibrium, and the component is one, the degree of freedom according to phase rule should be zero.

$$F = C - P + 2 = 1 - 3 + 2 = 0$$

The phases at different points of the Fig. 16.3 can be mentioned as

Stable points	B (95.6 °C)	... S_R, S_M, S_V (Three phases)
	C (120 °C)	... S_M, S_L, S_V (Three phases)
	E (151 °C)	... S_R, S_M, S_L (Three phases)
Metastable points	F (115 °C)	... S_K, S_L, S_V (Three phases)

The six curves formed from four boundaries of areas ABEG, ABCD, GECD, and BEC are labeled as solid S_R, vapor S_V, liquid S_L, and monoclinic S_M, respectively. Any point lying outside these curves lies in the areas. There is only one phase present in every area. To define the system completely any point in the area, it is necessary to specify the pressure and temperature. These areas have, therefore, two degrees of freedom. This also follows the phase rule equation.

$$\begin{aligned} F &= C - P + 2 \\ &= 1 - 1 + 2 \\ F &= 2 \end{aligned}$$

16.7 Eutectic Point

A solution of NaCl in water is when allowed to cool; pure ice first separates out at freezing point. As more and more ice separates out, the remaining solution becomes concentrated, and since the freezing point of a concentrated solution is lower than that of a dilute solution, there is a progressive lowering of freezing point. This phenomenon continues till a temperature is reached when the solution begins to freeze *as a whole*, i.e., at this point the solid that separates out is not pure ice but contains the salt also. At this point, there are four phases: the *solid salt, ice, solution*, and *vapor*. This temperature is known as cryohydric temperature or eutectic point and is different for different substances. For sodium chloride and water, it is about – 225 °C, and for a lead-silver system, it is 303 °C and corresponds to a fixed composition 2.4% Ag and 97.6% Pb. Hence,

> *"The lowest possible temperature at which the solution freezes enmass is called the eutectic point."*

It is an invariant point. This follows from phase rule when applied to a two component system containing four phases at this point.

$$F = C - P + 2$$
$$= 2 - 4 + 2 = 0$$

Since the eutectic point lies at a lower temperature than the melting point of either of the compounds, the study of eutectic point is very useful in case of alloys and the solids which mix in all proportions in the fused state.

16.8 Invariant System in Phase Rule

A system in equilibrium having *no* degree of freedom is called invariant system. In such a case, none of the factors like pressure, temperature, etc. can be changed; otherwise, one of the phases in equilibrium disappears.

Consider water system,

$$\underset{\text{solid}}{\text{Ice}} \rightleftharpoons \underset{\text{liquid}}{\text{Water}} \rightleftharpoons \underset{\text{gas}}{\text{Vapor}}$$

in which the three phases—solid, liquid, and gas—are in equilibrium at one definite temperature and pressure. This point in pressure-temperature curve in water system corresponds to triple point O which denotes definite temperature 0.0075 °C and definite pressure 4.58 mm. It is an invariable point that also follows from phase rule equation.

$$F = C - P + 2 = 1 - 3 + 2 = 0$$

Similarly the triple points B, C, and E in pressure temperature diagram for sulfur system are invariable.

16.9 Transition Point

It is the temperature at which the relative stability of two forms of the same element undergoes a change. For example,

- The rhombic form of sulfur is stable at ordinary temperature, but when heated above 95.6 °C and kept at this temperature for sometimes, it undergoes transition to monoclinic form. The reverse change can also take place on cooling. This point is denoted by B in the diagram (Fig. 16.3) for sulfur system.

$$S_R \overset{95.6\,^{\circ}\mathrm{C}}{\rightleftharpoons} S_M$$

- White tin undergoes transition into gray tin when cooled and kept for a long time below 18 °C.
- $Na_2SO_4 \cdot 10H_2O$ on heating to 234 °C undergoes transition to the anhydrous sodium sulfate.

16.10 Characteristics

The transition point is characterized by changes in density, solubility, viscosity, vapor pressure, and other physical properties. In fact, these changes in properties have been utilized for the determination of transition point.

Exercises

1. Write short notes on
   (a) Phase.
   (b) Degree of freedom.
   (c) Components of a system. In a system, the degree of freedom is equal to 2. If equilibrium involves three phases, determine the number of components in the system?
   (d) Eutectic point
   (e) Invariant system in phase rule studies.
   (f) Triple point.
   (g) Transition point.
2. What is phase rule? Explain the terms used. Give the number of components, number of phases, and degree of freedom in
   (a) Sodium chloride solution in water containing undissolved salt in equilibrium with water vapor.
   (b) A mixture of N_2 and H_2 contained in a vessel.
   (c) Ice, water, and vapor in equilibrium.

   Hint: (b) Mixture of nitrogen and hydrogen: A mixture of gases form a single gaseous phase. Thus, a mixture of N_2 and H_2 from a single phase ($P = 1$). It is one component system ($C = 1$) and hence $F = 2$.

$$F = C - P + 2$$

or

$$F = 1 - 1 + 2 = 2$$

3. An air-conditioner uses freon- 12 ($CCl_2 F_2$) as the coolant, and its boiling point is $-30\,°C$. The heat of vaporization of freon is 165 J/g, and the specific heats of gas and liquid freon are 0.61 J/g and 0.97 J/g. K, respectively. Determine the heat evolved when 10 kg of freon is cooled from $+40\,°C$ to $-40\,°C$.
4. A phase diagram obtained for Xe is presented below. Answer the following using the diagram.

   (a) In what phase xenon is normally found at temperature of 25 °C and 1 atm. pressure?
   (b) At $-114\,°C$ and 0.75 atm. pressure, in what phase does the xenon exist?
   (c) Calculate the vapor pressure of solid xenon at $-122\,°C$.
   (d) If the vapor pressure of liquid xenon is 380 mmHg, determine the temperature of liquid phase.
   (e) Which of the phase of Xe, solid or liquid, will be denser. Explain the reason.

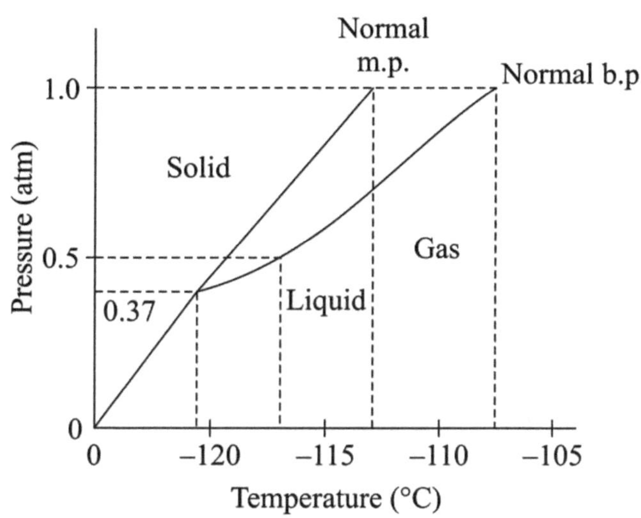

Phase diagram for xenon

5. State how many components are present in each of the following systems:

   (a) $H_2(g) + N_2(g)$
   (b) $NH_3(g)$
   (c) A solution of $Ca(NO_3)_2$ in water
   (d) An aqueous solution of NaCl and Na_2SO_4
   (e) An aqueous solution of $KCl + Na_2SO_4$

6. How many degrees of freedom will each of the systems has in question 5?

7. How many phases are there in.

   (i) A sealed bulb half filled with liquid water, the other half being air saturated with water vapor?
   (ii) A bulb containing small water and there is no air at 100 °C?
   (iii) A mixture of oil and water dispensed in an emulsion?

8. Determine the number of components in the systems composed of

   (i) $N_2(g) + O_2\ (g)$
   (ii) What is the answer of (i) if a catalyst is present to promote oxides of N_2
   (iii) Any salt + water

9. How many degrees of freedom are there in each of the following systems? Suggest variables that could correspond to these degrees of freedom.

   (i) Liquid water and water vapor in equilibrium at a pressure of 1 atm.
   (ii) Liquid water and water vapor in equilibrium
   (iii) An aqueous solution of H_3PO_4 and NaOH at 1 atm pressure.

10. Draw a rough phase diagram from the following data for SO_2. Normal boiling point and freezing point are − 10 °C and − 72.7 °C, respectively. The triple point is − 75.5 °C at 1.60×10^{-3} atm. The critical point is at 157.0 °C and 78 atm.
11. Explain why burns by steam at 100 °C are more severe than burns by water at 100 °C.
12. Explain why no substance can exist in a liquid state at a temperature above its critical temperature.
13. Calculate the amount of heat required to convert 74.6 g of water to steam at 100 °C.
14. Calculate the amount of heat (in kJ) needed to convert 866 g of ice at − 10 °C to steam at 126 °C. (Specific heat of ice and steam are 2.03 J/g °C and 2.0 J/g °C.)
15. Consider the phase diagram of CO and explain why solid carbon dioxide is called dry ice.

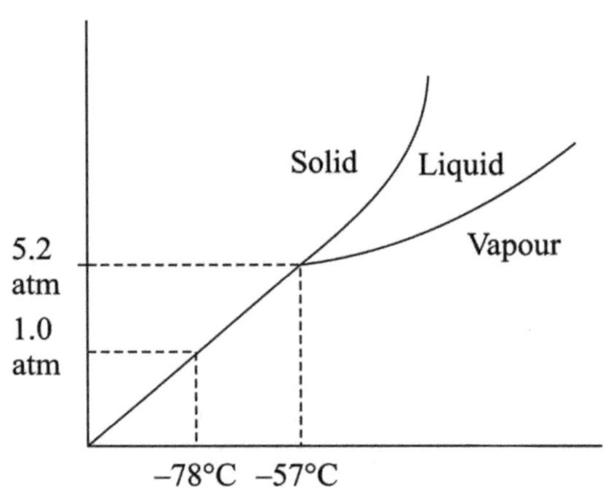

Chapter 17
Experiments in Engineering Chemistry

17.1 Experiment No. 1

17.1.1 Object

To determine the percentage of available chlorine in a given sample of bleaching powder

17.1.2 Chemicals and Apparatus

Burette, pipette, conical flask, 250 ml flask, bleaching powder, KI, glacial acetic acid, hypo solution, starch solution

17.1.3 Theory

The available chlorine is the amount of chlorine which is liberated out by the action of mineral acid.

The bleaching powder has the following constituents:

1. Calcium hypochlorite $Ca(OCl)Cl.4H_2O$
2. Basic calcium chloride $CaCl_2$. $Ca(OH)_2H_2O$
3. Calcium hydroxide $Ca(OH)_2$

From the above constituents, the calcium hypochlorite reacts with dilute mineral acid to liberate free chlorine

© The Author(s), under exclusive license to Springer Nature Switzerland AG 2026

R. N. Goyal, H. Goel, *Textbook of Engineering Chemistry*,
https://doi.org/10.1007/978-3-032-22068-4_17

$$Ca(OCl)Cl + 2CH_3COOH \longrightarrow Ca\,(COOCH_3)_2 + H_2O + Cl_2$$

The liberated Cl_2 reacts with water to form hypochlorous acid (HOCl) which is responsible for bleaching action and disinfection.

$$H_2O + Cl_2 \longrightarrow HOCl + HCl$$

The liberated Cl_2 is estimated by Iodometric method. In this method, bleaching powder reacts with dilute acid to liberate Cl_2 which reacts with KI to liberate I_2. The free I_2 liberated is titrated against hypo solution ($Na_2\,S_2O_3$) in the presence of starch indicator.

$$Cl_2 + 2KI \longrightarrow 2KCl + I_2$$

$$I_2 + \text{ Starch } \longrightarrow \text{ Blue complex}$$

$$2Na_2\,S_2O_3 + I_2 \longrightarrow 2NaI + Na_2\,S_4O_6 \text{ (Sodium tetra thionate)}$$

17.1.4 Procedure

5 g of bleaching powder sample is weighed and transferred to a pestle mortar. Prepare its paste in small amount of water and transfer the paste in a beaker using small amount of water. The milky solution obtained is allowed to settle. Transfer the supernatant to a 250 ml volumetric flask; repeat the process till entire solution is transferred to volumetric flask. Now make up the solution up to the mark and shake well for a homogeneous suspension.

Pipette out 25 ml of the above solution to a conical flask; add 1 g or a pinch of the KI solid or 25 ml of 10% KI solution. Now add 10 ml of glacial acetic acid followed by two drops of starch solution and titrate against 0.1 N Hypo ($Na_2\,S_2O_3$) solution. The end point is the disappearance of blue color of starch. Repeat the process to get concordant readings.

17.1.5 Observations

1. Weight of empty bottle $= \ldots\ldots W_1$ gm.
2. Weight of bottle + Bleaching powder $= \ldots\ldots W_2$ gm.
3. Weight of bleaching powder $= W_2 - W_1$ gm.
4. Titration of bleaching powder solution with standard hypo solution.

S.No.	Volume of bleaching powder solution taken (ml)	Burette reading		Volume of hypo solution consumed (ml)
		Initial	Final	
1.	25.0			
2.	25.0			
3.	25.0			

17.1.6 Calculation

(a) Calculation of strength of bleaching powder.

$$N_1 V_1 = N_2 V_2 \text{ if } V_H \text{ is volume of hypo used.}$$

then

$$0.1\text{ N} \times V_H = N_2 \times 25 \Rightarrow N_2 = \frac{V_H}{250}$$

(b) as

$$\text{Na}_2\,\text{S}_2\text{O}_3 \equiv \text{I}_2 \equiv \text{Cl}_2$$

then from normality × eq. wt. $= \text{g/l.}\ (N_2)$

$$\frac{V_H}{250} \times 35.5 = \text{g/l.} \Rightarrow \text{g/250 ml} = \frac{V_H \times 35.5}{250 \times 4}$$

(c)

$$\%\text{ of available chlorine} = \frac{V_H \times 35.5 \times 100}{250 \times 4 \times (W_2 - W_1)}\%$$

17.1.7 Result

The percentage of available chlorine in a given sample of bleaching powder = ...%

17.1.8 *Precautions*

(1) During weighing the bleaching powder should not be in the direct contact with weights or balance; otherwise, chlorine can corrode the materials.
(2) The solution should be well shaken before taken out for titration.
(3) Freshly prepared starch solution should be used.
(4) KI should be added first and then acid should be added; otherwise, some Cl_2 gas may be lost
(5) The glass apparatus should be washed first with chromic acid and then with distilled water.

17.2 Experiment No. 2

17.2.1 *Object*

To determine the ferrous content in the supplied sample of iron ore by titrimetric analysis against standard $K_2Cr_2O_7$ solution using $K_3Fe(CN)_6$ as an external indicator.

17.2.2 *Chemicals and Apparatus*

Burette, pipette, white tile, standard $K_2Cr_2O_7$ solution, $K_3Fe(CN)_6$ solution, Mohr's salt.

17.2.3 *Theory*

Fe_2O_3 or hematite is the common ore of iron. The ore is powdered and dissolved in conc. HCl to convert Fe to $FeCl_3$ and react with reducing agent $SnCl_2$ to convert $FeCl_3$ to $FeCl_2$. The excess of $SnCl_2$ is removed by adding mercuric chloride.

$$2FeCl_3 + SnCl_2 \longrightarrow 2FeCl_2 + SnCl_4$$

$$SnCl_2 + 2HgCl_2 \longrightarrow SnCl_4 + Hg_2Cl_2$$

The unknown solution of iron ore having $FeCl_2$ oxidizes with $K_2Cr_2O_7$ and dil. H_2SO_4

$$
\begin{array}{rl}
Cr_2O_7^{2-} + 14H^+ + 6e^- \longrightarrow & 2Cr^{3+} + 7H_2O \\
\{Fe^{2+} \longrightarrow & Fe^{3+} + e^-\} \times 6 \\
\hline
Cr_2O_7^{2-} + 14H^+ + 6Fe^{2+} \longrightarrow & 2Cr^{3+} + 6Fe^{3+} + 7H_2O
\end{array}
$$

The indicator $K_3Fe(CN)_6$ is not added into titration beaker; otherwise, it can form a permanent complex color with Fe^{2+}. The drops of indicator are placed on white tile, and a drop of titrate solution is mixed with drop of indicator to get blue color.

$$2\{Fe(CN)_6\}^{3-} + 3Fe^{2+} \longrightarrow Fe_3\,(Fe(CN)_6)_2$$

As soon as all the Fe^{2+} ions are oxidized to Fe^{3+} by $Cr_2O_7^{-2}$, the color of drop of indicator remains unchanged, and end point can be deduced.

17.2.4 Procedure

Take 25 ml of sample solution and add 10 ml of dil. H_2SO_4. Titrate this solution with $K_2Cr_2O_7$ using $K_3Fe(CN)_6$ as an external indicator. Take drop of the solution and touch it with one drop of $K_3Fe(CN)_6$ placed on a white tile. If blue color persists, then add more $K_2Cr_2O_7$, till no blue color is produced.

17.2.5 Observations

(a)

$$
\begin{aligned}
\text{weight of empty bottle} &= W_1 \text{ gm} \\
\text{weight of iron ore} + \text{bottle} &= W_2 \\
\text{weight of iron ore} &= W_2 - W_1
\end{aligned}
$$

(b) Titration of sample solution with $K_2Cr_2O_7$

S.No.	Volume of sample (ml)	Burette reading		Volume of $K_2Cr_2O_7$ used (ml)
		Initial	Final	

17.2.6 Calculation

$$N_1 V_1 = N_2 V_2 \quad \text{if } V_k \text{ is the Vol. of } K_2Cr_2O_7 \text{ used}$$

$$\text{(ore)} \quad (K_2Cr_2O_7)$$

$$N_1 \times 25 = N_2 \times V_k \quad N_2 \text{ is normality of } K_2Cr_2O_7$$

$$N_1 = \frac{N_2 \times V_k}{25} \quad \text{For Iron ore eq. wt. } = 56$$

250 ml sample contains

$$= N_1 \times 56 \times \frac{250}{1000} \text{ gm Fe}$$

or amount of Fe in $(W_2 - W_1)$ gm of

$$\text{Iron ore } = N_1 \times 56 \times \frac{250}{1000} \text{ gm Fe}$$

$$\% \text{ of Fe} = N_1 \times 56 \times \frac{250}{1000} \times \frac{100}{w}$$

$$\boxed{\% \text{ Fe} = \frac{N_2 \times V_k}{25} 56 \times \frac{250}{1000} \times \frac{100}{w}}$$

17.2.7 Result

The ferrous content in the sample ore $= \ldots\ldots$g/liter.

17.2.8 Precautions

(i) For preparation of Ferrous ammonium sulfate solution dil. H_2SO_4 should be added; otherwise, a clear solution will not be obtained.
(ii) The burette, pipette, and other glasswares should be washed with chromic acid and then with water. After washing, the burette should be rinsed with titrant.
(iii) The tile should be layered with a layer of grease.
(iv) The glass rod should be washed every time after withdrawing the solution from titration cell.

(v) Every time dil. H_2SO_4 should be added in titration flask as $K_2Cr_2O_7$ acts as an oxidant in acidic medium.
(vi) The withdrawal of solution from the titration flask should be least to increase the accuracy of results.

17.3 Experiment No. 3

17.3.1 Object

To determine the chloride content in supplied water sample using Mohr's method.

17.3.2 Apparatus and Chemicals

Burette, pipette, conical flask, $\frac{N}{50}$ standard $AgNO_3$ solution, potassium chromate solution as indicator.

17.3.3 Theory

The chloride ions are present in water due to different metal chloride dissolved in it. These ions cause corrosion in boilers, water pipes, etc.

When $AgNO_3$ solution is added to the sample solution, then a white precipitate of AgCl is obtained. On adding K_2CrO_4 in the solution, as soon as all the chloride ions have been precipitated out, then at the end point reddish brown precipitate of $AgCrO_4$ (silver chromate) is obtained.

$$NaCl + AgNO_3 \longrightarrow AgCl \downarrow + NaNO_3$$

$$MgCl_2 + AgNO_3 \longrightarrow 2AgCl \downarrow + Mg\,(NO_3)_2$$

$$K_2CrO_4 + 2AgNO_3 \longrightarrow \underset{\text{Brown Red.}}{Ag_2CrO_4 \downarrow} + 2KNO_3$$

17.3.4 Procedure

Take 25 ml of sample solution in a conical flask and add few drops of K_2CrO_4 indicator. Add $AgNO_3$ solution from burette drop wise. A white color will be obtained. Near the end point by the addition of a single drop of $AgNO_3$ solution, a brick red color precipitate will be obtained, which quickly disappears on shaking.

Continue the addition of $AgNO_3$ to get permanent red color. Repeat unless two concordant readings are obtained.

17.3.5 Observations

Titration of solution with standard $AgNO_3$ solution.

S.No.	Volume of sample solution (ml)	Burette Reading		Volume of $AgNO_3$ Solution (ml)
		Initial	Final	
1.	25 ml			
2.	25 ml			
3.	25 ml			

17.3.6 Calculation

$$N_1 V_1 = N_2 V_2$$

Chloride $AgNO_3$

Sample

$$N_1 \times 25 = \frac{N}{50} \times VAgNO_3 \Rightarrow N_1 = \frac{N}{50} \times \frac{VAgNO_3}{25}$$

$$\text{Strength of chloride ions g/l} = N_1 \times 35.5 \text{ (equivalent wt. of Cl)}$$

$$\text{Strength of chloride ions} = \frac{N}{50} \times \frac{VAgNO_3}{25} \times 35.5.\text{g}/1$$

17.3.7 Result

The amount of chloride content in given water sample $= \ldots$...mg/liter.

17.3.8 Precautions

(i) The silver nitrate solution should be prepared in Cl^- free water.
(ii) The pH of the solution should not be acidic; if the pH is acidic, it should be reacted with $CaCO_3$ to neutralize the acidity.

(iii) The same number of drops of indicator should be added each time.

17.4 Experiment No. 4

17.4.1 *Object*

To determine the constituents and amount of alkalinity of the supplied water sample.

17.4.2 *Apparatus and Chemicals*

Burette, pipette, comical flask, $\frac{N}{50}H_2SO_4$, methyl orange indicator.

17.4.3 *Theory*

The alkalinity in water is due to hydroxyl ion, carbonate ion, and bicarbonate ion. These ion can be estimated by titrating the water sample using phenolphthalein and methyl orange as an indicator. The alkalinity in water causes sludge formation in boiler and caustic embrittlement to the boiler wall. The reactions involved are given below:

(i)

$$OH^- + H^+ \longrightarrow H_2O$$

(ii)

$$CO_3^{2-} + H^+ \longrightarrow HCO_3^-$$

(iii)

$$HCO_3^- + H^+ \longrightarrow CO_2 + H_2O$$

On addition of phenolphthalein, the end point obtained is called phenolphthalein alkalinity (P) which is due to OH^- and CO_3^{2-} ions, and end point due to methyl orange is called methyl orange alkalinity (M) which is due to all the three types of ions. The values of P and M can estimate the type and quantity of alkalinity present, which is given in the table.

Result of P/M	OH^-	CO_3^{2-}	HCO_3^-
P = 0	NIL	NIL	[M]
P = M	P/M	NIL	NIL
$P = \frac{1}{2}M$	NIL	2P/M	NIL
$P > \frac{1}{2}M$	2P − M	2[M − P]	NIL
$P < \frac{1}{2}M$	NIL	2P	M-2P

17.4.4 Procedure

Take 25 ml of sample solution, and add few drops of phenolphthalein and titrate with $\frac{N}{50} H_2SO_4$ to get end point by decolorizing the pink color. This is the phenolphthalein alkalinity (P). Now add few drops of methyl orange in the same solution and again titrate with $\frac{N}{50} H_2SO_4$ to get color change from yellow to red. It is methyl orange alkalinity (M).

17.4.5 Observations

S.No.	Volume of water sample (ml)	Burette Reading		Volume of H_2SO_4 used
		Initial	Final	
1.	25 ml			
2.	25 ml			
3.	25 ml			

17.4.6 Calculation

$$N_1 V_1 = N_2 V_2 \quad \text{Say for P} = x \text{ ml } H_2SO_4 \text{ used}$$

Sample H_2SO_4 for M = y ml H_2SO_4 (total) used water.

(i) HpH alkalinity:

$$N_1 \times 25 = \frac{N}{50} \times V_{H_2SO_4} = \frac{N}{50} \times x$$

$$N_1 = \frac{N}{50} \times \frac{x}{25}$$

In terms of $CaCO_3$

$$\text{Strength} = N_1 \times 50 \text{ g/l}$$
$$= \frac{N}{50} \times \frac{x}{25} \times 50 \text{ g/l}$$
$$\text{P} = \frac{x}{25} \text{ g/l}$$

(ii) MeOH alkalinity:

$$N_1 \times 25 = \frac{N}{50} \times y$$
$$N_1 = \frac{N}{50} \times \frac{y}{25}$$

In terms of $CaCO_3$

$$\text{Strength} = N_1 \times 50 \text{ g/l}$$
$$\text{M} = \frac{y}{25} \text{ g/l}$$

Now HO^-, CO_3^{2-} and HCO_3^- alkalinity can be obtained from the table.

17.4.7 Result

The alkalinity in given sample of water is

CO_3^{2-} .... ppm, OH^- .... ppm and HCO_3^- = ... ppm

17.4.8 Precautions

(i) First add phenolphthalein indicator and titration be carried out; then methyl orange should be added.
(ii) The apparatus should be washed properly and rinsed with respective solutions.
(iii) The solution of H_2SO_4 must be prepared volumetrically from conc. H_2SO_4 and then standardized with standard Na_2CO_3 solution.

17.5 Experiment No. 5

17.5.1 Object

To determine the temporary and permanent hardness of water by complexometry.

17.5.2 Apparatus and Chemicals

Burette, pipette, conical flask, buffer solution of pH 10, EBT indicator sample water, EDTA (M/10).

17.5.3 Theory

Presence of hardness in water can cause problem of scaling. These salts are of calcium and magnesium metals. These salts form metal complexes with EDTA (ethylenediamine tetraacetic acid).

At pH10, Eriochrome Black T (EBT) gives wine red colored complex with Ca^{2+} and Mg^{2+} ions. When this complex reacts with EDTA solution, the color of complex changes from wine red to blue at the end point.

$$Ca^{2+} + \text{EBT} \longrightarrow \underset{\text{wine red.}}{\text{Ca EBT complex}} \xrightarrow{\text{EDTA}} \underset{\text{Blue.}}{\text{Ca–EDTA complex}} + \text{EBT}$$

EDTA is a chelating agent with four dent.

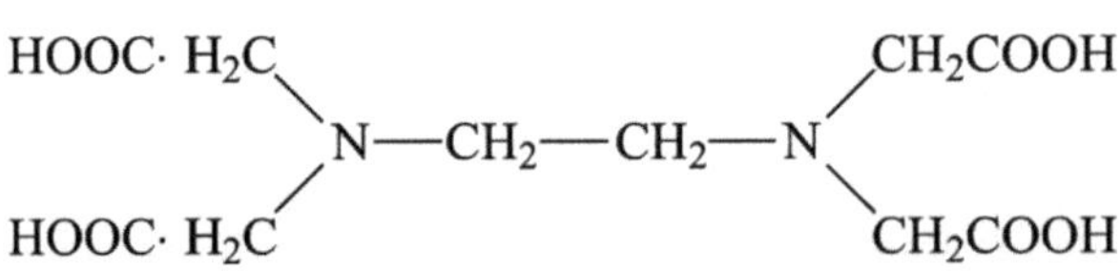

The two dents of EDTA combine with Ca^{2+} or Mg^{2+} to form complex.

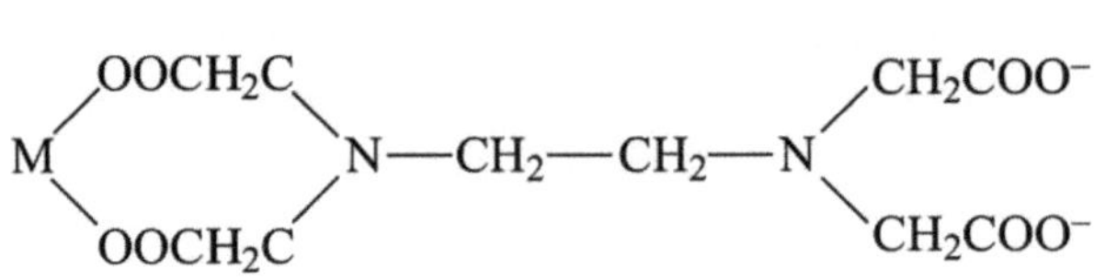

EBT (Eriochrome Black T) is chemically named as (Na-l-(l-hydroxy-2-naphthylazo)-6-nitro-2-napthol-4-sulfonate)

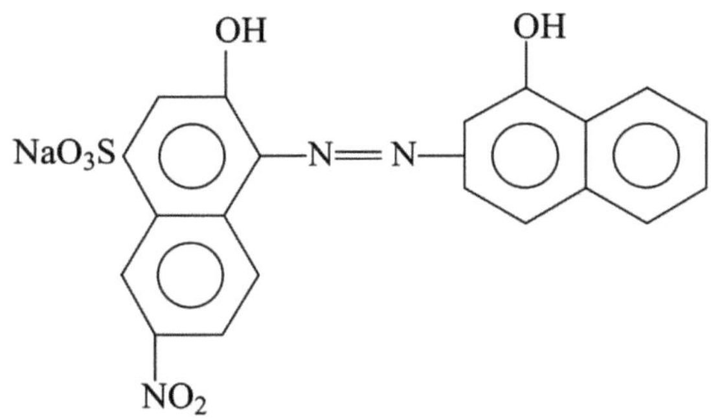

17.5.4 Procedure

Take 25 ml of sample water in a conical flask. Add about half a test tube of buffer solution. Make sure that the sample smells ammonia. Add few drops of EBT indicator. The color of solution will be wine red. Titrate the solution with $\frac{M}{10}$ EDTA solution till wine red color is changed to blue. Now take 250 ml of hard water in a beaker and boil it for half an hour; cool it and filter. Take 25 ml of this solution, titrate it as mentioned above, and find out end point. This value corresponds to permanent hardness.

17.5.5 Observations

(i) Readings for hard water

S.No.	Volume of water sample (ml)	Volume of EDTA		Amount of EDTA used
		Initial	Final	
				V_1 (Let)

(ii) Readings for boiled water

S.No.	Volume of boiled water sample (ml)	Volume of EDTA		Amount of EDTA used
		Initial	Final	
				V_2 (Let)

17.5.6 Calculation

$$\begin{aligned} 1 \text{ mole of EDTA } &= 1 \text{ mole of hardness} \\ 1 \text{ ml of M EDTA } &= 100 \text{ mg of } CaCO_3 \end{aligned}$$

Total Hardness

$$V_1 \text{ ml of } \frac{\text{M}}{10}\text{EDTA} = V_1 \times \frac{\text{M}}{10} \times 100 = 10V_1 \text{mg of } CaCO_3$$

$$\text{So total hardness in 25 ml of sample water } = 10V_1 \text{mg of } CaCO_3$$

$$\begin{aligned} \text{and in 1000 ml of sample water } &= 10V_1 \times \frac{1000}{25} \text{mg.} \\ &= 400V_1 \text{ mg/l} \quad \text{or} \quad \text{ppm of } CaCO_3 \end{aligned}$$

Permanent Hardness

$$V_2 \text{ ml of } \frac{M}{10} \text{ EDTA } = V_2 \times \frac{M}{10} \times 100 \text{ mg} = 10V_2 \text{ mg of } CaCO_3$$

$$\text{Permanent hardness in 25 ml of water } = 10V_2 \text{ mg of } CaCO_3$$

$$\begin{aligned} \text{In 1000 ml of water } &= \frac{10V_2 \times 1000}{25} \\ &= 400V_2 \text{ mg/l} \quad \text{or} \quad \text{ppm of } CaCO_3 \end{aligned}$$

Temporary Hardness

$$\begin{aligned} \text{Temporary hardness } &= \text{ Total } - \text{ Permanent.} \\ &= 400V_1 - 400V_2 \\ &= 400\,(V_1 - V_2) \text{ mg/l} \quad \text{or} \quad \text{ppm of } CaCO_3 \end{aligned}$$

17.5.7 Result

Total hardness = ... ppm.

Permanent hardness = ... ppm.

Temporary hardness = ... ppm.

17.5.8 Precautions

(i) The blue color at the end point is quite close to wine red hence; the change in color should be checked in good light.
(ii) The addition of ammonia buffer must be ensured to bring the pH10.
(iii) After boiling the sample water, the sample should be cooled and filtered to get permanent hardness.
(iv) The glass apparatus should be properly rinsed with distilled water.
(v) The amount of indicator added should be the same in all titrations.

17.6 Experiment No. 6

17.6.1 Object

To find the chemical oxygen demand (COD) of a wastewater sample using potassium dichromate.

17.6.2 Apparatus and Chemicals

Reflux apparatus, $\frac{N}{4}$ $K_2Cr_2O_7$ solution, H_2SO_4, Ag_2SO_4, standard $\frac{N}{4}$ $FeSO_4$ solution, Ferroin indicator, $HgSO_4$.

17.6.3 Theory

COD is the amount of O_2 used for oxidizing the organic matter present in wastewater sample using strong oxidant. The value of COD in water sample is the measure of pollution strength of domestic or industrial wastes.

The organic matter oxidized completely by $K_2Cr_2O_7$ in the presence of H_2SO_4 to convert it to CO_2 and H_2O. The excess of $K_2Cr_2O_7$ is titrated with ferrous ammonium sulfate solution. The$_2$ $K_2Cr_2O_7$ consumed gives the O_2 required for oxidation of organic matter.

$$6Fe^{2+} + Cr_2O_7^{2-} + 14H^+ \longrightarrow 6Fe^{3+} + 2Cr^{3+} + 7H_2O.$$

17.6.4 Procedure

Take 50 ml of the sample in a 500 ml round bottom flask and add 1 gm $HgSO_4$ and some pieces of porcelain. Immerse the flask in cold water and slowly add 75 ml of Ag_2SO_4–H_2SO_4 reagent with continuous shaking.

Now add 25 ml of $\frac{N}{4}$ $K_2Cr_2O_7$. Fix the reflux condenser and reflux the contents for 2 hours. Wash the condenser with distilled water into the flask. Cool and dilute by adding equal amount of water. Now add few drops of Ferroin indicator and titrate against $\frac{N}{4}$ ferrous ammonium sulfate (FAS) to get a color change from blue green to reddish brown.

A blank titration should also be carried out using distilled water in place of sample solution.

S.No.	Volume of sample (ml)	Burette reading		Volume of $\frac{N}{4}$ FAS
		Initial	Final	
1.	25 ml			
2.	25 ml			
3.	25 ml			

17.6.5 Calculation

$$\text{Volume of sample} = 25\text{ ml}$$

$$\text{Volume of } \frac{N}{4} \text{ FAS used with sample} = \text{A ml}$$

$$\text{Volume of } \frac{N}{4} \text{ FAS used with blank} = \text{B ml}$$

$$\text{Vol of } \frac{N}{4} \text{ FAS} \equiv K_2Cr_2O_7 \text{ used for COD} = \text{B} - \text{A}$$

$$N_1V_1\,(\text{sample}) = N_2V_2(\text{FAS})$$

$$N_1 \times 25 = \frac{1}{4} \times (\text{B} - \text{A})$$

$$N_1 = \frac{B - \text{A}}{100}$$

and

$$\text{COD} = \frac{B - \text{A}}{100} \times 8\text{ g/l}$$

17.6.6 *Precautions*

1. Addition of Ag_2SO_4 should be dropwise; otherwise, due to exothermic reaction, a blast may occur.
2. The addition of Ag_2SO_4 should be with constant shaking and cooling.

17.7 Experiment No. 7

17.7.1 *Object*

To determine iron concentration in the sample of water by spectrophotometric method.

17.7.2 *Apparatus and Chemicals*

Spectrophotometer, conc. HCl, standard solution of FAS, KCNS solution.

17.7.3 *Theory*

Fe^{3+} ion reacts with KCNS to give different red colored complexes.

$$Fe^{3+} + SCN^- = Fe(SCN)^{2+}$$

$$Fe^{3+} + 2SCN^- = [Fe(SCN)_2)]^+$$

$$Fe^{3+} + 6SCN^- = [Fe(SCN)_6]^{3-}$$

The large excess of SCN^- is generally used to increase the intensity of red color. In order to suppress the hydrolysis of Fe^{3+}, the highly acidic HCl is used. The detection of Fe^{3+} is carried out at $\lambda_{max} = 480$ nm.

The concentration of Fe is based on Beer's-Lambert's law

$$\text{Absorbance A} = \varepsilon \text{Cl}$$

$$\varepsilon = \text{ extinction coefficient.}$$

$$\text{C} = \text{ conc. of solution.}$$

$$\text{l} = \text{ length of solution.}$$

17.7.4 Procedure

1. Dissolve the weighed substance in conc. HCl. Evaporate nearly to dryness to remove excess of acid and oxidize the iron to iron (III) by adding dil. $KMnO_4$, and make up the solution to 250 ml. Take 40 ml of this solution in 50 ml flask; add 5 ml SCN^- solution and 3 ml HCl and make up to the mark with deionized water.
2. Prepare a blank using same quantity of reagents.
3. Take 5.0, 10.0, 15.0, and 20.0 ml of standard solution in 50 ml flasks. Add 5 ml dil. HCl and 5 ml 20% KCNS solution and make them up with deionized water.
4. Measure the absorbance of each solution at 480 nm, against blank after 5 min of color development, and plot a graph between the absorbance A and concentration C.
5. Now measure the absorbance of test solution (step 1), and determine the concentration of this solution using calibration curve.

17.7.5 Observations

S.No.	Volume of standard	Conc. of Fe^{3+} in (ppm)	Absorbance
1.			
2.			
3.			
4.			
Unknown.			

17.7.6 Result

Conc. of Fe^{3+} in given sample : =

17.7.7 Precautions

1. To prevent the interface of phosphate, arsenate, oxalate, and fluoride ions, a high conc. of HCl should be used.
2. H_2SO_4 is not used because SO_4^{2-} has a tendency to form complexes with Fe^{3+} ions.

3. Hg^{+} and Sn^{2+} salts, if present, should be converted to Hg^{2+} and Sn^{4+}; otherwise, the color will be destroyed.

17.8 Experiment No. 8

17.8.1 Object

To find out the velocity constant for the inversion of cane sugar in acidic medium and to show that inversion follows the first order kinetics:

17.8.2 Apparatus and Chemicals

Polarimeter, stopwatch, beaker, sugar solution, dil. HCl

17.8.3 Theory

The inversion of cane sugar in acidic medium is expressed as

$$C_{12}H_{22}O_{11} + H^{+} \rightleftharpoons \underset{\text{dextro}}{C_6H_{12}O_6} + \underset{\text{laevo}}{C_6H_{12}O_6}$$

The rate of this reaction depends upon the concentration of sugar solution. The dextrorotatory sugar solution decomposes to dextro and laevo rotatory mixture of glucose and fructose. This reaction is known as inversion.

The order of reaction can be measured by measuring the change in angle of rotation of plane polarized light during reaction using polarimeter. The rate constant k for first order kinetics will be

$$k = \frac{2.303}{t} \log \frac{a}{a - x} = \frac{2.303}{t} \log \frac{r_0 - r_\infty}{r_t - r_\infty}$$

where

$$r_0 = \text{Angle of rotation initially}$$

$$r_t = \text{angle of rotation at time } t.$$

$$r_\infty = \text{Angle of rotation at infinity/log time.}$$

17.8.4 Procedure

Weigh 20 g of sugar and dissolve it in water in 250 ml flask. Take 25 ml of this solution and add 25 ml of dil. HCl in a beaker. Now set up the polarimeter and get zero reading by using distilled water in polarimeter tube. Now place the mixture of sugar and dil. acid in polarimeter tube, and quickly take reading of r_0 and start stopwatch. Take reading of r_t at time 5, 10, 15, 20, 30, and 40 minutes.

Now take the remaining mixture and put in oven at 70 °C for half an hour, and take r_∞ after cooling the mixture.

17.8.5 Observations

S.No.	Time (min)	Rotations
1	0	
2	5	
3	10	
4	15	
5	20	
6	30	
7	40	
8	∞	

17.8.6 Calculation

$$k = \frac{2.303}{t} \log \frac{r_0 - r_\infty}{r_t - r_\infty}$$

Calculate k_1, k_2, k_3 at different values of r_t.

17.8.7 Result

The values of

$$k_1 = \ldots\ldots\ldots$$

$$k_2 = \ldots\ldots\ldots$$

$$k_3 = \ldots\ldots\ldots$$

which are expected to be practically constant and indicates the first order kinetics

17.8.8 Precautions

1. The mixture should be kept at room temperature
2. There should not be any bubble in polarimeter tube.
3. The rotation should be checked for +ve and −ve values.

17.9 Experiment No. 9

17.9.1 Object

To determine the molecular weight of polystyrene sample by using viscometric method.

17.9.2 Apparatus and Chemicals

Ostwald viscometer, stopwatch, oven, polystyrene, benzene, or toluene.

17.9.3 Theory

The coefficient of viscosity of a liquid can be calculated from the formula

$$\eta = \frac{P\pi r^4 t}{8vl} \quad \text{(i)}$$

where t is the time required by v ml of liquid to flow through a capillary tube of length l and radius r under the pressure of P dynes/cm^2.

An addition of a small amount of polymer to a solvent causes a sharp increase in viscosity. This increase in viscosity depends on the molecular weight of the polymer.

The ratio of coefficient of viscosity of solutions (η_s) and coefficient of viscosity of solvent η_0 can be obtained from Eq. i as,

$$\frac{\eta_s}{\eta_0} = \frac{d_s}{d_0}\frac{t_s}{t_0} \quad \text{(ii)}$$

where d_s and d_0 are the densities of solution and solvent, respectively, and t_s and t_0 are the time of flow for a constant volume of solution and solvent, respectively.

For a dilute solution $d_s = d_0$ hence,

$$\frac{\eta_s}{\eta_0} = \frac{t_s}{t_0}$$

increase in viscosity with respect to solvent

$$\eta_{\text{sp}} = \frac{\eta_s}{\eta_0} - 1 = \frac{t_s}{t_0} - 1$$

The ratio of the specific viscosity of the solution to its concentration C is called reduced specific viscosity

$$\eta_{\text{red}} = \frac{\eta_{sp}}{C}$$

The plot of $\frac{\eta_{sp}}{C}$ vs. C is a straight line. It is extrapolated to zero concentration to give limiting viscosity $|\eta|$ which is related to molecular weight as

$$|\eta| = KM^{\alpha}$$

where K is constant for a polymer solvent system and α is a function of molecule geometry. These are obtained from samples of known molecular weight polymer.

17.9.4 Procedure

Dissolve 0.5 g of polystyrene in 100 ml of benzene/toluene to prepare 0.5% solution of polystyrene. Using this conc. of 0.5%, prepare lower concentrations of 0.1%, 0.2%, 0.3%, and 0.4% by volumetric dilution. Now take pure solvent in Ostwald viscometer, and determine time of flow between the two marks of viscometer. It is the time for flow of pure solvent. Similarly find out the time of flow for solutions of different conc. of polystyrene.

17.9.5 Observation

Plot a graph between $\frac{\eta_{sp}}{C}$ versus C to get $|\eta|$ by extrapolating the straight line.

Conc. %	Flow time	$\frac{\eta_s}{\eta_0} = \frac{t_s}{t_0}$	$\eta_{sp} = \frac{\eta_s}{\eta_0} - 1 = \frac{t_s}{t_o} - 1$	$\eta_{\text{red}} = \frac{\eta_{sp}}{c}$
Solvent	t_0			
0.1	t_1			
0.2	t_2			
0.3	t_3			
0.4	t_4			
0.5	t_5			

17.9.6 Calculation

Polystyrene/benzene $K = 1.03 \times 10^4,\ \alpha = 0.74$

Polystyrene/toluene $K = 3.7 \times 10^4,\ \alpha = 0.62$

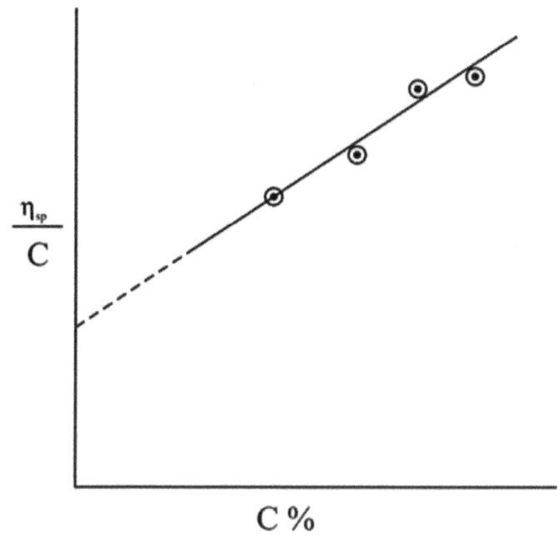

Taking log

$$\log|\eta| = \log K + \alpha \log M$$

$$\log M = \frac{\log|\eta| - \log K}{\alpha}$$

17.9.7 Result

The observed molecular weight of polystyrene $= \ldots\ldots\ldots$

17.9.8 Precautions

1. The viscometer should be bubble-free.
2. The time of flow of different solutions should be measured from lower concentration to higher concentration.
3. The viscometer should be dried in an oven after washing with chromic acid; otherwise, the water may form bubble due to non-dissolution of water in benzene or toluene.
4. Benzene and toluene are carcinogenus substances; hence, their inhalation should be avoided.

17.10 Experiment No. 10

17.10.1 Object

To determine pH of a solution using a pH meter and titration of such solution pH metrically.

17.10.2 Apparatus and Chemicals

pH meter, glass electrode, reference electrode, beakers, buffer tablets, unknown solution.

17.10.3 Theory

pH of a solution is the −ve logarithm of hydrogen ion concentration.

$$pH = -\log\left[H^+\right]$$

when alkali is added to an acid, the pH of solution increases slowly, but near the end point, there is large change in pH which indicates end point.

17.10.4 Procedure

Standardize the pH meter using buffers (acid and basic) of known pH. Now wash the electrodes with water.

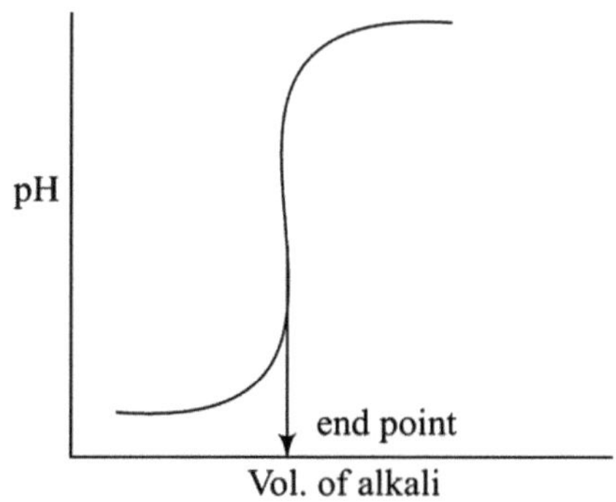

Dip the glass and reference electrode in 5.0 ml of acid solution and note the pH of pure acid. Add $\frac{N}{10}$NaOH solution dropwise and note the pH of solution at every 0.2 ml addition of $\frac{N}{10}$NaOH. Plot a graph between pH and volume of alkali added.

17.10.5 Observation

For 5 ml of acid taken.

Volume of alkali (ml)	0.2	0.4	0.8	1.0	2.0	3.0	4.0	5.0	5.5	5.6	5.7	5.8	5.9
pH													

17.10.6 Calculation

Using the plot of pH versus ml of alkali used, the middle of sharp increase in curve will give end point (say x ml)

$$N_1 V_1 = N_2 V_2$$

$$\text{Acid} = \text{Alkali}$$

$$N_1 \times 5 = \frac{N}{10} \times x$$

[Assume at end point volume of alkali $= x$ ml]

$$\text{Strength of HCl} = \frac{36.5 \times x}{10 \times 5}$$

17.10.7 Result

The strength of acid solution is $= \ldots$ g/litre.

17.10.8 Precautions

1. The glass electrode should not touch with the wall of glassware.
2. The glass electrode should not be rubbed with filter paper. It should be dried by simply touching onto the filter paper.
3. During titration, the solution should be stirred thoroughly.
4. The pH meter should be standardized by using acid and basic buffers.

17.11 Experiment No. 11

17.11.1 Object

To determine the calorific value of a fuel sample by using a Bomb/break calorimeter.

17.11.2 Apparatus and Chemicals

Bomb calorimeter, fuel sample, benzoic acid, glassware etc.

17.11.3 Theory

Calorific value of fuel is the amount of heat liberated by the complete combustion of a unit mass of fuel. The calorific value determines the quantity of a fuel and also helps in calculating the thermal efficiency. The higher calorific value can be calculated as

$$\text{HCV} = \frac{(W+w)\left(\Delta t + \begin{matrix}\text{cooling}\\ \text{correction}\end{matrix}\right) - \left(\text{Acid} + \begin{matrix}\text{fuse}\\ \text{correction}\end{matrix} + \begin{matrix}\text{thread}\\ \text{correction}\end{matrix}\right)}{m}$$

where

w = mass of water taken in calorimeter
W = Water equivalent of calorimeter

m = mass of fuel taken in bomb
Δt = difference of initial and final temperature of calorimeter water

and

$$\text{LCV} = \text{HCV} - (0.09\text{H} \times 587)$$

where H = % of Hydrogen in coal.

17.11.4 Procedure

Take 0.5 g of given fuel in a dry steel crucible. A fine platinum/nichron wire, touching the sample fuel, is stretched across the electrodes, and a cotton thread of known length is tied on fuse wire and inserted in sample inside. Place 10 ml of water inside the bomb and tightly close the bomb cover lid. Now charge the bomb by O_2 gas applying O_2 cylinder to get the pressure $\sim$ 25 atmosphere. Now bomb is placed inside the calorimeter. Add known mass of water, say 2 liters in calorimeter, so that the bomb is submersed in water but terminals remain above the water level. There should be no bubble or leakage in bomb. Connect the terminals with electricity. Switch on the stirrer and note the initial temperature of water. After burning the fuel, the water should be stirred for half an hour to get maximum temperature. Leave the bomb for 1 hour to settle down the fumes. Now release the oxygen of bomb and open the bomb, wash the residue of bomb with water, and transfer to a beaker. Use this washing to determine presence of HNO_3 and H_2SO_4.

Determination of H_2SO_4 **and** HNO_3 **(Acid Correction)** Due to the presence of sulfur in fuel, S is oxidized to SO_2 which dissolves in water to give H_2SO_4. This reaction is exothermic; hence, a necessary correction against H_2SO_4 formed should be made in calculation of HCV.

For every gram equivalent of H_2SO_4 (i.e., 49 gm H_2SO_4) 36 kcal of heat is librated.

$$1 \text{ ml of } \frac{N}{10} H_2SO_4 \equiv 49 \times \frac{N}{10} \times \frac{1}{100} \text{ g } H_2SO_4 \cong \frac{36}{10^4} \text{kcal} = 3.6 \text{ cal}$$

Similarly

$$1 \text{ ml of } \frac{N}{10} HNO_3 \equiv 6.3 \times \frac{N}{10} \times \frac{1}{1000} \text{ g } HNO_3 \cong 1.429 \text{ cal}$$

Dilute the washing from bomb to 100 ml, and boil to remove CO_2. Now titrate it with $\frac{N}{10}Ba(OH)_2/NaOH$ using methyl orange as an indicator. Add 20 ml $\frac{N}{10}Na_2CO_3$ solution. Warm and filter the solution. Wash the precipitate and titrate the supernatant with $\frac{N}{10}HCl$ using methyl orange as an indicator.

Let

$$\frac{N}{10}\text{HCl used} = \text{A ml}$$

$$\frac{N}{10}\text{Ba(OH)}_2\text{ used} = \text{B ml}$$

therefore

$$\frac{N}{10}\text{HNO}_3 = (20 - \text{A})\text{ml} \quad \Rightarrow 1.43(20 - \text{A})\text{cal}$$

$$\frac{N}{10}\text{H}_2\text{SO}_4 = (\text{A} + \text{B} - 20)\text{ml} \quad \Rightarrow 3.6(\text{A} + \text{B} - 20)\text{cal.}$$

Fuse Wire Correction For 1″ of fuse wire = 35 cal/g of nichrom wire.

Cotton Thread Correction For 1 g of cotton = 4140 cal/g.

Cooling Correction

$$t_{\text{cooling}} = nV + \left(\frac{V^1 - V}{t^1 - t}\right)\left[\Sigma(t) + \frac{t_0 + t_n}{2} - nt\right]$$

where

n = Number of minutes between firing and maximum temp.
V = Rate of fall of temp. during initial period before firing.
V^1 = Rate of fall of temp. during final period after maximum temp.
t = mean temp. during initial period.
t^1 = Mean temp. during final period.
$\Sigma(t)$ = Mean temp. during initial to final period.
t_0 = Temp. at the time of firing.
t_n = Temp. at the time of maximum value.

17.11.5 Observations

	Initial period		Combustion period		Final period	
S.No.	Time	°C	Time	°C	Time	°C
1.						
2.						
3.						
4.						
5.						
6.						

17.11.6 Result

The HCV of given fuel $=$ cal/g.

17.11.7 Precautions

1. The amount of sample taken should not be more than 1.0 g.
2. There should not be any leakage in bomb.
3. The fuel should be finally powdered before placing in bomb.
4. The bomb should not be opened just after the firing; otherwise, some fumes may blow out.
5. All the parts of bomb should be submersed in water except electrode terminals.
6. The water should be stirred all the time.

17.12 Experiment No. 12

17.12.1 Object

To analyze the coal sample by proximate analysis method.

17.12.2 Apparatus and Chemicals

Powdered coal, oven, muffle furnace, electric furnace, crucible, dessicators, furnace, tong.

17.12.3 Theory/Procedure

Proximate analysis involves the determination of moisture, volatile mater, ash, and fix carbon in a given sample of coal. The determination involves a stepwise procedure as under:

a. ***Moisture:*** Take 1 gram of finely powdered coal in a silica crucible and heat it in an oven at 105 °C for 1 hour; the loss in weight with respect to sample taken is called % of moisture.
b. ***Volatile matter:*** The moisture free above dried coal is heated in a closed crucible at 950 °C for 10 minutes in an electric furnace in absence of air. The crucible is cooled in dessicator and weighed and loss in weight is the % of volatile matter.

c. ***Ash:*** The moisture and volatile matter free coal sample is heated in an open crucible at 750 °C in a muffle furnace to get a constant weight. The residue is called ash on % basis.
d. ***Fix carbon:*** The fix carbon will be equal to [100 – % Moisture - % Volatile matter - % Ash].

17.12.4 Observations and Calculation

a. ***Moisture:***

$$\begin{aligned} \text{weight of empty crucible} &= w_1 \text{ g} \\ \text{crucible} + \text{sample} &= w_2 \text{ g} \end{aligned}$$

crucible + sample after heating at 105 °C = w_3

$$\% \text{ moisture} = \frac{w_2 - w_3}{w_2 - w_1} \times 100$$

b. ***Volatile matter:***

$$\begin{aligned} \text{weight of empty crucible} &= w_1 \\ \text{crucible} + \text{sample} &= w_2 \end{aligned}$$

crucible + sample after heating at 950 °C in absence of air $= w_4$

$$\% \text{ volatile matter} = \frac{w_2 - w_4}{w_2 - w_1} \times 100$$

c. ***Ash:***

$$\begin{aligned} \text{weight of emply crucible} &= w_1 \\ \text{crucible} + \text{sample} &= w_2 \end{aligned}$$

crucible + sample after heating at 750 °C in the presence of air $= w_5$

$$\% \text{ of Ash} = \frac{w_2 - w_5}{w_2 - w_1} \times 100$$

d. ***Fix carbon:***

% FC = 100 – % moisture - % volatile matter - % of Ash.

17.12.5 *Result*

(1) % Moisture =——— %
(2) % Volatile matter =——— %
(3) % Ash =——— %
(4) % Fix carbon =——— %.

17.12.6 *Precautions*

1. The silica crucible should be weighed after heating and cooling till a constant weight is achieved.
2. The crucible or sample should be first cooled in dessicator before weighing.
3. The furnace tong should be used for putting or removing the crucible.
4. The heating rate of sample should be adjusted at a slow rate; otherwise, some sample loss may occur due to rapid expulsion of volatile gases and moisture.

17.13 Experiment No. 13

17.13.1 *Object*

To determine equivalent weight of iron by the chemical displacement method. (The equivalent weight of copper is 63.5.)

17.13.2 *Chemicals and Apparatus*

Burette, pipette, standard $CuSO_4$, hypo solution, KI and starch solutions, iron strip.

17.13.3 *Theory*

When iron strip is allowed to dip in $\frac{N}{10}CuSO_4$ solution, then some iron ion goes in solution as $FeSO_4$ and some Cu is deposited over iron strip.

$$Fe + CuSO_4 \longrightarrow FeSO_4 + Cu$$

Thus, Fe displaces Cu from $CuSO_4$ then

$$\frac{\text{equivalent weight of Cu}}{\text{equivalent weight of Fe}} = \frac{\text{weight of copper deposited}}{\text{weight of Fe goes in solution}}$$

17.13.4 *Procedure*

Clean the iron strip and weigh accurately and dip it in 100 ml $\frac{N}{10}CuSO_4$ solution for 30 minutes. Take out the strip with the help of forceps and put in dessicator. Now titrate the remaining solution of $CuSO_4$ using $\frac{N}{10}$ hypo solution idometrically as follows.

Take 25 ml of $CuSO_4$ solution and add 1 g solid KI and allow to stand the mixture for 2–5 minutes in dark. The solution becomes dark brown due to I_2 libration. Now titrate with $\frac{N}{10}$ hypo solution using starch as an indicator. At end point the blue color disappears.

17.13.5 *Observations*

(i) Weight of iron strip $= x$ gms
(ii) Weight of iron strip + Cu after drying $= y$ gms
(iii) Weight of copper deposited $= z$ gms.

$$z = \text{by titration} = (\text{initial con. of Cu} - \text{final conc. of Cu})$$

(iv) Weight of Fe, goes into solution $= A = x + z - y$

17.13.6 *Iodometric Titration*

S.No.	Volume of $CuSO_4$	Burette Reading (ml) Initial	Burette Reading (ml) Final	Volume of hypo used (ml)
1.	25 ml			
2.	25 ml			
3.	25 ml			

17.13.7 *Calculation*

Final conc. of Cu.

$$N_1V_1 = N_2V_2$$

$CuSO_4$ Hypo

$$N_1 \times 25 = \frac{N}{10} \times \text{V hypo} \quad \Rightarrow N_1 = \frac{V_{\text{Hypo}}}{250}$$

$$\text{Strength g/l} = N_1 \times \text{ eq. wt. of Cu} = \frac{V_{Hypo}}{250} \times 63.5 \text{ g/l}$$

$$\text{Final strength of Cu} = \frac{V_{Hypo}}{250} \times \frac{63.5}{10} \text{ g}/100 \text{ ml.}$$

$$\text{z} = \text{ Initial conc. of Cu} - \text{ Final conc. of Cu}$$

$$\frac{\text{Eq. weight of Iron}}{\text{Eq. weight of Copper}} = \frac{A}{\text{z}}$$

$$\text{\% error of experiment} = \frac{\text{Eq. weight Iron experimentally}}{\text{eq. wt. of Iron}} \times 100.$$

17.13.8 Result

The eq. wt. of iron =———

% error =———.

17.13.9 Precautions

1. The starch indicator should not be added initially; otherwise, a permanent blue color may be obtained.
2. In dessicsator, dry silica gel should be used.
3. The starch solution should be freshly prepared.

17.14 Experiment No. 14

17.14.1 Object

To study the adsorption of acetic acid on dispersed charcoal.

17.14.2 Apparatus and Chemicals

Burette, pipette, six reagent bottles, $\frac{N}{10}$NaOH, activated charcoal, phenolphthalein indicator.

17.14.3 Theory

Adsorption is a physical property to hold the particles on the surface of a solid, without penetration into the bulk of the solid. The solid on which adsorption takes place is called adsorbent, and the material adsorbed is called adsorbate.

The phenomenon of adsorption depends on following factors.

1. ***Surface area of adsorbent:*** The larger the surface area, the larger the adsorption due to which charcoal and silica gel are excellent adsorbent because they have highly porous surface and hence large surface area is available. For increasing the surface area, finely divided substances are used for large adsorption power.
2. ***Nature of adsorbate:*** The gases which are easily liquefiable have more power of adsorption e.g., SO_2, HCl, Br_2, Cl_2 etc.
3. ***Temperature:*** Adsorption is exothermic process; therefore, the higher the temperature, the lesser will be the adsorption.
4. ***Pressure:*** If pressure is high, adsorption is more.
5. ***Concentration:*** The amount of solute adsorbed/g of adsorbent depends on the concentration of the solute in the solution.

The adsorption of solute follows Freundlich equation.

$$\frac{x}{m} = KC^n \Rightarrow \log\left(\frac{x}{m}\right) = \log K + n \log C$$

where $\frac{x}{m}$ is the amount of solute adsorbed/g of adsorbent. K and n are constants which depend on temperature and the nature of solute, solvent, and adsorbent, and C is the concentration of solute.

17.14.4 Procedure

(1) Take six stoppered bottles. Weigh accurately 2 gm of activated characoal and put in each bottle.
(2) Take 0.6 NCH_3COOH and distilled water in two separate burettes. Add different volumes of CH_3COOH and distilled water to make total volume 60 ml in each bottles as given in table.
(3) Shake the solutions with charcoal powder and put the bottles for 1 hour.

Bottle. Number	1	2	3	4	5	6
Volume of CH_3COOH	60	50	40	30	20	10
Volume of water	0	10	20	30	40	50

(4) Now filter the solution of Bottle No. 1 into a beaker and titrate 25 ml of it with $\frac{N}{10}$NaOH using phenolphthalein as an indicator. Repeat the process for other bottles

17.14.5 Observations

Volume of acetic acid in titration $= 25$ ml.

	Burette Reading			
Bottle No.	No.	Initial	Final	Volume of NaOH used
1	1			V_1
	2			
	3			
2	1			V_2
	2			
	3			
3	1			V_3
	2			
	3			
4	1			V_4
	2			
	3			
5	1			V_5
	2			
	3			
6	1			V_6
	2			
	3			

17.14.6 Calculation

(i) Bottle No. 1

$$\text{Initial con. of } CH_3COOH = 0.6\ N$$

After adsorption

$$N_1V_1 = N_2V_2$$
$$CH_3COOH \quad NaOH$$

$$N_1 \times 25 = \frac{N}{10}V_1 \text{ (from table)}$$

$$N_1 = \frac{V_1}{250}$$

Change in CH_3COOH concentration due to adsorption, $C = 0.6 - N_1$ and in 60 ml change in conc.

$$A = \frac{(0.6 - N_1) \times 60}{1000} \times 60 \text{ (mol. wt. of } CH_3COOH)$$

(ii) Similarity for Bottle No. 2

$$\text{Initial conc. of } CH_3COOH = \frac{0.6 \times 50}{60} = 0.5\ N$$

(iii) For Bottle No. 3

$$\text{Initial conc. of } CH_3COOH = \frac{0.6 \times 40}{60} = 0.4\ N$$

Similarly the change in conc. (A) in 60 ml volume for all bottles will be calculated and for each bottle $\log(A/m)$ is calculated. (Where $m = 2$ g of charcoal.)

Also plot $\log(A/m)$ versus $\log C$.

17.14.7 Result

The validity of Freundlich isotherm is shown by a straight line plot between $\log A/m$ vs $\log C$.

17.14.8 Precautions

1. There should be equal amount of adsorbent in each bottle.
2. Each bottle should be kept for equal time.

3. The temperature of experiment should remain constant.

17.15 Experiment No. 15

17.15.1 Object

Determination of equilibrium constant by applying distribution law.

17.15.2 Apparatus and Chemicals

Pipette, burette, reagent bottles, benzene, benzoic acid, NaOH, phenolphthalein.

17.15.3 Theory

When a substance distributes itself between two phases in equilibrium with each other, the ratio of the concentration of the substance in them is constant, at a constant temperature, provided the molecular state of the distribution of substance is same in both the phases.

If C_1 and C_2 are the conc. of substance in two phases, then

$$\frac{C_1}{C_2} = K \text{ (partition coefficient or distribution coefficient)}$$

If the solute associates or dissociates in any of the phases then, according to Nerst for association in one of the solvent (say for C_2), the new modification for distribution law is

$$K = \frac{C_1}{\sqrt[n]{C_2}}$$

where n is number of molecules associated for dissociation and the modification is

$$K = \frac{C_1}{C_2(1-\alpha)}$$

where α is degree of dissociation.

For benzoic acid in water-benzene system, the benzoic acid associates as dimer in benzene i.e., experimentally $\frac{C_1}{\sqrt{C_2}} = \text{constant}$.

17.15.4 Procedure

1. Dissolve 2.5 g benzoic acid in 250 ml of benzene.
2. Take four bottles numbered as 1, 2, 3, and 4 and add 50 ml, 40 ml, 30 ml, 20 ml of benzoic acid solution in bottles, respectively.
3. Now add 50 ml of water in each bottle and 10, 20, and 30 ml benzene in bottles 2, 3, and 4, respectively.
4. Shake well and leave the bottles for 1 hour. These should be vigorously shaken at every 10-minute interval.
5. After 1 hour, the two layers separate clearly; the upper layer is benzene and lower layer is water.
6. Now take 10 ml of benzene layer in conical flask and mix it with 20 ml of water with few drops of phenolphthalein. Titrate with $\frac{N}{10}$NaOH to get end point. Repeat the process to get two concordant readings.
7. Similarly titrate the benzene layer for remaining three bottles.
8. Now pipette out 10 ml water layer of one bottle and titrate it with $\frac{N}{10}$NaOH using phenolphthalein as indicator, and similarly titrate aqueous layer of the remaining three bottles.

17.15.5 Observations

1. Titration with benzene layer

Bottle no.	Burette reading		Volume of NaOH used (ml)
	Initial	Final	
1.	0.0		
2.	0.0		
3.	0.0		
4.	0.0		

2. Titration with water layer

Bottle no.	Burette reading		Volume of NaOH used (ml)
	Initial	Final	
1.	0.0		
2.	0.0		
3.	0.0		
4.	0.0		

17.15.6 Calculation

$$C_1 = \frac{\text{Volume of NaOH used } \times \frac{N}{10}}{\text{Volume of aquous layer (10 ml)}}$$

$$C_2 = \frac{\text{Volume of NaOH used } \times \frac{N}{10}}{\text{Volume of benzene layer (10 ml)}}$$

Calculate C_1 and C_2 for each bottle and put in $K = \frac{C_1}{\sqrt{C_2}}$. The value of K for each bottle should be constant. Also plot a graph between C_1 vs. $\sqrt{C_2}$ which should be straight line and shows the association of benzoic acid as 2 in benzene.

17.15.7 Result

1. The partition coefficient of benzoic acid between water and benzene is ———.
2. The association of benzoic acid in benzene is dimeric.

17.15.8 Precautions

1. The bottles should be properly stopped; otherwise, benzene evaporation may fluctuate the result.
2. The pipetting of aqueous layer should not possess benzene.
3. The droplets that stick on the wall of bottles should be involved in pipetting.

17.16 Experiment No. 16

17.16.1 Identification of Given Organic Compound

The given organic compound should first of all be tested for the special elements, i.e., nitrogen, sulfur, and halogens. These elements are detected by Lassaigne's test.

17.16.2 Lassaigne's Solution

Take a freshly cut sodium piece of very small size into an ignition tube, and a small amount of given compound is added. The tube is then held with a pair of tongs and heated genetly on the flame. Then tube is heated strongly till red not and then dropped in about 15.0 ml of distilled water in a small porcelain dish. The tube will itself break into pieces. Make three-four such ignitions. Any unreacted sodium will react with distilled water and make it alkaline. Boil the solution and filter, and the filterate is called sodium fusion extract. Now test the special elements as follows.

Test for Nitrogen

Mix 3.0 ml of sodium fusion extract with 1.0 ml of freshly prepared $FeSO_4$ solution. A dirty green precipitate of ferrous hydioxide will be obtained. Heat the solution, cool it, and then add dilute H_2SO_4 in excess. Appearance of green or blue color confirms the presence of nitrogen.

The appearance of blue/green color is due to the formation of $Fe_4[Fe(CN)_6]_3$ i.e., ferric ferrocyanide.

$$FeSO_4 + 2NaOH \longrightarrow \underbrace{Fe(OH)_2 + Na_2SO_4}_{\text{directly green ppt.}}$$

$$Fe(OH)_2 + \underbrace{6NaCN}_{\text{from compound}} \longrightarrow Na_4Fe(CN)_6 + 2NaOH$$

$$3Na_4Fe(CN)_6 + 2Fe_2\,(SO_4)_3 \longrightarrow Fe_4\,[Fe(CN)_6]_3 + 6Na_2SO_4$$

Test for Sulfur

- Mix 3.0 ml of sodium fusion extract with 3.0 ml of freshly prepared sodium nitroprusside. Appearance of violet color confirms the presence of sulfur.

  During sodium fussion, the sulfur present is converted to $Na_2\,S$. The reaction is

$$Na_2\,S + Na_2Fe(NO)(CN)_5 \longrightarrow \underbrace{Na_3\,Fe\,(O = N - S.Na)\,(CN)_5}_{\text{violet coloured complex}}$$

- Mix 3.0 ml of sodium fusion extract with acetic acid and lead acetate solution; formation of black precipitate (PbS) confirms the presence of sulfur.

Test for Bromine and Iodine

Take 2.0 ml of sodium fusion extract, and add few drops of dilute H_2SO_4. Now add 1.0 ml of chloroform and chlorine water in excess. Shake well the tube. If the color of the chloroform layer changes to

yellow or light brown	–Br present
violet	–I present.

The principle of the test is that chlorine liberates bromine from bromides and iodine from iodides. Thus, bromides or iodides formed during sodium fusion will librate corresponding halogen

$$2NaBr + Cl_2 \longrightarrow 2NaCl + Br_2$$

$$2NaI + Cl_2 \longrightarrow 2NaCl + I_2$$

Test for Chlorine

Take 2 to 3 ml of sodium fusion extract, and add 1 ml of conc. HNO_3. Boil the solution and cool it and add few drops of $AgNO_3$. Formation of white precipitate soluble in excess of NH_4OH indicates the presence of chlorine.

The students should be cautious with the appearance of white ppt. because if chlorine is present in the distilled water, a white ppt. may appear. It is advisable to compare the result of same test with the distilled water used.

The following groups should then be tested irrespective of the special elements present.

1. ***Phenols:*** Take 1 ml of aqueous or alcoholic solution of the given compound and add 1 ml of aqueous $FeCl_3$ solution. Appearance of blue, violet, and green-wine red color confirms the presence of phenolic compounds.
2. ***Carboxylic acids:*** Take 2 ml of 10% solution of $NaHCO_3$ in a test tube and then add a pinch of given compound. The effervescence of gas (CO_2) indicates the presence of carboxylic acid.
3. ***Aldehydes and Ketones:*** Dissolve a small amount of the compound in 2 – 3 ml of dil HCl in a tube. Similarly dissolve few crystals of 2, 4-dinitrophenylhydrazine in dil HCl in another tube. Mix the two solutions. The appearance of precipitate will confirm the presence of aldehydes or ketones.
4. ***Alcohols:*** To 1 ml of the aqueous solution of the given compound, add few drops of ceric ammonium nitrate. The appearance of red color will confirm the presence of alcohol.

$$(NH_4)_2\left[Ce\,(NO_3)_6\right] + 2R - OH \longrightarrow \underbrace{[Ce\,(NO_3)_4\,(R - OH)_2]}_{\text{red colour}} + 2NH_4NO_3$$

5. ***Carbohydrates:*** Take 2–3 ml of the aqueous solution of the given compound in a test tube. Add two to three drops of Molisch's reagent (10% solution of α-napthol in alcohol) and then add 2 ml of conc. H_2SO_4 from the sides of the test tube and allow to stand for few min. Appearance of red-violet ring at the junction of two layers confirms the presence of carbohydrate. If the tubc is shaken, the entire mixture will become red-violet.

If nitrogen is present, the following groups should be tested.

1. ***Nitro group:*** Take 0.1 g of the given compound in a test tube and add 2 ml of alcohol, a pinch of zinc dust, and two to three drops of $CaCl_2$. Boil the test tube and cool it. Filter the contents in another test tube containing Tollen's reagent. A white precipitate changing from gray to black confirms the presence of nitro group. In this test, nitro group is reduced to —NHOH and the test is known as Mulliken's test.
2. ***Primary Amino Group (Aromatic):*** First prepare fresh solution of $NaNO_2$ by dissolving small amount in water, and then 2 to 3 ml of dil HCl is added. The solution is mixed carefully. In a separate tube, take 2–3 ml of aqueous solution of compound and add 3 ml of dil. HCl and mix well. Now add cold $NaNO_2$ solution in the compound solution.

   If solution becomes green, violet, blue, red, and red-brown, amino group is present.

   In many cases, the above colors are not clear. In such cases, cool the mixture at 0 °C and add cooled solution of β-napthol in NaOH. The formation of intense color dye (azo dye) will confirm the presence of aromatic amino group in the given compound.
3. ***Amides:*** Take small amount of the given compound in a test tube and add 1 to 2 ml of aq. NaOH solution. Heat the tube. If ammonia gas is evolved, which can be characterized by its pungent smell, it confirms the presence of amide group in the compound.

If nitrogen and sulfur both are present, the compound should be tested as follows: Take 0.1 gm of the compound in a test tube and add 2 ml of NaOH solution. Heat the tube for 1 min, cool it, and then add few drops of aqueous lead acetate solution. The appearance of dark brown or black color will confirm the presence of thioureas. [It is advisable that students should write the experiment on identification in a tabular form in three columns—Experiment, Observation, and Inference.]

Some Common Reagents in the Chemistry Laboratory At many occassions, common reagents are not available in the laboratory. These reagents can be prepared as follows:

1. ***Benedict Solution***—Dissolve 1.70 g of $CuSO_4$ in 100 ml of water. Add to it a solution of 17.0 g of sodium citrate and 10 g anhydrous Na_2CO_3 prepared in 80 ml of water. Make the solution to 1.0 liter.
2. ***Chlorine Water***—Should be prepared carefully and kept in a tight bottle. The chlorine water can be prepared by bubbling chlorine gas in water till it gets saturated.
3. ***Ceric Ammonium Nitrate Solution***—Dissolve 40 g of ceric ammonium nitrate in 100 ml of 2 N nitric acid. If it does not dissolve completely, slightly warm the mixture.
4. ***2,4 -Dinitrophenylhydrazine reagent***—Dissolve 2.0 g of 2,4-dinitrophenylhydrazine in 10 ml of conc. H_2SO_4. Add to it slowly 200 ml of ethyl alcohol, and let the mixture stand for 1 h. Filter the mixture and keep it in a bottle.
5. ***Fehling Solutions***—

   – Fehling A – can be prepared by dissolving 17.0 g of crystalline $CuSO_4$ in 250 ml of water. Add few drops of dil H_2SO_4 to prevent hydrolysis.
   – Fehling B – can be prepared by dissolving 8.6 g of sodium potassium tartarate (Rochelle salt) and 35 g of NaOH in 250 ml of water.

   To use Fehling solution, equal volumes of Fehling solution A and B are mixed.
6. ***Ferric Chloride Solution*** - is prepared by dissolving 1 g of $FeCl_3$ in 100 ml of water.
7. ***Molisch's reagent*** - Dissolve 10 g of α-naphtol in alcohol. Filter it if the solution has some undissolved material in it.
8. ***Alkaline β-napthol solution*** - Dissolve 1 g of β-napthol in 100 ml of 10% NaOH solution.
9. ***Phenolphthalein*** - Dissolve 1 g of phenolphthalein in 100 ml alcohol. Add slowly 100 ml water with constant stirring.
10. ***Silver Nitrate Solution*** - Dissolve 1 g of silver nitrate in 100 ml of alcohol.
11. ***Tollen's Reagent*** - Mix 5 ml of silver nitrate solution with few drops of NaOH solution. A dirty white precipitate will be formed in the test tube due to AgOH. Dissolve the precipitate by drop by drop addition of NH_4OH.
12. ***Dilute acids or alkalies*** - Acids are diluted with water to get nearly 4 N solution. Similarly 4 N solution of NaOH or KOH is considered as dilute alkali.

17.17 Experiment No. 17

17.17.1 Object

Determination of heat of neutralization of a strong base by strong acid.

17.17.2 Apparatus

Beaker, calorimeter, thermometer.

17.17.3 Theory

The chemical change that occurs on mixing a strong acid with strong base is

$$H^+ + OH^- \longrightarrow H_2O$$

The value of ΔH for this reaction is called heat of neutralization. It can be determined by carrying out the reaction in a calorimeter and measuring the change in temperature using a sensitive thermometer. Initially the heat capacity of the calorimeter or water equivalent is calculated, and specific heat of the reacting solutions is assumed to be similar to that of water.

17.17.4 Procedure

Step 1. Take some water in a 400 ml beaker, and heat the beaker so that water temperature reaches to 50 °C. Use a sensitive thermometer to measure the temperature. Immediately transfer 50 ml of water into the calorimeter at fixed time, and again record temperature at every 30 s for 3 min. Now plot the graph between temperature and time, and draw a vertical line corresponding to the time of mixing cold water. Extrapolate the two branches of curve to this time and note the corresponding temperatures. This will provide information about initial temperature of calorimeter, with hot water and final temperature of mixture. If W is the water equivalent, T_1, T_2 and T_m are temperature of cold water, hot water and mixture, the water equivalent, and W can be calculated from the relation.

$$50\,(T_m - T_1) = (W + 50)\,(T_2 - T_m)$$

Step 2. Now empty the calorimeter and completely dry it using filter paper. Record the temperature of given acid solution [usually 1 M HCl solution]. Transfer 50 mL of 1M NaOH in calorimeter, and record temperature at every 30 s with stirring. At a known time, add 50 mL of acid in calorimeter and stir well and again record temperature at an interval of 30 s for about 3 min. Again plot cooling curve between temperature and time. Extrapolate the branches like in first case, and record the initial and final temperatures. Let the temperature be T_3 and T_4. Calculate heat evolved, ΔH, using relation

$$\Delta H = (50 + 50 + W)\,(T_4 - T_3)$$

17.17.5 Result

Report the heat of neutralization of the given 50 mL solution.

17.18 Experiment No. 18

17.18.1 Object

Paper chromatographic separation of mixture of different metal ions.

17.18.2 Apparatus

Strips of Whatmann paper - 20 of approximate size 15 cm × 5, 6 cm. Glass jar (30 × 10 × 20 cm^3) with a cover.

17.18.3 Theory

The distance traveled by solute and solvents is determined in a chamber saturated with vapors of solvent. The ratio of 2 is called R_f (Reference front). Thus,

$$R_f = \frac{\text{Distance traveled by the solute from the original spot}}{\text{Distance traveled by the solvent from the original spot}}$$

17.18.4 Procedure

Take the filter paper strip and draw a horizontal line nearly 2 cm from one end of strip. Now take solutions of different metal ions [such as Ag(I), Hg (II), Pb (II) etc.], and put a drop of each of them on the line using glass capillary. Mark them as A, B, C. Now put a drop of a mixture of three metal ions and mark it as D. Dry the strip in air or in an oven. Now put nearly 100 ml of suitable developing solvent in jar such as

(i) Methyl $- n$ propyl ketone + 10 N HCl(85 : 15) by volume

(ii) tert-Butanol + acetone + water + conc. $HNO_3(50 : 30 + 12 + 8)$ by volume
(iii) distilled or deionized water

The paper strip is now fixed vertically from the spot end, using a clip so that the spot does not dip in the solvent. Let the developing solvent move on the paper. When the solvent stops to move, remove the strip and dry it. The spots can be visualized by spraying them with ammonium sulfide solution. Now calculate the R_f of each metal ions (spots A, B, C) and compare them with mixture.

M/4 solution of potassium chromate can also be used for spray and imparts following colors; Ag (I)-orange, Hg (II)-pink, Pb (II)-light pink.

17.18.5 Result

Report R_f values for different metal ions.

Note: Different metal ions can be separated using paper chromatographic technique. The developing solvent and visualization solvent will be different. For example,

(a) For separation of Fe(III), Al(III), and Cr(III)-developing solvent is mixture of glacial acetic acid and methanol in ratio of 3 : 1. The visualizing solvent will be mixture of 1% alcoholic alizarin and 0.5% solution of benzidine in acetic acid.
(b) For separation of Ca(II), Ba(II), and Sr(II)-developing solvent may be 1 : 1 mixture of methanol and ethanol or 20% solution of KCNS. The developing solvent will be 0.1% aqueous solution of violuric acid.

17.19 Experiment No. 19

17.19.1 Object

Preparation of iodoform from propanone.

17.19.2 Apparatus

Measuring cylinder 50 ml, conical flask 250 ml, boiling tube, 250 ml, melting point apparatus, thermometer.

17.19.3 Theory

Iodoform can be prepared by the reaction of propanone with iodine and NaOH. The reaction is

$$CH_3COCH_3 + 3I_2 + 4NaOH \rightarrow CH_3COO^-Na^+ + CHI_3 + 3H_2O + 3NaI$$

17.19.4 Procedure

Transfer 1 ml propanone with a pipette to conical flask. Then add 40 ml of 40% KI solution and 16 ml of 2 M NaOH solution prepared in water. Finally add 40 ml of 10% solution of sodium chlorate in water. Mix the contents well by slowly swirling the conical flask. A yellow colored precipitate will form in about 4, 5 min. Keep the flask for 15 min with occasional shaking. Filter the product using Buchner funnel and wash the product with water. Allow the product to dry. Take a small quantity of product in boiling tube, and dissolve it in minimum amount of ethyl alcohol by keeping it in water bath. When the product dissolves completely, take the tube out of water bath and allow to cool. The yellow colored crystals will appear in the boiling tube. Filter them, dry, and determine melting point using block or melting point apparatus. The melting point should be between 119 and 121 °C.

17.19.5 Precaution

Don't heat boiling tube directly on burner. As ethyl alcohol is flammable, it may catch fire.

Chapter 18
Electrical Conductivity and Ionization

In metals, conductance is the reciprocal of resistance, and resistance is proportional to length l and inversely proportional to area "a" i.e.,

$$R \propto \frac{l}{a} \qquad \text{or} \qquad R = \rho \times \frac{l}{a}$$

where R is the resistance and "ρ" is a constant and called specific resistance. Similarly, like a metallic conductor, the resistance "R" is also shown by an electrolyte.

The reciprocal of specific resistance ρ, for a solution of electrolyte, is known as specific conductance, i.e.,

$$\frac{1}{R} = \frac{1}{\rho} \times \frac{a}{l}$$

i.e., conductivity = specific conductivity $\times \frac{a}{l}$, and specific conductivity (K_v) = conductivity $(C) \times \frac{l}{a}$.

So, specific conductivity may be defined as

"*The conductivity of* 1 cm^3 *of a solution of electrolyte is known as specific conductivity.*"

The specific conductivity is denoted by K_v, where V is the volume in ml of electrolytic solution. The unit of specific conductivity is reciprocal of ohms (r.o.) or mhos or Ω^{-1}.

The ratio l/a is called cell constant which depends upon the area of cross section of conductive solution and the length of conductive solution "l."

© The Author(s), under exclusive license to Springer Nature Switzerland AG 2026
R. N. Goyal, H. Goel, *Textbook of Engineering Chemistry*,
https://doi.org/10.1007/978-3-032-22068-4_18

18.1 Measurement of Conductivity of the Solution

The conductivity of a solution can be measured by measuring the resistance of the solution, which can be measured by using Wheatstone bridge arrangement.

The Wheatstone bridge method is used to measure the resistance of a solution, which has the following two problems:

1. On passing direct current (DC), the concentration of solution changes due to the electrolysis of electrolytic solution.
2. Due to the polarization of electrode, a back EMF is produced. To overcome these problems, the following changes are made:
   (i) In place of DC, if alternating current (AC) is used, the polarization problem is not developed. The polarization can further be minimized by using cell electrode of platinum having coating of finely divided platinum, to increase surface area of electrode which reduces the polarization.
   (ii) If AC is used then in place of galvanometer, headphone is used to detect the null point.

18.2 Procedure

The test solution is placed in the conductive cell as shown in Fig. 18.1 having platinum electrode coated with finely divided platinum and sealed in glass tubes and fixed in ebonite cover to fix the distance between electrodes. The cell is attached with circuit and AC is passed using induction coil. By changing the resistance in resistance box "*R*" and sliding the joystick on wire, the minimum sound in headphone is detected to get null point. At this point.

$$\frac{\text{Resistance of Solution}}{\text{“}R\text{'' of R. Box}} = \frac{\text{Length } BC}{\text{Length } AB}$$

or

$$\text{Resistance of solution} = \frac{BC}{AB} \times R$$

or

$$\text{Conductivity of solution} = \frac{AB}{BC} \times \frac{1}{R}$$

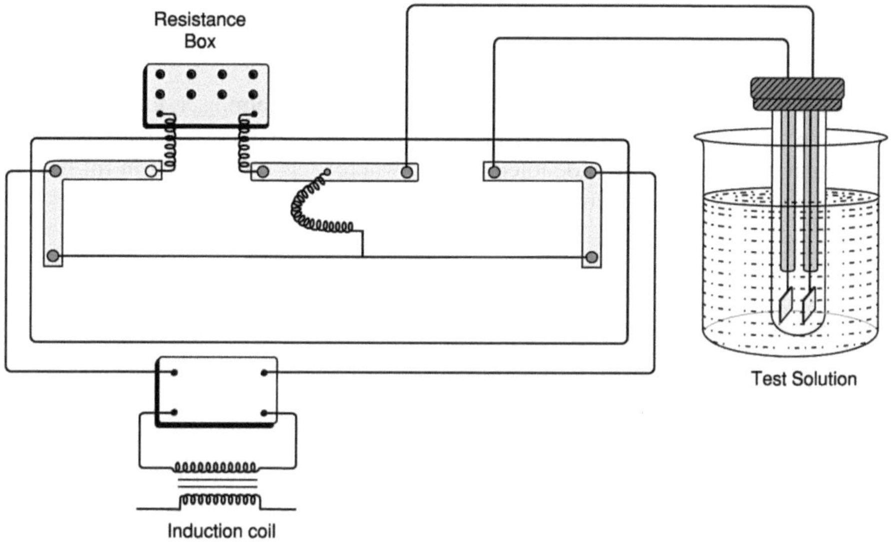

Fig. 18.1 Simple setup of a conductivity cell used to measure the conductance of a solution

18.3 Specific Conductance

The specific conductance K_v can be calculated by knowing the value of area "a" and length (l) between the electrodes in conductive cell. The value of l/a is called cell constant, and by measuring the cell constant, the value of K_v can be obtained using the formula.

$$K_v = C \times \frac{l}{a}$$

where C is the observed conductivity. Thus,

$$\text{Specific conductivity} = \text{Observed conductivity} \times \text{Cell constant}$$

18.4 Measurement of Cell Constant

The cell constant can be determined by using KCl solution of $N/50$ concentration and determining the conductivity at 25 °C. The specific conductivity of $N/50$ KCl at 25 °C was determined by Kohlrausch and was found to be 0.002765. Using these values, cell constant can be calculated as

$$\text{Cell constant} = \frac{0.002765}{\text{Conductivity of } N/50 \text{ KCl observed}}$$

Eq. Conductivity λ_v The conductivity of solution containing 1 gm. equivalent of an electrolyte when placed between two electrodes with 1 cm apart is called equivalent conductivity. It is denoted by Λ_v, where V is the volume in ml having 1 gm equivalent of electrolyte. Its unit is mohs or r.o. (revers of ohm). So,

$$\lambda_V = K_v \times V \quad (V \text{ is the volume in ml})$$

Thus, the eq. conductivity may also be defined as
"The product of specific conductivity and the volume of electrolytic solution in ml *having* 1 gm *equivalent of the electrolyte."*

Molecular Conductivity μ_v **or** Λ_v The conductivity of a solution having 1 gm-mol. of an electrolyte when placed between two electrodes 1 cm apart is known as molecular conductivity of the solution. It is denoted by μ_v or Λ_v. Its unit is mohs. or r.o. (reverse of ohm). So

$$\mu_v = K_v \times V \quad (V \text{ is the volume in ml})$$

18.5 Effect of dilution on C, K_v, λ_v and μ_v

(i) ***The conductivity "C":*** Conductivity of solution increases on dilution because the conductivity of solution depends upon the number of ions in solutions. On dilution, the degree of dissociation of electrolyte is increased to produce more ions. The dilution reduces the interionic interaction and reduces viscosity which also help in increasing the conductivity.

(ii) ***The specific conductivity "K_v":*** K_v decreases with dilution as the number of current carrying ions present per ml becomes less.

(iii) ***The equivalent "λ" and molecular conductivity "μ_v":*** λ_V and μ_v increase with dilution and reach to a maximum value at infinite dilution. The increase is due to the reason of $\mu_v = \boldsymbol{K_v} \times V$ or $\lambda_v = K_v \times v$. If $\boldsymbol{K_v}$ decreases, then on dilution V increases not in the same proportions as K_v decrease. So the values of λ_v and μ_v increase on dilution.
At infinite dilution, the values of λ_V and μ_v remain constant due to the fact that dissociation of electrolyte completes the numbers of ions per ml with dilution and hence the product of increase of v and decrease of K_v remains similar.

Example 18.1 Specific conductance of $N/10$ KCl at 25 °C is 0.0112 mhos: The resistance of cell containing the solution at the same temperature was found to be 65 ohms. Calculate the cell constant.

Solution Specific conductance $K_v = 0.0112$ mhos

$$\text{Conductance } C = \frac{1}{R} = \frac{1}{65}$$

$$K_v = C \times \frac{l}{a} \text{ (cell constant)}$$

$$\Rightarrow \quad 0.0112 = \frac{1}{65} \times \frac{l}{a} \text{ (cells constant)}$$

$$\frac{l}{a}\text{(cell constant)} = 65 \times 0.0112 = 0.728$$

Example 18.2 The resistance of an $N/10$ acid is 2.5×10^3 ohms. By using a cell having cell constant 1.150, calculate the equivalent conductivity.

Solution

$$\text{Resistance “}R\text{”} = 2.5 \times 10^3 \Omega$$

$$\text{Conductance “}C\text{”} = \frac{1}{2.5 \times 10^3} \Omega^{-1}$$

$$\text{Cell constant } (l/a) = 1.150$$

$$\text{Specific conductance, } K_v = C \times \text{ cell constant.}$$

$$K_v = \frac{1}{2.5 \times 10^3} \times 1.150$$

equivalent conductivity $\lambda_v = K_v \times V$.

Normal solution is obtained by dissolving 1 gm eq. in 1000 cc. (1 liter) $N/10$ solution is obtained by dissolving. 1 gm equivalent in = 1000 × 10 = 10000 c.c.

Hence

$$\lambda_v = \frac{1.150}{2.5 \times 10^3} \times 10000$$

$$= 4.6 \text{ mhos.}$$

Example 18.3 0.5 N solution of a salt has a resistance of 25 ohms with a cell having electrode 1.70 cm long and area 4.5 cm^2. Calculate the value of equivalent conductivity.

Solution Given $R = 25\Omega$

$$C = \frac{1}{25} \text{ r.o.}$$

$$l = 1.7 \text{ cm}$$

$$a = 4.5 \text{ cm}^2$$

and

$$K_v = C \times \frac{l}{a}$$

$$K_v = \frac{1}{25} \times \frac{1.7}{4.5}$$

Further

$$\lambda_v = K_v \times V \qquad \left(V = \frac{1000}{0.5} = 2000\right) \text{ml}$$

$$\lambda_v = \frac{1.7}{25 \times 4.5} \times 2000$$

$$= 30.22\Omega^{-1}$$

18.5.1 Relation Between Equivalent Conductivity and Degree of Dissociation

In a solution, the greater the number of ions, the greater is the conductivity. On dilution, the degree of dissociation of an electrolyte is increased, and the number of ions increased in solution and hence, the conductivity of solution should increase.

It is discussed previously that equivalent conductivity increases on dilution and when the electrolyte has completely dissociated into ions, the equivalent conductivity attained the maximum value. This is called equivalent conductivity at infinite dilution. Hence, the degree of dissociation may be related to equivalent conductivity "λ_v" for V ml dilution and "λ_∞" for infinite dilution.

$$\text{So, degree of dissociation } \alpha = \frac{\lambda_v}{\lambda_\infty}$$

When $\alpha = 1$ then $\lambda_v = \lambda_\infty$ which is the case for weak electrolytes. For strong electrolytes λ_v is much less than λ_∞, which is due to the relaxation and electrophoretic effects for strong electrolyte.

18.5.2 Relation Between μ_v and λ_v for Dibasic Acids

For monobasic acids, equivalent and molecular conductivity can be defined as

$$\lambda_v = K_v \times \text{V}$$

(where V is the volume having 1 gm equivalent of electrolyte), and

$$\mu_v = K_v \times \text{V}$$

(where V is the volume having 1 gm mol. of electrolyte).

For dibasic acids having $N/10$ solution then

$$\text{for } \lambda_v \text{ the value of } V = 10000 \text{ c.c.} \left(\frac{N}{10} \text{ sol.}\right)$$

and

$$\mu_v \text{ the value of } V = 20000 \text{ c.c.}$$

as for dibasic acids mol. wt. is double of the eq. weight. So,

$$\frac{\mu_v}{\lambda_v} = \frac{K_v \times 20000}{K_v \times 10000} = 2$$

So

$$\mu_v = 2\lambda_v$$

Example 18.4 The mol. conductivity of $\frac{N}{32}CH_3COOH$ is $9.625\Omega^{-1}$ and at infinite dilution is 385. Calculate the degree of dissociation of acid.

Solution For

$$CH_3COOH \text{ mol. wt. } = \text{Eq. wt.}$$

So

$$\lambda_v = \mu_v = 9.625\Omega^{-1}$$

and

$$\lambda_\infty = \mu_\infty = 385$$

and degree of dissociation

$$\alpha = \frac{\lambda_v}{\lambda_\infty}$$

$$\alpha = \frac{9.625}{385} = 0.025 \quad \text{or} \quad 2.5\% \text{ dissociation.}$$

18.6 Transport Number or Hittorff's Number

On passing current through an electrolytic solution, the ions of electrolyte move toward oppositely charged electrodes. Due to this movement of ions, the electric current flows through an electrolyte to show electrical conductivity. It is charge carriers like in solids. But the speed of ions usually are different. Thus, ion transport number is a quantity which can be defined as contribution of ions to the electric current in electrolytes due to different electrical mobility. This fraction/share of current carried by each type of ion is different which is named as transport number. "The fraction of the total current carried by an ions is termed it's transport number."

If n_a is the transport number of anion then,

$$n_a = \frac{\text{Current carried by anion}}{\text{Total current carried in solution}}$$

and if n_c is the transport number of cation, then

$$n_c = \frac{\text{Current carried by cation}}{\text{Total current carried in solution}}$$

If speed of cation and anion in solution are u_a and u_c, respectively, then

$$\text{Current carried by anion } \propto u_a = k \times u_a \tag{i}$$

and

$$\text{Current carried by cation } \propto u_c = k \times u_c \tag{ii}$$

Similarly total current carried by

$$\text{both ions } \propto u_a + u_c = k\,(u_a + u_c) \tag{iii}$$

Dividing (i) and (iii)

$$\frac{\text{Current carried by anion}}{\text{Total current carried}} = \frac{k \times u_a}{k\,(u_a + u_c)}$$

So,

$$n_a = \frac{u_a}{u_a + u_c} \tag{iv}$$

and similarly

$$n_c = \frac{u_c}{u_a + u_c} \tag{v}$$

adding equivalent (iv) and (v)

$$n_a + n_c = \frac{u_a}{u_a + u_c} + \frac{u_c}{u_a + u_c} = 1$$

18.6.1 Experimental Determination of Transport Number

The most common method used for the determination of transport number is Hittroff's method. It is based upon the principle that falls in concentration round any electrode is proportional to the speed of the ion moving away from it.

So

$$\text{Fall around cathode} \ \propto \ \text{speed of anion} \ \propto u_a$$

and

$$\text{Fall around anode} \ \propto \ \text{speed of cation} \ \propto u_c$$

again

$$\text{Fall around cathode} \ = k u_a \tag{vi}$$

and

$$\text{Fall around anode} \ = k u_c \tag{vii}$$

Dividing (i) and (ii),

$$\frac{\text{Fall around cathode}}{\text{Fall around anode}} = \frac{k u_a}{k u_c} = \frac{u_a}{u_c}$$

adding "1" on both side

$$1 + \frac{\text{Fall around cathode}}{\text{Fall around anode}} = 1 + \frac{u_a}{u_c}$$

or

$$\frac{\text{Fall around cathode} \ + \ \text{Fall around anode}}{\text{Fall around anode}} = \frac{u_a + u_c}{u_c}$$

or

$$\frac{\text{Total Fall around both electrode}}{\text{Fall around anode}} = \frac{u_a + u_c}{u_c}$$

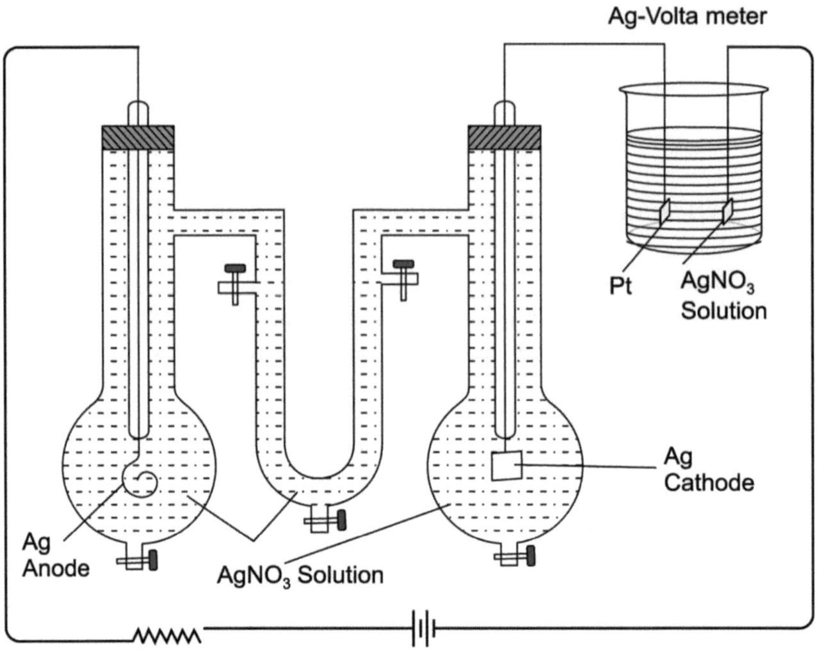

Fig. 18.2 Electrolysis apparatus with silver electrodes

Taking reciprocal

$$\frac{\text{Fall around anode}}{\text{Total fall around both electrode}} = \frac{u_c}{u_a + u_c}$$

again

$$\frac{\text{Fall around anode}}{\text{Total amount decomposed}} = \frac{u_c}{u_a + u_c} = n_c$$

To calculate the value of fall around anode and the total amount decomposed, a special type of electrolytic cell is used and is as shown below.

The cell consists of two glass flasks which are joined by a U-tube, as shown in Fig. 18.2. The flask and tubes are connected with stop cocks to draw out solution from different part of apparatus. One flask had silver anode and other has silver cathode. The apparatus is filled with standard silver nitrate solution and connected with silver voltameter in series. The current is passed for 2–3 hours and a suitable amount of solution is withdrawn from the anodic area. The solution is titrated with standard KCNS solution. Now the amount of Ag present before and after electrolysis is calculated.

The weight of silver deposited in silver voltameter is also noted. It is observed that the amount of silver increased in anode area, which is due to the fact that Ag^+ ions move from anode and NO_3^- ions move from cathode simultaneously. But Ag^+move slower than NO_3^- ions, and silver dissolved from anode and increased in amount of Ag near anode area. The amount of silver dissolved from anode by NO_3^- ions is equal to the amount of Ag^+ deposited in Ag-voltameter. Hence, if NoAg ion migrated from anode, the total increase in concentration of silver would have been equal to the Ag deposited in Ag-voltameter.

If W is the amount of Ag deposited in voltameter in series, which is equal to the amount of total current passed, and w is the increase in concentration of Ag round anode after electrolyis experiment, then $(W - w)$ is the decrease in concentration round anode, so

$$\text{Transport No. of Ag}^+ = \frac{\text{Fall in conc. around anode}}{\text{Total amount decomposed}} = \frac{W - w}{W}$$

and

$$\text{Transport No. of NO}_3^- = 1 - \frac{W - w}{W}$$

Example 18.5 A current deposited 20 mg of Ag in Ag voltameter in series. The increase in concentration of Ag-round anode was found to be 15.9 mg. Calculate transport number of Ag^+ and NO_3^- ions

Solution Total deposition in Ag-voltameter $W = 20.0$ mg

Increase in Ag concentration around anode = 15.9 mg.

$$\text{Total number of Ag+} = \frac{W - w}{W} = \frac{20 - 15.9}{20} = 0.205$$

and

$$\text{Total number of NO}_3^- = 1 - \text{T.No. of Ag}^+ = 1 - 0.205 = 0.795$$

Example 18.6 On passing current through Cu electrode dipped in $CuSO_4$, the total mass deposited in Cu-voltameter was 0.254 gm. The mass of $CuSO_4$ around cathode solution, before and after electrolysis, was 1.511 gms and 0.711 gm, respectively. Calculate total number of Cu^{++} and SO_4^{--} ions.

Solution Fall in wt. of $CuSO_4$ around cathode = 1.511 − 0.711 = 0.800 g.

Amount of $CuSO_4 \cong$ Cu deposited in Cu voltameter in series

$$\Rightarrow \quad \frac{\text{wt. of Cu deposited}}{\text{Wt. of CuSO}_4} = \frac{\text{At wt. of Cu}}{\text{Mol. wt. of CuSO}}$$

$$\Rightarrow \quad \frac{0.254}{\text{wt. of CuSO}_4} = \frac{63.5}{249.5}$$

$$wt.\text{ of CuSO}_4 = \frac{249.5 \times 0.254}{63.5} = 0.998(\text{ W})$$

As this value is for cathode so fall in conc. reflect SO_4^{--}

$$\text{T-No. SO}_4^{--} = \frac{W - w}{W} = \frac{0.800}{0.998} = 0.802$$

$$\text{T-No. Cu}^{++} = 1 - 0.802 = 0.198$$

18.7 Kohlrausch's Law of Ionic Mobilities

Kohlrausch's in 1875 observed that the difference between the equivalent conductivity at infinite dilution of pairs of salts, having an ion in common, is constant at a constant temperature.

For example,

	λ_∞ difference		λ_∞ difference
KBr	2.341	NaBr	20.6
NaBr		NaCl	
KCl	23.41	KBr	2.06
NaCl		KCl	

On the basis of above observation, he concluded that every ion at infinite dilution contributed a definite amount of equivalence conductivity irrespective of other ion of electrolyte, i.e.,

"The equivalent conductivity at infinite dilution is the sum of two values one depending upon the cation and other on the anion."

or in other words the equivalent conductivity at infinite dilution is equal to the sum of the conductance of anions and cations, i.e.,

$$\lambda_\infty = \lambda_a + \lambda_c$$

where λ_a and λ_c are the ionic mobilities or ionic conductances.

18.7.1 Application of Kohlrauch's Law

Equivalent Conductivity at Infinite Dilution: λ_∞ of weak electrolyte (like CH_3COOH)

As weak electrolytes are not completely ionized even at a very high dilution, then equivalent conductivity cannot be determined by measuring specific conductivity. According to Kohlrausch law, if the ionic mobilities of two ions of weak electrolyte are known, then the λ_∞ can be calculated for CH_3COOH

$$\lambda_\infty \text{ of } CH_3COOH = \lambda_{CH_3COO^-} + \lambda_H^+$$

The value of $\lambda_{CH_3COO} -$ and λ_{H^+} can be calculated by two methods.

1. By using strong electrolyte like HCl.CH_3COONa, NaCl.

$$\lambda_\infty \text{ of HCl} = \lambda_{H^+} + \lambda_{Cl^-} \quad \text{(viii)}$$

$$\lambda_\infty \text{ of } CH_3COONa = \lambda_{CH_3COO} + \lambda_{Na^+} \quad \text{(ix)}$$

$$\lambda_\infty \text{ of NaCl} = \lambda_{Na^+} + \lambda_{Cl^-} \quad \text{(x)}$$

On adding Equation (viii) (i) and (ix) and subtracting Equation (x), we can get the value of $\lambda_{\bar{C}H_3COO} + \lambda_{H^+}$.

2. The ionic mobility is proportional to the speed of an ion at constant temperature. Then

$$\lambda_a \propto u_a \quad \text{or} \quad \lambda_a = K u_a \quad \text{(xi)}$$

and

$$\lambda_c \propto u_c \quad \text{or} \quad \lambda_c = K u_c \quad \text{(xii)}$$

adding (i) and (ii)

$$\lambda_a + \lambda_c = K\,(u_a + u_c)$$

but

$$\lambda_a + \lambda_c = \lambda_\infty$$

so

$$\lambda_\infty = K\,(u_a + u_c) \quad \text{(xiii)}$$

dividing Eq. (i) by (iii),

$$\frac{\lambda_a}{\lambda_\infty} = \frac{ku_a}{k\,(u_a + u_c)}$$

$$\frac{\lambda_a}{\lambda_\infty} = n_a \quad \text{as} \quad na = \frac{u_a}{u_a + u_c}$$

therefore

$$\lambda_a = n_a \times \lambda_\infty$$

for CH_3COOH

$$\lambda_{CH_3COO^-} = \text{Total number of } CH_3COO^- \times \lambda_\infty CH_3COONa$$

and

$$\lambda_{H^+} = \text{TTotal number of } H^+ \times \lambda_\infty HCl$$

So by calculating λ_{CH_3COO} and λ_H^+, the λ_∞ for CH_3COOH can be calculated by adding two values.

$$\lambda_{\infty CH_3COO^-} = \lambda_{CH_3COO^-} + \lambda_H^+$$

This indicates that the ionic mobility is obtained by total number of ion and equivalent conductivity at infinite dilution of any electrolyte containing that ion.

Example 18.7 The equivalent conductivities of $\frac{N}{100}$ solution of NaCl, HCl, and CH_3COONa are 369.3, 101.7, and $71.7\Omega^{-1}$. Calculate the degree of dissociation and H^+ concentration of acid. Assume equivalent conductivity of CH_3COOH as $14.3\Omega^{-1}$.

Solution

$$\lambda_\infty \text{ for } CH_3COOH = \lambda_{CH_3COO^-} + \lambda_H^+$$

$$\lambda_\infty \text{ for HCl} = \lambda_H^+ + \lambda_{Cl} = 369.3 \qquad \text{(xiv)}$$

$$\lambda_\infty \text{ for NaCl} = \lambda_{Na}^+ + \lambda_{Cl} = 101.7 \qquad \text{(xv)}$$

$$\lambda_\infty \text{ for } CH_3COONa = \lambda_{CH_3COO^-} + \lambda_{Na}^+ = 71.7 \qquad \text{(xvi)}$$

Adding (xiv) and (xv) and subtracting (xvi), we get

$$\lambda_H^+ + \lambda_{Cl^-} - \lambda_{Na+} - \lambda_{Cl}^- + \lambda_{CH_3COO^-} + \lambda_{Na}^+ = 369.3 + 101.7 - 71.7$$

$$= 339.3$$

$$\Rightarrow \quad \lambda_{CH_3COO^-} + \lambda_{H^+} = 339.3$$

or

$$\lambda_{\infty} CH_3COOH = 339.3$$

Then

$$\text{Degree of dissociation } \alpha = \frac{\lambda_v}{\lambda_\infty} = \frac{14.3}{339.3} = 0.042$$

and

$$\begin{aligned} H^+\text{ion concentration } &= \frac{\alpha}{V} = \frac{0.042}{100 \times 1000}\text{gm (Kohlraush's law)} \\ &= 4.2 \times 10^{-4}\text{gm ion per litre} \end{aligned}$$

Example 18.8 The eq. conductivity at infinite dilution of HCl and CH_3COONa is 380 and 80, respectively. The transport numbers of H^+ and CH_3COO^- ions are 0.84 and 0.56, respectively. Calculate the equivalent conductivity at infinite dilution for CH_3COOH.

Solution Using

$$\begin{aligned} \lambda_a &= n_a \times \lambda_\infty \\ \lambda_{H+} &= n_{H+} \times \lambda_\infty^{HCl} \\ &= 0.84 \times 380 \\ &= 319.2\Omega^{-1} \\ \lambda_{CH_3COO^-} &= n_{CH_3COO^-} \times \lambda_\infty CH_3COONa \\ &= 0.56 \times 80 \\ &= 44.8\Omega^{-1} \\ \lambda_\infty^{CH_3COOH} &= \lambda_{CH_3COO^-} + \lambda_H^+ \\ &= 44.8 + 319.2 \\ &= 364.0\ \Omega^{-1} \end{aligned}$$

Example 18.9 The specific conductance of $N/10$ $AgNO_3$ is $9.46 \times 10^{-3}\Omega^{-1}$ at 25 °C. The mobilities of Ag^+ and NO_3^- at 25 °C are 55.7 and 60.8 r.o., respectively. Calculate the degree of dissociation of $AgNO_3$ in $N/10$ solution.

Solution

$$\text{Degree of dissociation } \alpha = \frac{\lambda_v}{\lambda_\infty}$$

and

$$\lambda_v = K_v \times V$$

$$K_v = 9.46 \times 10^{-3}\Omega^{-1} \quad \text{and} \quad V = 10 \times 1000\,\text{ml}$$

So

$$\lambda_v = 9.46 \times 10^{-3} \times 10000 = 94.7\Omega^{-1}$$

Now

$$\begin{aligned}\lambda_\infty^{AgNO_3} &= \lambda_{Ag}^{+} + \lambda_{NO_3}^{-} \\ &= 55.7 + 60.8 \\ &= 116.5\Omega^{-1}\end{aligned}$$

$$\alpha = \frac{94.7}{116.5} = 0.81 \quad \text{or} \quad 81\%$$

18.8 Solubility

It is the property of any solid, liquid, or gaseous substance called solute, which dissolves in a solvent to form a homogeneous solution. The solubility of a substance depends upon the physical and chemical properties of solute and solvent at a specific temperature, pressure, and the pH of the solution.

The extent of the solubility (at a temperature) of a substance in a solvent is measured by using its saturated solution, where adding of more solute does not increase the concentration of solution and begin to precipitate.

The extent of the solubility ranges widely from infinitely soluble to springly soluble, e.g., ethanol in water is infinitely soluble or fully miscible and $BaSO_4$ is poorly soluble in water.

18.8.1 Factors Affecting Solubility

There are several factors which can affect the extent of solubility. The solubility of one substance in another is determined by the balance of intermolecular forces between solute and solvent and the entropy change that accompanies solution.

Factors like temperature and pressure change the solute-solvent interaction, thus changing the solubility. Solubility also depends on the pressure of other species dissolved in the solution as common ion effect reduces the solubility. More clearly the solubility depends upon the ionic strength of solution and also depends on the physical size of solute. The solubility of solute may increase on increasing the surface area of solute.

18.8.2 Determination of Solubility of Sparingly Soluble Salts

Substances like $BaSO_4$, which are insoluble in water, still have some solubility in water. This small solubility can be determined by conductivity measurement of their salt solution at equilibrium (saturated condition).

By using conductive cell, the specific conductivity K_v of the saturated salt solution can be measured. Since the concentration of solute is very less, it will completely dissociate into ions. Thus, the value of equivalence conductivity, λ_v, will be the same as equivalent conductivity at infinite dilution, λ_∞.

Thus,

$$\lambda_v = \lambda_\infty = K_v \times V$$

where "V" is the volume in ml having 1 gm equivalent of solute.

By using Kohlrausch's law, λ_∞ can be obtained as

$$\lambda_\infty = \lambda_a \times \lambda_c$$

Hence, by putting the values of λ_∞ and K_v, the value of V can be calculated, which denotes 1 gm equivalent in V ml.

So solubility, $s = \frac{1000}{V} \times$ equivalent wt. g/litre

$$\text{More clearly solubility, } s = \frac{1000 \times \text{ equivalent wt} \times K_v}{\lambda_\infty} \text{g/litre}$$

18.8.3 Solubility Product

For a saturated solution of a normal binary salt (AB) in a solvent, at equilibrium,

$$AB \rightleftharpoons A^+ + B^-$$

on applying law of mass action on ionic equilibrium:

$$\frac{[A^+][B^-]}{[AB]} = \text{constant.}$$

For saturated solution, the active mass of dissolved substance AB is constant at constant temperature.

$$\frac{[A^+][B^-]}{\text{constant}} = \text{constant}$$

or

$$[A^+][B^-] = \text{constant "}S\text{"}$$

where "S" or K_{sp} is called solubility product which may be defined as
"The product of concentration of ions in a saturated solution of a binary electrolyte at a given temperature is constant and is known as solubility product."
The solubility product is only applied in cases where the solution is saturated; it cannot be applied to normal ionic solutions.

If "s" is the solubility of binary electrolyte (say AB),

$$AB \rightleftharpoons A^+ + B^-$$

then the concentration of $[A^+]$ and $[B^-]$ will be "s" as it is completely ionized. So,

$$\text{solubility product } S = [A^+][B^-]$$

and

$$S = s.s$$

or

$$s = \sqrt{S}$$

Hence, the solubility of a sparingly soluble salt (binary) is equal to the square root of the solubility product of that salt. It is expressed in gm. mol/ liter.

If the salts are not binary electrolytes, then the solubility product "S" will be calculated as

$$AxBy \rightleftharpoons xA^+ + yB^-$$

or

$$S = [A^+]^x \times [B^-]^y$$

For example,

$$Ag_2CO_3 \rightleftharpoons 2Ag^+ + CO_3^{--}$$

So

$$S = [Ag^{+}]^{2}[CO_{3}^{--}]$$

We must remember that solubility product applies only to saturated solutions, and unit of solubility product depends on the solubility product expression. For example, solubility product expression of barium sulfate is

$$S = \left[Ba^{2+}\right] + \left[SO_{4}^{2-}\right]$$

If each concentration is in moldm^{-3}, the unit of solubility product will be mol^{2} dm^{-6}. You must remember that

- If ionic concentration of a solution gives a value less than the solubility product, no precipitate will be formed-the solution is not saturated.
- If ionic concentration of a solution gives a value more than the solubility product, enough precipitate would be formed. The answer will be equal to solubility product.

18.8.4 Applications of Solubility Product

It is clear from the definition of solubility product that if the product of concentration of ions in a saturated solution of an electrolyte is greater than the solubility product of the electrolyte, then it will precipitate out. More clearly the precipitation will only occur when the concentration of ions exceeds the solubility product of the electrolyte. This concept is used in analytical chemistry for qualitative analysis of basic radicals.

(i) Application in third and fifth groups' qualitative analysis
The use of NH_4Cl in third and fifth group analysis along with NH_4OH and $(NH_4)_2CO_3$

(a) In third group NH_4Cl is used with NH_4OH then due to the presence of NH_4^+ as common ions, the concentration of OH^- is reduced, but it is sufficient to increase the solubility product of hydroxides of Fe, Al and Cr, while it is not sufficient for radicals of fourth group onward.

(b) In fifth group due to the addition of NH_4Cl and $(NH_4)_2CO_3$, again common ion effect of NH_4^+ developed to reduce concentration of CO_3^{--} only exceeding the solubility product of Ca, Sr, and Ba carbonates, not for Mg of sixth group. The solubility product of $MgCO_3$ is higher, which is not exceeded to get precipitated out.

(ii) Precipitation of sulfide in second and fourth groups:

Group II radicals are precipitated out by passing H_2S in presence of HCl, where as fourth group radicals are precipitated out by passing H_2S in the presence of NH_4OH. In the second group, due to the presence of HCl and H_2S, common ion effect due to H^+ ion occurs, which reduces the concentration of S^- ions. Due to this reason, the solubility product suppressed and only exceeded for second group sulfides only.
In the fourth group on addition of NH_4OH, the common ion effect does not develop and on passing H_2S the solubility product of sulfides, the fourth group exceeded due to which these sulfides get precipitated.

Example 18.10 The solubility of $CaCO_3$ at 25 °C is 2.8×10^{-4} g/100 ml. Calculate the solubility product of salt assuming it as completely ionized.

Solution For binary sparingly soluble salts.

$$CaCO_3 \rightleftharpoons Ca^{++} + CO_3^{2-}$$

Solubility of $CaCo_3$/liter will be $2.8 \times 10^{-4} \times 10 = 2.8 \times 10^{-3}$gm/litre.

Example 18.11 The solubility product of $PbBr_2$ is 8.1×10^{-4}. If the salt dissociated 60% in saturated solution, then calculate the solubility of $PbBr_2$.

Solution The dissociation of $PbBr_2$ will be

$$PbBr_2 \rightleftharpoons Pb^{++} + 2Br^-$$

at equilibrium, the solubility product

$$S = \left[Pb^{++}\right]\left[2 \times Br^-\right]^2$$

For 60% dissociation, the solubility s will be

$$\text{For } \left[Pb^{++}\right] = 0.6 \times s$$
$$\text{For } \left[Br^-\right] = 0.6 \times s$$

So,

$$\text{solubility product `}s\text{'} = [0.6 \times s][0.6 \times s]^2$$

$$8.1 \times 10^{-4} = [0.6 \times s]\left[0.36 \times s^2\right]$$
$$8.1 \times 10^{-4} = 0.216 \times s^3$$
$$s^3 = 13.5 \times 10^{-4}$$
$$s = 2.36 \text{ g/litre.}$$

Example 18.12 Specific conductance of a saturated solution of AgBr is $2.3 \times 10^{-6} \Omega^{-1}$. If the ionic conductivity of Ag^+ and Br^- are 58 and 70, respectively, then calculate the solubility in water and $N/10$ HBr

Solution Given K_v for

$$\text{AgBr} = 2.3 \times 10^{-6} \Omega^{-1}$$

As per Kohlrausch's law

$$\begin{aligned} \lambda_\infty \text{ for AgBr} &= \lambda_{\text{Ag}^+} + \lambda_{\text{Br}} \\ &= 58 + 70 \\ &= 128 \\ \lambda_v^\infty &= K_v \times V \\ 128 &= 2.3 \times 10^{-6} \times V \\ V &= 5.565 \times 10^7 \text{ml} \end{aligned}$$

Now

$$\begin{aligned} 5.565 \times 10^7 \text{ml contains } &= 1 \text{ gm eq. of AgCl} \\ 1000 \text{ ml will contain } &= \frac{1000}{5.565 \times 10^7} \\ &= 0.180 \times 10^{-4} \text{gm/litre} \\ \text{So, solubility `}s\text{'} &= 1.8 \times 10^{-5} \text{gm/litre} \end{aligned}$$

Solubility in $N/10$ HB*r:* Let the solubility of AgBr in $N/10$ HBr is s_1 gm/litre, hence, solubility product of AgBr will be

$$\begin{aligned} \text{AgBr} &\rightleftharpoons \text{Ag}^+ + \text{Br}^- \\ S &= \left[\text{Ag}^+\right]\left[\text{Br}^-\right] \\ S &= s \times s = s^2 \\ &= \left(1.8 \times 10^{-3}\right)^2 \text{ (in water)} \\ &= 3.24 \times 10^{-6} \end{aligned}$$

in $N/10$ HBr

$$\begin{aligned} &AgBr\ Ag^+ + Br^- \\ &HBr\ \ H^+ + Br^- \\ &S = (s_1)\,(s_1 + C) \end{aligned}$$

where C is the ionic concentration of Br^-

$$\text{given } C = 0.1 \text{ Normal}$$

and

$$S = 3.24 \times 10^{-6}$$

So,

$$3.24 \times 10^{-6} = s_1\,(s_1 + 0.1)$$

$$3.24 \times 10^{-6} = s_1^2 + 0.1s$$

as s_1 is very small; hence, s_1^2 is again very small. Thus, s_1^2 can be neglected. So,

$$0.1s_1 = 3.24 \times 10^{-6}$$

$$s_1 = 3.24 \times 10^{-5}\text{gm eq./litre}$$

Example 18.13 The solubility product of PbS is 4.4×10^{-26}. Calculate the concentration of sulfide ion in which PbS is just precipitated out. The concentration of $\left[Pb^{++}\right]$ is 0.01M.

Solution

$$\text{Given solubility product } = 4.4 \times 10^{-26}$$

$$\text{Solubility of } \left[Pb^{++}\right] = 0.01\text{M}$$

$$PbS \rightleftharpoons Pb^{++} + S^{--}$$

$$S = \left[Pb^{++}\right]\left[S^{--}\right]$$

$$4.4 \times 10^{-26} = [0.01] \times \left[S^{--}\right]$$

$$\left[S^{--}\right] = 4.4 \times 10^{-24}\text{M}$$

18.9 Solved Problems

18.9.1 Conductivity

1. The specific conductance of a solution was $3.0 \times 10^{-2} \Omega^{-1}\ \text{cm}^{-1}$. The resistance was 70Ω between two Pt-electrodes. If the area of electrodes is $1.86\ \text{cm}^2$, then find out the distance between the electrodes and determine cell constant.

   Solution Given

$$K_v = 3.0 \times 10^{-2}\ \text{r.o.},$$

$$R = 70\Omega$$

$$a = 1.86\ \text{cm}^2$$

$$l = ?$$

$$K_v = c \times l/a$$

$$3.0 \times 10^{-2} = \frac{1}{70} \times \frac{l}{1.86}$$

$$l = 70 \times 1.86 \times 3 \times 10^{-2}$$

$$= 3.906\ \text{cm}$$

$$\text{cell constant } = l/a = \frac{3.906}{1.86} = 2.1\ \text{cm}^{-1}$$

2. The resistance of a cell containing 0.02 N KCl was 400Ω. Calculate the equivalent conductivity of KCl if cell constant is $1.2\ \text{cm}^{-1}$

   Solution Given

$$\text{Cell constant } = 1.2$$

$$\text{Resistance “}R\text{” } = 400\Omega$$

$$V \text{ in ml} = \frac{1000}{0.02} = 50000\ \text{ml}$$

Then

$$K_v = C \times l/a$$

$$K_v = \frac{1}{R} \times \frac{l}{a}$$

$$K_v = \frac{1}{400} \times 1.2$$

$$= 3.0 \times 10^{-3}$$

New

$$\lambda_v = K_v \times V$$

$$\lambda_v = 3 \times 10^{-3} \times 50000$$

$$= 150\Omega^{-1}$$

3. The resistance of a decinormal solution was 28 ohms. The solution was placed between two electrodes, which were 1.5 cm apart, and area of the electrode was 2.5 cm^2. Calculate equivalent conductivity of solution.

   Solution Given

$$R = 28\Omega$$

V in ml for $N/10$ (decinormal solution) $= 10 \times 1000 = 10,000$ ml.

$$l = 1.5 \text{ cm}$$

$$a = 2.5 \text{ cm}^2$$

$$\lambda_v = ?$$

$$K_v = C \times \frac{l}{a}$$

$$K_v = \frac{1}{28} \times \frac{1.5}{2.5}$$

$$= 2.14 \times 10^{-2}\Omega^{-1}$$

again

$$\lambda_v = K_v \times V$$

$$= 2.14 \times 10^{-2} \times 10000$$

$$= 214 \text{ mohs.}$$

4. A solution having 0.1 N $AgNO_3$ is placed between two electrodes which are 4 cm apart and has 12 cm^2 area. The potential difference between electrodes was 20 volts and current passed was 0.2 amperes. Calculate specific and equivalent conductivity.

Solution Given

$$V = 10 \times 1000$$
$$= 10,000 \text{ ml}$$
$$l = 4 \text{ cm}$$
$$a = 12 \text{ cm}^2$$
$$\text{Voltage } V = 20 \text{ volts}$$
$$\text{Current } i = 0.2 \text{ A}$$
$$K_v = ?$$
$$\lambda_v = ?$$

as we know

$$K_v = \mathrm{C} \times \frac{l}{a}$$

and

$$V = iR \text{ (ohm's law)}$$
$$20 = 0.2 \times R$$
$$R = 100\Omega$$
$$C = \frac{1}{100}$$
$$= 0.01\Omega^{-1}$$

So

$$K_v = 0.01 \times \frac{4}{12}$$
$$= 3.33\Omega^{-1} \text{ cm}^{-1}$$

and

$$\lambda_v = K_v \times \text{V in ml}$$
$$\lambda_v = 3.33 \times 10,000$$
$$= 3.33 \times 10^3 \text{ mohs}$$

5. For a 0.01 N KCl solution, the specific conductance is $1.4 \times 10^{-3} \Omega^{-1}$ at 25 °C. The resistance of this solution is 500Ω. If $\frac{N}{1000}$ NaCl solution was filled in the same cell at the same temperature, then resistance was found as 6000Ω. Calculate the equivalent conductivity of NaCl solution.

Solution Given

$$K_v \text{ for KCl} = 1.4 \times 10^{-3} \Omega^{-1}$$
$$V_{\text{KCl}} = 100 \times 1000 \text{ ml}$$
$$R_{\text{KCl}} = 500\Omega \Rightarrow C = \frac{1}{500} \Omega^{-1}$$
$$R_{\text{NaCl}} = 6000\Omega$$
$$V_{\text{NaCl}} = 1000 \times 1000$$

Using first data constant can be calculated for KCl

$$K_v = \text{C} \times \frac{l}{a} \text{ (cell constant)}$$

So, K_v for KCl $= \frac{1}{500} \times$ cell constant

$$1.4 \times 10^{-3} = \frac{1}{500} \times \text{ cell constant}$$
$$\text{Cell constant } = 500 \times 1.4 \times 10^{-3}$$
$$= 700 \times 10^{-3}$$
$$= 0.7$$

Again for NaCl

$$K_v \text{ for NaCl} = C \times \text{ cell constant.}$$
$$K_v^{\text{NaCl}} = \frac{1}{6000} \times 0.7$$
$$K_v^{\text{NaCl}} = 1.67 \times 10^{-4} \Omega^{-1} \text{ cm}^{-1}$$

and

$$\lambda_v^{\text{NaCl}} = K_v \times V$$

$$\lambda_v^{\text{NaCl}} = 1.67 \times 10^{-4} \times 1000 \times 1000$$

$$\lambda_v \text{ for NaCl} = 167 \text{ mohs.}$$

6. 0.02 N solution of CH_3COOH has specific conductance of $3.1 \times 10^{-4}\ \Omega^{-1}$. If degree of dissociation of CH_3COOH is 0.045, then calculate the equivalent conductivity at infinite dilution.

Solution Given

$$V \text{ in ml} = \frac{1000}{0.02}$$

$$= 50,000 \text{ ml}$$

$$K_v = 3.1 \times 10^{-4} \Omega^{-1}$$

$$\alpha = 0.045$$

$$\lambda_\infty = ?$$

$$\lambda_v = K_v \times V \text{ in ml}$$

$$\lambda_v = 3.1 \times 10^{-4} \times 50,000$$

$$= 15.5 \text{ r.o.}$$

and

$$\alpha = \frac{\lambda_V}{\lambda_\infty}$$

$$0.045 = \frac{15.5}{\lambda_\infty}$$

$$\lambda_\infty = 344.4 \text{ r.o. or mohs.}$$

7. The resistance of a semi normal solution is 50Ω. If it is placed between a cell of length 1.5 cm and area 7.5 cm^2, calculate the equivalent conductivity of solution.

Solution Given

$$R = 50$$

$$V = 2 \times 1000 \text{ ml} \left(\text{semi normal } = \frac{N}{2}\right)$$

$$l = 1.5 \text{ cm}$$

$$a = 7.5 \text{ cm}^2$$

We know,

$$\lambda_v = K_v \times V$$

and

$$K_v = C \times \frac{l}{a}$$

$$K_v = \frac{1}{50} \times \frac{1.5}{7.5}$$

$$= 4 \times 10^{-3} \Omega^{-1} \text{ cm}^{-1}$$

So

$$\lambda_v = 4 \times 10^{-3} \times 2000 = 8 \text{ mohs}$$

8. The equivalent conductivity of H_2SO_4 at infinite dilution is $400\Omega^{-1}$. The concentration of solution is 20 gm/litre, and its specific resistance is $20.5\Omega^{-1}$. Find out the degree of dissociation for H_2SO_4.

   Solution Given

$$\lambda_v^{\infty} \text{ for } H_2SO_4 = 400\Omega^{-1}$$

$$\text{concentration } = 20 \text{ gm/ litre}$$

$$\rho \text{ for } H_2SO_4 = 20.5\Omega^{-1} \left(K_v = \frac{1}{\rho}\right)$$

$$\alpha = ?$$

We know

$$\alpha = \frac{\lambda_v}{\lambda_\infty}$$

and

$$\lambda_v = K_v \times V \text{ in ml}$$

For V in ml,

$$\text{Normality is required and normality} \times \text{equivalent wt.} = \text{g/l}$$

$$N \times 49 = 20$$

$$\text{Normality} = \frac{20}{49}$$

So

$$V = \frac{49}{20} \times 1000 \text{ ml}$$

and

$$\lambda_v = \frac{1}{\rho} V \text{ in ml.}$$

$$\lambda_v = \frac{1}{20.5} \times \frac{49}{20} \times 1000$$

$$= 119.5 \text{ mohs}$$

$$\alpha = \frac{\lambda_v}{\lambda_\infty} = \frac{119.5}{400}$$

$$= 0.299 \quad \text{or} \quad 29.9\%$$

9. A decinormal solution of a weak acid has the specific conductance of $4.7 \times 10^{-3} \Omega^{-1}$ cm^{-1} at room temperature. The equivalent conductance at infinite dilution of this acid is 564 r.o. Calculate the dissociation constant of acid and its equivalence conductance.

Solution Given

$$V \text{ in ml} = 10 \times 1000 \text{ ml} \left(\text{decinormal} = \frac{N}{10} \right)$$

$$K_v = 4.7 \times 10^3 \Omega^{-1} \text{ cm}^{-1}$$

$$\lambda_\infty = 564 \text{ r.o.}$$

$$K_a = ?$$

$$\lambda_v = ?$$

$$\lambda_v = K_v \times V \text{ in ml}$$

$$= 4.7 \times 10^{-3} \times 10,000$$

$$\lambda_v = 47 \text{ mohs or r.o.}$$

again

$$\alpha = \frac{\lambda_v}{\lambda_\infty} = \frac{47}{564} = 0.083$$

For weak electrolyte

$$K_a = \frac{\alpha^2}{(1-\alpha)V} = \frac{(0.083)^2}{(1-0.083)\times 10}$$

Ans.

$$K_a = 7.51 \times 10^{-4}$$

10. The equivalence conductivity of HI at infinite dilution is 404 mhos, and specific conductivity is 0.15 mhos at room temperature. If the concentration of solution is $N/2$, then determine the degree of dissociation.

Solution Given

$$\lambda_\infty^{\text{HI}} = 404 \text{ mhos}$$
$$K_v = 0.15 \text{ mhos}$$
$$V = 2 \times 1000$$
$$= 2000 \text{ ml}$$
$$\alpha = ?$$
$$\lambda_v = K_v \times V$$
$$\lambda_v = 0.15 \times 2000$$
$$= 300 \text{ mhos.}$$
$$\alpha = \frac{\lambda_v}{\lambda_\infty} = \frac{300}{404}$$
$$= 0.74 \quad \text{or} \quad 74\%$$

11. The equivalent conductance at infinite dilution for CH_3COONa, NaCl, and HCl are 11.2, 122.4, and 444.0, respectively. Calculate the degree of ionization and equivalent conductance of $\frac{N}{5}CH_3COOH$. (Given $K_a = 1.8 \times 10^{-5}$)

Solution Using Kohlrausch's law.

$$\lambda^{\infty}_{\text{CH,3COOH}} = \lambda^{\infty}_{\text{HCl}} + \lambda^{\infty}_{\text{CH}_3\text{COONa}} - \lambda^{\infty}_{\text{NaCl}}$$
$$= 444.0 + 11.2 - 122.4$$
$$= 332.8 \text{ mohs.}$$

Using Ostwald's Dilution law for weak acid

$$\begin{array}{ccc} \text{CH}_3\text{COOH} & \text{CH}_3\text{COO}^- & +\text{H}^+ \\ 1 & 0 & 0 \\ 1-\alpha & \alpha & \alpha \end{array}$$

$$K_a = \frac{\alpha/v \times \alpha/v}{(1-\alpha)/v} = \frac{\alpha^2}{v}, \text{ Neglecting } 1-\alpha \text{ (as } \alpha \text{ is very small)}$$

so

$$K_a = \frac{\alpha^2}{V}$$
$$1.8 \times 10^{-5} \times 5 = \alpha^2 (V \text{ is volume in liter } N/5 \text{ has } V = 5 \text{ litre})$$
$$\alpha^2 = (9 \times 10^{-5})$$
$$\alpha = 9.48 \times 10^{-3}$$

and

$$\alpha = \frac{\lambda_v}{\lambda_\infty}$$
$$9.48 \times 10^{-3} = \frac{\lambda_v}{332.8}$$
$$\lambda_v = 3.2 \text{ mhos.}$$

18.9.2 Transport Number

12. In a copper sulfate solution concentration of copper before electrolysis was 0.80 g/ml and after electrolysis, it was 0.90 g/ml in anode solution. The total mass of copper deposited in copper voltameter was 0.150 gm. Calculate transport number of Cu^{++} and SO_4^{--} ions.

    Solution Before electrolysis wt. = 0.80 gm copper
    After electrolysis wt = 0.90 gm copper
    (w) Increase in weight of copper = 0.90 − 0.80 = 0.10 gm copper

Total copper deposited in voltameter $= 0.15$ gm copper

$$\text{Thus transport no. of Cu}^{++} = \frac{W - w}{W}$$

$$= \frac{0.15 - 0.10}{0.15} = \frac{0.05}{0.15}$$

$$n_c = 0.333$$

$$n_a = 1 - 0.333$$

$$= 0.667$$

13. An electrolytic solution of $AgNO_3$ has 1.14 mg of Ag in unit gm of solution. After passing current through the solution, the concentration of anode solution was 39.70 mg of Ag in 20.10 gm of solution. The amount of Ag deposited in voltameter attached in series was 32.10 mg of Ag. Calculate transport number of Ag^+ and NO_3^- ions.

Solution Given
Before electrolysis
wt. of Ag in anode solution = 1.14 gm of Ag/g of solution
After electrolysis
wt. of Ag in anode solution = 39.70 mg of Ag/20 g of solution
Converting amount of Ag before electrolysis at the rate of solution amount after electrolysis 1 gm of anode solution before electrolysis containing = 1.14 g of Ag.
20.1 gm of anode solution will contain

$$= 1.14 \times 20.1 = 22.914 \text{ mg of Ag}$$

$$\text{Increase of amount of Ag} = 39.700 - 22.914$$

$$w = 16.79 \text{ mg}$$

Also, total Ag deposited in voltameter = 32.10mg(W). So,

$$\text{Transport number of Ag}^+ = \frac{W - w}{W}$$

$$= \frac{32.10 - 16.79}{32.10}$$

$$= 0.477$$

and transport number of $NO_3^- = 1 - 0.477 = 0.523$

14. A solution of $AgNO_3$ has 1.315 gm of Ag before electrolysis and 1.23 gm after electrolysis in 100 ml of cathode solution. On passing the same current in a copper voltameter in series, 0.05 gm of copper was deposited. Calculate the transport number of Ag^+ and NO_3^- ions.

Solution Before electrolysis

$$\text{cathode solution contains} = 1.315\ \text{gm}/100\ \text{ml}$$

After electrolysis

$$\text{cathode solution contains} = 1.23\ \text{gm}/100\ \text{ml}$$

So, fall in concentration of Ag round cathode = 1.315 − 1.23 = 0.085 gm of Ag.
In copper voltameter

$$\text{amount of copper deposited} = 0.05\ \text{gm of Cu}$$

Converting Cu to Ag by using formula

$$\frac{\text{wt. of Cu}}{\text{wt. of Ag}} = \frac{\text{Eq. wt. of Cu}}{\text{Eq. wt. of Ag}}$$

$$\frac{0.05}{\text{wt. of Ag}} = \frac{63.5/2}{108/1}$$

$$\text{wt. of Ag} = \frac{108 \times 0.05}{31.75}$$

$$= 0.1701\ \text{gm.}$$

$$\text{So, Transport number of } NO_3^- \text{ ion} = \frac{0.085}{0.1701} = 0.4997$$

$$\text{and Transport number of } Ag^+ \text{ ion} = 1 - 0.4997 = 0.5003$$

15. In $AgNO_3$ solution 0.008 gm. $AgNO_3$ is dissolved in one gm of water. During electrolysis 0.079 gm of Ag was deposited in a Ag voltameter and after electrolysis 25 gm of anode solution had 0.26 gm of $AgNO_3$. Calculate transport number of Ag^+ and NO_3^- ions.

Solution Before electrolysis 1 gm of water has 0.008 gm. of $AgNO_3$, after electrolysis 25 gm of water has 0.26 gm of $AgNO_3$ (consider the solution is very dilute so $AgNO_3$ is dissolved in water). Converting before electrolysis solution to 25 gm of water (as after electrolysis)
wt. of $AgNO_3$ in 1 gm water = 0.008 gm.
wt. of $AgNO_3$ in 25 gm of water = 0.008 × 25 = 0.2 gm in 25 g of water.

Increase in concentration of $AgNO_3$ after electrolysis

$$= 0.26 - 0.2 = 0.06 \text{ gm of } AgNO_3(w)$$

Amount of Ag deposited in Ag voltameter = 0.079 gm of Ag converting Ag to $AgNO_3$ by equivalence formula

$$\frac{\text{wt. of } AgNO_3}{\text{wt. of } Ag} = \frac{\text{eq. of } AgNO_3}{\text{eq. of } Ag}$$

$$\frac{\text{wt. of } AgNO_3}{0.079} = \frac{170}{108}$$

$$\text{wt. of } AgNO_3 = \frac{170 \times 0.079}{108}$$

$$= 0.124 \text{ gm of } AgNO_3(W)$$

Then

$$\text{Total number of } Ag^{+}\text{ion} = \frac{W - w}{W}$$

$$= \frac{0.124 - 0.06}{0.124}$$

$$= 0.516$$

and

$$\text{Total number of } NO_3^{-}\text{ion} = 1 - 0.516$$

$$= 0.484$$

16. The amount of silver before electrolysis was 1.31 gm. and after electrolysis was 1.13 gm at cathode. The same current was passed through a copper voltameter which deposited 0.05 gm of copper. Calculate the transport number of Ag^{+} and NO_3^{-} ion.

Solution Before electrolysis

$$\text{Amount of Ag present} = 1.31 \text{ gm. at cathode sol.}$$

After electrolysis

$$\text{Amount of Ag present} = 1.13 \text{ gm. at cathode sol.}$$

So, fall in concentration of silver round cathode

$$= 1.31 - 1.23 = 0.08\ \text{gm.}(W - w)$$

At copper voltameter amount of copper deposited = 0.05 gm.
Convert Cu to Ag by equivalence formula.

$$\frac{\text{wt. of silver}}{\text{wt. of copper}} = \frac{\text{Eq. wt. of silver}}{\text{Eq. wt. of Cu}}$$

$$\frac{\text{wt. of silver}}{0.05} = \frac{108}{63.5/2}$$

$$= \frac{108}{31.75}$$

$$\text{wt. of silver} = \frac{108 \times 0.05}{31.75}$$

$$= 0.17\ \text{gm. (W)}$$

So

$$\text{Transport number of } NO_3^- = \frac{W - w}{W}$$

$$= \frac{0.08}{0.17}$$

$$= 0.417$$

and

$$\text{Transport number of } Ag^+ = 1 - 0.471$$

$$= 0.529$$

17. During electrolysis of a solution of HCl, increase in concentration of solution is 0.00665 gm for chlorine. If 0.12 gm of Ag is deposited in silver voltameter, then find the transport number of H^+ and Cl^- ions.

Solution As conc. of chlorine is increasing around anode, so it is w for $Cl^- =$ 0.00665 gm. Convert in terms of Ag using equivalent formula.

$$\frac{\text{wt. of chlorine}}{\text{wt. of silver}} = \frac{\text{eq. wt. of chlorine}}{\text{eg. wt. of silver}}$$

So

$$\frac{0.00665}{\text{wt. of silver}} = \frac{35.5}{108}$$

$$\text{wt. of silver} = \frac{0.00665 \times 108}{35.5}$$

$$= 0.0202 \text{ gm. silver } (w)$$

and

$$W = 0.12$$

So

$$\text{transport number of } H^+ = \frac{W - w}{W}$$

$$= \frac{0.12 - 0.0202}{0.12}$$

$$= 0.832$$

$$\text{Total number } Cl^- = 1 - 0.832 = 0.17$$

18. A solution of KOH (0.12%) was electrolyzed. After electrolysis, the conc. of KOH was 0.095 gm, per 60 gm. of cathode solution. If 0.03 gm of KOH was decomposed in silver voltameter, then calculate the transport No. of Na^+ and OH^- ions.

Solution Before electrolysis 0.12 gm. of KOH was in 100 gm. of cathode solution.
After electrolysis 0.095 gm. of KOH was in 60 gm. of cathode solution.
Convert before electrolysis to 60 gm. 100 gm of cathode solution having 0.12 gm KOH

$$\text{60 gm of cathode solution will have} = \frac{0.12 \times 60}{100}$$

$$= 0.072 \text{ gm.}$$

$$\text{Increase of KOH around cathode} = 0.095 - 0.072$$

$$= 0.023 \text{ gm.(w)}$$

wt. of KOH decomposed in Ag voltamin = 0.03 gm. (w)

$$\text{Transport No. of OH}^- = \frac{W - w}{W}$$
$$= \frac{0.08 - 0.23}{0.03}$$
$$= 0.233$$
$$\text{Transport No. of Na}^+ = 1 - 0.233$$
$$= 0.767$$

18.9.3 Kohlrausch's Law

19. The conductance of NaCl, HCl, and CH_3COONa is 125.9, 425.9 and 111.0, respectively, at infinite dilution. Calculate λ_∞ for acetic acid, and if equivalent conductance of acetic acid is 50.10, also calculate the degree of dissociation.

    Solution As per Kohlrausch's law

$$\lambda_\infty^{\text{CH}_3\text{COOH}} = \lambda_{\text{CH}_3\text{COOH}^-} + \lambda_{\text{H}}^+$$
$$\lambda_\infty^{\text{HCl}} = \lambda_{\text{H}}^+ + \lambda_{\text{Cl}^-} \quad \text{(i)}$$
$$\lambda_{\text{NaCl}}^\infty = \lambda_{\text{Na}^+} + \lambda_{\text{Cl}^-} \quad \text{(ii)}$$
$$\lambda_\infty \text{CH}_3\text{COONa} = \lambda_{\text{CH}_3\text{COO}^-} + \lambda_{\text{Na}}^+ \quad \text{(iii)}$$

applying Eq. (i) +, Eq. (ii) −, and Eq. (iii)

$$\lambda_\infty^{\text{HCl}} - \lambda_\infty \text{NaCl} + \lambda_\infty \text{CH}_3\text{COONa} = \lambda_{\text{H}^+} + \lambda_{\text{CH}_3\text{COO}^-}$$
$$425.9 - 125.9 + 111.0 = \lambda_{\text{CH}_3\text{COOH}}^\infty$$
$$\lambda_{\text{CH}_3\text{COOH}}^\infty = 411.0$$
$$\alpha = \frac{\lambda_v}{\lambda_\infty} = \frac{50.1}{411.0}$$
$$= 0.122$$

20. The conductivities at infinite dilution of CH_3COONa and HCl are 85 and 410, respectively. If the transport numbers for acetate and hydrogen ions are 0.60 and 0.90, respectively, then calculate λ_∞ for acetic acid.

    Solution We know

$$\lambda_{\infty}^{CH_3COOH} = \lambda_{CH_3COOH^-} + \lambda_{H}^{+}$$

and

$$\lambda_c = n_c \times \lambda_\infty \text{ (for cation)}$$

or

$$\lambda_{H}^{+} = n_{H}^{+} \times \lambda_\alpha \quad \text{or} \quad \text{HCl}$$

$$\lambda_{H}^{+} = 0.9 \times 410 = 369$$

Similarly

$$\lambda_{CH_3COO^-} = n_{CH_3COO^-} \times \lambda_{\infty \text{ for } CH_3Na}$$

$$\lambda_{CH_3COO^-} = 0.6 \times 85 = 51$$

So,

$$\lambda_{\infty}^{CH_3COOH} = 369.0 + 51.0 = 420.0 \text{ mhos}$$

21. The specific conductivity of $N/10$ $AgNO_3$ solution is 9.46×10^{-3} mohs at 25 °C. If the mobilities of Ag^+ and NO_3^- ions are 59.0 and 61.1 mohs, respectively, then calculate the degree of dissociation for $AgNO_3$ solution.

Solution Degree of dissociation

$$\alpha = \frac{\lambda v}{\lambda \infty}$$

and

$$\lambda_v = K_v \times \text{V in ml}$$

given

$$K_v = 9.46 \times 10^{-3}$$

$$V = 10 \times 1000 \text{ ml.} (N/10 \text{ solution})$$

So,

$$\lambda_v = 9.46 \times 10^{-3} \times 10 \times 1000$$
$$= 94.6 \text{ mohs.}$$
$$\lambda_v \text{AgNO}_3 = \lambda_{\text{Ag}^+} + \lambda_{\text{NO}_3^-}$$
$$= 59.0 + 61.1$$
$$= 120.1$$

So,

$$\alpha = \frac{\lambda v}{\lambda \infty}$$
$$= \frac{94.6}{120.1}$$
$$= 0.788$$

22. The specific resistance of 2 N solution of an acid was 3.12 ohms. If the equivalent conductivity at infinite dilution for acid is 380 r.o., calculate the degree of dissociation of acid.

Solution Given

$$\rho = 3.12\Omega \left(Kv = \frac{1}{\rho} = \frac{1}{3.12} \right) \Omega^{-1}$$
$$\text{strength} = 2\,\text{N} \left(V = \frac{1000}{2} = 500 \text{ ml} \right)$$
$$\lambda_v^{\infty} = 380 \text{ r.o.}$$
$$\lambda_v = K_v \times V$$
$$\lambda_v = \frac{1}{3.12} \times 500 \text{ r.o.}$$
$$\alpha = \frac{\lambda_v}{\lambda_\infty}$$
$$= \frac{\frac{1}{3.12} \times 500}{380} = 0.422$$

23. A decinormal solution of NaCl has specific conductivity of 9.2×10^{-3} r.o. If ionic mobilities for Na^+ and Cl^- ions are 43.4 and 66.0, respectively, calculate the degree of dissociation of NaCl.

Solution Given

$$K_v = 9.2 \times 10^{-3} \text{ r.o.}$$

$$\text{Conc.} = \frac{N}{10}(\text{decinormal})(V = 10 \times 1000 \text{ ml})$$

$$\lambda_{Na^+} = 43.3 \text{ r.o.}$$

$$\lambda_{Cl^-} = 66.0 \text{ r.o.}$$

$$\lambda_\infty \text{NaCl} = 43.3 + 66.0 = 109.4 \text{ r.o.}$$

$$\lambda_v = K_v \times V$$

$$= 9.2 \times 10^{-3} \times 10 \times 1000 = 920 \text{ r.o.}$$

$$\alpha = \frac{\lambda_v}{\lambda_\infty} = \frac{92.0}{109.4} = 0.841$$

18.9.4 Solubility Product

24. The specific conductivity of AgCl is 1.44×10^{-6} r.o. The ionic mobilities of Ag^+ and NO_3^- ions are 61.0 and 69.4, respectively. Calculate the solubility of AgCl.

Solution Given

$$\lambda_{Ag^+} = 61.0$$

$$\lambda_{NO_3^-} = 69.4$$

$$K_v = 1.44 \times 10^{-6} \text{ r.o.}$$

$$\lambda_v = K_v \times V \text{ (for spraingly soluble solts } \lambda_v \cong \lambda_\infty)$$

$$\lambda_\infty = K_v \times V \text{ and } \left(\lambda_\infty = \lambda_{Ag^+} + \lambda_{NO_3^-}\right)$$

$$130.4 = 1.44 \times 10^{-6} \times V$$

$$V = 9.06 \times 10^7 \text{ml}$$

The V explains the volume in ml containing 1 gm equivalent of salt in Vml of solution.

So, 9.06×10^7ml of solution has 1 gm equivalent AgCl

$$1000 \text{ ml of solution will have } = \frac{1 \times 1000}{9.06 \times 10^7}\text{ml}$$

$$= \frac{1}{9.06} \times 10^{-7}/\text{litre}$$

$$= 0.11 \times 10^{-7}\text{gm mole/litre}$$

$$\text{and in g/litre } = 0.11 \times 10^{-7} \times 143.5$$

$$= 1.58 \times 10^{-6}\text{gm/litre}$$

25. At 25 °C, the K_v value of AgBr is $3 \times 10^{-6}\Omega^{-1}$, and the λ_∞ of AgBr at 25 °C is 105 r.o. Calculate the solubility product of AgBr.

Solution

$$K_v = 3.0 \times 10^{-6} \text{ r.o.}$$

$$\lambda_\infty = 105 \text{ r.o.}$$

$$\lambda_v = \lambda_\infty = K_v \times V \text{ (for spraingly soluble salt } \lambda_v = \lambda_\infty)$$

$$105 = 3 \times 10^{-6} \times V$$

$$V = 35 \times 10^6\text{ml (having 1 g meg. of AgBr)}$$

$$\text{As} = 35 \times 10^6\text{ml has 1 gm eq. of AgBr}$$

$$1000 \text{ ml has} = \frac{1000}{35 \times 10^6}$$

$$= 28.57 \times 10^{-6} \text{ gm mole/liter.}$$

or

$$\text{solubility of AgBr} = 28.57 \times 10^{-6} \times 187.9$$

$$= 5.368 \times 10^{-3} \text{ g/1}$$

$$\text{solubility product } = s^2 = \left(5.36 \times 10^{-3}\right)^2 = 2.88 \times 10^{-5}$$

Exercises

1. The resistance of a solution having $3.0 \times 10^{-2}\Omega^{-1}$ cm^{-2} was 170.5Ω. The area of electrodes was 1.96 cm^2. Calculate the length of solution between electrodes.

2. $N/10$ solution of a salt is placed between two platinum electrodes of 2.1 cm length and 4.0 cm^2 area. The resistance was found to be 25Ω. Calculate equivalent conductivity of solution
3. A solution of specific conductance $2.8 \times 10^{-3}\Omega^{-1}$ has a resistance of 500Ω. Calculate the cell constant of cell.
4. For solutions having concentration $N/10$ KCl. and $N/10$ $AgNO_3$, the resistances were 337.3 and 366.6 ohms, respectively, at 35 °C. The specific conductance of $N/10$ KCl is $1.3 \times 10^{-3}\Omega^{1}$ cm^{-1} at the same temperature. Calculate the cell constant and equivalent conductivity of $N/10$ $AgNO_3$ solution.
5. KCl solution ($N/50$) has a resistances of 5000 ohms. If the cell constant of the cell is 1.2, then find the equivalent conductivity of solution.
6. The specific conductivity of $N/10$ acid is $3 \times 10^{-4}\Omega^{-1}$. Calculate the equivalent conductivity at infinite dilution. The degree of dissociation of acid is 0.045.
7. The molar conductivity at infinite dilution of a monobasic acid is $411\Omega^{-1}$. The specific conductivity of this acid is $3.3 \times 10^{-4}\Omega^{-1}$. If concentration of acid is $N/50$, then find the degree of dissociation of acid.
8. A decinormal salt solution is placed between platinum electrodes which are at a distance of 1.8 cm and have an area 5 cm^2. Calculate equivalent conductivity of solution.
   (*Hint:* Decinormal is $N/10$)
9. A $N/50$ acid has equivalent conductivity $50\Omega^{-1}$. If the equivalent conductivity at infinite dilution is $350\Omega^{-1}$, then calculate the degree of ionization of acid.
10. In a conductive cell, the distance between electrodes is 1.8 cm and area is 5.4 cm^2. The concentration of salt solution is $N/20$, and the resistance obtained is 30 ohms. Calculate equivalent conductivity.
11. The equivalent conductivity of hydrogen iodide at infinite dilution is $394\Omega^{-1}$, and for $N/20$ solution, the specific conductivity is $1.3 \times 10^{-1}\Omega^{-1}$. Calculate degree of dissociation of HI.
12. The specific conductivity of ($N/15$) KCl solution was $0.12\Omega^{-1}$. Calculate the equivalent conductivity.
13. Calculate equivalent conductivity and degree of dissociation of $N/5$ acetic acid. The equivalent conductances at infinite dilution for NaCl, HCl, and CH_3COONa are 120.5, 425.2, and 9.1, respectively, and $K_a = 1.8 \times 10^{-5}$
   (*Hint:* Ostwald's dilution formula $K = \frac{\lambda^2}{V}$ for weak acid)
14. The equivalent conductivities at infinite dilution of HCl, CH_3COONa, and NaCl are 346, 78, and 111 r.o., respectively. Calculate equivalent conductivity of acetic acid at infinite dilution. The degree of association of λ_v for CH_3COOH is 35 mohs.
15. During an electrolysis of $CuSO_4$ solution between copper electrodes, the total mass of copper deposited at cathode was 0.23 g. Before electrolysis, the mass of copper/unit volume of anode liquid was 1.185 gm and after electrolysis was 1.375 gm. Calculate the transport numbers of Cu^{++} and SO_4^{--}.
16. An aqueous solution of $CuSO_4$ has 1 gm copper per unit volume. The amount of copper deposited in copper voltameter in series was 0.22 gm. If after

electrolysis the concentration of copper at anode solution is 1.165 gm/ unit volume, then calculate the transport number of Cu^{++} and SO_4^{--} ions.

17. Before electrolysis, the concentration of $CuSO_4$ solution was 5.047 gm of $CuSO_4$ in 94.95 gm of water. The silver deposited in a coulometer in series was 1.548 gm. After electrolysis, 137.7 gm of anode solution had 7.702 gm $CuSO_4$. Calculate (i) transport number of Cu^{++} and SO_4^{--} ions and (ii) relative velocity of migration of Cu^{++} and SO_4^{-}.
Hint:

$$\left(\frac{u_c}{u_a} = \frac{\text{Fall around anode}}{\text{Fall around cathode}}\right)$$

18. 100 gm of $CuSO_4$ solution having 10.16 gm of $CuSO_4$ was electrolyzed. After electrolysis 59.54 gm of anode solution had 5.8 gm. of $CuSO_4$. The series Ag voltameter deposited 0.5010 gm of Ag. Calculate transport number of ions.
19. A solution of $AgNO_3$ contains 0.265 gm. of silver in 25 gm of anode solution. After electrolysis 30 gm solution had 0.043 gm of silver. In series voltameter, the deposited silver was 0.032 gm. Calculate transport number of ions.
20. Before electrolysis one gm solution had 1.05 gm silver, whereas after electrolysis 25 gm solution had 4.23 gm silver. In the Ag voltameter in series, 31.80 gm silver was deposited. Calculate transport number of Ag^+ and NO_3^- ions.
21. A current which deposited 0.047 gm of Cu in a copper voltameter was also passed through $AgNO_3$ solution. The solution obtained from cathode contained 1.315 gmAg before electrolysis and after electrolysis has 1.231 gm of Ag. Calculate transport number of ions.
22. A solution of NaOH (0.1605%) was electrolyzed. After electrolysis 55.25 gm of cathode solution had 0.095 gm of NaOH. In series of Ag voltameter deposited the equivalent of 0.0290 gm of NaOH. Calculate transport number of Na^+ and OH^- ions.
23. At 15 °C, the conductivities at infinite dilution of HCl and CH_3 COONa are 380 and $90\Omega^{-1}$, respectively. The transport numbers of H^+ and CH_3COO^- ions are 0.84 and 0.56, respectively. Calculate equivalent conductivity at infinite dilution for CH_3COOH.
24. The equivalent conductivities at infinite dilution for HCl, NaCl, and CH_3 COONa are 380, 109, and $79\Omega^{-1}$, respectively. Calculate equivalent conductivity at infinite dilution for CH_3COOH.
25. The K_v of $AgNO_3$ for $N/10$ solution is 9.46×10^{-3} r.o. at 18 °C. The mobilities of Ag^+ and NO_3^- at the same temperature are 55.8 and 61.0 r.o., respectively. Calculate the degree of dissociation of $AgNO_3$ solution.
26. The equivalent conductivity of NH_4Cl at infinite dilution is $148\Omega^{-1}$, and ionic conductancies for Cl^- and OH^- are 76.3 and 198 r.o., respectively. Calculate equivalent conductivity at infinite dilution for NH_4OH.
27. The equivalent conductivity of HCl at infinite dilution is $390\Omega^-$, and specific conductivity of $N/2$ solution is $0.133\Omega^-$. Calculate degree of dissociation.

28. The strength of a HCl solution is 1.4 N at 25 °C. If its specific resistance was found to be 2.6 Ω, then calculate degree of dissociation of HCl if equivalent conductivity at infinite dilution is 384 r.o.
29. The transport numbers of Ag^+ and NO_3^- ions at 18 °C are 9.470 and 0.534, respectively. At 18 °C, the molecular conductivity of $AgNO_3$ is 115.8 omhos. Calculate ionic mobilities of Ag^+ and NO_3^-.
    (*Hint:* $\lambda_a = n_a \times \lambda_\infty, \lambda_c = n_a \times \lambda_\infty$)
30. The specific conductivity of saturated solution of AgCl is 1.33×10^{-6} mhos. The ionic mobilities of Ag^+ and NO_3^- ions are 56.9 and 68.5, respectively. Calculate the solubility of AgCl.
31. A monobasic acid has concentration $N/20$ and at 15 °C it is dissociated 3.5%. Calculate the dissociation constant of this acid.

$$k = \frac{\alpha^2}{(l - \alpha)V}$$

32. The $N/100$ solution of an acid has degree of dissociation of 0.0387. Calculate the H^+ ion concentration of solution
    (*Hint:* H^+conc. $= \alpha/V$)
33. The equivalent conductivity of a decinormal solution of a weak acid is $7.0\Omega^{-1}$. If λ_∞ of acid is 400 mhos, calculate the dissociation constant of the acid.
34. The solubility of $BaSO_4$ at 25 °C is 2.33×10^{-4}gm/100 ml. Calculate the solubility product of $BaSO_4$, assuming the ionization is 100%.
35. The solubility product of $PbBr_2$ is 6×10^{-5}. If the salt of 78% is dissociated, then find the solubility of $PbBr_2$.
36. At 18 °CAgCl is soluble in water as 1.4×10^{-3}gm/liter. Calculate the solubility of AgCl at the same temperature.
37. The solubility product of $Mg(OH)_2$ is 3.4×10^{-10}; calculate the solubility of $Mg(OH)_2$.

Index

© The Author(s), under exclusive license to Springer Nature Switzerland AG 2026
R. N. Goyal, H. Goel, *Textbook of Engineering Chemistry*,
https://doi.org/10.1007/978-3-032-22068-4

MIX
Papier aus verantwortungsvollen Quellen
Paper from responsible sources
FSC® C105338

If you have any concerns about our products,
you can contact us on
ProductSafety@springernature.com

In case Publisher is established outside the EU,
the EU authorized representative is:
Springer Nature Customer Service Center GmbH
Europaplatz 3, 69115 Heidelberg, Germany

Printed by Libri Plureos GmbH
in Hamburg, Germany